Y0-COV-265

THE PROCEEDINGS OF THE IAU 8TH ASIAN-PACIFIC REGIONAL MEETING VOLUME I

COVER ILLUSTRATION:

S106 taken with CISCO and the Subaru telescope.

By the courtesy of the National Astronomical Observatory, Japan.

A SERIES OF BOOKS ON RECENT DEVELOPMENTS IN ASTRONOMY AND ASTROPHYSICS

Publisher

THE ASTRONOMICAL SOCIETY OF THE PACIFIC

390 Ashton Avenue, San Francisco, California, USA 94112-1722
Phone: (415) 337-1100 **E-Mail: orders@astrosociety.org**
Fax: (415) 337-5205 **Web Site: www.astrosociety.org**

A listing of all of the ASP Conference Series Volumes and IAU Volumes published by The ASP may be found at the back of this volume

THE ASTRONOMICAL SOCIETY
OF THE PACIFIC – CONFERENCE SERIES

Volume 289

THE PROCEEDINGS OF THE IAU 8TH ASIAN-PACIFIC REGIONAL MEETING VOLUME I

A meeting held at
National Center of Sciences, Hitotsubashi Memorial Hall, Tokyo, Japan
2-5 July 2002

Edited by

Satoru Ikeuchi
Department of Astrophysics, Nagoya University, Chikusa-ku, Nagoya, Japan

John Hearnshaw
University of Canterbury, Department of Physics & Astronomy
Christchurch, New Zealand

and

Tomoyuki Hanawa
Department of Astrophysics, Nagoya University, Chikusa-ku, Nagoya, Japan

Library of Congress Cataloging in Publication Data
Main entry under title

Card Number: 2003101696
ISBN: 1-58381-134-6

ASP Conference Series - First Edition

Printed in United States of America by Sheridan Books, Ann Arbor, Michigan

Contents

Preface xv

Program xvii

Participants xxix

Part 1. Large and New Facilities: Science and Development

Steps Toward the Next Generation CFHT 3

Gregory G. Fahlman

HERCULES: A High-resolution Spectrograph for Small to Medium-sized Telescopes 11

J. B. Hearnshaw, S. I. Barnes, N. Frost, G. M. Kershaw, G. Graham, and G. R. Nankivell

A Detailed Analysis of the Short- and Long-term Precision of Stellar Radial Velocities Obtained Using HERCULES 17

Jovan Skuljan

Square Kilometre Array (SKA) 21

R. D. Ekers

The Giant Metrewave Radio Telescope 29

Rajaram Nityananda

A Proposal for Constructing a New VLBI Array, Horizon Telescope . . 33

Makoto Miyoshi, Seiji Kameno, and Heino Falcke

Solar-B: Status of Project 37

Yoshinori Suematsu and the Solar-B team

On Astronomy Cooperation between, and New Facilities from, Indonesia and Japan 45

Hakim L. Malasan

Astronomy Against Terrorism: an Educational Astronomical Observatory Project in Peru 49

Mutsumi Ishitsuka, Hernán Montes, Takehiko Kuroda, Masaki Morimoto, and José Ishitsuka

Part 2. Extrasolar Planets: Discovery and Formation

Techniques for the Detection of Planets beyond the Solar System 55

John Hearnshaw

A Transit Search for Extrasolar Planets with the 0.5m Automated Patrol Telescope 65

M. G. Hidas, J. K. Webb, M. C. B. Ashley, M. Hunt, J. Anderson, and M. Irwin

Microlensing at High Magnification: Extra-solar Planets 69

Nicholas Rattenbury, Ian Bond, Jovan Skuljan, and Phil Yock

Searches for Extrasolar Planets with the Subaru Telescope: Companions and Free-Floaters 73

Motohide Tamura, Yoichi Itoh, Yumiko Oasa, and CIAO/CISCO/AO/SUBARU Telescope Teams

Low Mass Companions to Stars: Implications for Models of the Formation and Evolution of Binary and Planetary Systems 77

David C. Black and Tomasz F. Stepinski

Investigation of the Physical Properties of Protoplanetary Disks around T Tauri Stars by a High-resolution Imaging Survey at $\lambda = 2$ mm . . 85

Munetake Momose, Yoshimi Kitamura, Sozo Yokogawa, Ryohei Kawabe, Motohide Tamura, and Shigeru Ida

A Dynamical Analysis of the GJ 876 and HD 82943 Systems 89

Ji Jianghui and Li Guangyu, Liu Lin

Part 3. Large Scale Survey

The 6dF Galaxy Survey 97

Ken-ichi Wakamatsu, Matthew Colless, Tom Jarrett, Quentin Parker, William Saunders, and Fred Watson

Radio Sources in the 2dF Galaxy Redshift Survey: AGN, Starburst Galaxies and Their Cosmic Evolution 105

Elaine M. Sadler, Carole A. Jackson, and Russell D. Cannon

A Large Scale Extragalactic Survey with an Infrared Camera Onboard ASTRO-F 113

Hideo Matsuhara and Hiroshi Murakami

Infrared All-Sky Survey with ASTRO-F 117

Hiroshi Shibai

ISOGAL Survey of the Intermediate Galactic Bulge: Mass-losing AGB Stars in Galactic Bulge ISOGAL Fields. 121

D. K. Ojha, A. Omont, and ISOGAL Team

A Sensitive VLA Image Covering the SIRTF First-Look Survey Area . . 125

Q. F. Yin, J. J. Condon, and W. D. Cotton

The Virgo High-Resolution CO-Line Survey 129

Y. Sofue, J. Koda, H. Nakanishi, S. Onodera, K. Kohno, A. Tomita, and S. K. Okumura

Part 4. Education and Popularization of Astronomy in Asia

Education and Understanding of Astronomy through Total Solar Eclipses 137

E. Hiei, N. Takahashi, and Y. Iizuka

The Establishment of an Astrophysics Course in the Philippines through the IAU TAD 145

Cynthia P. Celebre

New Trends in Astronomy Education: A "Mapping" Strategy in Teaching and Learning Astronomy 151

Sergei Gulyaev

Astronomical Education in Azerbaijan: Existing Heritage, Current Status and Perspectives 157

Elchin S. Babaev

Part 5. Star Formation and ISM

The Completion and Release of the AAO/UKST Hα Survey 165

Q. A. Parker, S. Phillipps, and the Hα survey consortium and associates

Star Formation in the Southern Hemisphere: A Millimetre-Wave Survey of Dense Southern Cores 173

Paul A. Jones, Maria R. Hunt-Cunningham, John B. Whiteoak, and Graeme L. White

Star Formation Studies with SIRIUS – JHK Simultaneous Near-infrared Camera 177

Yasushi Nakajima, Chie Nagashima, Takahiro Nagayama, Daisuke Baba, Daisuke Kato, Mikio Kurita, Tetsuya Nagata, Shuji Sato, Motohide Tamura, Takahiro Naoi, Hidehiko Nakaya, and Koji Sugitani

Evolution of Spatial Structure of Star Clusters 181

W. P. Chen and C. W. Chen

Molecular Line Studies of Galactic Young Stellar Objects 185

Ji Yang, Ruiqing Mao, Zhibo Jiang, Tao Geng, and Yi-ping Ao

Observation of Interstellar H_3^+ Using Subaru IRCS: the Galactic Center 189

Miwa Goto, Benjamin J. McCall, Thomas R. Geballe, Tomonori Usuda, Naoto Kobayashi, Hiroshi Terada, and Takeshi Oka

The Formation of Molecular Clouds: the Origin of Supersonic Turbulence 195

Shu-ichiro Inutsuka and Hiroshi Koyama

Shocked Atomic and Molecular Gas in Supernova Remnants 199

Bon-Chul Koo

Discovery of Overionized Plasma in the Mixed-Morphology SNR IC 443 207

Masahiro T. Kawasaki, Masanobu Ozaki, and Fumiaki Nagase

On the Relation between Abundances in Planetary Nebulae and Their Central Star Evolution 211

Mezak A. Ratag

Part 6. Cosmology, Galaxy Formation and Evolution

The Two Degree Field Galaxy Redshift Survey (2dFGRS): The Instrument, Data and Science 219

Bruce A. Peterson

The Deep Near-Infrared Universe Seen in the Subaru Deep Field 227

Tomonori Totani

The Cluster 'Sphere of Influence': Tracking Star Formation with Environment via Hα Surveys 235

Warrick J. Couch, Michael Balogh, Richard Bower, and Ian Lewis

A Search for Galaxy Clustering at z=4 Using an SDSS Quasar Pair . . 243

Osamu Nakamura, Masataka Fukugita, Mamoru Doi, Nobunari Kashikawa, and Donald P. Schneider

The Discovery of Submillimeter Galaxies 247

A. W. Blain

Prospects for Detecting Neutral Hydrogen Using 21-cm Radiation from Large Scale Structure at High Redshifts 251

J. S. Bagla and Martin White

Cosmological Constant with Sunyaev-Zel'dovich Effect towards Distant Galaxy Clusters 255

Masato Tsuboi, Takashi Kasuga, Atsushi Miyazaki, Nario Kuno, Akihiro Sakamoto, and Hiroshi Matsuo

An Ellipsoid Model for the Halo Density Profile 259

Y. P. Jing

Star Formation History in Galaxies Inferred from Stellar Elemental Abundance Patterns 263

T. Shigeyama and T. Tsujimoto

Cosmic Star Formation History Associated with QSO Activity: An Approach by the Black Hole to Bulge Mass Correlation 267

Y. P. Wang, T. Yamada, and Y. Taniguchi

Part 7. Compact Objects and High Energy Astrophysics

Evolution of Low- and Intermediate-Mass X-Ray Binaries 273

Xiang-Dong Li

Raman Scattering and Accretion Disks in Symbiotic Stars 281

Hee-Won Lee

X-ray Observations of the X-ray Binary Pulsar Centaurus X-3 with *RXTE* 285

Takayoshi Kohmura and Shunji Kitamoto

Peculiar Characteristics of a Hyper-luminous X-ray Source in M82 . . . 291

Hironori Matsumoto, Takeshi Go Tsuru, Ken-ya Watarai, Shin Mineshige, and Satoki Matsushita

X-ray Study of Thermal Composite Supernova Remnants 3C391 and G349.7+0.2 295

Yang Chen and Patrick Slane

Time-Dependent Properties of Black-Hole Accretion Flow 301

S. Mineshige

Results in Gamma-Ray Burst Observations and the Use of GRBs to Explore the Early Universe 305

Toshio Murakami, Hisayo Izawa, and Daisuke Yonetoku

A Deep Optical Observation of an Enigmatic Unidentified Gamma-Ray Source 3EG J1835+5918 311

Tomonori Totani, Wataru Kawasaki, and Nobuyuki Kawai

Magnetic Reconnection in the Accretion Disk Corona of Compact Stars 315

I. V. Oreshina, A., V., Oreshina, and B. V. Somov

General Relativistic Simulation of Magnetohydrodynamic Energy Extraction of a Rotating Black Hole 319

Shinji Koide

High Energy Astrophysical Neutrinos 323

H. Athar

Part 8. QSOs, AGNs and IGM

Associated Absorption in Radio-loud Quasars and the Growth of Radio Sources 329

Richard W. Hunstead and Joanne C. Baker

Very Extended Emission-Line Region around the Seyfert 2 Galaxy NGC 4388 337

M. Yoshida, M. Yagi, S. Okamura, K. Aoki, Y. Ohyama, Y. Komiyama, N. Yasuda, M. Iye, N. Kashikawa, M. Doi, H. Furusawa, M. Hamabe, M. Kimura, M. Miyazaki, S. Miyazaki, F. Nakata, M. Ouchi, M. Sekiguchi, K. Shimasaku, and H. Ohtani

Evidence for the Evolutionary Sequence of Blazars: Different Types of Accretion Flows in BL Lac Objects 341

Xinwu Cao

3–4 μm Spectroscopy of 17 Seyfert 2 Nuclei: Quantification of the Compact Starburst Contribution 345

Masatoshi Imanishi

A High Resolution Imaging Survey of CO, HCN and HCO^+ Lines towards Nearby Seyfert Galaxies 349

Kotaro Kohno

Starburst-AGN Connections from High Redshift to the Present Day . . 353

Yoshiaki Taniguchi

A New Picture of QSO Formation 363

Masayuki Umemura and Nozomu Kawakatu

Distribution of the Faraday Rotation Measure in AGN Jets as a Clue for Deciding a Valid Theoretical Model 367

Y. Uchida, M. Nakamura, H. Kigure, and S. Hirose

Cosmic Rays in the Large Scale Structure of the Universe 371

Dongsu Ryu

Highlights from the First Five Years of the VSOP Mission 375

H. Hirabayashi, P. G. Edwards, Y. Murata, Y. Asaki, D. W. Murphy, H. Kobayashi, M. Inoue, S. Kameno, and T. Umemoto

Part 9. Solar and Stellar Activities, Binaries

Progress on Numerical Simulations of Solar Flares and Coronal Mass Ejections 381

Kazunari Shibata

An Investigation of Loop-Type CMEs with a 3D MHD Simulation . . . 389

J. Kuwabara, Y. Uchida, and R. Cameron

Basic Principles and Examples of Solar-type Flare Modelling 393

Boris V. Somov, Takeo Kosugi, and Taro Sakao

Energy Accumulation for the Bastille Flare 397

A. I. Podgorny, I. A. Bilenko, S. Minami, M. Morimoto, and I. M. Podgorny

Fine Structures Observed by the Chinese Solar Radio Broadband Fast Dynamic Spectrometers 401

Yihua Yan, Qijun Fu, Yuying Liu, Zhijun Chen, Hongao Wu, Fuying Xu, Zhihai Qin, Min Wang, and Zhiguo Xia

Flux Tube Oscillations and Coronal Heating 409

Y. Sobouti, K. Karami, and S. Nasiri

Binary Evolution – Problems and Applications 413

Zhanwen Han

Annual Parallax Measurements of Mira-Type Variables with Phase-Referencing VLBI Observations 421

Tomoharu Kurayama and Tetsuo Sasao

Magnetic Reconnection in the Solar Lower Atmosphere 425

C. Fang, P. F. Chen, and M. D. Ding

Part 10. Gravitational Lens

Future Microlensing Observations 431

Cheongho Han

A Perturbative Approach to Astrometric Microlensing due to an Extrasolar Planet 441

Hideki Asada

Detecting Astrometric Microlensing with VERA (VLBI Exploration of Radio Astrometry) 445

Mareki Honma and Tomoharu Kurayama

The Dark Matter Halo of the Gravitational Lens Galaxy 0047-2808 . . . 449

Randall B. Wayth and Rachel L. Webster

The New Era of Precision Microlensing 453

Andrew Gould

Secular Component of Apparent Proper Motion of QSOs Induced by Gravitational Lens of the Galaxy 461

Kouji Ohnishi, Mizuhiko Hosokawa, and Toshio Fukushima

Chandra Spectroscopy and Mass Estimation of the Lensing Cluster of Galaxies CL0024+17 465

Naomi Ota, Makoto Hattori, Etienne Pointecouteau, and Kazuhisa Mitsuda

Towards Direct Detection of Substructure around Galaxies – Quasar Mesolensing 469

Atsunori Yonehara, Masayuki Umemura, and Hajime Susa

An Analytical Model for the Distribution of Image Separations in Gravitational Lensing 473

Premana W. Premadi and Hugo Martel

Part 11. Business Sessions

Summary of Business Sessions 479

Satoru Ikeuchi

First Author Index 485

Preface

The IAU 8th Asian-Pacific Regional Meeting (APRM) was held from 2 to 5 July 2002 at the Hitotsubashi Memorial Hall, Tokyo. This meeting was organized by the National Committee for the IAU in Japan, the Astronomical Society of Japan and the National Astronomical Observatory of Japan. The meeting aimed to bring together astronomers in the Asian-Pacific region, with the goals (1) to exchange the results of recent research, (2) to promote research exchanges and collaborations in the region, (3) to encourage young astronomers' activity and hence to nurture the next generation of researchers, and (4) to discuss ways of driving research activities in the region and of popularizing astronomy.

Since this is the first APRM to be held since the Pusan Meeting in Korea in 1996, many astronomers have consequently attended and discussed a wide variety of topics. The total number of participants was 462 astronomers, including 148 foreign astronomers from 23 different countries in the region.

In this meeting, four plenary sessions, six parallel sessions, one special lecture and two business sessions, as well as poster presentations, were held. The total number of invited talks was 28, and there were 79 contributed oral presentations and 423 poster papers. As is seen, almost all participants presented their research results. This shows that the APRM is a very important oppotunity for astronomers in the region to meet and discuss their results. It is very pleasing that this meeting was so successful.

Because so many papers were presented, the proceedings of the 8th APRM had to be split into two volumes: Volume 1 includes invited and contributed oral talks, and is published by the Astronomical Society of the Pacific, while Volume 2 includes poster papers and is published by the Astronomical Society of Japan.

Finally the editors would like to express their deepest gratitude to the Monbu-Kagakushou (MEXT), to the Astronomical Society of Japan and to the International Astronomical Union (IAU) for their financial support on behalf of all participants. This meeting was finanically supported in part by Kambayashi Scholarship Foundation and personal donations from more than hundred Japanese astronomers. Also, we are grateful to the SOC members for preparing the scientific program and to the LOC members for organizing this meeting. We are indebted to Hirohisa Nagata for his assistance to editing of these proceedings.

Satoru Ikeuchi
John Hearnshaw
Tomoyuki Hanawa

The Scientific Organizing Committee:
L. Bronfman (Chile), K. Chang (Korea), S. M. Chitre (India), C. Fang (China Nanjing), B. Hidayat (Indonesia), J. Hearnshaw (New Zealand), S. Ikeuchi (Japan, chair), N. Kaifu (Japan), I. S. Kim (Russia), H. M. Lee (Korea), K. Y. Lo (China Taipei), J.V.Narlikar (India), M. Raharto(Indonesia), V. Rubin (USA),

Oral Session Program

PL: Plenary Sessions

(All plenary sessions are held in Room A)

Opening Session

July 2 (Tuesday) 9:30 – 10:00 Chair: T. Hasegawa

Opening Address	S. Ikeuchi	(SOC Chair)
Speeches	N. Kaifu	(IAU Vice-President)
	H.-M. Lee	(SOC)
Logistical Information	T. Hasegawa	(LOC Chair)

PL1: Large and New Facilities: Science and Development

July 2 (Tuesday) 10:00 - 18:15
Session 1 Chair: M. Ishiguro

10:00-10:40
Karoji, H.* (National Astronomical Observatory, Japan)
Status Report and Future Prospects of Subaru Telescope

10:40-10:55
Kinoshita, D. (European Southern Observatory, Chile) et al.
Wide-Field Survey near the Ecliptic with Subaru Telescope

10:55-11:10
Yamada, T. (National Astronomical Observatory, Japan)
Subaru Observations of Galaxies at High Redshift: When the Hubble Sequence were Formed?

11:10-11:50
Fahlman, G. G.* (Canada-France-Hawaii Telescope Corporation, USA)
Steps toward a Next Generation Canada-France-Hawaii Telescope

11:50-12:05
Hearnshaw, J. (University of Canterbury, New Zealand) et al.
HERCULES: A High-Resolution Fibre-fed Echelle Spectrograph for Small to Medium-sized Telescopes

12:05-12:20
Skuljan, J. (University of Canterbury, New Zealand) A Detailed Analysis of the Short- and Long -term Precision of Stellar Radial Velocities Obtained using the HERCULES spectrograph
12:20-13:50
Lunch

Session 2 Chair: N. Kaifu

13:50-14:30
Ishiguro, M.* (National Astronomical Observatory, Japan)
ALMA and Prospect for Japanese Participation

14:30-15:10
Ekers, R.* (Australia Telescope National Facility, Australia)
Square Kilometer Array

15:10-15:50
Nityananda, R.* (National Center for Radio Astronomy, India) The Giant Metrewave Radio Telescope

15:50-16:20
Coffee Break

16:20-16:35
Kobayashi, H. (National Astronomical Observatory, Japan) and the VERA team
Status of VERA (VLBI Exploration of Radio Astrometry) Project

16:35-16:50
Miyoshi, M. (1), Kameno, S. (1) and Falcke, H. (2) ((1) National Astronomical Observatory, Japan, (2) Max-Planck Institute, Germany)
A Proposal for Constructing a New VLBI Array, Horizon Telescope

16:50-17:05
Daishido, T. et al. (Waseda University, Japan) Pulsar Huge Array

Session 3 Chair: R. Ekers

17:05-17:45
Suematsu, Y.* (National Astronomical Observatory, Japan)
Solar B

17:45-18:00
Malasan, Hakim (Institute of Technology Bandung, Indonesia) On Astronomy Cooperation between, and New Facilities from, Japan and Indonesia

18:00-18:15
Ishitsuka, M.[†] (Instituto Geofisico del Peru, Peru) et al.
Astronomy against Terrorism: An Educational Astronomical Observatory Project in Peru
[†]Paper presented by José K. Ishitsuka (Univ. of Tokyo, Japan)

PL2: Extrasolar Planets: Discovery and Formation

July 3 (Wednesday) 9:30 - 13:00

9:30-11:00 Chair: D. C. Black

9:30-10:00
Hearnshaw, J. B.* (University of Canterbury, New Zealand)
Techniques for the detection of planets beyond our solar system

10:05-10:20
Hidas M. G. (1), Webb, J. K. (2, 1), Ashley, M. C. (1), Anderson, J. (2) and Irwin, M. (3) ((1)University of New South Wales, Australia, (2)University of California, Berkeley, USA, (3)Institute of Astronomy, University of Cambridge,UK)
A transit search for extrasolar planets with the 0.5-m automated patrol telescope

10:20-10:35
Rattenbury, N. J. (1), Bond, A. (1, 2), Skuljan, I. A. (2) and Yock, P. C. M. (1) ((1) University of Auckland, New Zealand, (2) University of Canterbury, New Zealand)
Detecting extrasolar planets via high magnification microlensing events

10:35-10:50
Tamura, M. and CIAO, AO, CISCO, Subaru Telescope teams (National Astronomical Observatory, Japan) Search for extrasolar planets with the Subaru telescope

11:00-11:30
coffee break

11:30-12:25 Chair: J. B. Hearnshaw

11:30-12:00
Black, D. C.* (Lunar and Planetary Institute, Houston, USA)
Low-mass companions to stars: Implications for the formation and evolution of binary and planetary systems

12:05-12:20
Momose, M. et al. (Ibaraki University, Japan)
Investigation of the physical properties of protoplanetary disks around T Tauri stars by a high-resolution imaging survey at lambda = 2 mm

12:20-12:35
Ji, J. (1, 2), Li, G. (1, 2), Zhao, H. (1, 2) and Liu, L. (3) ((1)Purple Mountain Observatory, Nanjing, PRC, (2)National Astronomical Observatory, Beijing, PRC, (3)Nanjing University, PRC)
The dynamical simulations of the planets orbiting GJ 876

PL3: Large Scale Survey

July 4 (Thursday) 9:30 - 13:00

9:30-11:30 Chair: H. Shibai

9:30-10:00
Turner, E.* (Princeton University, USA)
SDSS-FAQ and Favorites

10:00-10:30

Wakamatsu, K.* (Gifu University, Japan)
Galaxy Redshift Survey with 6dF

10:30-11:00
Sadler, E.* (University of Sydney, Australia)
Radio sources in the 2dF Galaxy Redshift Survey: AGN, starburst galaxies and their cosmic evolution

11:00-11:30
Boyle, B.* (Anglo Australian Telescope, Australia)
Future Large Scale Spectroscopic Surveys (tentative)

11:15-11:45
Coffee Break

11:45-13:00 Chair: E. M. Sadler

11:45-12:00
Matsuhara, H.† (ISAS, Japan) and ASTRO-F/IRC team Large Scale Extragalactic Survey with an Infrared Camera onboard ASTRO-F
†Paper presented by H. Murakami (ISAS, Japan)

12:00-12:15
Shibai, H. (Nagoya University, Japan) and ASTRO-F team
Infrared All-Sky Survey with ASTRO-F

12:15-12:30
Ojha, D. K. (Tata Institute of Fundamental Research, India)
ISOGAL Survey of the Outer Galactic Bulge: Mass-Losing AGB Stars in Galactic Bulge ISOGAL Fields

12:30-12:45
Yin, Q. F., Condon, J. J. and Cotton, W. D. (NRAO, USA)
The VLA Review of the SIRTF First Look Survey

12:45-13:00
Sofue, Y. (The University of Tokyo, Japan) et al.
High-Resolution CO-Line Survey of Virgo Spirals Using the Nobeyama mm-wave Array (NMA)

PL4: Education and Popularization of Astronomy in Asia

July 5 (Friday) 9:30 - 11:00

9:30-11:00 Chair: I. S. Kim

9:30-9:55
Hiei, E.* (Meisei University, Japan)
Education and Understanding of Astronomy at the Total Solar Eclipse

9:55-10:20
Celebre, C.* (PAGASA, Philippines)
The Establishment of an Astrophysics Course in the Philippine through the IAU/TAD

10:20-10:35
Matsumoto, R. (1), Fukuda, N., (1,2) Yokoyama, T. (3) and net laboratory team ((1) Chiba University, Japan, (2) Japan Science and Technology Corporation, (3) Nobeyama Radio Observatory, Japan)
Education and Popularization of Numerical Astrophysics by using a Coordinated Astronomical Numerical Softwares

10:35-10:50
Gulyaev, S. (Auckland University of Technology, New Zealand) New Trends in Astronomy Education: a "Mapping" Strategy in
Teaching and Learning Astronomy

10:50-11:00
Babayev, E. S. (Shamakhy Astrophysical Obs., Azerbaijan)
Astronomical Education in Azerbaijan: Existing Heritage, Current Status and Perspective (abstract PL(O)-4)

Special Lecture

July 5 (Friday) 11:00 - 12:00 Chair: K. Chang

11:00-12:00
Refsdal, S. (Institute of Theoretical Astrophysics, University of Oslo, Norway)
Gravitational Lensing as an Astrophysical Tool Closing Session

July 5 (Friday) 12:00 – 12:30 Chair: T. Hasegawa

Introduction to the IAU GA 2003 in Sydney	R. Ekers (IAU President-Elect)
Reports of Business Sessions	M. Raharto M. H. Lee
Closing Address	S. Ikeuchi (SOC Chair)

PS: Parallel Sessions

PS1: Star Formation and ISM

July 3 (Wednesday) 14:30 - 18:30 Room A

14:30-16:15 Chair: B. Hidayat

14:30-15:00
Parker, Q.* (University of Macquarie Sydney, Australia)
The AAO/UKST H-alpha Survey

15:00-15:15
Jones, P. A. (1) and Hunt, M. R. (2) ((1)Australia Telescope Facility, Australia, (2)University of New South Wales, Australia)
Star Formation in the Southern Hemisphere: A Millimeter-wave Survey of Dense Southern Cores

15:15-15:30
Ueno, M. (University of Tokyo, Japan) and the ASTRO-F team

Observations of Star Formation and Star Forming Region with ASTRO-F

15:30-15:45
Nakajima, Y. (Nagoya University, Japan) et al.
Star Formation Studies with SIRIUS-JHK Simultaneous Near Infrared Camera

15:45-16:00
Chen, W. P. and Chen, J. W. (National Center University, Taiwan)
Spatial Structure Evolution of Star Clusters

16:00-16:15
Yang, Ji et al. (Purple Mountain Observatory, China)
Molecular Line Studies of Galactic Young Stellar Objects

16:15-16:45
Coffee Break

16:45-18:30 Chair: Q. Parker

16:45-17:00
Goto, M. (Subaru Telescope, NAOJ, Hilo, USA) et al.
Observation of Interstellar $H3^+$ Using Subaru IRCS

17:00-17:15
Hong, S. S. (1), Lee, S. M. (1,2) and Kim, J. (3, 4) ((1) Seoul National University, Korea, (2) Korea Institute of Science Technology, Korea, (3) Korea Astronomy Observatory, Korea, (4) University of Notre Dame, USA)
Parker Instability under External and Self Gravities

17:15-17:30
Inutsuka, S. (1) and Koyama, H. (2) ((1) Kyoto University, Japan, (2) National Astronomical Observatory, Japan)
The Formation of Molecular Clouds: An Origin of Supersonic Turbulence

17:30-18:00
Koo, B. C.* (Seoul National University, Korea)
Shocked Atomic and Molecular Gas in Supernova Remnants

18:00-18:15
Kawasaki, T. M., Ozaki, M., and Nagase, F. (ISAS, Japan)
Discovery of the Overionized Plasma in the Mixed-Morphology SNR IC 443

18:15-18:30
Ratag, M. A. (Indonesian National Institute of Aeronautics & Space, Indonesia)
On the Relation between Abundances in Planetary Nebulae and their Central Star Evolution

PS2: Cosmology, Galaxy Formation and Evolution

July 3 (Wednesday) 14:30 - 18:30 Room B
14:30 – 16:15 Chair: K.-Y. Lo

14:30 – 15:00
Peterson, B.* (Australian National University, Australia) and the 2dF team

The 2dF Galaxy Redshift Survey: The Instrument, Data, and Science

15:00 – 15:30
Totani, T.* (National Astronomical Observatory, Japan)
Deep NearInfrared Universe Seen in the Subaru Deep Field

15:30 – 15:45
Chen, H. W. (Carnegie Observatories, USA) et al.
The z$\sim$1.2 Galaxy Luminosity Function from the Las Campanas Infrared Survey

15:45 – 16:15
Couch, W.* (University of NSW, Australia)
The Cluster Sphere of Influence : Tracking Star Formation with Environment via H-alpha Survey

16:15 – 16:45
Coffee Break

16:45 – 18:30 Chair: B. A. Peterson

16:45 - 17:00
Nakamura, O. (1), Fukugita, M. (1) and Doi, M. (2) ((1)Institute for Cosmic Ray Research, Japan, (2)The University of Tokyo, Japan) Search for Galaxy Clustering at z=4 using an SDSS Quasar Pair

17:00 - 17:15
Blain, A. (Calfornia Institute of Technology, USA)
The Discovery of Submillimeter Galaxies

17:15 - 17:30
Bagla, J. S. (1) and White, M. (2) ((1) Harish-Chandra Research Institute, Allahabad,India, (2) University of Calofornia, Berkeley, USA)
Redshifted 21 cm from the Large Scale Structure at High Redshifts

17:30 - 17:45
Tsuboi, M. (1) and Kasuga, T. (2) ((1) Ibaraki University, Japan, (2) Hosei University, Japan) Cosmological Constant with Sunyaev-Zel'dovich Effect of Distant Clusters 17:45 - 18:00

Jing, Y. P. (1) and Suto, Y. (2) ((1) Shanghai Astronomical Observatory, China, (2) The University of Tokyo, Japan)
An Ellipsoid Description for Halo Density Profile

18:00 - 18:15
Shigeyama, T. (1), Tsujimoto, T. (2) and Yoshii, Y. (1) ((1) The University of Tokyo, Japan, (2) National Astronomical Observatory, Japan)
Star Formation History in Galaxies Inferred from Stellar Elemental Abundance Patterns

18:15 - 18:30
Wang, Y. P. (1), Yamada, T. (2) and Taniguchi, Y. (3) ((1) Purple Mountain Observatory, Nanjing, China, (2) National Astronomical Observatory, Japan, (3)Tohoku University, Japan)
X-ray Viewing the Cosmic Star Formation History: An Approach by the Black Hole to Bulge Mass Correlation

PS3: Compact Objects and High Energy Astrophysics

July 3 (Wednesday) 14:30 - 18:30 Room C
14:30-16:15 Chair: T. Murakami

14:30-15:00
Li, X. D.* (Nanjing University, PRC)
Evolution of Low- and Intermediate Mass X-ray Binary

15:00-15:15
Lee, H. W.(Sejong University, Seoul, Korea)
Raman Scattering and Accretion Disk in Symbiotic Stars

15:15-15:30
Kohmura, T. and Kitamoto, S. (Rikkyou University, Tokyo, Japan)
X-ray Observation of X-ray Binary Pulsar Centaurus X-3 with RXTE

15:30-15:45
Matsumoto, H. et al. (Kyoto University, Japan)
Peculiar Characteristics of Hyperluminous X-ray Source M82 X-1

15:45-16:00
Chen, Y. (1) and Slane, P. (2) ((1) Nanjing University, PRC, (2) Harvard-Smithsonian Center for Astrophysics, USA)
X-ray Study of Thermal Composite Supernova Remnants 3C391 and G349.7+0.2

16:00-16:15
Mineshige, S. (Kyoto University, Japan) Time-Dependent Properties of Black Hole Accretion Flow

16:15-16:45
Coffee Break

16:45-18:30 Chair: X.-D. Li

16:45-17:15
Murakami, T.* (Kanazawa University, Japan)
The Most Recent Results in Gamma-Ray Burst Observations

17:15-17:30
Totani, T. (1, 2), Kawasaki, W. (3) and Kawai, N. (4) ((1) Princeton University Observatory, USA, (2) National Astronomical Observatory, Japan, (3) The University of Tokyo, Japan, (4) Tokyo Institute of Technology, Japan)
A Deep Optical Observation for an Enigmatic Unidentified Gammma-Ray Source 3EG1835+5918

17:30-17:45
Kawai, N. and HETE Science Team (Tokyo Institute of Technology and RIKEN, Saitama, Japan)
Results from High Energy Transient Explorer 2 (HETE-2) 17:45-18:00

Oreshina, I. V., Oreshina, A. V. and Somov, B. V. (Shternberg Astronomical Institute Moscow, Russia)
Magnetic Reconnection in the Accretion Disc Corona of a Compact Star

18:00-18:15
Koide, S. (1), Shibata, K. (2), Kudoh, T. (3) and Meier, D. L. (4) ((1) Toyama University, Japan, (2) Kyoto University, Japan, (3) National Astronomical Observatory, Japan, (4) Jet Propulsion Laboratory,USA)
General Relativistic Simulation of Magnetohydrodynamic Energy Extraction of Rotating Black Hole

18:15-18:30
Athar, H. (National Center for Theoretical Sciences, Hinchu, Taiwan)
High Energy Astrophysical Neutrinos

PS4: QSOs, AGNs and IGM

July 4 (Thursday) 14:30 - 18:30 Room A
14:30-16:15 Chair: R. Webster

14:30-15:00
Hunstead, R. W.* (University of Sydney, Australia) and Baker, J.C. (Oxford University, UK)
Associated absorption in radio quasars and the growth of radio sources

15:00-15:15
Yoshida, M. (National Astronomical Observatory, Japan) et al.
Very Extended Emission-Line Region around the Seyfert 2 Galaxy NGC 4388

15:15-15:30
Cao, X. (Shanghai Astronomical Observatory, China)
Evidence for the evolutionary sequence of blazars: different types of accretion flows in BL Lac objects

15:30-15:45
Jauncey, D.L.† (Australia Telescope National Facility, Australia) et al.
Microarcsecond Radio Imaging of AGN with Interstellar Scintillation †Paper presented by J. Lovell (ATNF, Australia)

15:45-16:00
Imanishi, M. (National Astronomical Observatory, Japan)
3-4 micro-m Spectroscopy of Seyfert 2 Nuclei to Quantatively Asses the Energetic Importance of Compact Nuclear Starbursts

16:00-16:15
Kohno, K. (The University of Tokyo, Japan) et al.
High Resolution Imaging Survey of CO, HCN, and HCO+ Lines toward Nearby Seyfert Galaxies using NMA and RAINBOW Interferometer

16:15-16:45
Coffee Break

16:45-18:30 Chair: R. W. Hunstead

16:45-17:15
Taniguchi, Y.* (Tohoku University, Japan)
Starburst-AGN Connection from High Redshift to the Present Day

17:15-17:30
Umemura, M. (Tsukuba University, Japan)
A New Picture of QSO Formation

17:30-17:45
Nishikawa, K. I. (Rutgers University, USA) et al.
3-D General Relativistic MHD Simulations of Jet Formation

17:45-18:00
Uchida, Y., Nakamura, M., Kigure, H. and Hirose, H. (Science University of Tokyo, Japan)
Distribution of Faraday Rotation Measure in AGN Jets as a Clue for Deciding Valid Theoretical Model

18:00-18:15
Ryu, D. (Chungnam National University, Korea)
Cosmic Rays in the Large Scale Structure of the Universe

18:15-18:30
Hirabayashi, H. (ISAS, Japan) et al.
Highlights from the first 5 years of the VSOP Mission

PS5: Solar and Stellar Activities, Binaries

July 4 (Thursday) 14:30 - 18:30 Room B
14:30-16:15 Chair: C. Fang
14:30-15:00
Shibata, K.* (Kyoto University, Japan)
Progress on Numerical Simulation of Solar Flare and Coronal Mass Ejections

15:00-15:15
Kuwabara, J., Uchida, Y. and Cameron, R. (Science University of Tokyo, Japan)
Investigation of Loop-type CME's with 3D MHD Simulations

15:15-15:30
Somov, B., Kosugi, T. and Sakao, T. (ISAS, Japan)
Basic Principles and Examples of Solar-type Flare Modeling

15:30-15:45
Podgorny, A. I. (1), Bilneko, I. A. (2), Minami, S. (3) and Podgorny, I. M. (4) ((1) Lebedev Physical Institute, Russia, (2) Moscow State University, Russia, (3)Osaka City University, Japan, (4) Institute for Astronomy, Russia)
Energy Accumulation for the Bastille Solar Flare

15:45-16:15
Yan, Y.* (National Astronomical Observatory, China)
Fine Structures Observed by the Chinese Solar Radio Broadband Fast Dynamic Spectrometers

16:15-16:45
Coffee Break

16:45-18:30 Chair: B. Somov

16:45-17:00
Sobouti, Y., Karami, K. and Nasiri, S. (Institute for Advanced Studies in Basic Science, Iran) Flux tube oscillations and coronal heating

17:00-17:30
Han, Z. W.* (Tunnan Observatory, PRC))
Binary Evolution: Problems and Applications

17:30-17:45
Pazhouhesh, R. (1) and Edalati, M. T. (Ferdowsi University, Iran)
Light curve analysis of the new eclipsing binary LD355

17:45-18:00
Edalati, M. T. and Salehi, F. (University of Ferdoesi, Iran)
AN analysis of GSC0008.324,newly Discovered binary star

18:00-18:15
Kurayama, T. (1, 2) and Sasao, T. (2) ((1) The University of Tokyo, Japan, (2) National Astronomical Observatory, Japan)
Annual Parallax Measurements of Mira-type Variables with Phase- Reference VLBI Observation

18:15-18:30
Fang, C., Chen, P. F., and Ding, M. D. (Dept. of Astronomy, Nanjing Univ., China)

Numerical Simulation on Magnetic Reconnection in the Solar Lower Atmosphere (abstract PS5-6)

PS6: Gravitational Lens

July 5 (Friday) 14:30 - 18:00 Room A
14:30-16:15 Chair: A. Gould
14:30-15:00
Han, C.* (Chungbuk National University, Korea)
Astrometric Microlensing Observations

15:00-15:15
Asada, H. (Hirosaki University, Japan)
Perturvative Approach to the Astrometric Microlensing due to an Extra Solar Planet

15:15-15:30
Honma, M. (National Astronomical Observatory of Japan, Japan)
Detecting Astrometric Microlensing with VERA (VLBI Exploration of Radio Astrometry)

15:30-15:45
Wayth, R. B. and Webster, R. L. R. B. (University of Melbourne, Australia
Mapping Dark Matter Distribution in Galaxies

15:45-16:00
Premadi, P. (Bandung Institute of Technology, Indonesia), et al. Lens Image Separation Statistics in Inhomogeneous Universe Models (abstract PS6-4)

16:00-16:15
Discussion (Leader: Andrew Gould)

16:15-16:45
Coffee Break

16:45-18:30 Chair: S. Refsdal

16:45-17:15
Gould, A.* (Ohio State University, USA)
The New Era of Precision Microlensing

17:15-17:30
Mitsuda, K. (1), Oshima, T. (1), Yonehara, A. (2), and Hattori, M. (3) ((1) ISAS, Japan, (2) Tsukuba University, Japan (3) Tohoku University, Japan)
X-Ray Observations of Lensed Quasars

17:30-17:45
Ohnishi, K. (1), Hosokawa, M. (2), Fukushima, T. (3) ((1) Nagano National College of Technology, Japan, (2) Communication Research Laboratory, Japan, (3) National Astronomical Observatory, Japan)
Secular Component of Apparent Proper Motion of QSOs induced by Gravitational Lens of our Galaxy

17:45-18:00
Ota, N. (1), Hattori, M. (2), Pointecouteau, E. (2) and Mitsuda, K. (3) ((1) Tokyo Metropolitan University, Japan, (2) Tohoku University, Japan, (3) ISAS, Japan)
Chandra Spectroscopy and Mass Estimation of the Lensing Cluster of Galaxies CL0024+17

18:00-18:15
Yonehara, A., Susa, H. and Umemura, M. (University of Tsukuba, Japan)
Toward direct detection of substructure around galaxies

18:15-18:30
Discussion (Leader: Sjur Refsdal)

BS: Business Sessions

BS1: Future Arrangement of APRM

July 4 (Thursday) 14:30 - 16:15 Room C Chair: M. Raharto

16:15-16:45
Coffee Break

BS2: Regional Publication and Network

July 4 (Thursday) 16:45 - 18:30 Room C Chair: H.-M. Lee

Participants

AHN, SANG-HYEON, Korea Institute for Advanced Study, Korea

AKABANE, KENJI, Nobeyama Radio Observatory, Japan

ALLEN, ANTHONY, Academia Sinica Institute of Astronomy and Astrophysics, China Taipei

AOKI, SEIICHIRO, Kwasan Observatory, Japan

AOYAMA, HIROKO, Nagoya Univ., Japan

ARIMOTO, JUN'ICHI, Kyoto Municipal Tounan High School, Japan

ASADA, HIDEKI, Hirosaki University, Japan

ASADA, KEIICHI, The Graduate University for Advanced Studies, Japan

ASAI, AYUMI, Kwasan and Hida observatories, Kyoto University, Japan

ASAYAMA, SHIN-ICHIRO, Osaka Prefecture University, Japan

ATHAR, HUSAIN, Institute of Physics, National Chiao Tung University, China Taipei

ATHREYA, RAMANA, Pontificia Universidad Catolica, Chile

AWAKI, HISAMITSU, Ehime University, Japan

AWANO, YUMI, Okayama Astronomical Museum, Japan

AYANI, KAZUYA, Bisei Astronomical Observatory, Japan

BABA, HAJIME, Center for Planning and Information Systems, ISAS, Japan

BABA, NAOSHI, Dept. Appl. Phys., Fac. Eng., Hokkaido Univ., Japan

BABAYEV, ELCHIN SAFARALY-OGLU, Shamakhy Astrophysical Observatory, Azerbaijan National Academy of Sciences, Azerbaijan Republic

BAEK, CHANG HYUN, Pusan National University, Korea

BAGLA, JASJEET, Harish-Chandra Research Institute, India

BAMBA, AYA, Kyoto University, Japan

BEKKI, KENJI, School of Physics, University of New South Wales, Australia

BLACK, DAVID, Lunar and Planetary Institute, U.S.A.

BLAIN, ANDREW, Department of Astronomy, Caltech, U.S.A.

BOFFIN, HENRI M.J., Royal Observatory of Belgium, Belgium

BOYLE, BRIAN, Anglo-Australian Observatory, Australia

BUAT, VERONIQUE, Laboratoire d'Astrophysique de Marseille, France

BURGARELLA, DENIS, Laboratoire d'Astrophysique de Marseille, France

BYUN, DO-YOUNG, Astronomy Program, SEES, Seoul National University, Korea, Korea

CAO, XINWU, Shanghai Astronomical Observatory, Chinese Academy of Sciences, China Nanjing

CELEBRE, CYNTHIA, Philippine Atmopsheric, Geophysical and Astronomical Services Administration (PAGASA), Philippines

CHANG, KYONGAE, Department of Physics, Chongju University, Korea

CHANG, HSIANG-KUANG, Institute of Astronomy, National Tsing Hua University, Taiwan, China Taipei

CHANG, HEON-YOUNG, Korea Institute for Advanced Study, Korea

CHATIEF, KUNJAYA, Dept of Astronomy, Institut Teknologi Bandung, Indonesia

CHEN, YANG, Department of Astronomy, China Nanjing

CHEN, DONGNI, Shanghai Astronomical Observatory of CAS, China Nanjing

CHEN, HSIAO-WEN, Carnegie Observatories, U.S.A.

CHEN, LI, Dept. of Astronomy, Beijing Normal University, China Nanjing

CHEN, P. F., Kwasan Observatory, Kyoto University, Japan

CHEN, WEN-PING, National Central University, Taiwan, China Taipei

CHEN, LI, Shanghai Astronomical Observatory, China Nanjing

CHERNENKO, ANTON, Space Research Institute, Russia

CHIBA, MASASHI, National Astronomical Observatory, Japan

CHOI, CHUL-SUNG, Korea Astronomy Observatory, Korea

CHOI, MINHO, ASIAA, China Taipei

CHOI, YOUNG-JUN, Tel-Aviv University, Israel, Israel

CHOUDHURY, MANOJENDU, Tata Institute of Fundamental Research, India

COUCH, WARRICK, The University of New South Wales, Australia

DAISHIDO, TSUNEAKI, Institute of Astrophysics, Waseda University, Japan

DJAMALUDDIN, THOMAS, National Institute of Aeronautics and Space (LAPAN), Indonesia

DOBASHI, KAZUHITO, Tokyo Gakugei University, Japan

DODDAMANI, VIJAYAKUMAR, Deptt. of Physics, Bangalore University, Bangalore-560 056, India

DODSON, RICHARD, University of Tasmania, Australia

DOI, MAMORU, Institute of Astronomy, School of Science, Univ. of Tokyo, Japan

DOTANI, TADAYASU, ISAS, Japan

EDALATI SHARBAF, MOHAMMAD TAGHI, University of Ferdowsi, Mashhad, Iran

EDWARDS, PHILIP, Institute of Space and Astronautical Science, Japan

EKERS, RON, Australia Telescope National Facility, CSIRO, Australia
FAHLMAN, GREGORY, Canada-France-Hawaii Telescope (CFHT), U.S.A.
FAN, JUNHUI, Center for Astrophysics, Guangzhou University, China Nanjing
FANG, CHENG, Department of Astronomy, Nanjing University, China Nanjing
FUJIMOTO, RYUICHI, Institute of Space and Astronautical Science, Japan
FUJISAWA, KENTA, Yamaguchi university, Japan
FUJISHITA, MITSUMI, Kyushu Tokai University, Japan
FUJITA, YUTAKA, National Astronomical Observatory, Japan, Japan
FUKUDA, NAOYA, Chiba University, Japan
FUKUE, JUN, Osaka Kyoiku University, Japan
FUKUSHIGE, TOSHIYUKI, University of Tokyo, Japan
FUKUSHIMA, TOSHIO, National Astronomical Observatory of Japan, Japan
FURUYA, IZUMI, Kobe University, Japan
GOUDA, NAOTERU, National Astronomical Observatory of Japan, Japan
GOULD, ANDREW, Ohio State University, U.S.A.
GULYAEV, SERGEI, Auckland University of Technology, New Zealand
HABE, ASAO, Hokkaido University, Japan
HACHISU, IZUMI, Dept. of Earth Science & Astronomy, College of Arts & Siences, University of Tokyo, Japan
HAN, CHEONGHO, Chungbuk National University, Korea
HAN, ZHANWEN, The Yunnan Observatory, National Astronomical Observatories of China, China Nanjing
HANAWA, TOMOYUKI, Nagoya University, Japan
HANDA, TOSHIHIRO, Institute of Astronomy, University of Tokyo, Japan
HASEGAWA, TETSUO, National Astronomical Observatory of Japan, Japan
HASEGAWA, TAKASHI, Gunma Astronomical Observatory, Japan
HASHIMOTO, OSAMU, Gunma Astronomical Observatory, Japan
HATSUKADE, ISAMU, Miyazaki University, Japan
HAYASHI, MASAHARU, Chiba University, Japan
HEARNSHAW, JOHN, University of Canterbury, New Zealand
HERDIWIJAYA , DHANI, Institute Technology of Bandung, Indonesia
HIDAS, MARTON, University of New South Wales, Australia
HIDAYAT, BAMBANG, Bosscha Observatory and Dept.of Astronomy, ITB, Indonesia
HIDAYAT, TAUFIQ, Institut Teknologi Bandung, Indonesia
HIEI, EIJIRO, Meisei University, Japan
HIGUCHI, ARIKA, Kobe University, Japan
HIKAGE, CHIAKI, School of science, the University of Tokyo, Japan

HIRABAYASHI, HISASHI, ISAS, Japan

HIRANO, KOICHI, Dept. of Physics, Tokyo University of Science, Japan

HIRANO, NAOMI, Institute of Astronomy & Astrophysics, Academia Sinica, China Taipei

HIROI, KUMIKO, Center for Computational Physics , University of Tsukuba, Japan

HIROSE, SHIGENOBU, Department of Physics, Tokyo University of Science, Japan

HIROSE, YOSHIYASU, University of Tsukuba, Japan

HOJAEV, ALISHER S., Ulugh Beg Astronomical Institute of the Uzbek Academy of Sciences & Uzbek Space Agency, Uzbekistan

HONDA, YASUKO, Kinki University Technical College, Japan

HONG, SEUNG SOO, Seoul National University, Korea

HONMA, MAREKI, National Astronomical Observatory of Japan, Japan

HOSOKAWA, MIZUHIKO, Communications Research Laboratory, Japan

HOSOKAWA, TAKASHI, Yukawa Institute for Theoretical Physics, Japan

HOU, JINLIANG, Shanghai Astronomical Observatory, Chinese Academy of Science, China Nanjing

HU, YOUQIU, University of Science & Technology of China, China Nanjing

HUNSTEAD, RICHARD, University of Sydney, Australia

HUNT, MARIA, University of New South Wales, Australia

HWANG, W-Y. PAUCHY, Department of Physics, National Taiwan University, China Taipei

ICHIKI, KIYOTOMO, National Astronomical Observatory of Japan, Japan

IDETA, MAKOTO, Department of Astronomy, University of Tokyo, Japan

IKEUCHI, SATORU, Nagoya University, Japan

IMAEDA, YUSUKE, NAOJ, Japan

IMANISHI, KENSUKE, Faculty of Science, Kyoto University, Japan

IMANISHI, MASATOSHI, National Astronomical Observatory of Japan, Japan

INOUE, AKIO, Department of Astronomy, Kyoto University, Japan

INOUE, HAJIME, Institute of Space and Astronautical Science, Japan

INOUE, MAKOTO, National Astronomical Observatory of Japan, Japan

INUTSUKA, SHU-ICHIRO, Kyoto University, Japan

ISHIGAKI, TSUYOSHI, Department of Applied Physics, Hokkaido University, Japan

ISHIGURO, MASATO, National Astronomical Observatory of Japan, Japan

ISHII, TAKAKO, Kwasan and Hida Observatories, Kyoto University, Japan

ISHISAKI, YOSHITAKA, Department of Physics, Tokyo Metropolitan University, Japan

ISHITSUKA, JOSE K., University of Tokyo, Japan

ISOBE, HIROAKI, Kwasan and Hida Observatories, Kyoto University, Japan

ISOBE, SYUZO, National Astronomical Observatory, Japan

ITO, TAKASHI, National Astronomical Observatory, Japan

ITOH, MASAYUKI, Kobe University, Japan

IWAMOTO, SHIZUO, Departmant of Earth and Space Science, Graduate School of Science, Osaka University, Japan

IWATA, IKURU, Department of Astronomy, Kyoto University, Japan

IZUMIURA, HIDEYUKI, Okayama Astrophysical Observatory, NAOJ, Japan

JI, JIANGHUI, Purple Mountain Observorary, Chinese Academy of Sciences, China Nanjing

JIANG, XIAOJUN, National Astronomical Observatories, Chinese Academy of Sceinces, China Nanjing

JING, YI PENG, Shanghai Astronomical Observatory, China Nanjing

JONES, PAUL, Australia Telescope National Facility, Australia

JUGAKU, JUN, Research Institute of Civilization, Japan

KAIFU, NORIO, NAOJ, Japan

KAMAYA, HIDEYUKI, Department of Astronomy, School of Science, Kyoto University, Japan

KAMEGAI, KAZUHISA, Department of Physics, The University of Tokyo, Japan

KAMENO, SEIJI, National Astronomical Observatory of Japan, Japan

KAMEYA, OSAMU, National Astronomical Obaservatory of Japan, Japan

KANAI, TSUNETO, Research Center for the Early Universe, University of Tokyo, Japan

KANDORI, RYO, Tokyo Gakugei University, Japan

KAN-YA, YUKITOSHI, National Astronomical Observatory of Japan, Japan

KAROJI, HIROSHI, Subaru Telescope, NAO Japan, U.S.A.

KASUGA, TAKASHI, Hosei University, Japan

KATAZA, HIROKAZU, Institute of Space and Astronautical Science, Japan

KATO, YOSHIAKI, Chiba University, Japan

KATO, SEIICHI, Cybermedia Center, Osaka Universtiy, Japan

KAWABATA, SHUSAKU, , Japan

KAWABATA, KOJI, National Astronomical Observatory of Japan, Japan

KAWABATA, KIYOSHI, Tokyo University of Science, Japan

KAWAGUCHI, NORIYUKI, National Astronomical Observatory, Japan

KAWAI, NOBUYUKI, Tokyo Institute of Technology, Japan

KAWAKATU, NOZOMU, Center for Computational Physics, University of Tsukuba, Japan

KAWAKITA, HIDEYO, Gunma Astronomical Observatory, Japan

KAWANO, YOZO, Nagoya University, Japan

KAWASAKI, MASAHIRO, ISAS, Japan

KAWASAKI, WATARU, Institute of Astronomy and Astrophysics, Academia Sinica, China Taipei

KHOSROSHAHI, HABIB, Institute for Advanced Studies in Basic Sciences (IASBS), Iran

KIGUCHI, MASAYOSHI, Kinki University, Japan

KIGURE, HIROMITSU, Department of Physics, Science University of Tokyo, Japan

KIM, IRAIDA, Sternberg State Astronomical Institute of Moscow State University, Russia

KIM, JIN HYUNG, Kyungpook National University, Korea

KIM, KAP-SUNG, Department of Astronomy & Space Science, Kyung Hee University, Korea

KIM, JONGSOO, Korea Astronomy Observatory, Korea

KIM, SOO HYUN, Kyungpook National University, Korea

KIM, SUNGSOO S., Space Science Telescope Institute, U.S.A.

KINOSHITA, HIROSHI, National Astronomical Observatory, Japan

KINOSHITA, DAISUKE, European Southern Observatory, Chile

KINUGASA, KENZO, Gunma Astronomical Observatory, Japan

KITABATAKE, ETSUKO, Osaka Kyoiku University, Japan

KITAMOTO, SHUNJI, Rikkyo University, Japan

KITAMURA, MASATOSHI, National Astronomical Observatory of Japan, Japan

KO, CHUNG-MING, Department of Physics and Institute of Astronomy, National Central University, China Taipei

KOBAYASHI, HIDEYUKI, National Astronomical Observatory of Japan, Japan

KODA, JIN, National Astronomical Observatory of Japan, Japan

KODAIRA, KEIICHI, Graduate University for Advanced Studies, Japan

KOGURE, TOMOKAZU, Bisei Astronomical Observatory, Japan

KOHMURA, TAKAYOSHI, Rikkyo University, Japan

KOHNO, KOTARO, Institute of Astronomy, University of Tokyo, Japan

KOIDE, SHINJI, Toyama University, Japan

KOKUBO, EIICHIRO, National Astronomical Observatory, Japan

KOMIYA, ZEN, Tokyo Univ. of Science, Japan

KOO, BON-CHUL, Seoul National University, Korea

KOYAMA, KATSUJI, Department of Physics, Japan

KOYAMA, HIROSHI, National Astronomical Observatory Japan, Japan

KOZAI, YOSHIHIDE, Gunma Astronomical Observatory, Japan

KOZU, HIROMICHI, Kwasan Observatory, Kyoto Uvinerstiry, Japan

KUBO, YUKI, Hiraiso Solar Observatory, Communications Research Laboratory, Japan

KUNIYOSHI, MASAYA, Waseda Univ., Japan

KUNO, NARIO, Nobeyama Radio Observatory, Japan

KURAYAMA, TOMOHARU, National Astronomical Observatory of Japan, Japan

KURTANIDZE, OMAR, Abastumani Observatory, Georgia

KUWABARA, TAKUHITO, Graduate Institute of Astronomy, National Central University, China Taipei

KUWABARA, JOJI, Science University of Tokyo, Japan

LANZAFAME, GIUSEPPE, Osservatorio Astrofisico di Catania, Italy

LEE, HYUNG MOK, Seoul National University, Korea

LEE, HEE-WON, Sejong University, Korea

LEE, HSU-TAI, Institute of Astronomy, National Central University, Taiwan, China Taipei

LEE, SANG HYUN, Pusan National University, Korea

LEE, JOUNGHUN, University of Tokyo, Japan

LEE, YOUNGUNG, Korea Astronomy Observatory, Korea

LEE, SANG MIN, Korea Institute of Science Technology and Information, Korea

LEE, OH KYUN, Kyungpook Nation University, Korea

LEE, JEONG AE, Kyungpook National University, Korea

LI, XIANGDONG, Department of Astronomy, Nanjing University, China Nanjing

LI, JIAN_PING, Department of Astronomy, Nanjing University, China Nanjing

LI, KEJUN, Yunnan Observatory, China Nanjing

LI, HUI, Purple Mountain Observatory, Chinese Academy of Sciences, China Nanjing

LIU, BIFANG, Yukawa Institute for Theoretical Physics, Kyoto Uni., Japan

LIU, LIN, Department of astronomy, Nanjing University, China Nanjing

LO, KWOK-YUNG, Academia Sinica Insitute of Astronomy & Astrophysics, China Taipei

LOVELL, JAMES, Australia Telescope National Facility, Australia

MA, YUEHUA, Purple Mountain Observatory, Academia Sinica, China Nanjing

MACHIDA, MAMI, Chiba University, Japan

MACHIDA, MASAHIRO, National Astronomical Observatory, Japan

MAEDA, HIDEKI, Waseda university, Japan

MAKITA, MITSUGU, Osaka-Gakuin Junior College, Japan

MALASAN, HAKIM L., Department of Astronomy, Institute of Technology Bandung, Indonesia

MARLOW, ANNIE, Ehime University, Japan

MASAKI, YOSHIMITSU, National Astronomical Observatory, Japan, Japan

MATSUDA, TAKUYA, Department of Earth and Planetary Sciences, Kobe University, Japan

MATSUMOTO, RYOJI, Chiba University, Japan

MATSUMOTO, TOMOAKI, Hosei University, Japan

MATSUMOTO, HIRONORI, Kyoto University, Japan

MATSUO, HIROSHI, National Astronomical Observatory, Japan

MATSUSHITA, EIKO, Tokyo Univ. of Science, Japan

MCGRUDER, CHARLES, Western Kentucky University, U.S.A.

MELEK, MAHER, Astronomy Department, Faculty of Science, Cairo University, Giza, Orman, Egypt., Egypt.

MICHIKO, OHKUBO, Astronomical Institute, Osaka kyoiku University, Japan

MIKAMI, TAKAO, Osaka Gakuin University, Japan

MINAMI, SHIGEYUKI, Dept. Electrical Engineering, Osaka City University, Japan

MINESHIGE, SHIN, Yukawa Institute for Theoretical Physics, Kyoto University, Japan

MITSUDA, KAZUHISA, ISAS, Japan

MIURA, HITOSHI, Center for Computational Physics, University of Tsukuba, Japan

MIYAGOSHI, TAKEHIRO, National Astronomical Observatory of Japan, Japan

MIYAJI, SHIGEKI, Chiba University, Japan

MIYAWAKI, RYOSUKE, Fukuoka Univ. of Education, Japan

MIYOSHI, MAKOTO, NAOJ, Japan

MIZUNO, YOSUKE, Department of Astronomy, Kyoto University, Japan

MIZUNO, NORIKAZU, Department of Astrophysics, Nagoya University, Japan

MIZUSHIMA, HIROFUMI, Hiroshima University, Japan

MIZUTANI, KOHEI, Department of Physics, Saitama University, Japan

MOCHIZUKI, NANAKO, ISAS, Japan

MOMOSE, MUNETAKE, Institute of Astrophysics and Planetary Sciences, Ibaraki Univeristy, Japan

MOON, NANMO, Osaka Kyoiku University, Korea

MORI, MASAO, Institute of Natural Science, Senshu university, Japan

MORIGUCHI, YOSHIAKI, Department of Astrophysics, Nagoya University, Japan

MORINO, JUN-ICHI, Subaru Telescope NAOJ, Japan, U.S.A.

MORITA, KOH-ICHIRO, National Astronomical Observatory, Japan, Japan

MORIWAKI, KAZUMASA, Department of Earth and Planetary Sciences, Kobe University, Japan

MURAKAMI, NAOSHI, Department of Applied Physics, Hokkaido University, Japan

MURAKAMI, TOSHIO, Fuculty of Science, Knazawa University, Japan

MURAKAMI, HIROSHI, ISAS, Japan

MURAKAMI, HIROSHI, The Institute of Space and Astronautical Science , Japan

MURATA, YASUHIRO, The Institute of Space and Astronautical Science, Japan

MURAYAMA, TAKASHI, Tohoku University, Japan

NAGAO, TOHRU, Astronomical Institute, Tohoku University, Japan

NAGASHIMA, MASAHIRO, National Astronomical Observatory, Japan

NAITOH , SEIICHIROU , Institute of Astrinomy, The University of Tokyo, Japan

NAKAGAWA, YOSHITSUGU, Kobe University, Japan

NAKAHARA, JUNJI, Graduate School of Science, Hiroshima University, Japan

NAKAI, NAOMASA, Nobeyama Radio Observatory, Japan

NAKAJIMA, YASUSHI, Nagoya University, Japan

NAKAJIMA, HIROSHI, Physics, Kyoto Univercity, Japan, Japan

NAKAMURA, YASUHISA, Department of Education, Fukushima University, Japan

NAKAMURA, OSAMU, institute for cosmic ray research, Japan

NAKAMURA, KENJI, Matsue College of Technology, Japan

NAKANISHI, HIROYUKI, Institute of Astronomy, University of Tokyo, Japan

NAKANO, TAKENORI, Department of Physics, Faculty of Science, Kyoto University, Japan

NAKANO, MAKOTO, Ōita University, Japan

NAKASATO, NAOHITO, Department of Astronomy, University of Tokyo, Japan

NAKATA, FUMIAKI, National Astronomical Observatory of Japan, Japan

NAKAZATO, TAKESHI, Center for Computational Physics, University of Tsukuba, Japan

NARUKAGE, NORIYUKI, Kwasan and Hida Observatories, Kyoto Univ., Japan

NISHIHARA, EIJI, Gunma Astronomical Observatory, Japan

NISHIKORI, HIROMITSU, Graduate School of Science and Technology, Chiba University, Japan

NISHIOKA, HIROAKI, Hiroshima University, Japan

NITYANANDA, RAJARAM, National Centre for Radio Astrophysics, Tata Institute of Fundamental Research, India

NORO, AYATO, Graduate School of Science and Technology, Chiba University, Japan

NOUNO, KIMIHIKO, School of Science, Kyoto Univ., Japan

OASA, YUMIKO, NASDA, Japan

OBAYASHI, HITOSHI, Gunma Astronomical Observatory, Japan

OCHI, YASUHIRO, Institute for Theoretical Astrophysics in Nagoya University, Japan

OGAWA, HIDEO, Osaka Prefecture University, Japan

OGURA, KATSUO, Kokugakuin University, Japan

OGURI, MASAMUNE, University of Tokyo, Japan

OHASHI, TAKAYA, Tokyo Metropolitan University, Japan

OHISHI, MASATOSHI, National Astronomical Observatory of Japan, Japan

OHNISHI, KOUJI, Nagano National College of Technology, Japan

OHSUGA, KEN, Yukawa Institute for Theoretical Physics, Kyoto University, Japan

OHTANI, HIROSHI, Kyoto University, Japan

OHYAMA, YOUICHI, Subaru Telescope, NAOJ, Japan

OJHA, DEVENDRA KUMAR, Tata Institute of Fundamental Research, India

OKA, KAZUTAKA, Department of Earth and Planetary Sciences, Kobe University, Japan

OKA, TAKESHI, Department of Astronomy and Astrophysics, The University of Chicago, U.S.A.

OKADA, YOKO, Department of Astronomy, University of Tokyo, Japan

OKAMOTO, YOSHIKO, Institute of Physics, Center for Natural Science, Kitasato University, Japan

OKAMURA, SADANORI, University of Tokyo, Japan

OKAZAKI, AKIRA, Gunma University, Japan

OKAZAKI, WATARU, Science and Engineering, Ibaraki University, Japan

OKIZONO, RYOUSUKE, Osaka Kyoiku University, Japan

OKUDA, TORU, Hakodate College, Hokkaido Univ. of Education, Japan

OKUDA, HARUYUKI, Gunma Astron. OBS., Japan

ONAKA, TAKASHI, Department of Astronomy, University of Tokyo, Japan

ONODERA, SACHIKO, Institute of Astronomy, Faculty of Science, University of Tokyo, Japan

ORESHINA, ANNA, Sternberg Astronomical Institute, Russia

ORIHARA, SHIHO, Osaka Kyoiku University, Japan

OSHIMA, TAI, ISAS, Japan

OTA, NAOMI, Tokyo Metropolitan University, Japan

OYAMA, TOMOAKI, Tokyo University, Japan

OZAWA, TOMOKO, Osaka Kyoiku University, Japan

PARK, MYEONG-GU, Kyungpook National University, Korea

Parker, Quentin, University of Macquarie, Australia

Paul, Bikash Chandra, Physics Department, North Bengal University, India

Paul, Biswajit, Tata Institute of Fundamental Research, India

Pazhouhesh, Reza, Mashad University, Iran

Peterson, Bruce A., Australian National University, Australia

Phan, Van Dong, Faculty of Physics, Ha Noi Pedagogical University, Vietnam

Podgorny, Igor, Institute for Astronomy, Russia

Podgorny, Alexander, Lebedev Physical Institute, Russia

Premadi, Premana, Institut Teknologi Bandung, Indonesia

Raharto, Moedji, Bosscha Observatory, Department of Astronomy, Institut Teknologi Bandung, Indonesia

Rao, A. R., Tata Institute of Fundamental Research, India

Ratag, Mezak Arnold, Indonesian National Institute of Aeronautics and Space (LAPAN), Indonesia

Rattenbury, Nicholas, University of Auckland, New Zealand

Refsdal, Sjur, Institute of Theoretical Astrophysics, University of Oslo, Norway, Norway

Ryu, Dongsu, Chungnam National University, Korea

Sadler, Elaine, University of Sydney, Australia

Saigo, Kazuya, Center for Computational Physics University of Tsukuba, Japan

Saito, Tomoki, Nobeyama Radio Observatory, Japan

Saitou, Masashi S., Department of Science Education. Gunma University, Japan

Saitou, Takayuki, Depertment of Physics. Faculty of Science. Hokkaido University., Japan

Sakamoto, Tsuyoshi, Department of Astronomical Science, The Graduate University for Advanced Studies, Japan

Sakamoto, Seiichi, National Astronomical Observatory of Japan, Japan

Sakurai, Takashi, National Astronomical Observatory, Japan, Japan

Sasaki, Toshiyuki, Subaru Telescope, NAOJ, U.S.A..

Sato, Shuji, Department of Astrophysics, Nagoya University, Japan

Sato, Fumio, Tokyo Gakugei University, Japan

Sato, Yasuhiko, Tokyo University of Science Dept. of Physics, Japan

Sato, Bun'ei, Univ. of Tokyo, Japan

Sawa, Takeyasu, Department of Physics and Astronomy, Aichi University of Education, Japan

Sawada, Tsuyoshi, National Astronomical Observatory of Japan, Japan

SAWADA-SATOH, SATOKO, ISAS, Japan

SETA, MASUMICHI, Communications Research Laboratory, Japan

SHAO, ZHENGYI, Shanghai Astronomical Observatory, China Nanjing

SHARMA, MANGALA, Indian Institute of Astrophysics, India

SHEN, ZHI-QIANG, The Institute of Space and Astronautical Science, Japan

SHIBAI, HIROSHI, Nagoya Uniersity, Japan

SHIBATA, SHINPEI, Department of Physics, Yamagata University, Japan

SHIBATA, KAZUNARI, Kwasan Observatory, Kyoto University, Japan

SHIBATA, KATSUNORI, M., National Astronomical Observatory Japan, Japan

SHIGEYAMA, TOSHIKAZU, Research Center for the Early Universe, Graduate School of Science, University of Tokyo, Japan

SHIMADA, NOBUE, Communications Research Laboratory (CRL), Japan

SHINKAI, HISA-AKI, RIKEN, Comp. Sci. Div., Japan

SHIOYA, YASUHIRO, Astronomical Institute, Tohoku University, Japan

SKULJAN, JOVAN, University of Canterbury, New Zealand

SOBOUTI, YOUSEF, Institute for Advanced Studies in Basic Sciences, Iran

SOFUE, YOSHIAKI, Institute of Astronomy, University of Tokyo, Japan

SOHN, YOUNG-JONG, Center for Space Astrophysics, Yonsei University, Korea

SOMOV, BORIS, Institute of Space and Astronautical Science (ISAS), Japan

STEIN, ROBERT, Michigan State University, U.S.A..

SUDA, HIROSHI, University of Tokyo, VERA Project office, NAOJ, Japan

SUDOU, HIROSHI, Astronomical Institute, Tohoku University, Japan

SUGIMOTO, KANAKO, Nagoya University, Japan

SUGIMOTO, MASAHIRO, National Astronomical Observatory of Japan, Japan

SUGITANI, KOJI, Nagoya City University, Institute of Natural Sciences, Japan

SUGIYAMA, NAOSHI, National Astronomical Observatory Japan, Japan

SUSA, HAJIME, University of Rikkyo, Japan

SUSUKITA, RYUTARO, RIKEN (The Institute of Physical and Chemical Research), Japan

SUTO, YASUSHI, University of Tokyo, Japan

SUWA, TAMON, Hokkaido Univ., Japan

SUZUKI, ISAO, Department of Physics, Science University of Tokyo, Japan

SUZUKI, TAKERU, National Astronomical Observatory of Japan, Japan

TACHIHARA, KENGO, Max-Planck-Institut fuer extraterrestrische Physik, Germany

TAGUCHI, HIKARU, Gunma Astronomical Observatory, Japan

TAJIMA, YUKIKO, Chihaya Nature and Astronomy Museum, Japan

TAKAGI, SHIN-ICHIRO, Department of Physics, Kyoto University, Japan

TAKAHASHI, MASAAKI, Aichi University of Education, Japan

TAKAHASHI, JUNKO, Meiji Gakuin University, Japan

TAKAHASHI, NORITSUGU, Meisei University, Japan

TAKAHASHI, HIDENORI, National Astronomical Observatory of Japan, Japan

TAKAHASHI, ROHTA, Yukawa Institute for Theoretical Physics, Japan

TAKAYAMA, JUN, Sony Corp., Japan

TAKEDA, TAKAAKI, National Astronomical Observatory of Japan, Division of Theoretical Astrophysics, Japan

TAKEI, YOH, Institute of Space and Astronautical Science, Japan

TAKEUCHI, TSUTOMU T., Optical and Infrared Astronomy Division, National Astronomical Observatory of Japan, Japan

TAKEUCHI, AKITSUGU, Yonago National College of Technology, Japan

TAMURA, MOTOHIDE, National Astronomical Observatory of Japan, Japan

TANAKA, YASUO, Ibaraki University, Japan

TANIGAWA, TAKAYUKI, Nagoya University, Dept. of Earth and Planetary Sciences, Japan

TANIGUCHI, YOSHIAKI, Astronomical Institute, Tohoku University, Japan

TANUMA, SYUNITI, Hida and Kwasan Observatoris, Kyoto University, Japan

TARUYA, ATSUSHI, Research Center for the Early Universe(RESCEU), School of Science, Japan

TATEMATSU, KEN'ICHI, National Astronomical Observatory of Japan, Japan

TOMISAKA, KOHJI, National Astronomical Observatory, Japan

TOMITA, KOICHIRO, Advanced Engineering Services Co, Ltd, Japan

TOMITA, AKIHIKO, Wakayama University, Japan

TOSAKI, TOMOKA, Gunma Astronomical Observatory, Japan

TOTANI, TOMONORI, Princeton University, U.S.A.

TSUBOI, MASATO, Ibaraki University, Japan

TSUJIMOTO, MASAHIRO, Kyoto University, Japan

TSUJIMOTO, TAKUJI, National Astronomical Observatory, Japan

TSURIBE, TORU, Osaka University, Japan

TSURU, TAKESHI, Physics, Kyoto University, Japan

TURNER, EDWIN L., Princeton University Observatory, U.S.A.

UCHIDA, YUTAKA, Physics Department, Science University of Tokyo, Japan

UEHARA, HAYATO, Tokyo Gakugei University, Japan

UEMATSU, SACHIKO, School of Science and engineering, Ibaraki Univ., Japan

UENO, MUNETAKA, University of Tokyo, Japan

UJIHARA, HIDEKI, Naotional Astronomical Observatory, Japan

UMEKAWA, MICHIHISA, Chiba University, Japan

Umemoto, Tomofumi, National Astronomical Observatory of Japan, Japan

Umemura, Masayuki, Center for Computational Physics, University of Tsukuba, Japan

Usui, Fumihiko, Dept. of Earth Science & Astronomy, Grad. School of Arts and Sciences, Univ. of Tokyo, Japan

Uyeda, Chiaki, Institute of Earth and Space Science, Graduate School of Science, Osaka University, Japan

Wada, Keiichi, University of Colorado, U.S.A.

Wakamatsu, Ken-ichi, Gifu Univ., Japan

Wang, Yiping, Purple Mountain Observatory, Academia Sinica., China Nanjing

Wang, Jia-Ji, Shanghai Astronomical Observatory, Chinese Academy of Sciences, China Nanjing

Watabe, Yasuyuki, University of Tsukuba, Japan

Watanabe, Jun-ichi, National Astorn. Obs. Japan, Japan

Watarai, Ken-ya, Kyoto University, Japan

Wayth, Randall, University of Melbourne, Australia

Webster, Rachel, University of Melbourne, Australia

Wiramihardja, Suhardja D., Department of Astronomy / Bosscha Observatory, Institute of Technology, Bandung, Indonesia

Wu, Xuebing, Dept. of Astronomy, Peking University, China Nanjing

Xia, Yifei, Department of Astronomy, Nanjing University, China Nanjing

Xiao, Naiyuan, Department of Astronomy, Nanjing University, China Nanjing

Xiao-Ma, Gu, Yunnan Observatory, National Astronomical Observatories, CAS, Kunming, China, China Nanjing

Yahagi, Hideki, Institute of Astronomy, University of Tokyo, Japan

Yaji, Kentaro, Kawabe Cosmic Park, Japan

Yamada, Masako, Osaka University, Japan

Yamada, Toru, NAOJ, Japan

Yamagata, Tomohiko, Ministry of Education, Culture, Sports, Science and Technology, Japan

Yamamoto, Hiroaki, Department of Astrophysics, Nagoya University, Japan

Yamamoto, Naotaka, Science University of Tokyo, Japan

Yamamoto, Fumio, the University of Tokyo, Japan

Yamamoto, Mikio, Miyazaki University, Japan

Yamaoka, Hitoshi, Dept. of Physics, Kyushu Univ., Japan

Yamasaki, Atsuma, Department of Earth and Ocean Sciences, National Defense Academy, Japan

Yamauchi, Shigeo, Iwate University, Japan

YAMAUCHI, MAKOTO, Miyazaki University, Japan

YAMAZAKI, RYO, Department of Physics, Kyoto University, Japan

YAN, YIHUA, National Astronomical Observatories, Chinese Academy of Sciences, China Nanjing

YANAGISAWA, KENSHI, Okayama Astrophysical Observatory, Japan

YANG, ZHILIANG, Dept. of Astronomy, Beijing Normal University, China Nanjing

YANG, JI, Purple Mountain Observatory, China Nanjing

YANO, TAIHEI, National Astronomical Observatory, Japan

YAO, YONGQIANG, Purple Mountain Observatory, Academia Sinica, China Nanjing

YEH, CHIN-TEH, Department of Astronomy, Nanjing University, China Nanjing

YIN, QI FENG, The National Radio Astronomy Observatory, U.S.A.

YOKOGAWA, SOZO, National Astronomical Observatory of Japan, Japan

YOKOO, TAKEO, Osaka Kyoiku University, Japan

YOKOO, HIROMITSU, Kyorin University, School of Health Sciences, Japan

YOKOSAWA, MASAYOSHI, Ibaraki University, Japan

YOKOYAMA, TAKAAKI, National Astronomical Observatory of Japan, Japan

YONEHARA, ATSUNORI, Center for Computational Physics, University of Tsukuba, Japan

YONEKURA, YOSHINORI, Osaka Prefecture University, Japan

YOON, TAE SEOG, Kyungpook National University, Korea

YOSHIDA, FUMI, National Astronomical Observatory of Japan, Japan

YOSHIDA, MICHITOSHI, National Astronomical Observatory of Japan, Japan

YOSHII, YUZURU, Institute of Astronomy, School of Science, Univ. of Tokyo, Japan

YOSHIKAWA, MAKOTO, Institute of Space and Astronautical Science, Japan

YOSHIKAWA, KOHJI, Research Center for the Early Universe, The University of Tokyo, Japan

YOSHIOKA, SATOSHI, Tokyo University of Mercantile Marine, Japan

YUAN, QIRONG, Nanjing Normal University, China Nanjing

ZHAO, GANG, National Astronomical Observatories of China, China Nanjing

ZHU, ZI, Shaanxi Astronomical Observatory, China Nanjing

Large and New Facilities: Science and Development

IAU 8th Asia-Pacific Regional Meeting
ASP Conference Series, Vol. 289, 2003
S. Ikeuchi, J. Hearnshaw & T. Hanawa, eds.

Steps Toward the Next Generation CFHT

Gregory G. Fahlman

Canada-France-Hawaii Telescope Corporation, P. O. Box 1597, Kamuela, HI 96743

Abstract. The 3.6-m Canada-France-Hawaii Telescope was started as a project in 1974 and saw first light in 1979. It was the first international project to commit to a location on the summit of Mauna Kea and, from the outset, was dedicated to the best exploitation of its superb site. The context in which CFHT operates has changed with the presence of four 8-m-class telescopes on the Mauna Kea summit. The CFHT site can easily accommodate a larger aperture facility and the recent designation of our site by the University of Hawaii as one that may be "recycled" has opened the real possibility of constructing an ngCFHT. Here I will describe the progress toward this end.

1. Introduction

The CFHT is a partnership between the National Research Council of Canada, the Centre National de la Recherche Scientifique of France, and the University of Hawaii, with these member agencies holding 42.5%, 42.5% and 15% interests in the Corporation. The project was started in 1974 with the signing of a Tripartite Agreement and first light was obtained in 1979. Under the direction of René Racine, the CFHT became noted for the superb image quality delivered to the instruments, vindicating the belief of the early proponents of the site that sub-arc second imaging should be a routine occurrence given the altitude and favorable wind patterns that pertain to Mauna Kea. Indeed, CFHT was instrumental in establishing Mauna Kea unequivocally as the best site in the northern hemisphere, if not the world, for ground-based optical and infra-red astronomy. A recent citation analysis (Benn & Sanchez, 2001) showed that CFHT was the leading 4-m-class telescope in the world in the past decade.

At the present time, CFHT is completing its wide-field imaging initiative. MegaCam, a 340-megapixel CCD camera with a 1° x 1° field of view is being commissioned now, and WIRCam, the first of a new generation of mosaicked infrared imagers is under development for deployment in 2004. These instruments, in particular, will allow CFHT to remain competitive in forefront astronomy to about the end of this decade. However, looking at the 2010 horizon, the member agencies have encouraged the Corporation and the CFHT astronomical community to consider the replacement of the present CFHT with a new facility.

The first important step in that direction was the internal *Report on Options for use of the Existing Pier with a New Telescope* prepared by Walter Grundman in 1997. This study, which was based mainly on the emerging experience with 8-m-class telescopes, concluded that an 8-m–12-m telescope could be accommodated on the existing pier. This technical report was evidently not acted upon, largely because the two major partners, Canada and France, were already engaged in 8-m-class projects (Gemini and VLT respectively). A science case was needed to identify the requirements for a new CFHT.

The CFHT Board then commissioned a study by a select committee of astronomers from the three communities. The committee was chaired by Harvey Richer and Claude Catala. This group, known as the NGC, focussed their attention on science issues and the potential synergies between a new optical facility and other ambitious projects, like ALMA and NGST, that will operate at other wavelengths. The NGC report (1998, available from the CFHT Web site: www.cfht.hawaii.edu) identified five key questions for any new astronomical facility: (1) Are we alone in the universe? (2) What is the universe made of and what is its overall geometry? (3) How did our and other solar systems form and evolve? (4) What were the first sources of light in the universe and how did galaxies like our own come into being? (5) Are there things in the universe that we haven't yet dreamed of? These questions informed a detailed examination of the capabilities of current and proposed facilities. The conclusions of the NGC were: (1) The existing CFHT should be operated with most of its current suite of instruments for about another decade. (2) Beginning almost immediately the CFHT Board should initiate a Phase A study of a large optical telescope, in the range of 25 m+. This could be built on the current CFHT site. (3) If it is decided to build this large telescope, construction could begin in about 2008.

While these conclusions have not been acted upon, the idea that the next step in OIR facilities should be a large aperture telescope has certainly been endorsed *de facto* by the astronomical community in general. The NGC set the bar at 25 m to be competitive with, and complementary to, an 8-m NGST. As NGST is now baselined at 6 m, a 20-m aperture would satisfy the science case developed by the NGC. The idea of rebuilding on the CFHT site predated the adoption of the new Master Plan for the Mauna Kea summit, which imposes certain guidelines for redevelopment that are discussed briefly below. Finally, the intent expressed by the proposed construction date is clear: to move quickly so that discoveries made by new facilities like NGST and ALMA can be followed up at optical and near-infrared wavelengths in a timely way. This basic argument remains a key motivation behind all proposals for next generation large telescopes.

2. The Master Plan for the Mauna Kea Science Reserve

The summit of Mauna Kea is the highest point of land in the entire Pacific basin and is considered sacred ground by the indigenous Hawaiian people. The construction of telescopes on the summit has therefore always been a contentious issue. The present generation of facilities were all constructed under guidelines stipulated in a previous Master Plan. The site is managed by the University of Hawaii under a lease granted by the State of Hawaii. A state audit completed in 1998 revealed deficiencies in fulfilling the environmental and cultural goals contained in the then existing Master Plan. As a result of this, and recognizing that further development might be anticipated, the University initiated a process to develop a new Master Plan that would address all the issues raised in a comprehensive way. After a long public debate, the new Master Plan was finally adopted on 2000 June 16. (The full text is available from the University of Hawaii site: www.hawaii.edu/maunakea/.)

For the purposes here, the most significant part of the Master Plan is that those sites on the summit ridge now occupied by 4-m-class telescopes (or smaller) may be "recycled"; i.e., redeveloped with new telescopes of larger aperture. The real estate now occupied by these *small* telescopes is very precious and, especially given the sensitivity of the site, should be used for the highest purposes. Of additional note is that a new site for a "next generation large telescope" (defined as one with an aperture in excess of 25 m) has been designated in a location just below the summit ridge.

In response to concerns about the visual impact of the telescope enclosures, the Master Plan sets forth a number of guidelines for the enclosures that might be placed on the redeveloped sites. The most critical guidelines specify that the height and width should be limited to approximately 130 feet (about 40 m). These restrictions impose a severe constraint on the maximum aperture that could be contemplated for a ridge site. (The published guidelines do allow for buried structures but that is of doubtful practicality.) An aggressive optical design might specify a primary f-ratio of f/1, and thus with a conventional hemispherical dome, the aperture would be limited to about 20 m. With an even faster f-ratio, innovative design of the enclosure and perhaps elevation restrictions on pointing, a somewhat larger aperture could be accommodated within the guidelines, but probably not all the way to 30 m.

3. Concept Studies

In order move the process towards a next generation CFHT (ngCFHT) forwards, the Observatory, in consultation with its Science Advisory Council, issued a call for concept studies. The concepts were to be constrained by the Master Plan guidelines and were to include a science case, initial instrumentation ideas and at least a discussion of the feasibility of the proposed design. Three teams were selected from among the responders, one in each of the three communities served by the CFHT. Mid-study results were first presented in December 2000 at a meeting held to celebrate the first 21 years of CFHT operations , and the final reports were presented in 2001 May, in time for a discussion at the tri-annual CFHT Users Meeting held in Lyons. More detailed information about each of these concepts may be found by following links from the CFHT Web site. I will only briefly summarize the principle features of each concept here.

3.1. University of Hawaii

Jeff Kuhn served as the leader of a large group of UH astronomers who contributed to a study of a High Dynamic Range Telescope (HDRT) as a solution for the ngCFHT. The concept was derived from an pre-existing study to replace the NASA funded IRTF on Mauna Kea with a 6.5-m New Planetary Telescope. The innovative aspect of the NPT was that it would be an off-axis design with an un-obscured aperture. For the ngCFHT, the group presented a design that consists of six circular NPT type segments, 6.5-m off-axis mirrors that are figured

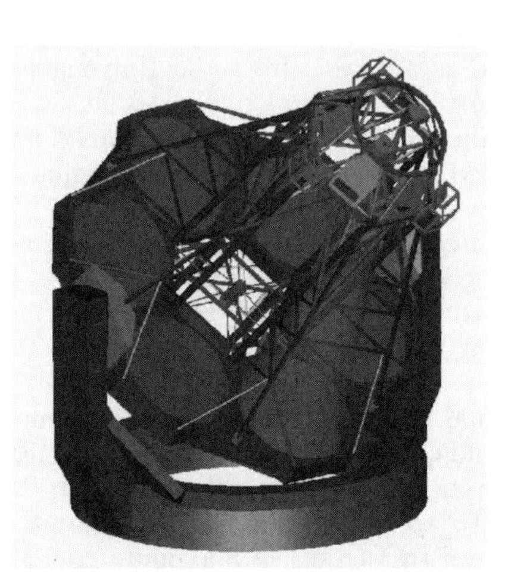

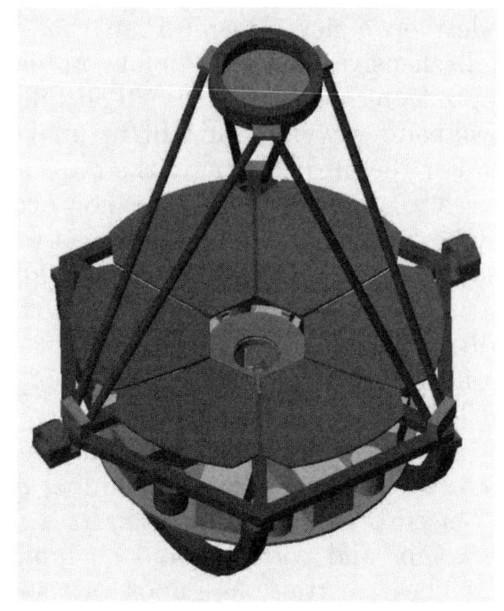

Figure 1. The HDRT is shown on the left, the LPT on the right.

from a parent 22-m paraboloid, and arranged around a stiff central structure. Two distinct optical configurations are possible with this design: (1) the wide-field option has 6 deployable 2.3-m secondary mirrors that direct light downwards to a monolithic 7-m tertiary, in a Paul-Baker configuration, that yields a $1^\circ \times 1^\circ$ field of view; and (2) the narrow-field mode in which the wide-field secondaries are retracted and the light passes to a Gregorian focus employing 6 small secondary mirrors, yielding a $3' \times 3'$ field. This design, in which the wave front incident on a segment from an on-axis target suffers no obscuration, was chosen to minimize the light diffracted from the core of the PSF. A schematic view of the telescope is shown in Figure 1.

3.2. Marseille-Paris

A group centered at the Laboratoire d'Astrophysique in Marseille and the Observatoire de Paris headed by Denis Burgarella presented a concept, now called the Large Petal Telescope (LPT) that features a filled aperture consisting of six large segments. This design was motivated by the desire to minimize the degradation of the pupil PSF caused by the scattering of light from edge effects associated with segment boundaries. The primary segments are nominally 8-m aspherics, shaped, as shown in Figure 1, to fill the pupil plane. The study noted that the use of large, thin segments has the further advantage that figuring tolerances can be relaxed (compared to small segments) because the active support system provides compensation. The optical design was aimed at providing good image quality over a large focal plane to be located behind the primary.

3.3. Canada

The concept from Canada was the work of a consortium headed by Raymond Carlberg of the University of Toronto that involved contributions from the university community, industry, and the Herzberg Institute of Astrophysics (HIA), a unit of the National Research Council of Canada. This study was undertaken within the larger context of the Long Range Plan (LRP): *The origins of Structure in the Universe*, a document that sets out development goals for astronomy in Canada over the next decade (see: www.casca.ca/lrp/). Among the priorities identified in the LRP is work directed towards Canadian participation in a next generation, large optical telescope (LOT). The HIA has established a project office to support on-going efforts in this area and a science steering committee has been formed to provide the top-level requirements for the Canadian design.

The concept presented to CFHT was for a 20-m primary aperture constructed of hexagonal segments, generically similar to the Keck telescopes but using smaller pieces. Considerable effort was devoted to achieving a good understanding of the segment edge-effects on the telescope PSF and to develop concepts for the telescope structure and enclosure design that met the Mauna Kea guidelines. Work in these and related areas has advanced considerably over the past year, and many elements of the original concept have undergone substantial evolution. A recent summary of the project status has been presented by Roberts et al. (2002).

4. Present Status

Of the three concepts summarized above, only that from Canada has been significantly advanced over the past year for the reasons mentioned above. The work is not uniquely oriented toward ngCFHT, since it is applicable to any LOT and, prudently, Canada will continue to explore all reasonable partnership arrangements. The French astronomical community will be conducting a prospective study in 2003 May that will include consideration of the future of CFHT. In the meantime, some work on ngCFHT is continuing in France, mainly in the context of examining the issue of providing adaptive optics for a (generic) LOT.

Recently, the HIA and the Marseille-Paris groups have joined the CFHT in sponsoring a fairly comprehensive study of the manufacturing issues related to primary segmentation. This study, to be carried out under contract by Sagem, will consider segments ranging from small (1 m) to large (8 m), and is aimed partly at the issue of whether the use of large segments is a cost-effective solution for a filled LOT primary.

The Canadian and French teams have agreed to adopt the baseline Canadian optical design for the purposes of this study. The key elements of the design include: A classical R-C design with a 20-m primary at f/1 and a 2.5-m secondary with a final focal ratio of f/15. The back focal length is 18 m. The field of view is $20'$ and will be limited to $11'$ by a 1-m refractive corrector with an ADC for applications that demand the highest image quality. A tertiary will take the beam out to flanking Nasmyth platforms.

In the meantime, The Canadian firm, AMEC Dynamic Systems (ADS) has advanced work on the telescope structure and enclosure design. The telescope concept, shown in figure 2, resembles the LBT design, with a hexapod support structure for the secondary mirror. The proposed enclosure is a hemi-spherical calotte that satisfies the Mauna Kea Master Plan guide lines and offers great protection against wind and other deleterious weather conditions.

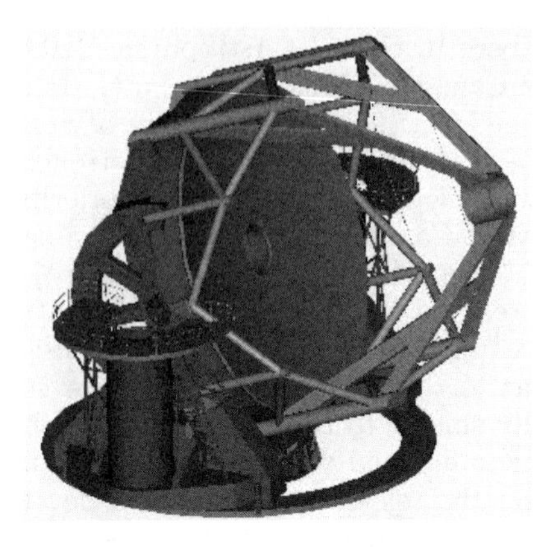

Figure 2. The proposed Canadian VLOT and Calotte enclosure.

5. International Context

The evolution in optical astronomy from 4-m-class facilities to 8-m-class facilities was begun with the segmented design embodied in the twin telescopes of the William O. Keck Observatory. The first of these came on-line in 1992. Since then, several new general purpose telescopes in the same class have been completed to an operational state, including the four unit telescopes of the VLT, the Gemini twins, and Subaru. These new-comers all employ very thin monolithic primaries with active support systems to maintain their optical figure. Others are still under construction, notably the Gran Telescopio Canaries, superficially similar to the Keck design, but with many detailed improvements, and the Large Binocular Telescope (LBT), a design that uses a pair of 8-m primaries mounted on a common support structure. There are also some intermediate-sized facilities in use: the twin Magellan telescopes at the Las Campanas site of the Carnegie Institution and, the refurbished MMT telescope, each with 6.5-m stiff borosilicate primaries. In addition, there are two telescopes, the HET in Texas and the SALT, with large segmented primaries that sacrifice pointing flexibility to achieve a large collecting area at relatively low cost. The most extreme example of that particular approach are the liquid mirror telescopes that are restricted to a field of view around the zenith. The largest of these, an experimental 6-m, is under construction by Paul Hickson (see www.astro.ubc.ca/LMT/).

The upgrading of world-wide optical facilities has seen an astounding increase in collecting area that has also been accompanied by a leap forward in delivered image quality under natural atmospheric conditions and, of course, the deployment of adaptive optics has further expanded the information obtained from using our telescopes. The 8-m-class telescopes are, with the exception Keck, only now becoming fully operational with a reasonably complete instrumentation complement and will certainly be the principal tools used in OIR astronomy in the next decade or so.

The large number of 8-m-class facilities has side-effects that bear some consideration. The first is that the sheer number of these facilities has levelled the astronomical playing field – competitiveness is not so much a matter of aperture anymore, but rather in the domains of instrumentation and operational efficiency. The scope for differentiation between different observatories is small, except that the cost of maintaining a range of instruments fully to cover the spectral resolution-wavelength plane is very high. These instrumentation costs are likely to drive cooperative developments that would limit instrumentation competition through time exchange programs, further levelling the playing field. However, this equilibrium is unlikely to remain static for too long. One result of the proliferation of 8-m-call telescopes is that the science will reach the point of *needing* a bigger aperture telescope much sooner than past historical trends might suggest.

A second point is more relevant to the issue of recycling the Mauna Kea site: the large number of 8-m-class telescopes has made the prospect of replacing the small telescopes with 8-m-class (plus a minus a modest number) telescopes unappealing in that any science case that might be made is going to be matched against the (largely unexploited) capabilities of existing facilities. It is therefore more likely that the redevelopment will be directed either towards special purpose facilities of moderate aperture, or to a telescope with a significantly larger aperture than any existing today. As we we have already noted, the present design guidelines will limit the aperture on the summit ridge sites to about 20 m. That is probably the minimum size that one would consider a worthwhile challenge for a next generation telescope.

There are several groups around the world that are developing concepts for a next-generation large optical telescope. The most prominent projects are the 30-m CELT, developed by the University of California and Cal Tech partnership, and the 100-m OWL studies sponsored by ESO. The OWL project is especially interesting because its development has been couched in terms of what is *feasible* rather than what is *needed*, relying on the obvious point that, all else being equal, a bigger aperture is better than a smaller one. The 30-m aperture of CELT has been selected largely for similar considerations. Both choices provide some verification of the sense that technical feasibility has been the main driver in facility development. At the time of this writing, only the CELT partnership has, with publication of the Green Book, demonstrated feasibility and argued that they are in a position to move forwards to a construction phase.

6. Concluding Remarks

In recent history, telescopes have passed through three generations defined by factors of about two in aperture. However, it is difficult to know for sure the extent to which science cases are shaped by technical feasibility. Technology reflects past discoveries of a more fundamental nature and further, does not advance equally in all areas. A huge leap forward with today's technology may land us in tomorrow's dead-end. Another consideration is that our science cases are necessarily rooted in what we know today. A case for a huge change in capability is not hard to make, of course, but does get harder to justify as the cost rises. Some new discovery could point us in an unanticipated direction that has been precluded by a prejudice in our design arising from some unrecognized assumption. Both of the above arguments present risks which are acceptable at some cost point – that point being a judgment of how future options may be limited by spending in this way today. The historical factor of two in telescope aperture may therefore also reflect an implicit assessment of scientific knowledge in the field, quite apart from any technology issues.

A problem with relying on the "bigger is better" argument is that all else is rarely equal. Clearly cost is a major consideration and, traditionally, cost is traded against benefit. Now pretty much anything that can be done on a smaller telescope can be done faster and *cheaper* on a bigger telescope, but this still begs the question: why build a big telescope if a smaller telescope will accomplish the key scientific goals? Public funding agencies ultimately place constraints which are logically (often, practically) independent of technical feasibility issues.

An ngCFHT will be expensive. At the present time, we have no reliable way to estimate the likely cost. Simple scaling arguments, based on the estimates published by the CELT partnership, suggest that a functioning telescope in an enclosure could be constructed for about $150 M. This would *not* include an AO system or any instruments. The cost of an AO system is particularly difficult to estimate (CELT has allocated close to $150 M for an AO system) and, as yet, there are no well-developed instrumentation proposals for an ngCFHT. It isn't difficult to imagine that a suite of three instruments, together with some form of AO could add $100 M to the cost of a bare telescope The price of a functioning ngCFHT is about 10 times the cost of the original CFHT that opened in 1979. Some fraction of this is simple inflation (about one third) with the rest attributed to the higher cost of being competitive in today's research environment.

The key points in favor of an ngCFHT include:

- it would be located on one of the premier sites in the world, with a proven, accessible, and well-developed infrastructure in place;
- the current conceptual designs for both the telescope and enclosure are robust and compatible with the development guidelines applicable to Mauna Kea;
- the mechanical and structural requirements of a fast-20-m design are a modest extension from the LBT;
- the science case for a 20-m telescope is already strong and meets the key requirement of competitiveness with the 6-m NGST;
- the cost, while high, is not outrageous, and is within reach of a consortium of 3–5 partners with ambition but relatively modest means

Neither Canada nor France is likely to assume the same proportional financial interest in ngCFHT as they now have in CFHT. At least one, and probably more, new partners will be needed to bring this concept to reality.

References

Benn, C. B. & Sánchez, S. F. 2001, PASP, 113, 385

Roberts, S. et al. 2002, to be published in the proceedings SPIE Conference 4840, Future Giant Telescopes.

IAU 8th Asia-Pacific Regional Meeting
ASP Conference Series, Vol. 289, 2003
S. Ikeuchi, J. Hearnshaw & T. Hanawa, eds.

HERCULES: A High-resolution Spectrograph for Small to Medium-sized Telescopes

Hearnshaw, J. B., Barnes, S. I., Frost, N., Kershaw, G. M., Graham, G.

University of Canterbury, Department of Physics and Astronomy, Private Bag 4800, Christchurch 8020, New Zealand

Nankivell, G. R.

formerly Lower Hutt, Wellington New Zealand; now deceased

Abstract. The High Efficiency and Resolution Canterbury University Large Echelle Spectrograph (HERCULES) is a fibre-fed échelle spectrograph that has been in operation at the Mount John University Observatory for just over one year. HERCULES can capture the spectrum from 380 nm to 880 nm on a single 50-mm square CCD. Resolving powers of 40 000 to 80 000 are possible using different fibres.

The spectrograph is designed to achieve high efficiency when the seeing is well matched to the image scale on the fibre input (up to 20% in 1 arcsec seeing), and high stability is achieved by having the spectrograph installed inside a vacuum tank in a thermally isolated environment. Radial velocities with a precision of $\leq 10\,\mathrm{ms}^{-1}$ are routinely possible.

1. Introduction

A new fibre-fed spectrograph has recently been commissioned at the Mt John University Observatory. This instrument replaces the échelle spectrograph for high resolution stellar spectroscopy that has being operating at the Mt. John University Observatory (MJUO) since 1977 (Hearnshaw 1977).

A much more versatile and efficient instrument has been designed. Our new spectrograph is known as HERCULES (High Efficiency and Resolution Canterbury University Large Echelle Spectrograph). It is used with the MJUO's 1-m telescope. Details of the philosophy behind the design and a complete description of the instrument can be found in Hearnshaw et. al (2002).

2. The Optical Design

The goals of the HERCULES optical design were: (a) to capture the entire wavelength range 380–880 nm in a single exposure on a 50 mm square CCD; (b) to maximize stability and reproducibility of the spectra; (c) to achieve high efficiency even in the mediocre seeing ($2''$–$3''$) experienced at MJUO; (d) to deliver resolving powers in the range $R = 40$–$80\,000$.

It was decided early in the design process that the spectrograph would have no moving parts. This greatly simplifies the mechanical design while ensuring maximum stability and reproducibility of spectra. Motivation for this design is given in Hearnshaw et al. (2002). HERCULES uses a single 31.6 gr/mm échelle

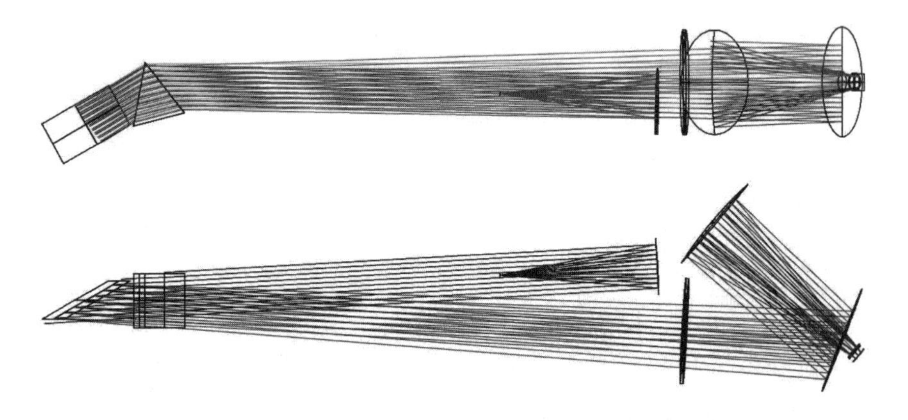

Figure 1. Optical design of HERCULES

grating with a blaze angle $\theta_B = 64.33°$ ($\tan\theta_B = 2.08$) This grating is a replica from the Richardson Grating Laboratory's master ruling MR152. The grating substrate is made from the low-expansion ceramic AstroSital.

The Littrow angle θ is made as small as possible so as to give high peak efficiency while still separating the incident and diffracted beams. HERCULES uses $\theta = 3.0°$ which ensures a modest spectrograph length of 4.5 m.

The collimator an on-axis paraboloid. This means that the fibre exits inside the return beam; however the obstruction is minimal. The collimator has a focal length of $f_{coll} = 783$ mm, and a diameter of 210 mm. While the elliptical footprint of the incident beam overfills the échelle grating by 14%, the larger collimator allows a larger fibre angular size on the sky and hence a net efficiency gain (Diego & Walker 1985). The collimator is made from Zerodur and is 35 mm thick.

Prism cross-dispersion was chosen because it allows significantly greater efficiency over a broad wavelength range than gratings. Cross-dispersion is provided by a single BK7 prism with an apex angle of $\alpha_P = 49.50°$. The prism height is 276 mm, base 258 mm, and length 255 mm and the mass is 23 kg. Even though the prism is used in double pass, over 80% of photons are transmitted at all wavelengths. The use of a prism in double pass causes the angle of illumination relative to the normal plane of the échelle facets to vary from one order to the next. This introduces a small wavelength-dependent line tilt. However, this has negligible effect on resolving power. See Hearnshaw et. al (2002) for further details.

HERCULES uses three different fibres in order to achieve the desired resolving powers (Table 1). All fibres are Ceramoptec step index fibres with enhanced UV transmission.

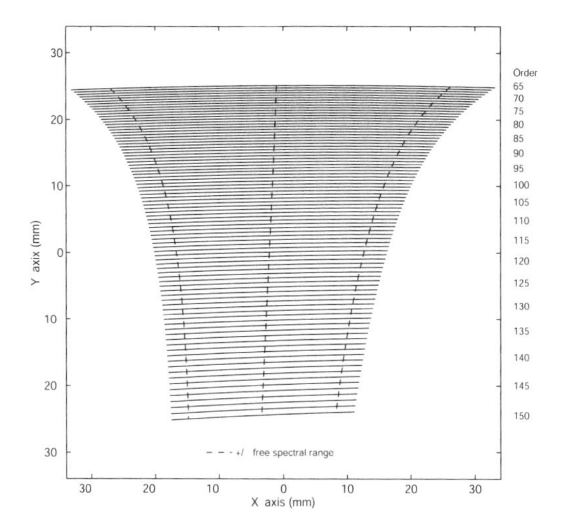

Figure 2. The spectral format of HERCULES. Shown are orders $m = 65$ ($\lambda_B = 875.1\,\text{nm}$) to $m = 150$ ($\lambda_B = 379.3\,\text{nm}$). The dashed lines show the positions of the order centre and the limits of the free spectral range. The orders are traced as far as the half-maximum intensity on each side of the order centre.

Table 1. Fibres and resolving powers.

Fibre #	core diam. (μm)	microslit (μm)	R
1	100	–	41 000
2	100	50	70 000
3	50	–	82 000

The f/13.5 beam of the MJUO 1-m telescope is converted to f/4.5 by microlenses which are cemented onto each of the fibre's input faces. This means that each of the 100-μm fibres accepts 4.5″ and the 50-μm half that. Fibre option 2 has a 50-μm microslit aluminized on the output face. Options 2 and 3 were designed to give the same resolving powers, although option 3 would give superior throughput in good seeing ($\theta_* \leq 1.8''$).

A Schmidt camera is used to give very good performance over a large wavelength range. The design is folded by a 550-mm diameter flat at an angle of 19.0°. The 100-mm diameter perforation that this necessitates leads to 23% light loss at some wavelengths, but gives the advantage of a focal plane which is external to the camera. A 525-mm BK7 corrector plate with a 15-mm thickness forms the camera entrance. The camera mirror of HERCULES is 500 mm in diameter. The optical surface is spherical giving a focal length of 973 mm. The last optical component is a small field correcting lens. The rms spot size ranges from 4.5 μm in the UV to below 1 μm near Hβ, 3 μm at Hα, rising to 6 μm near the ends of the red orders. All mirrors are made from Zerodur and have an enhanced silver coating applied. All refracting surfaces are overcoated with single layer MgF_2 broad-band anti-reflection coatings. The HERCULES optical layout is shown in Figure 1.

Presently a 25×25-mm square CCD with 24-μm pixels is in use. This necessitates three exposures to cover the entire available wavelength range. It is anticipated that we will upgrade to a larger CCD with smaller pixels in the the near future, The entire spectral format is shown in Figure 2.

3. Mechanical Design

The entire spectrograph is housed inside a vacuum tank which is maintained at a pressure of 2-3 torr. This is to ensure minimal susceptibility of the spectrograph to the effects of changing air pressure, temperature stratification and convection currents. All these effects would compromise precise radial-velocity measurements. The vacuum also ensures an extremely clean dust-free environment for the optical components. The camera's field-flattening lens serves as the window between the externally mounted CCD and the vacuum-encased spectrograph. Experience has shown that the vacuum can be maintained for months between successive evacuations. The HERCULES vacuum tank is shown in Figure 3.

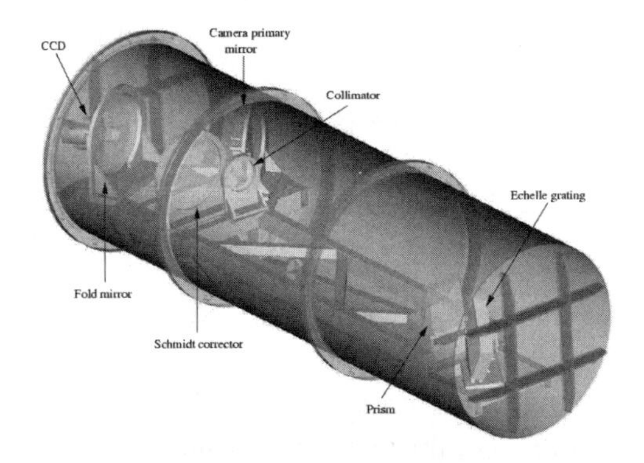

Figure 3. The HERCULES spectrograph inside the vacuum tank. The tank is formed in three sections. The first section, which encompasses the camera optics, is rigidly connected to the spectrograph bench. The lid (on which the camera is mounted) and the other two sections are free to roll away on rails.

4. Performance

Theoretical predictions of the efficiency of HERCULES have been made to find the fraction of photons falling on the primary mirror of the telescope which are finally detected as a function of wavelength and seeing. Table 2 shows the results.

Table 2. Efficiency of HERCULES.

	seeing λ (nm)	$1''$	$2''$	$3''$
red	650	20%	17%	12%
green	500	17%	15%	11%
blue	420	11%	9%	7%

Measurements of the actual efficiency of HERCULES were made using the star ϵ Eri (HR 1084) with spectrophotometric data taken from Breger (1976). While the spectrograph's performance comes close to that predicted, there was some discrepancy, with actual values being only half those calculated in the green. This is likely due to cumulative guiding errors. An improved auto-guiding system will improve the throughput. Uncorrected atmospheric dispersion and larger atmospheric extinction than adopted may also be contributing factors.

Currently a radial-velocity precision of less than $10\,\mathrm{ms}^{-1}$ is routinely possible. Observations of a thorium-argon emission lamp before and after stellar exposures are used to provide wavelength calibration, and velocity measurements are made using standard cross-correlation techniques. Observations of blue-sky solar spectra and late-type dwarf stars have produced radial velocity measurements with an rms scatter of around $4\,\mathrm{ms}^{-1}$. Bright stars are less precise. Continuous autoguiding will soon be implemented, and we expect throughput will improve significantly, and stellar radial-velocity precision will also be down to about 3 or $4\,\mathrm{ms}^{-1}$.

5. Conclusion

The HERCULES spectrograph has been in continual operation at MJUO for over one year. The efficiency and stability of HERCULES have proven to be exceptional, although some improvements are expected. The acquisition of a single large CCD which will remain attached to the instrument will lead to considerable gains. Upgraded auto-guiding will result in increased efficiency as well as radial-velocity precision.

References

Breger, M. 1976, A&A, 32, 7

Diego, F. & Walker, D. D. 1985, MNRAS, 217, 347

Hearnshaw, J. B. 1977, Proc. Astron. Soc. Australia, 3, 102

Hearnshaw, J. B., Barnes, S. I., Kershaw, G. M., Nankivell, G. R, Frost, N., Ritchie, R., & Graham, G. 2002, submitted to Exper. Astron.

Murdoch, K. A., Hearnshaw, J. B., Clark, M. 1993, ApJ, 413, 349

Skuljan, J., Hearnshaw, J. B., Cottrell, P. L. 1999, in ASP Conf. Ser. Vol. 185 (IAU Coll. 170), Precise Stellar Radial Velocities, ed. J. B. Hearnshaw & C. D. Scarfe (San Francisco: ASP), 91

IAU 8th Asia-Pacific Regional Meeting
ASP Conference Series, Vol. 289, 2003
S. Ikeuchi, J. Hearnshaw & T. Hanawa, eds.

A Detailed Analysis of the Short- and Long-term Precision of Stellar Radial Velocities Obtained Using HERCULES

Jovan Skuljan

Department of Physics and Astronomy, University of Canterbury, Private Bag 4800, Christchurch 8020, New Zealand

Abstract. An analysis has been made to examine the overall stability and final precision of relative stellar radial velocities obtained using the Hercules spectrograph. The velocities are measured by cross-correlation between one-dimensional spectra in logarithmic wavelength space. Some delicate reduction steps are performed in order to achieve the highest precision possible. A typical precision of a few metres per second is easily obtained during one observing night. Stellar velocities are somewhat affected by insufficient scrambling by the optical fibre. The long-term velocities also exhibit some systematic run-to-run differences. A preliminary analysis shows that these problems can be eliminated by making some minor modifications to the equipment and also by careful comparison of thorium images. This is expected to bring the overall precision to a level of several metres per second, which would make Hercules fully capable of detecting extra-solar planets.

1. Introduction

A new high-resolution fibre-fed échelle spectrograph called HERCULES has been installed at Mt John University Observatory, to be used on the 1-m telescope (Hearnshaw et al. 2002). A 1024×1024 SITe CCD camera is used as a detector in four different spectral regions.

Over the past twelve months a large number of stellar spectra has been collected and a dedicated reduction software package has been developed. Some preliminary analysis of the results has been made in order to examine the overall stability and final precision of relative stellar radial velocities obtained with this spectrograph.

2. Reduction Procedure

A dedicated computer program called HRSP (Hercules Reduction Software Package) has been developed in order to achieve the maximum efficiency in the reduction of Hercules spectra. The program has been written in C and has been optimized for Hercules échelle spectra. A single command is needed to reduce a given stellar spectrum and the reduction typically takes less than a minute. Standard FITS images are used both as input and output.

Three types of images are obtained with Hercules: the white lamp (used for flat-fielding), the thorium lamp (used for wavelength calibration) and the stellar spectra. A typical CCD image contains about 40–50 orders, depending on the spectral region. For high-precision radial velocity work we use orders 90–126, between $m\lambda_1 \approx 567\,000$ and $m\lambda_2 \approx 571\,000$ Å.

A standard échelle reduction procedure is used, including: the background and cosmic ray subtraction, order extraction, flat-fielding (continuum straightening), continuum normalization and wavelength calibration (rebinning).

Relative radial velocities are obtained by cross-correlation between two spectra of the same star. Every order is cross-correlated separately and the arithmetic mean velocity is calculated. Spectra are prepared for cross-correlation by first subtracting the mean flux and then by applying a cosine-bell window to the edges (Simkin 1974).

3. Possible Sources of Errors

There are numerous factors affecting the precision of stellar radial velocities (see e.g. Griffin & Griffin 1973). We shall focus here on a special case of a Cassegrain échelle spectrograph in a vacuum, with fibre feed, a thorium lamp for wavelength calibration and cross-correlation technique used for velocity determination. Some possible sources of errors are: difference in illumination of the optical fibre by stellar and reference sources (poor scrambling), possible physical instabilities (shifts, vibrations) in the spectrograph and CCD camera, poor (unstable) wavelength calibration, line smearing due to long exposures combined with the Earth's rotation, presence of telluric lines, poor barycentric correction, problems associated with discrete sampling and rebinning and, finally, the photon noise. A careful design and construction of the spectrograph, combined with some delicate steps involved in the data reduction process, have been used to eliminate (or minimize) most of these effects.

The photon noise is unavoidable and it sets a theoretical limit to the velocity precision. It can be demonstrated (Murdoch & Hearnshaw 1991) that the standard deviation of measured radial velocities is inversely proportional to the S/N ratio. This result can be verified easily by means of numerical simulations, if a set of random images is generated from a single observed spectrum (each pixel value can be varied randomly, in accordance with the Poisson statistics) and then all spectra are cross-correlated with the original. It was found from such a test that a precision of about 3 $\mathrm{m\,s^{-1}}$ should be expected at $S/N \sim 100$.

4. Blue Sky Spectra

Blue sky spectra are used to check the stability of radial velocity determinations. There are many advantages of using the blue sky as a calibration source and these include the fact that the sky is uniformly bright, which eliminates any problems with poor fibre scrambling. Also, the radial velocity of the Sun is known to extremely high accuracy.

A typical short-term precision of Hercules sky spectra is about $3 - 4$ $\mathrm{m\,s^{-1}}$, which is close to the photon-noise limit. Our long-term precision is somewhat lower ($5 - 6$ $\mathrm{m\,s^{-1}}$), as seen in Figure 1a. There seem to exist some systematic

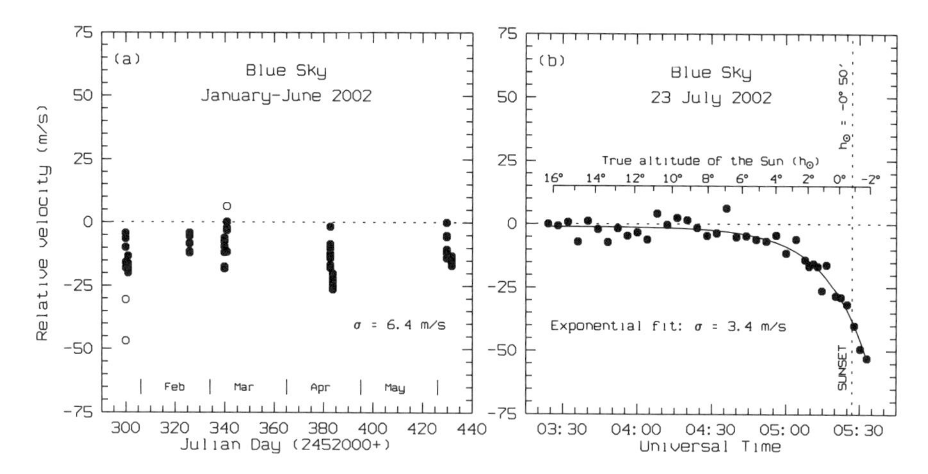

Figure 1. (a) Long-term radial velocities of the blue sky. (b) The effect of the differential atmospheric extinction at low altitudes.

shifts in the zero point between different observing runs (even from one day to another), but these effects can be eliminated by introducing some improvements to the observing and reduction procedures. For example, it has become obvious that the blue sky velocities depend very much on the apparent altitude of the Sun (Figure 1b). This is caused by the differential extinction in the Earth's atmosphere, combined with the Sun's rotation. The upper solar limb appears slightly brighter than the lower one and this produces an apparent shift in the average radial velocity of as much as $40 - 50$ m s^{-1}, when the Sun is on the horizon. This effect can be minimized by observing the sky only when the altitude of the Sun is greater than $8 - 10$ degrees.

5. Stellar Spectra

Stellar radial velocities seem to exhibit a significantly larger scatter. This is almost certainly a result of poor scrambling by the optical fibre, so that the whole spectrum appears shifted across the CCD chip, depending on the actual position of the stellar image (seeing) with respect to the fibre. Although the telescope is repositioned for each exposure, the star is never found exactly at the same spot. This effect becomes especially prominent for the brightest stars, when the exposure times are extremely short (~ 10 s), so that the illumination of the fibre is highly non-uniform. At longer exposures, the situation is slightly different, due to frequent adjustments of the telescope position (guiding) and usual distortions of the seeing due to scintillation in the Earth's atmosphere. These effects tend to reduce the non-uniformity of the fibre illumination and hence give better radial velocities.

A typical plot of radial velocities for a bright star (α Cen) is shown in Figure 2a. In spite of careful repositioning of the telescope before each exposure, the differences in individual radial velocities can be as high as $40 - 50$ m s^{-1}. An interesting experiment was made on 2002 June 4th by taking two separate

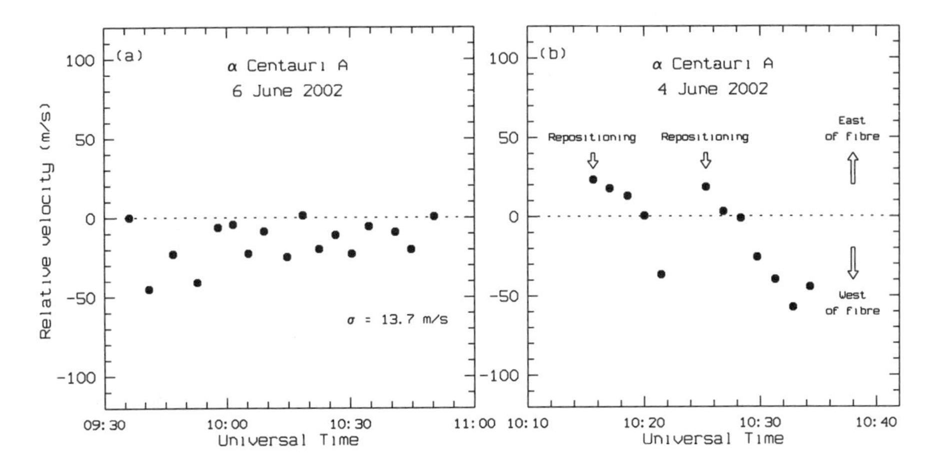

Figure 2. (a) Radial velocities of α Cen A over a couple of hours. (b) The effect of poor scrambling by the optical fibre.

sequences of exposures. The telescope was repositioned before each sequence so that the star was found slightly to the east of the optical fibre, for the first exposure. Then the telescope was not repositioned while a series of exposures was taken. Also, the sidereal rate was made slightly worse, so that the star drifted westwards, very slowly. The resulting radial velocities are seen in Figure 2b. There is an obvious red shift when the star is found east of fibre and an opposite, blue shift when the star is west of fibre. The difference can be as high as $\sim 70\,\mathrm{m\,s^{-1}}$.

6. Conclusion

Observations collected so far demonstrate that the radial velocity precision of Hercules spectra is somewhat limited by insufficient scrambling properties of the optical fibre. Blue sky spectra give a short-term precision of $3 - 4\,\mathrm{m\,s^{-1}}$, which is close to the photon-noise limit. The long-term precision can be brought to the same level by careful reduction and by avoiding observing the sky while the Sun is low. A similar high precision in stellar radial velocities can be achieved by improving the telescope positioning and guiding and, possibly, by introducing a double scrambler.

References

Griffin, R., & Griffin, R. 1973, MNRAS, 162, 243

Hearnshaw, J. B., Barnes, S. I., Frost, N., Kershaw, G. M., & Nankivell, G. R. 2002, ASP Conference Series, (this volume)

Murdoch, K., & Hearnshaw, J. B. 1991, Ap&SS, 186, 137

Simkin, S. M. 1974, A&A, 31, 129

IAU 8th Asia-Pacific Regional Meeting
ASP Conference Series, Vol. 289, 2003
S. Ikeuchi, J. Hearnshaw & T. Hanawa, eds.

Square Kilometre Array (SKA)

R. D. Ekers

ATNF, CSIRO, PO Box 76, Epping, NSW, 2121, Australia

Abstract. Our first glimpse of the epoch of re-ionization will be the most dramatic of the predictable outcomes of the SKA, but the history of unexpected discoveries at radio wavelengths suggests that the SKA will do much more. In its short life, radio astronomy has had an unequaled record of discovery, including three Nobel prizes: big-bang radiation, neutron stars, and gravitational radiation. Radio telescopes have followed the pattern of exponential growth generally seen in flourishing areas of science and technology. There is no technical reason for this not to continue, but to do so will require a telescope of a square kilometre aperture (the SKA) and a shift in technology that will set new challenges. Being sensitive to neutral and ionized hydrogen gas, high-energy particles and magnetic fields, the SKA will complement other planned instruments in the optical, infrared and millimetre wavebands.

1. Astronomy Before the Light - The Epoch of Re-Ionization

By the end of this decade one of the biggest questions remaining in astronomy will be the state of the universe when the neutral hydrogen which recombined after the big bang is being re-ionized by the first sources of ionizing radiation. This epoch of the universe is totally opaque to optical radiation but can be probed by the 21-cm H line. The first structures will appear as inhomogeneities in the primordial hydrogen, heated by infalling gas or the first generation of stars and quasars. A patchwork of either 21-cm emission or absorption against the cosmic background radiation will result. This structure and its evolution with z will look completely different for different re-ionization sources. A large population of low mass stars will be completely different from a small number of QSOs. For $z \sim 6$ we expect to see a growing 'cosmic web' of neutral hydrogen and galaxy halos forming and evolving (e.g. Tozzi et al. 2000). A radio telescope with a square kilometre of collecting area operating in the $100 - 200$ MHz frequency range will have the sensitivity to detect and study this web in HI emission!

2. The Development of Radio Astronomy

Our first glimpse of the epoch of re-ionization will be the most dramatic of the predictable outcomes of the SKA but the history of unexpected discoveries at radio wavelengths suggest that the SKA will do much more.

2.1. The Exponential Growth of Radio Telescope Sensitivity

It is well known that most scientific advances follow technical innovation, Harwit (1981). De Solla Price (1963), had also reached this conclusion from his application of quantitative measurement to the progress of science. His analysis also showed that the normal mode of growth of science is exponential and he gave examples from many areas. Moore's law, describing the 18-month doubling of transistor density on semiconductor chips, is a recent example. A plot of the continuum sensitivity of telescopes used for radioastronomy shows this exponential character (Figure 1) with an increase in sensitivity of 10^5 since 1940, doubling every three years. To maintain the extraordinary momentum of discovery of the last few decades a very large new radio telescope will be needed in the next decade.

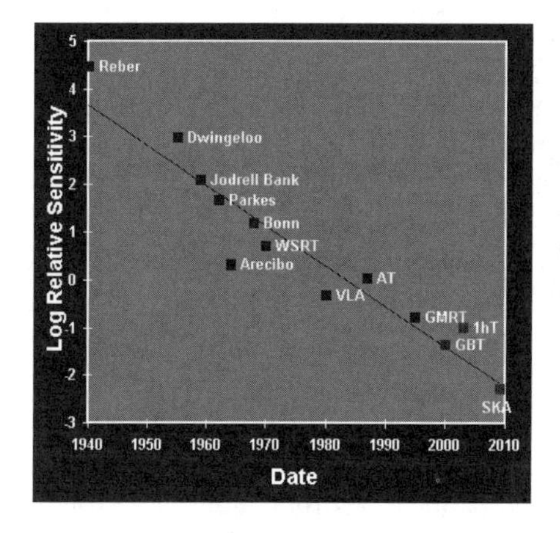

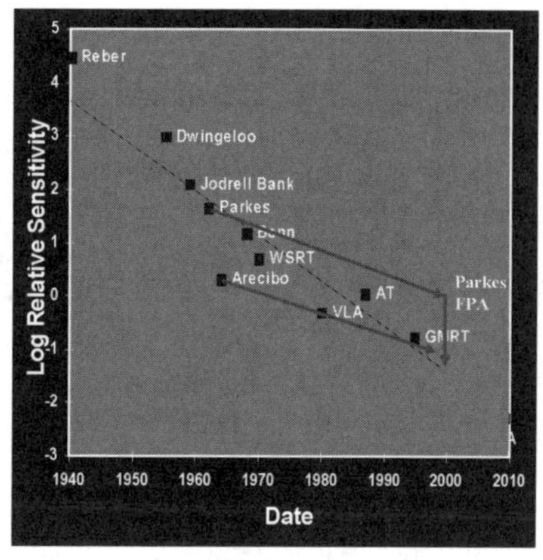

Figure 1. (a) Radio telescope sensitivity (b) Upgrade examples

2.2. The 3 Nobel Prizes in Radio Astronomy

This exponential development of radio astronomical instrumentation has generated an impressive list of discoveries with three of the five Nobel prizes awarded for discoveries in astrophysics coming directly from radio astronomy.

1. Cosmic microwave background radiation (1965); Nobel prize awarded to Penzias and Wilson, Bell Telephone Labs, for this technology driven serendipitous discovery.

2. Discovery of neutron stars (1974) through their radio pulsations (pulsars) by Jocelyn Bell and Tony Hewish – another technology enabled serendipitous discovery. This Nobel prize was shared with Sir Martin Ryle for the development of the aperture synthesis technique which has been the basis for all high resolution imaging in radio astronomy.

3. Verification of Einstein's prediction of gravitational radiation (1993); Nobel prize to Taylor and Hulse using Arecibo to make precise time measurements of a binary pulsar to test the theoretical prediction.

3. Radio Telescopes Make Discoveries!

Radio telescopes have produced a stream of discoveries including an unusually high fraction which are serendipitous (shown in italics in the following compilation).

- *Quasars and radio galaxies*
- 21cm HI line
- *Cosmological evolution required to explain steep slope of radio source counts*
- *Super-luminal motions of the jets in radio galaxies*
- *First evidence for the need for dark matter in spiral galaxies from the 21-cm HI rotation curves*
- *Masers and megamasers*
- Mass of the black hole in NGC4258 using H_2O megamaser emission
- *Pulsars*
- *Gravitational lenses*
- Gravitational relativistic time delay measured by planetary radar
- Gravitational radiation
- *First extra-solar planetary system*

3.1. Why Has Radio Astronomy Had So Much Impact?

The fraction of the energy output in the radio wavelength range of the electromagnetic spectrum is negligibly small but radio waves are easy to generate and they propagate with little absorption so provide unique information about the universe.

- Radio astronomy provided the first evidence of non-thermal processes in astronomy.

  The discovery of the galactic nuclei and quasars drove the paradigm shift to gravitational astrophysics, black holes and AGNs.

- Radio galaxies and quasars are easily seen at high red shift and they dominate radio catalogues. The counts of these radio sources had a profound impact on cosmology giving the first evidence for strong cosmic evolution.

- The long radio wavelengths and benign atmosphere made it possible to build interferometers with very long baselines (VLBI) providing the highest angular resolution in astronomy.

- The low opacity has let us probe regions deep in the centres of galaxies which are often highly obscured by dust at optical and even infrared wavelengths.

- Finally, the microwave band is well suited to long distance communications and so is one of the most interesting wavelength for SETI searches.

4. The Square Kilometre Array (SKA)

The exponential improvement in Figure 1 is obtained by the successive introduction of new design concepts. Upgrades to existing systems (illustrated for Arecibo and Parkes in Figure 1(b)) are impressive but it's hard to maintain the same exponential improvement. An increase in sensitivity of the order needed to maintain this exponential growth until 2010 cannot be achieved by improving the electronics or receiver systems in existing telescopes but only by *increasing the total effective collecting area of radio telescopes to about a million square metres.* The project has therefore acquired the appellation, the Square Kilometre Array.

4.1. The Concept

The SKA is a unique radio telescope now being planned by an international consortium. Extensive discussion of the science drivers and of the evolving technical possibilities led to a set of design goals for the Square Kilometre Array (Taylor & Braun 1999). Some of the basic system parameters required to meet these goals are summarized in Table 1.

Table 1. **Instrumental Design Goals**

Parameter	Design Goal
Total Frequency Range	0.03 – 20 GHz
Imaging Field of View	1 square deg. @ 1.4 GHz
Angular Resolution	0.1 arcsec @ 1.4 GHz
Surface Brightness Sensitivity	1 K @ 0.1 arcsec (continuum)
Instantaneous Bandwidth	$0.5 + \nu/5$ GHz
Number of Spectral Channels	10^4
Number of Instantaneous Pencil Beams	100 (at lower frequencies)

5. Sensitivity

The most obvious impact of the SKA will be its sensitivity, almost 100 times that of any existing radio telescope.

5.1. Line

At 1.4 GHz for a moderate resolution spectral line observations ($\nu/d\nu = 10^4$, corresponding to 30 km s^{-1}), the system sensitivity is 1.6 μJy after 8 hours observation. Taylor & Braun (1999) make an informative comparison of the spectral line sensitivity compared with those of many existing and planned facilities using a simulated spectra of the spiral galaxy M101 as it would appear at high red-shifts. The striking result is that even a normal galaxy like M101 (with an atomic mass $M_{\rm HI} = 2 \times 10^{10} M_\odot$ and molecular gas mass $M_{\rm H_2} = 3 \times 10^9 M_\odot$) could be detected efficiently at arbitrary red-shift. The HI line can be tracked to red-shifts up to 4, while the CO lines becomes accessible and easily detectable at red-shifts of 4 and larger.

5.2. Continuum

At 1.4 Ghz, for the broad-band continuum ($\sim$ 600 MHz) the system sensitivity is 23 nanoJy after 8 hours' observation.

The non-thermal continuum emission of normal galaxies can be detected out to arbitrary distances and gives a direct measurement of the star formation rate. It is interesting to note that, just as for the Hubble deep field, a continuum image at the sensitivity of the Square Kilometre Array is dominated by normal galaxies – the radio galaxies and quasars have already been seen at higher flux levels out to the beginning of the universe and no longer dominate the radio sky.

6. Field of View and Angular Resolution

The Square Kilometre Array will be the world premier instrument for astronomical imaging. No other instrument, existing or currently planned, on the ground or in space, at any wavelength, will provide simultaneously: a wide instantaneous field of view (1 square degree) and exquisite and well defined angular resolution (0.1" – 0.001"); and wide instantaneous bandwidth ($\Delta\nu/\nu > 50\%$), coupled with high spectral resolution ($\nu/d\nu > 10^4$) for detecting small variations in velocity. This translates to a very powerful survey mode which could even be conducted in parallel with other targeted observations.

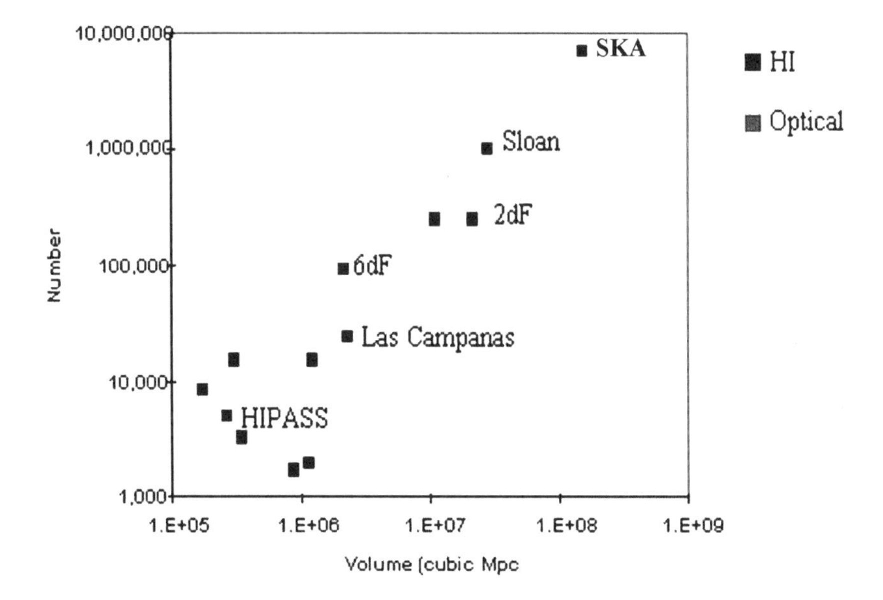

Figure 2. Effective volume and number of galaxies in various planned and completed surveys

6.1. Redshift Surveys and Large Scale Structure

The Square Kilometre Array will be able to perform surveys covering a large area of the sky with very uniform sensitivity. In 3 months of observing time one could cover 1000 square degrees and be able to detect L_* galaxies out to redshift of $z = 1.3$. Assuming the Zwaan et al. (1997) HI mass function, one expects to detect 10^7 galaxies in a volume of 3×10^8 Mpc^3. This is an order of magnitude more than in optical surveys, such as the Sloan Digital Sky Survey (see e.g. Lahav 1999) and the AAO 2dF Survey (Colless 2000). Following the methodology of Lahav, we have included the HI surveys in Fig. 2.

7. Multiple Beams

The reduction in the cost and size of the electronics in telescopes of the future will allow radio astronomers to take increasing advantage of multibeaming through either focal plane or aperture plane arrays. The Parkes multibeam HI system, now being replicated at Jodrell Bank and Arecibo, is an example of the impact of focal plane arrays. In the extreme aperture plane array with element size comparable to a wavelength it is even possible to generate simultaneous independent beams anywhere in the sky changing the whole sociology of big telescope astronomy.

8. Some Surprising Science

8.1. Searching for Redshifted CO

If an upper frequency of 22 GHz can be achieved the $1-0$ transition of CO is redshifted into the band for $z \geq 4$. Even though the $1-0$ line may be as much 100 times weaker than the higher order lines seen at mm wavelengths, the larger field of view, sensitivity and relative bandwidth makes the SKA $1-200$ times faster than even the ALMA array for a blind CO survey of galaxies.

8.2. CMB Anisotropy and S-Z at cm Wavelengths

Although the SKA will only operate at cm wavelengths, where discrete source confusion dominates the CMB anisotropy, its extreme sensitivity to point sources will make it possible to subtract the source contamination at these wavelengths and thereby image the low surface brightness CMB anisotropies on small angular scales. The SKA, operating at 10–20 GHz, will be able to make high-l observations of the CMB anisotropy spectrum and survey the sky for Sunyaev-Zeldovich decrements with unprecedented sensitivity. (Subrahmanyan & Ekers 2002).

8.3. High Energy Neutrinos

A wide range of astrophysical models predict the existence of neutrinos in the energy range $10^{19} - 10^{22}$ eV, e.g. Protheroe (1999). These high energy neutrinos produce an observable pulse of Cerenkov emission at radio frequencies if they interact in the front surface of the Moon (Dagkesamanskii & Zhelenznyk 1989). A patch of SKA antennas operating in the continuum between 1 – 2 GHz could be configured to detect such pulses with sensitivity 3 orders of magnitude better than any existing experiment (Hankins et al. 1996, Gorham et al. 2001).

8.4. Observing Transients before They Happen

Entirely new ways of doing astronomy may be possible with the SKA. With an array that is pointed electronically, the raw, "undetected" signals can be recorded in memory. These stored signals could be used to construct beams pointing anywhere in the sky. Using such beams astronomers could literally go back in time and use the full collecting area to study pulsar glitches, supernovae and gamma-ray bursts or SETI candidate signals, following a trigger from a sub-array or other wavelength domain.

9. Building the Square Kilometre Telescope

Costs of major astronomy facilities have now reached $US 1 billion levels. International funding is unlikely to exceed this value implying it has to be built at a cost < $US1000 per square metre for 10^6 square metres. If we compare the costs per square metre of existing radio telescopes (Table 2) we see that we will need an innovative design to reduce costs.

Table 2. **Cost / sq metre**

Telescope	$US/sqm	ν_{max}
GBT	10,000	100 GHz
VLA	10,000	50 GHz
ATA	3,000	11 GHz
GMRT	1,000	1 GHz

New technologies involving large scale integration of transistor amplifiers into complex systems which can be duplicated inexpensively, and the rapidly increasing capability to apply digital processing at high bandwidth now make it possible to obtain this aperture at an affordable cost. This will require unconventional techniques involving the large scale replication of individual elements. The time frame during which a new radio facility is needed to complement other planned instruments will be in the years around 2010.

Some aspects of the technology needed are still in the development stage. Institutions participating in the SKA are now designing and building prototype systems and the key technologies will be determined from these. Both planar phased arrays and reflectors/refractors are being considered for the antennas.

10. Radio Frequency Interference (RFI)

Ironically the very developments in communications that drive Moore's Law and make these radio telescopes possible also generate radio interference at levels far in excess of the weak signals detectable with an SKA. The future of observations with this high sensitivity will depend on our ability to mitigate against interference but I am optimistic that a combination of adaptive cancellation, regulation and geographic protection will let us access the faint redshifted HI signals from the early universe (Ekers & Bell 2001).

11. International Collaboration

Even with the dramatic reduction in cost of unit aperture, the SKA will be expensive. One path to achieving this vision is through international collaboration, to build facilities which no single nation can afford. While the additional overhead of a collaborative project is a penalty, the advantages are also great. It can avoid wasteful duplication and competition; provide access to a broader knowledge base; generate innovation through cross fertilization; and create wealth for the nations involved. An analysis of successful international collaborations has shown the importance of starting the collaboration early. This has been the case for the SKA, starting with an URSI/IAU Working Group on large facilities in 1993, and an agreement between a number of observatories for joint technology development.

An MOU to establish an SKA International Steering Committee was signed at the IAU General Assembly in Manchester, 2000 August. Current members are: UK, Germany, Netherlands, Sweden, Italy, Poland, United States, Canada, Australia, China and India. Russia, South Africa and Japan are participating as associate members.

References

Colless M. M. 2000, in Dunk Island Conference, Redshift Surveys and Cosmology, Publications of the ASA, *17*, 215-226

Dagkesamanskiia, R. D., & Zhelenznyk, I. M. 1989, JETP *50*, 233

de Solla Price, D. J. 1963 'Little Science, Big Science' (town: Columbia University Press)

Ekers, R. D. & Bell, J. F. 2001, in IAU Symp. 196, Preserving the Astronomical Sky, Vienna, eds R.J. Cohen & W.T. Sullivan (San Francisco: ASP)

Gorham, P. W., Liewer, K. M., Naudet, C. J., Salzburg, D. P. & Williams, D. 2001, in Radhep 2000, Los Angeles, eds. D. Salzburg & P. Gorham, AIP Conference Proceedings *579*, 177-188

Hankins, T. H., Ekers, R. D. & O'Sullivan, J. D. 1996, MNRAS, *283*, 1027

Harwit, M. 1981, 'Cosmic Discovery' (New York: Basic Books, Inc)

Lahav, O. 1999, in ESO Astrophysics Symposia, Looking Deep in the Southern Sky, eds. R. Morganti & W. J. Couch (Berlin: Springer-Verlag) 42

Protheroe, R. J. 1999, in Neutrino 98, Takayama 1998. Nuclear Physics B (proc. Suppl.), *77*, 465-473

Subrahmanyan, R. & Ekers, R. D. 2002, URSI General Assembly, Maastricht

Taylor, A. R. & Braun, R. 1999, 'Science with the Square Kilometre Array' http://www.skatelescope.org/ska_science.shtml

Tozzi, P., Madau, P., Meiksin, A. & Rees, M. J. 2000, ApJ, *528*, 597-606

Zwaan, M. A., Briggs, F. H., Sprayberry, D. & Sorar, E. 1977, ApJ, *490*, 173

IAU 8th Asia-Pacific Regional Meeting
ASP Conference Series, Vol. 289, *2003*
S. Ikeuchi, J. Hearnshaw & T. Hanawa, eds.

The Giant Metrewave Radio Telescope

Rajaram Nityananda

National Centre for Radio Astrophysics, TIFR, Post Bag 3, Ganeshkhind, Pune University Campus, Pune 411007, India

Abstract. The Giant Metrewave Radio Telescope (GMRT) of the National Centre of Radio Astrophysics (NCRA) of the Tata Institute of Fundamental Research (TIFR) at Khodad, India, has been operational in the band 0.2 to 2 metres for the last two and a half years. The system characteristics and performance and recent results from the group will be presented. Details of use over the last six months by scientists from other observatories under the GMRT Time Allocation Committee (GTAC) and future plans will be also be reviewed in this paper. Areas which have been studied include observations made in the GMRT band of neutral hydrogen, nearby galaxies, supernova remnants, the Galactic Centre, pulsars, the Sun and others.

1. Background

The early plans for the GMRT are described in a 1991 paper (Swarup et al. 1991). Briefly, fourteen 45-metre antennas stand in a central region about a kilometre square, providing good surface brightness sensitivity, while the remaining sixteen are spread out over three 14-km long arms, providing the resolution to study (and remove if needed) compact sources in the field. The system provides five frequency bands, centred at approximately 1250, 610, 325, 235, and 150 MHz. The L-band feed allows coverage from 1000 (or even 850 with some loss) MHz to 1450 MHz, and was built at the Raman Research Institute in Bangalore, India as was an array combiner for pulsar work. Optical fibre links carry control and local oscillator signals and the RF data between each telescope and the central building. A major piece of in-house built special purpose hardware, the FX correlator, takes in sixty signals each of 16 MHz + 16 MHz bandwidth (two sidebands) with 128 spectral channels, and forms 435 baselines. The overall cost of the GMRT was about 12 million dollars (a number which has stayed more or less constant from the planning stage to completion, while the number increased in rupee terms). In 1994, the TIFR radio astronomy group was reorganized as the National Centre for Radio Astrophysics.

2. Status and Use

After two years of testing and use by the NCRA group, the Raman Research Institute (RRI) group, and a few outsiders, the telescope was opened from

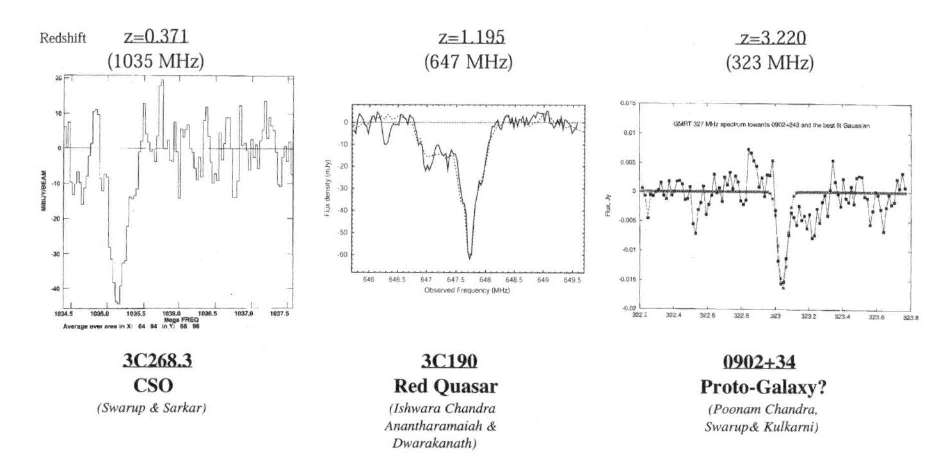

Figure 1. GMRT observations of associated HI absorption in radio galaxies and quasars: A few typical results

2002 January to experienced observers from observatories and astronomy departments worldwide, with formal review of proposals and time allocation being handled by a GMRT Time Allocation Committee, GTAC (see NCRA web page http://www.ncra.tifr.res.in for more details). Currently, one of the proposers (who can but need not be a collaborator at NCRA) is required to be present at the GMRT during the observations.

About half a dozen PhD theses, and about twenty publications have emerged in the last two years, and about an equal number must be currently in the pipeline. (A list is maintained in the Library section of the NCRA page). The topics include the Sun, pulsars, supernova remnants, HII regions, nearby galaxies, active galaxies, clusters of galaxies, and the intergalactic medium (damped Lyman alpha systems). Since there can be a significant time delay between observation and publication, a more up-to-date indicator of the areas is given by the list of observations taken since 2002 January, which is maintained at the website.

The following table gives the specifications of the GMRT as presented to prospective observers by the GTAC, in consultation with the observatory. More details are available on the web page. Notice that while some of the numbers, such as synthesized beam-widths, are straightforward, some, like recommended observing times and dynamic ranges, are indicative, being based on current experience. They could, and indeed should, evolve as more work is done on and with the telescope.

It is hard to give the flavour for all the work being done, since it is so diverse, so I have chosen just two examples. The first, (Fig. 1) is a collage of associated absorption studies for AGNs at different redshifts. It is appropriate, because it shows neutral hydrogen, and also because it was given to me by Swarup, who started the GMRT project and saw it through many crucial milestones. The picture reminds us that he continues to take an active interest in using the telescope. It also brings out the great advantages of good collecting area and

	Frequency (MHz)			
	235	325	610	1420
Primary beam (Degrees HPBW)	2.5	1.8	0.9	0.4 $*(1400/f)$
System temperature (K)	180	100	90	70
Antenna temp. (K/Jy/antenna)	0.3	0.35	0.3	0.25
Synthesised beam (arcsec)				
Whole array	13	9	5	2
Central square	270	200	100	40
Largest detectable source (arcmin)	44	32	17	7
Sensitivity (rms image noise mJy/MHz/min)				
Whole array	3.9	1.2	1.2	1.2
Central square	7.8	2.4	2.4	2.4
Usable frequency range (MHz)				
Reliable	232 to 238	315 to 335	590 to 630	1000 to 1450
With some luck	230 to 240	305 to 360	580 to 640	850 to 1450
Fudge factor(actual to estimated time)				
Short observations	10	10	5	5
Long observations*	5	4	3	3
Best rms sensitivities while imaging (mJy)	3	1	0.5	0.2
Typical dynamic ranges	200	500	500	500

* For spectral observations fudge factor is close to 1

Table 1. System parameters of GMRT. (The 150 MHz band is not in routine use)

frequency coverage, which are bound to drive many more absorption studies at GMRT. Of course, emission measurements were part of the original GMRT manifesto. Dwarakanath, just before this meeting, informed me that the record for the highest redshift ($z = 0.1887$) detection of cold HI in emission is held jointly by the VLA and the GMRT, since it emerged from a merged data cube made by him, Verheijen & van Gorkom (unpublished). One is sure this record will not last very long.

Fig. 2, courtesy of Gupta and Gangadhara, is a pulsar profile which shows some special features – the largest number of components seen to date, which required both high sensitivity and dynamic range as well as some novel data processing.

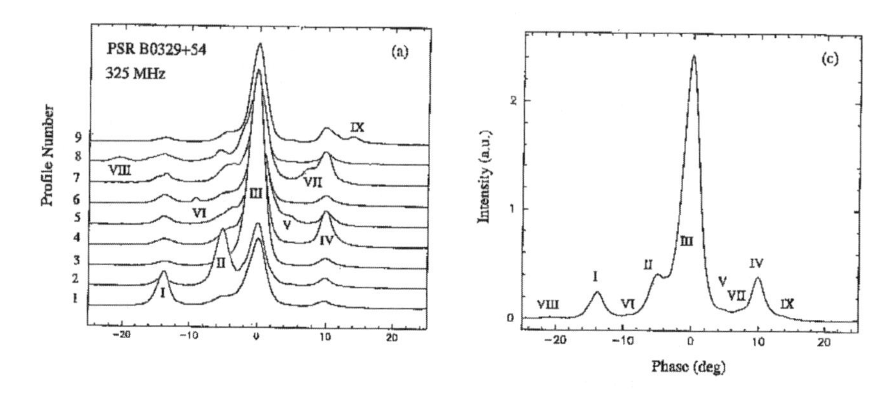

Figure 2. GMRT observations of B0329+54 showing nine pulse components (Gangadhara and Gupta 2001)

3. Outlook

The GMRT represents an important opportunity for astronomers everywhere, and certainly in India and the Asia Pacific Region, to use a modern facility which occupies a special and interesting niche in radio astronomy. Allowing for maintenance, ionospheric conditions, and other factors, there are still about four thousand hours of observing time at four different frequencies available per year. This can be thought of as dozens of PhD theses! Indeed, the experience has been that graduate students, with their willingness to travel, stay in new places, and learn new things, have been an important force at the GMRT. Even those for whom the radio data are only a part of their overall work have been able to use the telescope effectively in a few months time, especially in the shorter wavelength bands. Hopefully, many of these students should develop appetites for using LOFAR and SKA later in their careers. And for the black belt radio astronomer, Table 1 shows that there is still the challenge of reaching the limiting performance at the longest wavelengths. This involves addressing many issues relating to the ionosphere, the telescopes, receiving system, observing strategy, and data analysis, which will be important for future projects as well.

Acknowledgments. I thank all the the members of the GMRT academic group at NCRA who have shared their experience, knowledge and results with me over the last two years. The material presented here has received major inputs from S. Ananthakrishnan and A.P. Rao, Observatory Director and Chief Scientist at the GMRT. This paper is presented as a spokesman of the entire group, which built and now operates the telescope, and I would particularly like to mention the large and dedicated engineering team.

References

Swarup, Ananthakrishnan, Kapahi, Rao, Subrahmanya & Kulkarni, 1991, Current Science, 60, 95

Gangadhara & Gupta, 2001, ApJ, 555, 31

IAU 8th Asia-Pacific Regional Meeting
ASP Conference Series, Vol. 289, 2003
S. Ikeuchi, J. Hearnshaw & T. Hanawa, eds.

A Proposal for Constructing a New VLBI Array, Horizon Telescope

Makoto Miyoshi

National Astronomical Observatory, 2-21-1, Osawa, Mitaka, Tokyo, Japan

Seiji Kameno

National Astronomical Observatory, 2-21-1, Osawa, Mitaka, Tokyo, Japan

Heino Falcke

Max-Planck-Institute für Radioastronomie, Auf dem Hügel, 69, 53121, Bonn, Germany

Abstract. The existence of a black hole in the universe has become very clear and is now one of our common sense in astronomy. But the direct image of a black hole showing relativistic phenomena around the event horizon was still beyond our reach in the previous century. The apparent sizes of black holes are too small to observe. Sagittarius A* is the closest massive black hole at our galactic center. Even the Schwarzschild radius of SgrA* is only about 6μarcseconds. Early in the 21st century, however, the development of VLBI techniques and millimeter and sub-millimeter radio astronomy will soon reach the point to make such observations of black holes possible. We here propose to construct a new VLBI array that should be named as the Event Horizon Telescope.

1. The Existence of Black Holes Was Confirmed Last Century

The existence of black hole in the universe is now really confirmed from observations using the Hubble Space Telescope and the VLBA and ground based IR telescopes (Miyoshi et al. 1995; Macchetto et al. 1997; Herrnstein et al. 1999; Ghez et al. 2000). One of best examples is in NGC4258, which has a massive black hole with mass of $3.9 \times 10^7 M_\odot$ at its center, while another good case is for SgrA* at our galactic center, whose mass is measured to be $2.6 \times 10^6 M_\odot$. In both cases, the masses are measured from dynamical motion around the black holes, namely from proper motions of a rotating molecular gas disk, and from orbiting stars with a velocity of more than 1000 km/s, respectively. It is too difficult to deny the existence of black holes in the universe today.

2. Can We Watch Black Holes?

Though some observations suggest the existence of a surrounding disk or matter at the parsec scale around central black holes (Dayton et al. 2001; Kameno et al. 2001), the vicinities of black holes are still veiled observationally. We now know of their existence, but nothing about their real nature. Only theorists have investigated how black holes look like at a scale of several Schwarzschild radii (Fukue & Yokoyama 1988; Usui et al. 1998; Takahashi & Mineshige 2002).

3. Apparent Sizes of Black Holes

Once we get a new telescope with higher resolution, which object is the best candidate so that we can observe the black hole vicinity? From the mass and distance of black holes we can calculate the apparent angular size of their Schwarzschild radii, and the diameter of the shadow. The Schwarzschild radius of a stellar black hole with 1 solar mass at 1 pc is only 0.02μarcseconds (20 nano-arcseconds), and hence a stellar black hole would be too small to observe even in 21-st century. Most of massive black holes at several Mpc also show very small apparent sizes, less than 1μas.

4. Best Candidate, SgrA*

The black hole of SgrA* at our galactic center has the largest apparent size ($R_S = 6\mu$as, $D_{\rm shadow} = 30\mu$as), and hence SgrA* is the best candidate we select to observe a black hole vicinity.

Though SgrA* is quite a low luminosity AGN ($L \sim 10^{-8.5} L_{\rm Edd}$), recent X-ray observations reveal its short-time burst with a timescale of a few hours (Baganoff et al. 2001) and further Zhao et al. (2001) reported the periodic change about $T \sim 106$ days in radio flux density; both of them suggest that SgrA* is really the closest massive black hole worthy of monitoring deeply.

Previous VLBI observations at the centimeter to millimeter region were already performed to watch the nature of SgrA*, but in vain. This is because the plasma gas surrounding SgrA* washed away the true image of the black hole (Lo et al. 1999). At sub millimeter wavelengths, however, the effect of the plasma is reduced with λ^{-2} and the true face of the black hole can be seen (Falcke et al. 2000)

In this century, we should construct a sub-mm VLBI array for observing the black hole vicinity of SgrA*. The Horizon Telescope will testify general relativity at strong gravity, and at the same time make a new field of observational black-hole astronomy.

5. Horizon Telescope for Monitoring the Black Hole at SgrA*

In order to obtain the black hole image of SgrA*, the Horizon Telescope must be a sub-millimeter VLBI system. Below we show the minimum specifications of the Horizon Telescope.

Figure 1. Horizon Telescope in the Near Future

- Observing Frequency: 350 GHz to 800 GHz (sub-millimeter) – to escape from the scattering effect of circum-nuclear plasma;
- Observing site: (1) space, or (2) southern hemisphere highlands or the Antarctic – to receive sub-mm radio emission from SgrA* (low declination);
- Stations: more than 10 like VLBA – for getting sufficient uv coverage to obtain high dynamic range in image;
- Array size: more than 8000 km at 500 GHz – to attain less than 10μas resolution.

6. A New Possibility for SgrA*

Recently Miyoshi et al. (2002) found a jet activity in SgrA* with 43-GHz VLBA observations at one burst epoch, at the end of 2001 July. The jet elongated during the observing time of 7 hours with apparent velocity $v \sim 0.1c$. They confirmed the reliability of the detection by means of phase referencing mapping of the in-beam SiO maser source, IRS10EE. Then the detection means that we can begin the watching the black hole and its related activity of SgrA* with ground-based mm-VLBI.

References

Baganoff, F. K., Bautz, M. W., Brandt, W. N., Chartas, G., Feigelson, E. D., Garmire, G. P., Maeda, Y., Morris, M., Ricker, G. R., Townsley, L. K., & Walter, F. 2001, Nature, 413, 45-48

Dayton J. L., Wehrle, Ann E., Glenn, Piner, B., & Meier, David L. 2001, ApJ, 553, 968-977.

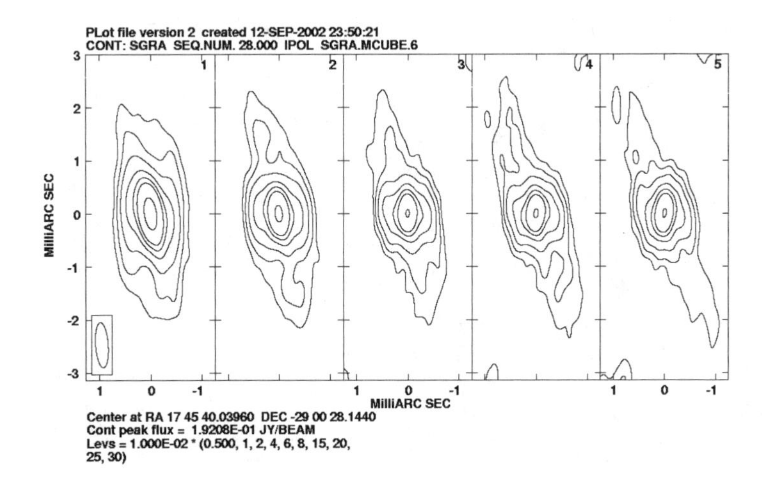

Figure 2. SgrA* observed at a burst epoch (from Miyoshi et al. 2002). Every map is integrated for 3 hours, overlapping 2 hours with adjacent maps.

Falcke, H., Melia, F., Agol, E. 2000, ApJ, 528, L13-L16

Fukue, J., Yokoyama, T. 1988, PASJ, 40, 15-24

Ghez, A. M., Morris, M., Becklin, E. E., Tanner, A., & Kremenek, T. 2000, Nature, 407, 349

Herrnstein, J. R., Moran, L. M., Greenhill, L., Diamond, P., Inoue, M., Nakai, N., Miyoshi, M., Henkel, C., & Riess, A. 1999, Nature, 400, 539-541

Kameno, S., Sawada-Satoh, S., Inoue, M., Shen, Zhi-Qiang, & Wajima, K. 2001, PASJ, 53, 169-178

Lo, K. Y., Shen, Z., Zhao, J.-H., Ho, P. T. P. 1999, in The Central Parsecs of the Galaxy, ASP Conf. Ser. Vol. 186. Ed. by Heino Falcke, Angela Cotera, Wolfgang J. Duschl, Fulvio Melia, & Marcia J. Rieke., 72

Macchetto, F., Marconi, A., Axon, D. J., Capetti, A., Sparks, W., & Crane, P. 1997, ApJ, 489, 579

Miyoshi, M., Moran, J., Herrnstein, J., Greenhill, L., Nakai, N., Diamond, P. & Inoue, M. 1995, Nature, 373, 127-129

Miyoshi, M., Imai, H., Nakashima, J., & Deguchi, S. et al., in preparation.

Takahashi, R., & Mineshige, S. 2002, in preparation

Usui, F., Nishida, S., & Eriguchi, Y. 1998, MNRAS, 301, 721-728

Zhao, Jun-Hui, Bower, G. C., & Goss, W. M. 2001, ApJ, 547, L29-L32

IAU 8th Asia-Pacific Regional Meeting
ASP Conference Series, Vol. 289, 2003
S. Ikeuchi, J. Hearnshaw & T. Hanawa, eds.

Solar-B: Status of Project

Yoshinori Suematsu[1] and the Solar-B team

[1] *National Astronomical Observatory, Mitaka, Tokyo 181-8588, Japan*

Abstract. The Solar-B spacecraft, currently under development for a planned launch in the summer of 2005, carries the Solar Optical Telescope (SOT) to make precise measurements of magnetic fields of the solar photosphere with a high spatial resolution, the X-Ray Telescope (XRT) to observe the dynamics of the high temperature corona, and the EUV Imaging Spectrometer (EIS) to observe plasma motions in the transition region and corona. The aim of Solar-B is to investigate the physical coupling between the photosphere (engine) and the corona (dissipater) to ultimately understand the mechanism of coronal dynamics and heating. The magnetic field maps with 0.2–0.3″resolution, the images of the high temperature corona with 1″resolution, and the precise coronal velocity maps provided by these telescopes will all be new, and unprecedented scientific outcomes are expected.

1. Introduction

Solar-B is an ISAS (Institute of Space and Astronautical Science) solar space mission following YOHKOH (Solar-A). The YOHKOH (1991–2001), SoHO (1995 – present) and TRACE spacecraft (1998 – present) have been providing exciting new results on the dynamical behaviour of the corona and flares (e.g. Schrijver et al. 1999). For example, they revealed that the corona is extremely dynamic, with restructuring (Tsuneta et al. 1992), rapid heating (e.g. active region transient brightenings by Shimizu et al. 1992, nanoflares by Shimizu and Tsuneta 1997, coronal flashes in polar coronal holes by Koutchmy et al. 1997) and mass acceleration (e.g. X-ray jets from bright points by Shibata et al. 1992; Shimojo et al. 1996) being common phenomena. In addition, it turned out that the cores of many coronal bright points including small flares are too small in size to surmise their energy release mechanisms.

It is clear that magnetic fields play the most important role in coronal phenomena and they are deeply rooted in the photosphere and controlled by plasma motions there and beneath. It is well known, however, that it is very difficult to observe the precise and detailed behaviour of the photospheric magnetic fields with ground-based instruments, because the fields are concentrated in the sub-arcsec structures which is much smaller than an atmospheric seeing-limited resolution. Especially, the magnetic components transverse to the line of sight, which give the measure of excess magnetic energy, are difficult to observe, and cannot be measured with any degree of accuracy if the field is not spatially resolved. Among mechanisms, the magnetic reconnection seems to play the central role in the energy release of most solar dynamical phenomena (e.g. Shibata 1997). Therefore, understanding the detailed physical process of magnetic reconnection is one of the most important objectives of the Solar-B mission.

The Solar-B is also aimed at comprehensive understanding of the solar photosphere and corona as a system. To do this, we have to determine the evolution of the photospheric vector magnetic field, which is an indicator of free energy build-up. We also need to determine the topology, location and timing of energy release. The determination of these boundary conditions will require high spatial resolution and high-cadence imaging.

Therefore, the Solar-B should be capable of coronal imaging better than YOHKOH/SXT and photospheric vector magnetic field measurements in a sub-arcsec resolution. From these requirements, the Solar-B was designed to carry three instruments: a solar optical telescope (SOT) to observe the photosphere to chromosphere and the photospheric magnetic fields with 0.2–0.3″resolution, an X-ray telescope (XRT) to observe the dynamical behavior of the corona with 1″resolution, and an EUV imaging spectrometer (EIS) to observe the precise plasma motion in the transition region and corona (Suematsu 1998; Shimizu 2002).

These observational instruments are being developed under close collaboration between Japan (ISAS/NAOJ), the United State (NASA), and the United Kingdom (PPARC). Regarding the SOT, the largest of the three telescopes, the 50-cm aperture optical telescope assembly (OTA) is developed by Japan and the Focal Plane Package (FPP) is developed by NASA. Regarding the XRT, NASA is in charge of the X-ray optics and Japan is in charge of the X-ray CCD camera and observation control system. The EIS is developed mainly by US and UK. NAOJ (National Astronomical Observatory of Japan), in collaboration with ISAS, is playing a role in designing and testing the interface between each instrument and the spacecraft and in developing the Mission Data Processor (MDP), which is one of the key components of the Solar-B mission; the MDP works for bit compressions and 8/12 bit JPEG compressions with lossy DCT and lossless DPCM to reduce stored data amount, and for onboard analysis such as flare onset detection for XRT.

In the following, Solar-B observing instruments and their test status are described.

2. Solar-B Observing Instruments

The overview of Solar-B is given in Table 1 and Figure 1. The cubic box in the bottom is the bus module which contains common electrical devices including the attitude control system. Mounted on both sides of the bus are the solar array panels, each of which will be expanded to 4.5 m in orbit. The cylindrical Optical Bench Unit (OBU) made of a low expansion CFRP (carbon fiber reinforced plastics) with 1m diameter and 1m height is attached on the bus module. The optical telescope (OTA) partly inside the OBU occupies an central part of the spacecraft structure. The OBU is the base of accurate pointing (targeted ∼1″) of three independent telescopes of SOT, XRT and EIS.

Table 1. Overview of Solar-B satellite.

Orbit	630 km solar-synchronous polar orbit
Weight	$\sim$770 kg (dry) + $\sim$130 kg (thruster gas)
Launch	summer/2005 by ISAS M-V rocket
Lifetime	$\geq$ 3 years
Attitude control	3-axis stabilized body control
Power consumption	$\sim$1000 W
Data rate	300$\sim$500 kbps , max 2 Mbps
Data compression	Bit compression and 8/12 bits JPEG compression
Amount of data	$\sim$3 Gbits per orbit (8 Gbits recorder capacity)
Telemetry rate	$\sim$4 Mbps

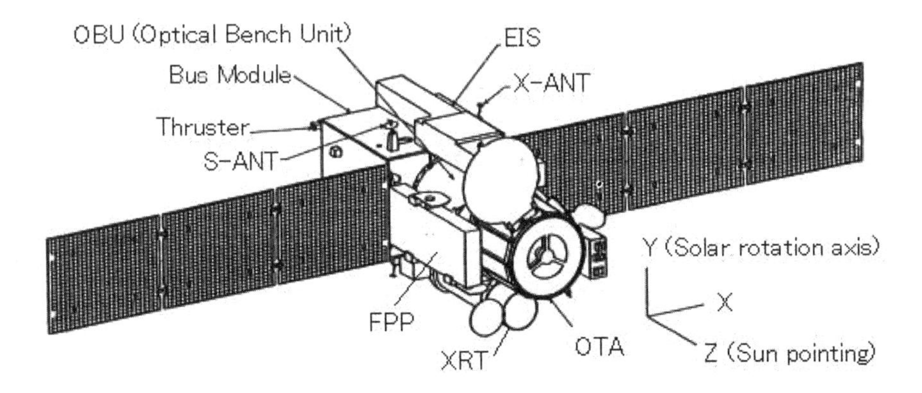

Figure 1. The Solar-B spacecraft layout.

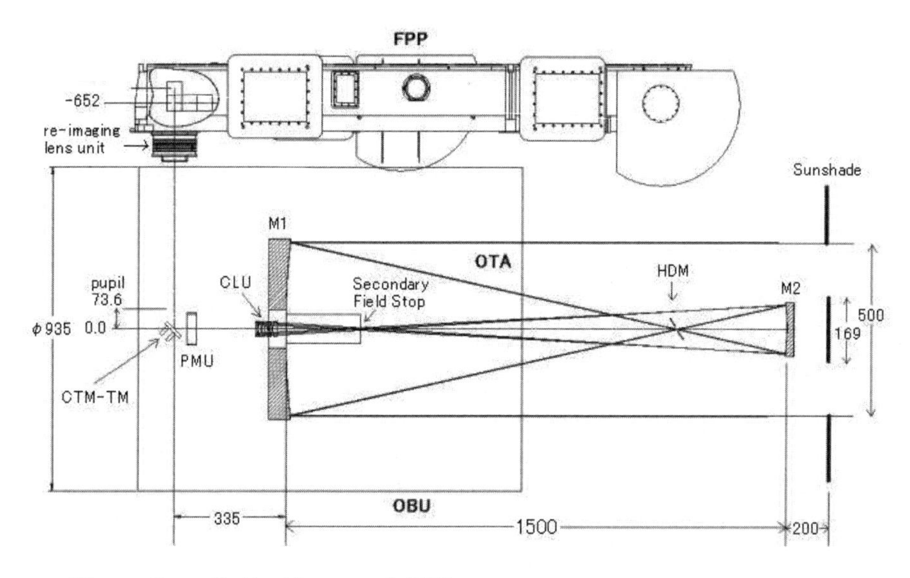

Figure 2. Optical layout of OTA.

2.1. SOT

The solar optical telescope (SOT) is designed to give diffraction limited images and to be capable of high precision polarimetry for the vector magnetic measurements; target spatial resolution is 0.2–0.3″ depending on wavelength and the S/N ratio is ~1000 to detect fine-scale magnetic elements. The OTA is an aplanatic Gregorian (1.5 m in length, 50 cm aperture) with a heat dump oblique mirror at the primary focus (the mirror works as a field stop of diameter ~500″). A collimator lens unit (CLU) comes after the secondary focus to yield a pupil image around a polarization modulator unit (PMU) and a tip-tilt mirror behind the primary mirror and also to relax a positional tolerance of the following focal plane package. The optics layout of OTA is shown in Figure 2.

The FPP consists of a monochromatic imager tunable to several spectral lines and continua useful for atmospheric diagnostics, and a spectrometer for magnetic sensitive lines (Table 2).

The OTA framework is composed of truss of CFRP, and is an adhesively bonded structure. The primary mirror, which is only 14 kg weight, and the secondary mirror are made of lightweight ULE.

An integration of OTA proto model, an alignment of its optical components and random vibration/shock test have been carried out with ISAS facility from 2001 August through 2002 April. In order to perform the optical test, the OTA proto-model was integrated with actual optical components quite similar to flight ones except for their wavefront error accuracies; the flight model CFRP truss was used. The proto-model primary and secondary mirrors are also intended as backups for the flight by re-polishing in a contingency case. The proto-model CLU consists of two groups of two lenses to give an optimal performance in a test wavelength of 632.8 nm (flight model CLU is almost aberration-free and temperature insensitive, made of six-group six-lens). The tip-tilt mirror is made of ULE glass supported and tilt adjusted by three piezo actuators for image stabilization.

All the optical components of OTA are fixed at their best optical performance positions during the integration and test phase, since the OTA has no adjustment mechanism on orbit; only focus adjustment is made with a re-imaging lens of FPP in giving the diffraction-limited performance; mounting mechanism of optical components, especially of the primary mirror, should not degrade their optical qualities and truss should maintain the separation between the primary and secondary mirror in the accuracy of several microns.

The alignment of OTA by using an interferometer was verified in the beginning of the test; comatic aberration due to decentering and tilt misalignment of the secondary mirror and defocus due to its despace misalignment can be made negligibly small. This first measurement is affected by the deformation caused by gravity. We then carried out wavefront error measurements in both the upward ($+1\,g$) and downward ($-1\,g$) pointing telescope configurations to estimate a zero gravity despace position. Based on these measurements, the secondary mirror was adjusted to the zero gravity focus position. The integration and optical tests were made using the tower (3.2 m high) which is capable to mount the OTA, orienting it upside down and to set the 60 cm flat mirror at the top or bottom depending on the OTA orientation. To prevent the OTA from being contaminated, all these activities were done in class 100 clean booth.

In addition to static misalignment, dynamic misalignment of OTA optics was a more important concern. To check the healthiness of optics against the satellite launch, the most violent mechanical environment, the random vibration and shock test simulating the launch condition was performed using the ISAS facility. By comparing the wavefront errors of OTA before and after the test, it was demonstrated that the change of coma and defocus aberration of OTA is within the budget to achieve the diffraction-limit. This was one of the most important milestones for the flight model.

2.2. XRT

The X-ray telescope (XRT) will provide the detailed information of the high temperature material to permit the study of both the dynamics of fine scale coronal phenomena, such as magnetic reconnection, and coronal heating mechanisms. Grazing-incidence-type optics was designed to meet the requirement on angular resolution ($\sim 1''$) and to be capable of wide range temperature diagnostics from 1 MK (as observed with TRACE and SoHO/EIT) through several MK. Table 3 summarizes the characteristics of XRT.

The $2\text{k}\times 2\text{k}$ back-side illuminated X-ray CCD is cooled down to -45°C$\sim$-70°C using radiator, and its position can be controlled in a range of ± 1 mm to accommodate any change of focus of the 2.7 m long telescope and to optimize the focus position for various fields of views of observation program. NAOJ has been performing the calibration and evaluation of the CCD camera using the medium size vacuum chamber and X-ray monochromator installed in the Advanced Center for Technology, NAOJ. The extensive activities for fabrication and testing of the flight CCD camera will continue until its completion in 2003 March. The CCD will then be delivered to the US to be combined with the

Table 2. SOT instrument

Telescope	Aplanatic Gregorian, 50 cm aperture, 1.5 m length
Polarization modulator	Rotating waveplate, frequency = 0.625 Hz, retardance = $n + 0.35\lambda$ at 630 nm
Image stabilizer	Tip-tilt mirror with correlation tracker
Two-channel filtergraph	
narrow-band channel	Tunable Lyot (passband = 100 mÅat 630 nm) for Mg I 5173, Fe I 5250, Fe I 5576, Fe I 6302, H I 6563, etc.
wide-band channel	interference filter (passband= 3–10Å) for CN 3883, Ca II 3968, CH 4305, 4505, 5550, 6684
detector	one 2k×4k CCD
image scale	0.0533″/pixel (wide-band) 0.08″/pixel (narrow-band)
Spectrograph	Littrow-type, Image Scanning Mirror
spectral coverage	630.08 – 630.32 nm Fe I 6301.5 Å and Fe I 6302.5 Å
spectral resolution	35 mÅ
sampling	20 mÅ
image scale	0.16″/pixel
field of view	164″(along slit), ±164″(transverse to slit)
detector	dual 192×1024 CCD
polarimetry	simultaneous I,Q,U,V/ orthogonal linear (dual beam)

X-ray telescope at NASA and after final testing the XRT will be delivered back to Japan to be installed to the Solar-B spacecraft in 2004 January.

2.3. EIS

The EUV imaging spectrometer (EIS) is a spectrograph with a slit/slot and a scanning mirror and designed to provide detailed physical condition of high temperature plasmas such as density, velocity and temperature. It should be noted that observing wavelength regions (170–210 Å and 250–290 Å) were carefully selected to make both normal corona and flare plasma diagnostics possible. EIS specification is given in Table 4 together with CDS/SoHO and its optics layout is shown in figure 3.

Though the CCD and the grating of EIS are under fabrication in UK and US, it is noted that a significant contribution from Japanese technology was made to improve their performance in resolution and sensitivity.

3. Test Schedule

After extensive study and design work under the strict constraints of mass, size, cost, time and environmental conditions, the Mechanical Test Model (MTM) of the Solar-B was completed in 2002 May. The MTM was used to verify the

Table 3. Characteristics of X-ray telescope (XRT).

		Remarks
Telescope	modified Walter I type of diameter 34 cm and focal length 2.7 m	polynomial surface Ir mirror
	co-centered aspect sensor of diameter 5 cm and focal length 2.7 m	
Temperature	$\geq 10^6$ K	several filters for T-diagnostics
	$\sim$6000 K	visible wavelength for aspect sensor
Detector	Back-illuminated CCD (2048×2048)	13.5 μm pixel
Pixel size	1″	FOV: 34′×34′
Exposure	0.01 sec (flare) - 200 sec (coronal hole)	
Flare detection	onboard analysis of flare patrol images	by Mission data processor (MDP)
Flare onset	pre-storage buffer for high temporal obs.	by MDP
Onboard control	automatic region selection and exposure control	by MDP

Table 4. Specification of EIS and its comparison with CDS aboard SoHO.

	EIS (Solar-B)	CDS (SoHO)
Optics	Normal incidence	Grazing Incidence
collective area	ϕ15 cm Mo/Si multi-layer half for 170-210 Å, half for 250-290 Å	34.3 cm^2
spatial res.	1″	3″
Wavelength Band	170–210 Å, 250–290 Å	308–381 Å, 513–633 Å
Detector	two back-illuminated CCDs of 1k×2k format	MCP+ Front-Illuminated CCD
Pixel size	1″, 20 mÅ (@270 Å)	1.68″, 70 mÅ (@360 Å)
FOV (slit)	$\sim$8.5′×8.5′	240″
FOV (slot)	60″×512″	
Method of data analysis	Line-ratio Line-profile (Vel res. $\leq$10 km s^{-1})	Line-ratio

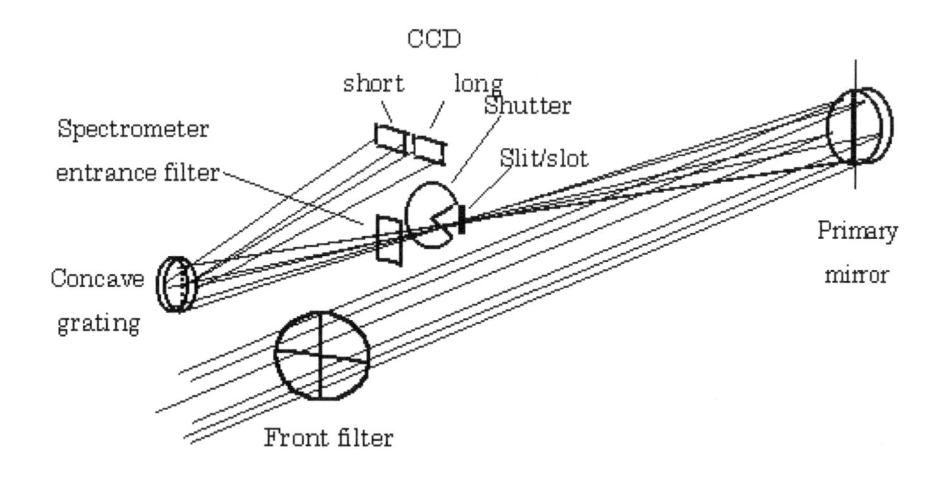

Figure 3. Layout of EIS optics.

mechanical validity of the spacecraft and each instrument against the launch condition; its fidelity to the flight model was quite high.

The system level mechanical test will be followed by the system level thermal vacuum test in 2002 October. The integration of the flight OTA will start in 2003 April. The integration and optical performance tests of the flight model OTA will be performed in the large clean room of the environment test facility newly constructed in NAOJ. After electrical/mechanical interface check starting in 2004 January and the final integration testing, the Solar-B is planed to be launched in the summer of 2005 by ISAS M-V rocket.

References

Koutchmy S. Hara H., Suematsu Y. & Reardon K. 1997, A&A, 320, L33

Schrijver C. J. et al. 1999, Solar Phys., 187, 261

Shibata K. et al. 1992, PASJ, 44, L173

Shibata K. 1997, in The Corona and Solar Wind near Minimum Activity, the Fifth SOHO Workshop, ESA SP-404, 103

Shimizu T. et al. 1992, PASJ, 44, L141

Shimizu T. & Tsuneta S. 1997, ApJ, 486, 1045

Shimizu T. 2002, Advances in Space Research, 29, 2009

Shimojo M. et al. 1996, PASJ, 48, 123

Suematsu Y. 1998, in High Resolution Solar Physics: Theory, Observations, and Techniques, ASP Conference Series, 183, 198

Tsuneta S. et al. 1992, PASJ, 44, L211

IAU 8th Asia-Pacific Regional Meeting
ASP Conference Series, Vol. 289, 2003
S. Ikeuchi, J. Hearnshaw & T. Hanawa, eds.

On Astronomy Cooperation between, and New Facilities from, Indonesia and Japan

Hakim L. Malasan

Bosscha Observatory, Department of Astronomy, Institut Teknologi Bandung, Jl. Ganesa 10 Bandung 40132, Indonesia

Abstract. The long-lasting activities in astronomy and astrophysics in Japan and Indonesia have provided the basis of cooperation in astronomy, which started in 1978. In the course of the cooperation, several instruments have been set up, especially at Bosscha Observatory in Lembang, Indonesia. The endeavour culminated in the erection of a new 45-cm Cassegrain reflector equipped with modern instruments for photometry and spectroscopy in 1989. A new impetus has been received from the cooperation with public observatories in Japan. Several new modern instruments were donated and installed in Indonesia in the period of 2000–2001, leading to new methods of astrophysical observation. In this context, networking of small observatories provides a better basis in pursuing modern scientific researches. This talk will review the activities in the past cooperation in astronomy between the two countries, and discuss new aspects of future collaboration.

1. Introduction

Cooperation in astronomy between Indonesia and Japan started in 1978 under auspices of Japan Society for Promotion of Sciences (JSPS) and Directorate General of Higher Education (DGHE) took the form of exchange of astronomers and joint researches in the fields galactic structure, cosmology and stellar physics. The wide spectrum of collaboration has been published in three proceedings (Kogure & Hidayat 1985, Ishida & Hidayat, 1989 and Ishida & Hidayat, 1992).

Among many fields that have been collaborated between two countries, activities in astronomical instrumentations have been undertaken for the purpose of enhancing the capability of the existing facilities at Bosscha Observatory. These activities yielded in procurement of a new microdensitometer (1982), construction of a DC photoelectric photometer (1980), installation of a sub-beam prism (1984) and employment of interference filters for the 51-cm Schmidt telescope (1984), and a hypersensitizing system for photographic plates (1986).

These new items of apparatus have been very indispensable mostly for survey observations of the southern hemisphere. The astronomical community in Indonesia has witnessed since then the development of persons interested in instrumentation and its utilization in astronomical observations.

2. The 45-cm Reflector at Bosscha Observatory

Background and motivation behind the erection of a modern reflector for photometry and spectroscopy of variable stars in the southern hemisphere have been described by Hidayat et al. (1992). In the earlier phase of its operation, the telescope has been primarily used for photometric observations, at which a complete set of *UBV* light curves of V505 Sgr has been secured and analyzed (Okazaki et al. 1994). The low-resolution spectrograph (94 Å/mm) furnished with photographic camera has been used primarily for educational purposes. Consistent photometric observations since 1993 has been also conducted to determine transformation coefficients and sky brightness of Lembang using Landolt standard stars and open cluster such as IC 4665 (Malasan et al. 2001). Motivated by the finding of an increase of aerosol content in the stratosphere over the observatory (Malasan & Raharto 1993), a multidisciplinary approach to the atmospheric extinction problem and its relation to meteorological aspects using the long-term data collected through photometric observation, is, at present, being undertaken. This research is participated actively by colleagues of Department of Geophysics and Meteorology and Department of Mathematics of Institut Teknologi Bandung.

3. Cooperation in Instrumentation

Although the telescope was optimized for photoelectric photometry and low-dispersion spectroscopy, it also has degree of versatility for another purposes such as experiment and imaging with CCD camera (Widjanarko et al. 1995, Kunjaya 2000). The spectrograph has also been used as the base instrument for fiber optic experiments with five-star configuration (Handojo & Fachrizal 1995). These activities have stimulated consciousness and awareness of self-supporting technical development of astronomical instrumentation.

In the period 1995 through 1997, mini-workshops organized by Bisei Astronomical Observatory, a public observatory in Japan, have been actively participated by Indonesian astronomers (cf. Kogure 1997). These workshops have been very valuable to astronomical community in Indonesia, since it opened a new vista of collaboration with public observatories in Japan, in which public education could be incorporated.

In the period 1998 through 2000, the author has been engaged in the construction and development of a new public observatory in Gunma Prefecture. Among many instrumentations that have been procured and installed for the observatory, a compact spectrograph, manufactured by Genesia Corporation, Mitaka, Tokyo, has an attractive feature and versatility for both education and research purposes. This versatile instrument was donated to Bosscha Observatory in 2000. Detailed features of this spectrograph have been described by Malasan et al. (2001). One striking feature is its flexibility to be attached to various kinds of telescopes at Bosscha Observatory. Although optimized for the 45-cm GOTO reflector, it also can be used with the 60-cm twin-refractor as shown in Figure 1 (*left*). A high resolution Hα spectrogram of Sirius (taken with 1200 grooves/mm grating), a typical double star observed regularly at Bosscha Observatory, is also displayed in the same figure (*right*). Since October this

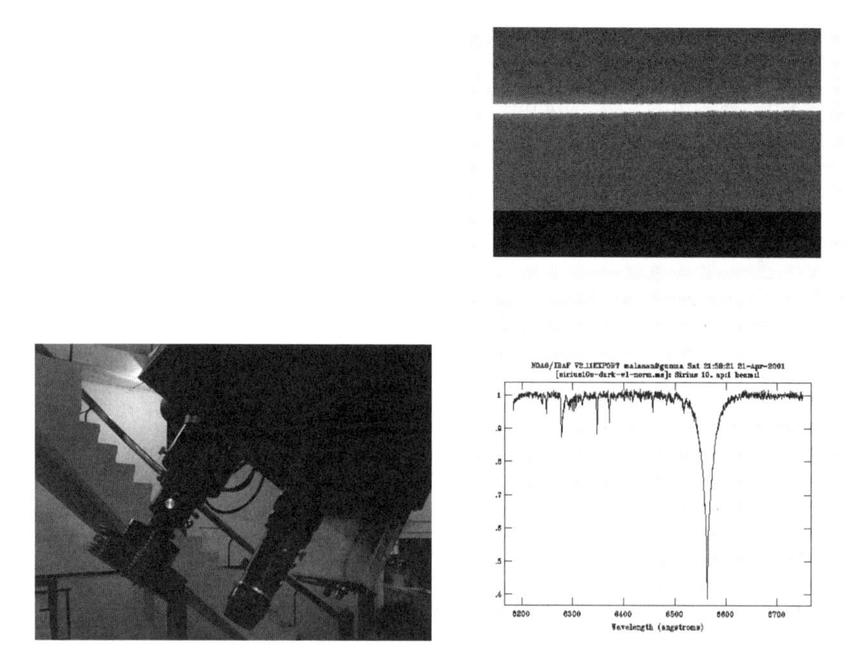

Figure 1. The Bosscha Compact Spectrograph attached to the 60-cm twin refractor at the Bosscha Observatory (left); High resolution Hα spectrum of Sirius (right)

year, the spectrograph has been put into regular operation and mostly used for stellar spectroscopic observations.

In the spirit of enhancing capability of telescope, so that it can be put at the same level with those in a modern observatory, a proposal has been prepared to add a new CCD camera, to renew the control system and to install an autoguider system to the 45-cm GOTO reflector. It is expected that with the new instruments, coordinated observations could be pursued more efficiently.

4. Closing Remarks

As has been stated in Hidayat & Arifyanto (1998), cooperation among institutions and countries should be beneficial for the respective parties and more importantly for a larger circle of the scientific community. Cooperative effort, as illustrated in the case of the Institute of Technology Bandung as a common higher learning center in Indonesia and Gunma Astronomical Observatory as a unique public observatory in Japan, would ensure development of man-power through practical exchange of science and technology.

Acknowledgments. It is a pleasure for me to thank the organizing committee for inviting and providing me support to attend the meeting. I am grateful to Profs B. Hidayat & Y. Kozai for constant encouragement, Dr. O. Hashimoto & Mr T. Kurata for invaluable discussions on the content of collaboration. Thanks are also due to colleagues at the Department of Astronomy ITB and at Gunma Astronomical Observatory, who have agreed to cooperate through the assignment of a memorandum of understanding on 2002 July 1st.

References

Handojo, A. & Fachrizal, N. 1985, in ASP Conf. Ser. 84, The Future Utilisation of Schmidt Telescopes, ed. J. M. Chapman, R. D. Cannon, S. J. Harrison & B. Hidayat (San Fransisco: ASP), 81

Hidayat, B., Mahasenaputra, McCain, C. & Malasan, H. L. 1992, in Proceedings of the Three-Year Cooperation in astronomy between Indonesia and Japan, Evolution of Stars and Galactic Structure, ed. K. Ishida & B. Hidayat (Tokyo: The University of Tokyo), 10

Hidayat, B. & Arifyanto, I. 1998, in Astrophysics at Gunma Astronomical Observatory, ed. O. Hashimoto, H. L. Malasan, A. Shimoda & T. Hasegawa, (Maebashi: Gunma Astronomical Observatory) 70

Ishida, K. & Hidayat, B. (eds.) 1989, Evolution of Stars and Stellar Systems, Proceedings of the Six-Year Cooperation in Astronomy between Indonesia and Japan 1986–1989 (Tokyo: The University of Tokyo)

Ishida, K. & Hidayat, B. (eds.) 1992, Evolution of Stars and Galactic Structure, Proceedings of the Three-Year Cooperation in Astronomy between Indonesia and Japan 1989–1991 (Tokyo: The University of Tokyo)

Kogure, T. & Hidayat, B. (eds.) 1985, Galactic Structure and Variable Stars, Proceedings of The Six-Year Cooperation in Astronomy between Indonesia and Japan 1979–1984 (Kyoto: Kyoto University)

Kogure, T. (editor) 1997 Proceedings of the Third Mini-Workshop held at Bisei Astronomical Observatory, Toward Amateur-Professional Cooperation in Observations and Popularization (Okayama: Bisei Astronomical Observatory)

Kunjaya, C. 2000, Jurnal Matematika & Sains, 5, 53

Mahasenaputra, Okazaki, A. & Hidayat, B. 1994, Southern Stars, 36, 65

Malasan, H. L. & Raharto, M. 1993, in Proceedings of the 5th International Symposium on Equatorial Atmosphere Observation over Indonesia (Jakarta: BPP Teknologi)

Malasan, H. L., Senja, M. A., Hidayat, B. & Raharto, M. 2001, in IAU Symp. 196, Preserving the Astronomical Sky, ed. R. J. Cohen & W. T. Sullivan (Dordrecht: Reidel), 147

Malasan, H. L., Yamamuro, T., Takeyama, N., Kawakita, H., Kinugasa, K., Hidayat, B., Setyanto, H. & Ibrahim, I. 2001, in Proceedings of the Indonesian-German Conference on Instrumentation, Measurements and Communication for the Future, ed. M. Djamal, O. Kanoun, E. Sesa & H.-R. Tränkler (Bandung: ITB), 159

Widjanarko, T., Handojo, A., and Malasan, H. L. 1995, in Ground Based Astronomy in Asia, Proceedings of The Third East-Asian Meeting on Astronomy, ed. N. Kaifo, (Tokyo: NAOJ), 378

IAU 8th Asia-Pacific Regional Meeting
ASP Conference Series, Vol. 289, 2003
S. Ikeuchi, J. Hearnshaw & T. Hanawa, eds.

Astronomy Against Terrorism: An Educational Astronomical Observatory Project in Peru

Mutsumi Ishitsuka, Hernán Montes

Instituto Geofísico del Perú, La Molina, Lima, Perú

Takehiko Kuroda, Masaki Morimoto

Nishi-Harima Astronomical Observatory, Sayo-gun, Hyogo, Japan

José Ishitsuka

National Astronomical Observatory of Japan, Mitaka-shi, Tokyo, Japan

Abstract. The Cosmos Coronagraphic Observatory was completely destroyed by terrorists in 1988. In 1995, in coordination with the Minister of Education of Peru, a project to construct a new Educational Astronomical Observatory has been executed. The main purpose of the observatory is to promote an interest in basic space sciences in young students from school to university levels, through basic astronomical studies and observations. The planned observatory will be able to lodge 25 visitors; furthermore an auditorium, a library and a computer room will be constructed to improve the interest of people in astronomy. Two 15-cm refractor telescopes, equipped with a CCD camera and a photometer, will be available for observations. Also a 6-m dome will house a 60-cm class reflector telescope, which will be donated soon, thanks to a fund collected and organized by the Nishi-Harima Astronomical Observatory in Japan. In addition a new modern planetarium donated by the Government of Japan will be installed in Lima, the capital of Peru. These installations will be widely open to serve the requirements of people interested in science.

1. Introduction

The idea to build a coronagraph and a solar observatory in the Peruvian Andes, was born in the Kyoto University in 1951. Professor Joe Ueta travelled in 1956 to the Huancayo Observatory of the Instituto Geofísico del Perú (IGP), with a small experimental coronagraph and observed successfully the green coronal line of Fe XIV (5303 Å). IGP was founded by the Carnegie Institution of Washington in 1922 as a Geomagnetic Observatory, and many great discoveries such as the Forbush Effect were carried out. Also in 1936 the Observatory became famous with the detection of the combined occurrence of solar flares and magnetic storms. One of the most important contributions of this observatory was the discovery of the equatorial electro-jet.

I was sent to the Huancayo Observatory in 1957 in order to establish the Department of Solar Activity and to find the place to install a new coronagraph in Peru. Indeed it was possible to observe from Huancayo the so-called green coronal Line, but not enough for good quality observations. Since then I have searched for clear skies from many possible places along the Peruvian Andes with an Evans-type sky photometer. Finally in 1965 I found nearby the Huancayo Observatory, about 70 km away, appropriate sky conditions in a place called Cosmos. Meanwhile I was searching for the site for the observatory, a new equatorial mount for the coronagraph, which I had designed, was constructed in Japan, and arrived at Huancayo in 1966. The construction of the Cosmos Coronagraphic Observatory began in 1972, at the altitude of 4 600 meters above sea-level. The design of the coronagraphic laboratory was done with a half-cylindrical sliding roof (see figure 1). A guest house for 12 people was also built, together with two geomagnetic laboratories, one meteorological facility and an electric power house. The observatory was inaugurated on 1978 October 22,

Figure 1. The Coronagraphic Laboratory and the half-cylindrical sliding roof (1978).

when representatives from both the Peruvian and the Japanese governments were present. Then finally in 1988 July, we took the first picture of the green coronal line of 5303 Å, and clear coronal details were registered, using 2415 Technical Kodak film. But unfortunately, a few weeks after that, terrorists occupied the observatory and destroyed everything.

2. The Educational Astronomical Observatory

In 1995 the Minister of Education of Peru suggested to the Instituto Geofísico del Perú to construct an Educational Astronomical Observatory, that should be relatively near the capital Lima, in order to give astronomical knowledge to students of schools and universities. Unfortunately in Lima we cannot expect favorable weather conditions for observations, and we searched for good astronomical sites around the capital. After inspecting several possible sites, a place called Cerro Jahuay (Jahuay Hill) was chosen as the best place for the future observatory. The selected site is at the distance of about 270 km from Lima, and 5 km from the Southern Pan-American Highway. The coordinates are $13°57'$ south and $76°03'$ west. The highest place is 458 meters above sea level. A 3 km $\times$ 3 km area of desert will be occupied by the whole observatory and at the present time, the respective proceedings to get the land have been almost finished.

The Universidad Nacional de Ingeniería designed the buildings that include three domes to house two 15-cm refractor telescopes and one 60-cm telescope. The government of Japan donated the two refractor telescopes in 1999. The 60-cm telescope is due to the donation the Nishi-Harima Astronomical Observatory in Japan, and currently the telescope is under construction. The observatory will have a maximum capacity to lodge 25 persons, with the availability of an astronomical discussions auditorium, internet facilities and a comfortable library. The estimated budget for the whole construction of the observatory is about one million US dollars.

3. Other Projects

Near the Educational Astronomical Observatory, there is a National University, called Universidad de Ica. Here we installed in 2001 a small refractor telescope to make solar observations. The telescope in this small observatory is furnished with a digital camera to monitor the Sun, and also it is planned to connect to the Internet and have real time images of the Sun.

As one of the other projects to expand astronomical education, we are developing optical parts of telescopes in the Ancon Observatory. To this idea, the Pontificia Universidad Católica del Perú is kindly cooperating with valuable suggestions. The aim of the project is to make optical mirrors for telescopes, and to teach students how to make their own telescope at school.

To keep the high technical quality during the construction of the Cosmos Observatory, we are planning to reconstruct a miniature equatorial mount similar to that lost previously at Cosmos. It will be used for future solar observations. People who worked at the Cosmos Observatory, will be deeply involved in this project.

4. Conclusions

The Instituto Geofísico del Perú is deeply involved in astronomy. Even though the Cosmos Coronagraphic Observatory was destroyed, they are still supporting me in my projects. Also universities in Peru are interested in developing astronomy. The construction of the Educational Astronomical Observatory in Peru will promote an interest in space sciences and astronomy amongst young students from school to university levels. The installation of the new planetarium at the capital of the country, will be widely open for the whole population, and hence it will be a good means to promote the sciences.

Acknowledgments. For the construction of the Cosmos Coronagraphic Observatory local people worked hard, and also the population resisted for many years the internal violence due to terrorism. My congratulations and acknowledgement go to them for this heroic attitude, and it will remain forever in my soul and mind. I would like to mention the continuous support given by the headquarters of the Instituto Geofísico del Perú, in particular I wish to acknowledge Dr. R. Woodman. In the course of preparing this report, we would like to express our thanks to Professor M. Kitamura for his kind help. Thank you very much to the governments of Perú and Japan for their support, and for the unconditional help that the people of Japan are giving me. And a special thanks for all the Peruvian people that gave me and to my family their warmest friendship all these years.

Extrasolar Planets: Discovery and Formation

IAU 8th Asia-Pacific Regional Meeting
ASP Conference Series, Vol. 289, *2003*
S. Ikeuchi, J. Hearnshaw & T. Hanawa, eds.

Techniques for the Detection of Planets beyond the Solar System

John Hearnshaw
University of Canterbury, Department of Physics and Astronomy, Private Bag 4800, Christchurch 8020, New Zealand

Abstract. In this review I will discuss the techniques for the detection of extrasolar planets. These are broadly the Doppler, astrometric, microlensing, direct imaging, interferometric and photometric transit methods. The successes or otherwise of these techniques to date will be outlined and the prospects for future successes will be described, in particular those based on proposed space missions. The paper emphasizes both historical aspects of extrasolar planet detection as well as future prospects over the next few decades.

This review was completed in 2002 late June, and only covers material available in the literature or on the internet up to that time.

1. Introduction

In the first half of the twentieth century there was a widespread belief that planetary systems comparable to our solar system were rare. This view was expressed, for example, by James Jeans (1917) who believed that the solar system was the product of a chance stellar encounter: "It is quite possible that our [solar] system may have been produced by events of an exceptional nature, that there are only a very few systems similar to ours in existence. It may even be that our system is something quite unique in the whole of space." Both the scarcity of such encounters and the widespread presence of binary stars appeared to favour against planetary system formation.

This view was completely reversed in the second half of the twentieth century, for two reasons. First there was a new realization of the vast extent of the universe, and the non-uniqueness of both the Sun as a star and the Milky Way as a galaxy, which seemed to make a multiplicity of planetary systems all the more plausible. Secondly, from the 1940s, a number of claims for the astrometric detection of planets around nearby stars were made. None of these claims have subsequently been robust enough to pass the test of further scrutiny, but they heightened interest in the feasibility of the method. Thus claims of massive planet detection (in the mass range 1 to 16 $M_{\rm J}$ were made by Strand (1943, 1957) for 61 Cygni, by Reuyl and Holmberg (1943) for 70 Ophiuchi, by van de Kamp (1944, 1963) for Barnard's star, by van de Kamp and Lippincott (1944, 1960) for Lalande 21185 and by Deich (1960) for 61 Cygni.

Henry Norris Russell seized on these claims of new planetary systems and declared: "On the basis of this new evidence, it therefore appears probable that among the stars at large, there may be a very large number which are attended by bodies as small as the planets of our own system" Russell (1943). And in a lecture at Vanderbilt University in the spring of 1958, Harlow Shapley declared that "Millions of planetary systems must exist. Whatever the methods of origin, planets may be the common heritage of all stars ..." (Shapley 1958). This view of the prevalence of planets was further enhanced by Struve's belief that the slow rotational velocities of many later type stars must somehow be linked to planet formation and the removal of the star's angular momentum. "Can it be that a large fraction of those late-type stars which would normally possess rapid rotations have, by some as yet unknown process, produced planetary systems, and in this manner relieved themselves of a large fraction of their angular momenta?" Struve (1946).

A fuller account of the history of the search for extrasolar planets is given by Dick (1996) in his book, *The Biological Universe.*

2. Planetary Detection Techniques

Planetary detection techniques can be classified into indirect and direct methods. In the former no signal in the form of radiation is detected from the planet itself, but the presence of a planetary body is inferred by its effect on the radiation emitted by the parent star or (in the microlensing technique) by the effect of the planet on the light from another star. In the direct technique, on the other hand, light or radiation comes from the planet itself.

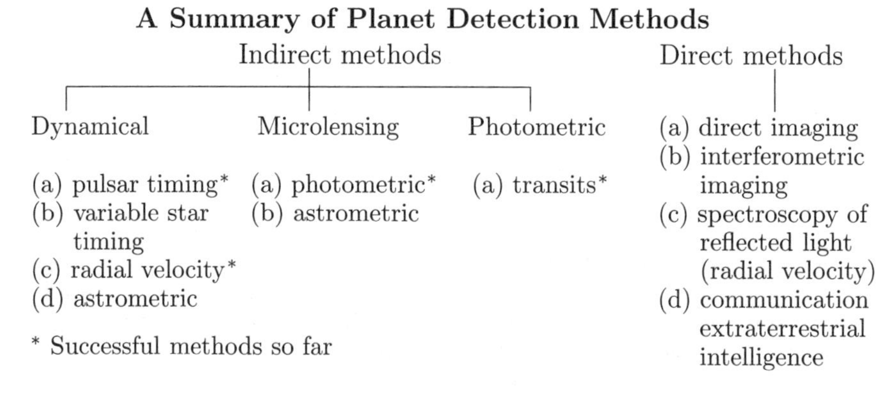

Figure 1: Planet detection methods

Figure 1 summarizes the various techniques. There are three main categories of indirect technique: these are the dynamical, microlensing and photometric methods.

It is remarkable that at the present time, although successful detection of extrasolar planets is extremely challenging using any of these techniques, nevertheless most of them have either just been producing their first successful results in the last few years, or they have every chance of doing so in the relatively near future. All require the implementation of state-of-the art technology, and these different methods all appear to be coming to fruition roughly simultaneously. Certainly none of the techniques listed are any longer in the domain of pure speculation, and all are the subject of practical experimentation at the present time.

In general, direct detection of planets is probably more difficult than indirect detection. However even here the first direct detections could be imminent. One direct technique which will not be discussed here further, but which should not be ignored, is the possibility of a successful SETI search leading is to information about an extrasolar planet as the source of intelligent signals.

2.1. Dynamical Techniques

A variety of dynamical methods are possible, but all rely on the gravitational reflex motion of the star when a planet is in orbit. The size of the star's orbit is then in inverse proportion to its mass relative to the planetary mass. One dynamical method relies on the timing of some precisely periodic event from the star in its orbit, where the delay in the time for a signal to reach Earth corresponds to the light travel time across the star's orbit about the common centre of mass. Both the pulsar timing technique and the variable star technique (for example periodic oscillations of a white dwarf) use this principle. In the Doppler technique the line-of-sight component of velocity of the star's motion is detected.

These techniques are all favoured by orbital inclinations more nearly edge-on ($i \simeq 90°$) than face-on. On the other hand the astrometric technique is more or less insensitive to inclination, and may even slightly favour face-on orientations. In the astrometric technique the amplitude of the positional wobble of a star is proportional to the size a of the planet's orbit, and given that $P \propto a^{\frac{3}{2}}$, clearly the longer period orbits have the best chance of detection. The projected displacement of a star of mass $M_\star$ and distance d in the plane of the sky, when being orbited by a planet of mass m_p is $\delta\theta = \frac{m_\mathrm{p}}{M_\mathrm{L}} \cdot \frac{r}{d}$, which for the Jupiter-Sun system observed from 10 pc would amount to a displacement of amplitude just 0.5 mas on a period of 11.9 yr.

2.2. The Doppler Technique

The Doppler technique has in recent years been by far the most successful of the techniques. Campbell and Curtis speculated on the feasibility of detecting massive dark objects in orbit around other stars as early as 1905 using Doppler methods. In 1952 Otto Struve had the foresight to realize the potential of the method to detect Jupiter-mass objects in small orbits, and he advocated searching for such planets which he surmised might have $a \sim 0.02$ AU and $P \sim 1$ d.

The basic relations based on Newtonian dynamics for interpreting a star's radial-velocity variations in terms of the mass of an orbiting but unseen planet are well known. They are:

$$K = \left(\frac{2\pi G}{P}\right)^{1/3} \frac{m_\mathrm{p} \sin i}{(m_\mathrm{p} + M_\star)^{2/3}} \frac{1}{(1 - e^2)^{1/2}}$$

For circular orbits ($e = 0$) this becomes $K = 28.4P^{-1/3}(m_{\rm p} \sin i)M_\star{}^{-2/3}$ where K is in m/s, P is in years, $m_{\rm p}$ is in Jupiter masses and $M_\star$ is in solar masses. In addition Kepler's third law gives for the period

$$P = a^{3/2} M_\star^{-1/2}$$

with P in years, a in AU and $M_\star$ in solar masses. In these equations K is the measured amplitude of the velocity variation imparted to the parent star by the reflex motion of the orbiting planet. Clearly only orbits inclined to the line of sight (not face on) have the factor $\sin i$ large enough for the Doppler method to be viable.

2.3. Microlensing Techniques

Microlensing techniques represent another category of indirect planetary detection. A microlensing event may be detected as the result of the chance alignment of a massive object, normally a star, along the line of sight to another star, known as the source. The result is the temporary amplification of the light of the source star on a timescale typically of days to weeks. The classical amplification curve, corresponding to a point source and point-mass lens, can be perturbed by the presence of a planet accompanying the lens star. The Einstein radius around the lens has a radius given by

$$R_{\rm E} = \sqrt{\frac{4GM_{\rm L}}{c^2}} \times \sqrt{\frac{D_{\rm OL}.D_{\rm LS}}{D_{\rm OS}}}$$

where $M_{\rm L}$ is the lens mass and the $D_{\rm OL}$, $D_{\rm LS}$, $D_{\rm OS}$ are respectively the distances from observer to lens, observer to source and lens to source. In a typical case of a source in the galactic bulge $D_{\rm OS} \sim 8$ kpc, and the lens also being a less distant bulge red dwarf star with $M_{\rm L} \sim 0.3M_\odot$ and $D_{\rm OL} \sim 6$ kpc, $R_{\rm E}$ would be approximately 1.9 A.U. The angular radius of the Einstein ring is therefore approximately 0.3 mas, well below the resolution limit for ground-based astrometry. Any planet lying fairly close to this distance from the lens nevertheless has a high chance of being detected as a result of brief perturbations in the amplification light curve, especially in high amplification events (Gould and Loeb 1992; Rattenbury et al. 2002). The timescale of the perturbation goes as $\sqrt{\epsilon} = \sqrt{m_{\rm p}/M_\star}$, and is typically a day for a Jupiter-mass planet, but of the order of an hour for one of Earth-mass.

Another type of microlensing technique is based on astrometry of the source star in a microlensing event (Safizadeh et al. 1999). The lens splits the source image into two normally unresolved components, one just inside and the other just outside the Einstein ring. The photometry measures the sum of the brightnesses of these images, while astrometry records the position of the photocentre, a point on the line between the images but in general not coincident with the unperturbed source star position. The photocentre in a microlensing event describes an elliptical path in the sky in a classical microlensing event, with semi-major axis of the order of a fraction of a milli-arc second. When the lens is accompanied by a planet, then an astrometric excursion of the photocentre from its regular elliptical path may occur, of duration hours to days and amplitude of tens to a hundred micro-arc seconds (for $\epsilon = m_p/M_\star \sim 10^{-3}$). This is in principle detectable using interferometry or in astrometry from space. Although the astrometric microlensing signatures are tiny, they occur relatively more often than photometric microlensing events, as the requirements for lens-source alignment are not so stringent.

2.4. Photometric Transit Technique

The photometric transit technique detects a planet by the diminution of the light from the parent star when the planet transits across the face of the star in a near-edge on orbit. The light loss goes as the square of the radius ratio, that is as $(R_p/R_\star)^2$, which is about 1 per cent for the Jupiter-Sun system. However the probability of Jupiter transits occurring for an external observer of the Sun is very small; the inclination must be $i > 89.94°$ and for random orbital orientations the probability of transits being possible would be only 9.8×10^{-4}. The transit duration for Jupiter would be just 32.6 hours, once every 11.9 years! However the detection probability improves rapidly for smaller orbits (in inverse proportion to the orbit dimensions), so Jupiter-size planets with short periods may well be detectable.

The transit technique has been explored by numerous authors, notably by Borucki and Summers (1984). It was also explicitly discussed by Struve in 1952 and also by T.J.J. See as early as 1897.

2.5. Direct Detection Techniques

Direct planet detection of the light scattered from a planetary surface is extremely difficult because of the proximity of star to planet in angular measure and the many orders of magnitude lesser intensity of the planet than the star. For Jupiter at 10 pc from the observer the angular separation $\theta = r/d$ would never exceed 0.5 arc seconds and the visual intensity ratio is of the order of 10^{-9}, giving a visual apparent magnitude of the planet of about $m_V \simeq 27.3$. In general $I_p/I_\star \simeq (\frac{\eta}{16}).(\frac{D_p}{a})^2$ where η is the planet's albedo, D_p is its diameter and a is the planet's distance from the parent star. In the thermal infrared ($\lambda \sim 10$ to 60 μm) this intensity ratio improves to somewhere between 10^{-4} to 10^{-7} (Black 1995), making this the favoured observational regime. For Jupiter at 60 μm the ratio is 10^{-4}; for the Earth or Jupiter at 10 μm, it is about 10^{-7}.

Apart from making the observations in the thermal IR, a number of further measures have been proposed for facilitating the direct detection of planets. One or several of the following could in principle be implemented: (1) reduction of scattered light in telescope optics by using super smooth mirrors; (2) use of adaptive optics to reduce stellar image size (Angel 1994); (3) observations made from space, so as further to improve image quality; (4) use of apodized mirrors so as to reduce diffracted light in outer part of a stellar image (Jacquinot and Roizen-Dossier 1964; Angel et al. 1986); (5) use of a coronograph, which can reduce the intensity in a stellar image by as much as 10^9 for observations made in space; (6) use of a very large aperture telescope with adaptive optics (Angel 1994) (e.g. OWL will have $D \sim 100$ m; CELT, $D \sim 30$ m), given that planet detectability $\propto D^4$. CELT with adaptive optics will give 7 mas resolution at 1μm; (7) use of a nulling interferometer (Bracewell 1978).

It seems probable that the first successful direct extrasolar planet detections will be made from space, even though ground based nulling interferometry from large ground-based instruments such as Keck or VLT may also have the potential to achieve this goal. Brown (1988) has considered the potential for direct planet detection from HST, and in a simulation showed that Jupiter observed from 5 pc at 550 nm in a 50 nm bandpass would have a peak in the diffraction image whose intensity was just 4 orders of magnitude below the intensity level from the Sun at 1 arc second (this is the intensity of the 0.02 arc s wide peak in Jupiter's image, not the integrated intensity).

3. Current Status of Different Planet Detection Techniques

At the present time (2002 June) by far the greatest number of successful planetary detections have come from the Doppler technique. Ninety-nine planets in 86 systems had been found by this method. Pioneering observations were made by Campbell et al. (1988) using a hydrogen fluoride cell as a way of imposing stable velocity reference lines on the stellar spectrum. The first discovery of an extrasolar planet using the Doppler method was that of 51 Peg b by Mayor and Queloz (1995). The short period of only 4.23 days was a major surprise and has led to much speculation about the origin of such "hot Jupiters". Approximately a dozen planets annually have been found since then, and many of these also have much shorter periods than any of the solar system planets. Even before the discovery of 51 Peg b, Latham et al. (1989) announced the presence of a low-mass companion to the F-type dwarf star HD114762 (a radial-velocity standard star!) from its low amplitude velocity variations of period 84 days and semi-amplitude 0.6 km/s. The unseen object could be a giant planet, a brown dwarf or even a red dwarf. The value of $m_{\rm p} \sin i$ is about $11 M_{\rm J}$.

Several other groups are pursuing this type of work using moderately high resolving power, including Marcy and Butler (1998), the McDonald Observatory group in Texas (Hatzes and Cochran 1998), the AFOE group (Korzennik et al. 2000), and a group based at ESO (Kürster et al. 1999). The present author has initiated a search using the Hercules spectrograph at Mt John (Hearnshaw et al. 2002).

Several multiple planetary systems have been successfully detected by the Doppler method. There are 11 such systems currently known, including υ And (Butler, Marcy et al. 1999) with three Jupiter-mass planets at 0.059, 0.83 and 2.50 AU; 47 UMa (Fischer, Marcy and Butler et al. 2002) has two giant planets and 55 Cnc (Marcy, Butler, Fischer et al. 2002) also has two, including one at 5.9 AU from the parent star, comparable to the size of Jupiter's orbit.

The transit technique has had a notable success when a transit was detected for the star HD209458 (Charbonneau et al. 2000), already known to have a giant planet from Doppler observations. The 2 per cent dip in the light curve indicates a planet of 1.54 Jupiter radii and an orbital inclination of $i = 85.2^\circ$ has been deduced. The planet mass can then be calculated to be $0.69 M_{\rm J}$ and the density to be 0.23 times that of water (Mazeh et al. 2000).

Recently the OGLE group monitored about 5 million stars photometrically in the Galactic Bulge and detected multiple periodic transit events in as many as 42 stars with periods 0.8 to 8.6 days and depths in the photometry of a few per cent (Udalski et al. 2002). These companion objects could be giant planets, brown dwarfs, white dwarfs or M-type red dwarfs, all of which have similar dimensions.

In 1991, even before the announcement of 51 Peg b by the Doppler method, Wolszczan and Frail (1992) found two planets orbiting the millisecond pulsar PSR1257+12 ($P = 6.2$ ms) from precise timing of pulse arrivals. The $m_{\rm p} \sin i$ values in Earth masses are 3.4 and 2.8 $M_{\rm E}$, much less than any planetary masses from the Doppler method. Since then a third probable planet has been found orbiting this pulsar, at $m_{\rm p} \sin i \simeq 0.015 M_{\rm E}$, or about a lunar mass. The nature and origin of these objects is still unresolved.

One other probably successful planet detection has been by microlensing. In 1999 the MOA (Microlensing in Astrophysics) and MPS (Microlensing Planet Search) groups jointly analysed the data for event MACHO 98-BLG-35 and found a low mass planet could account for the the perturbation of the light curve near maximum light of this event, which had $A_{\rm max} \sim 80$ (Rhie et al. 2000). Subsequently the data were reanalysed by the MOA group using difference imaging software, and the conclusions pertaining to planet detection have been much strengthened: a planet of mass between 0.4 and 1.5 $M_{\rm E}$ near the Einstein ring around the lens star can account for the observations (Bond et al. 2002).

4. Proposals for Extrasolar Planet Detection from Space

The prospects for planet detection from space are much better for most of the known techniques than they are from the ground. Only the Doppler technique can be done almost as well from the ground, even though it too is hampered by poor weather and seeing. However the best Doppler information for solar-type stars is in the green spectral region ($\lambda \sim 490$ to 550 nm), which is readily accessible from the ground. This is the result of falling flux and detection efficiency levels in the blue and the paucity of lines in the red.

Table 1 lists eleven future proposed space missions which have the potential to detect planets.

Although it is probable that not all these missions will be funded or launched, the majority of them probably will be. The space environment offers the possibility for the near continuous or repeated monitoring of many millions of stars for photometry or astrometry on a routine basis; astrometric precisions of GAIA and SIM are planned to be at the micro-arc second level. Several of the spacecraft, including TPF (Terrestrial Planet Finder) and Darwin, will not merely detect planets, but follow up on the discoveries by SIM and GAIA and other missions to determine the characteristics of many of the planets discovered, including those of about Earth mass and lying in the habitable zone around the parent stars where water is liquid. Space also offers the opportunity for direct planet detections, notably from missions such as NGST and TPF operating in the infrared.

5. Future Prospects for Extrasolar Planet Detections

At the time of writing, about 100 extrasolar planets are known, mainly from ground-based Doppler searches. Other ground-based techniques such as pho-

Table 1. Table of some planned and proposed space missions capable of detecting extrasolar planets

Mission	search method	launch	comments
Corot	transits	2005	3×10^4 stars $V = 11$ to 15.5
Kepler	transits	2007	10^5 stars $V < 14$ monitored continuously for 4 years
Eddington	transits	2008	5×10^5 stars, 5 yr mission
DIVA	astrometry	2004	10^4 stars surveyed for planets
FAME	astrometry (postponed)	2004	funding not supported by US Navy (Jan. 2002)
GAIA	astrometry	2012	astrometry of $> 10^9$ stars to $V = 20$; 5 yr mission. Several $\times 10^5$ nearby stars searched for planets
SIM	astrometry and nulling interferometry	2009	5 yr mission. All m_{J} planets within 1 kpc; m_{E} ones for 50 stars; $10 m_{\mathrm{E}}$ ones for 200 stars
Darwin	nulling interferometry	2014	6×1.5-m telescopes; survey 10^3 closest stars for m_{E}-planets
TPF	nulling interferometry or coronograph imaging	2010	5 yr mission; detailed characteristics of m_{E}-planets within 150 pc
NGST	direct imaging	2010	direct imaging of m_{J}-planets and brown dwarfs within 20 pc in near IR
GEST	photometric microlensing	?	10^8 stars in galactic bulge; not currently supported by NASA

tometric transits, microlensing and pulsar-timing should also come to fruition in the next few years. The number of detections from the ground is likely to increase, perhaps at 30 to 40 new detections annually from all these techniques.

From space, the missions DIVA, Corot, Kepler and Eddington are likely to give a greatly increased inventory of Jupiter-mass planets. Reasonable estimates are 250 such objects from DIVA astrometry by 2008, a further 100 from Corot transits by 2009, 1000 from Kepler by 2011 and 5000 from Eddington by 2013. Direct detections by NGST should follow by 2013. SIM, Darwin and TPF all have the potential for detecting Earth-mass planets – perhaps SIM will have found 50 by 2013, while Darwin and TPF will have a further 50–100 by 2015. In terms of planetary numbers, the most dramatic results will come from GAIA, with 20 000 to 30 000 giant planet discoveries predicted, and possibly as many as 30 new planets a day for 5 years, most of them within 100 pc.

References

Angel, J. R. P. 1994, Nature 368, 203

Angel, J. R. P., Cheng, A. Y. S. and Woolf, N. J. 1986, Nature, 322, 341

Bond, I. A. et al. 2002, MNRAS, 333, 71

Borucki, W. J. and Summers, A. L. 1984, Icarus, 58, 121

Bracewell, R. N. 1978, Nature, 274, 780

Brown, R. A. 1988, in IAU Coll. 99, Bioastronomy (ed. G. Marx), 117

Butler, R. P., Marcy, G. W., Fischer, D.A. et al. 1999, ApJ, 526, 916

Campbell, B., Walker G. A. H. and Yang, S. 1988, ApJ, 331, 902

Charbonneau, D., Brown, T. M., Latham, D. W., Mayor, M. 2000, ApJ, 529, 45

Deich, A. N. 1960, Izvestia Pulkovo Astron. Observ., 166, 1; also A. N. Deich and O. N. Orlova 1977, Soviet Ast., 21, 182

Dick, S., 1996, The Biological Universe, Camb. Univ. Press. See p.160 et seq.

Fischer, D .A., Marcy, G. W., Butler, R.P. et al. 2002, ApJ, 564, 1028

Gould, A. and Loeb, A. 1992, ApJ, 396, 104

Hatzes, A. P. and Cochran, W. D. 1998, ASP Conf. Ser. 134, 169

Hearnshaw, J. B. et al. 2003, Exp. Astron. (submitted)

Jacquinot, P. and Roizen-Dossier, B. 1964, Prog. Optics, 3, 29

Jeans, J. 1917, MmRAS, 62, 1

Korzennik, S. et al. 2000, ApJ, 533, L147

Kürster, M., Hatzes, A. P., Cochran, W. D., Dennerl, K., Döbereiner, S., Endl, M. 1999, IAU Coll. 170, ASP Conf. Ser. 185, 154

Latham, D. W. et al. 1989, Nature, 339, 38

Lippincott, S. L. 1960, AJ, 65, 445; see also P. van de Kamp 1944, S&T, 4, 5

Marcy, G. W. and Butler, R. P. 1998, Ann. Rev. Astron. Astrophys. 1998, 36, 57

Marcy, G. W., Butler, R. P., Fischer, D. A. et al. 2002, abstract 2002astro.ph..7294M

Mayor, M. and Queloz, D. 1995, Nature, 378, 355

Mazeh, T. et al. 2000, ApJ, 532, 55

Rattenbury, N. J., Bond, I. A., Skuljan, J., Yock, P. C. M. 2002, MNRAS, 335, 159

Reuyl, D. and Holmberg, E. 1943, ApJ, 97, 41

Rhie, S. H. et al. 2000, ApJ, 533, 378

Russell, H. N. 1943, Sci. Amer., July 1943, p.18

Safizadeh, N., Dalal, N. and Griest, K. 1999, ApJ, 522, 512

Shapley, H. 1958, Lecture on 'Religion in an age of science' presented at Vanderbilt Univ., Nashville, TN, Spring 1958, recorded by M.S. Snowden

Strand, K. A. 1943, PASP, 55, 29

Strand, K. A. 1957, AJ, 62, 35

Struve, O. 1946, S&T, 6, 3

Struve, O. 1952, Observ., 72, 199

Udalski, A., et al. 2002, Acta Astron. 52, 1

van de Kamp, P. 1944, Proc. Amer. Phil. Soc., 88, 372

van de Kamp, P. 1963, AJ, 68, 515

Wolszczan, A. and Frail, D. A. 1992, Nature, 355, 145

IAU 8th Asia-Pacific Regional Meeting
ASP Conference Series, Vol. 289, 2003
S. Ikeuchi, J. Hearnshaw & T. Hanawa, eds.

A Transit Search for Extrasolar Planets with the 0.5m Automated Patrol Telescope

M. G. Hidas, J. K. Webb, M. C. B. Ashley, M. Hunt

University of New South Wales, Sydney 2052 NSW, Australia

J. Anderson

University of California, Berkeley, USA

M. Irwin

Institute of Astronomy, Cambridge, UK

Abstract. We are beginning a search for transiting extra-solar planets using the 0.5-m Automated Patrol Telescope (APT) at Siding Spring Observatory, Australia. The main challenge has been overcoming the effects of intra-pixel sensitivity variations in the CCD, coupled with undersampling, to achieve high photometric precision. Using a modified version of software developed by Anderson and King (2000), we can now measure stars with $V \lesssim 14$ to 1% precision from a single 150-second APT image. Dedicating 50% of the available APT observing time to this project, we estimate that we should detect ~ 7 new short-period, Jupiter-sized planets per year.

1. Introduction

To date, radial-velocity searches have identified over 100 extrasolar planets. The transit method for detecting extrasolar planets involves photometric monitoring of a large sample of stars, in search of a periodic dip ($\sim$ 1%) in a star's light curve as an orbiting planet passes in front of it. This method has the potential to achieve a higher detection rate than the radial-velocity searches.

For a transiting planet, the size can be measured from the light curve (given an estimate of the host star's radius from its spectral type). Follow-up radial-velocity measurements give the actual mass (since the inclination is known). If the host star is sufficiently bright, the atmosphere of the planet can also be studied by high-resolution spectroscopy during a transit (e.g. Charbonneau et al. 2002; Webb & Wormleaton 2001).

We are searching for transiting extrasolar planets using the 0.5-m Automated Patrol Telescope (APT) at Siding Spring Observatory, Australia. The current CCD has a $2^\circ \times 3^\circ$ field of view and $9.4''$ pixels. For further technical details, see Ashley (2002) and Hidas et al. (2002). Regular observations for the project will begin in 2002 September. An upgrade to a pair of CCDs with $4.2''$

pixels, covering the entire useful field of view of the telescope ($\sim 6^\circ$ in diameter) is planned for early 2003.

2. High-precision Photometry from Undersampled Images

For stars with $V \lesssim 14$ in our 150-second APT images, photometric precision is limited by systematic errors. The most significant of these is due to intra-pixel variations (IPV) in the sensitivity of the CCD, coupled with under-sampling (the FWHM of a star in the image is ~ 1.3 pixels). A star centered on the more sensitive central part of a pixel appears $\sim 3\%$ brighter than the same star falling on a boundary between pixels. This constitutes a systematic error dependent on the *pixel phase* of the star (i.e. its exact position within a pixel). Because the distribution of flux among the pixels in a star's image is also affected by IPV, astrometry also suffers from pixel phase errors.

In a set of images offset from each other by non-integer numbers of pixels (a *dithered* set), each star samples a range of pixel phases. Therefore, when measurements from these images are combined, the average values are less biased by IPV effects.

Anderson & King (2000, herein referred to as AK) have developed a new technique to deal with this problem. They define the *effective* point-spread function (ePSF) as the instrumental PSF convolved with the sensitivity function of a single pixel. A model of the ePSF can be constructed from a dithered set of images. Using this model, the effect of IPV on photometric (and astrometric) measurements is considerably reduced.

The algorithm described in AK was originally developed to perform high-precision astrometry on images from HST's WF/PC2, which are also under-sampled. The needs of our project differ in two important aspects. Firstly, because we are using a ground-based telescope, with atmospheric seeing and tracking errors, the PSF varies slightly from image to image. We therefore need to derive a PSF for each individual image from the image itself. This is possible once the accurate positions and magnitudes of the stars (unbiased by IPV effects) have been calculated by averaging measurements from a large, dithered set of images.

The second difference is that our project requires precise photometry, not astrometry. This places different demands on the PSF. Astrometry requires the model PSF to accurately describe the distribution of flux among the central pixels, as a function of the pixel phase. Photometry only requires knowledge of how the *total* flux within the central pixels varies with pixel phase. Because of this difference, we are able to derive the PSF using a simpler algorithm than that described in AK. Stars are first measured with a PSF assuming *no* intra-pixel variation. Comparing these measurements to the average values, their variation with pixel phase can be determined, and used as a constraint for the final model of the ePSF.

Using our modified version of the AK algorithm, we can now measure stellar magnitudes to 1% precision down to $V \sim 14$ in a single 150-second image (Fig. 1). We will monitor 4 fields simultaneously, obtaining 5 images of a field per hour. The box-fitting algorithm described by Kovács, Zucker, & Mazeh (2002) requires a combined signal-to-noise ratio (for all in-transit measurements) of 6 for a

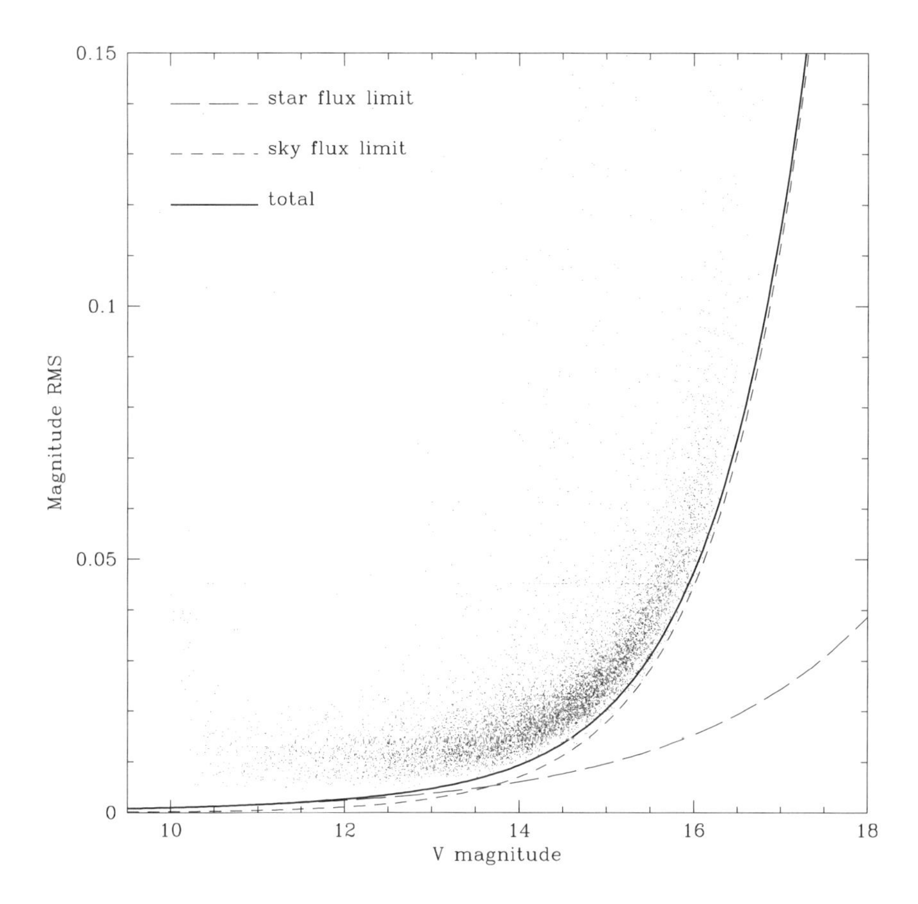

Figure 1. Photometric precision for a series of 150-second exposures taken with the APT. The field is centered on the open cluster NGC6633, close to the Galactic plane. The limits due to photon shot noise in the star and sky flux are shown. Readout noise is negligible. Systematic errors limit the precision for bright stars. Stars brighter than $V \sim 10.5$ are saturated.

significant detection of a transit from a light curve. For a typical "hot Jupiter" causing 3 hour-long, 1% deep transits every 3–5 days, we can reach this by observing three transit events. Considering the observational parameters of our search, including the properties of the upgraded detector, we estimate we will find ~ 7 planets per year (Hidas et al. 2002).

References

Anderson, J., & King, I. R. 2000, PASP, 112, 1360

Ashley, M. C. B. 2002, http://www.phys.unsw.edu.au/~mcba/apt.html

Charbonneau, D., Brown, T. M., Noyes, R. W., & Gilliland, R. L. 2002, ApJ, 568, 377

Hidas, M. G. et al. 2002, astro-ph/0209388

Kovács, G., Zucker, S., & Mazeh, T. 2002, A&A, 391, 369

Webb, J. K., & Wormleaton, I. 2001, PASA, 18, 252

IAU 8th Asia-Pacific Regional Meeting
ASP Conference Series, Vol. 289, 2003
S. Ikeuchi, J. Hearnshaw & T. Hanawa, eds.

Microlensing at High Magnification: Extra-solar Planets

Nicholas Rattenbury

Department of Physics, The University of Auckland, Auckland, New Zealand

Ian Bond

The Royal Observatory, Edinburgh, United Kingdom

Jovan Skuljan

Department of Physics and Astronomy, The University of Canterbury, Christchurch, New Zealand

Phil Yock

Department of Physics, The University of Auckland, Auckland, New Zealand

Abstract. Extra-solar planets can be efficiently detected in gravitational microlensing events of high magnification. High accuracy photometry is required over a short, well-defined time interval only, of order 10–30 hours. Most planets orbiting the lens star are revealed by perturbations in the microlensing light curve in this time. Consequently, telescope resources need be concentrated during this period only. Here we discuss some aspects of planet detection in these events.

1. Introduction

The presence of a planet in the lens system of a high magnification microlensing event results in a small deviation from the single lens light curve. This deviation occurs near the peak of the microlensing event and is detectable in events of high magnification (Liebes 1964; Griest & Safizadeh 1998). Heavier mass planets induce a larger perturbation, and planets closer to the Einstein ring of the main lens also produce larger perturbations. The Einstein ring radius, $R_{\rm E}$, and crossing time, $t_{\rm E}$, give the typical spatial and temporal scales for microlensing events:

$$R_{\rm E} \simeq 1.9\sqrt{\frac{M_{\rm L}}{0.3M_{\odot}}}\ \text{AU} \quad \text{t}_{\rm E} \simeq 16.6\sqrt{\frac{\text{M}_{\rm L}}{0.3\text{M}_{\odot}}}\ \text{days.}$$

The observer-lens and lens-source distances are assumed to be 6 kpc and 8 kpc respectively. The projected transverse velocity of the source is assumed to be 220 kms^{-1}. The time of full-width at half maximum for a high magnification event is: $t_{\rm FWHM} = \frac{3.5t_{\rm E}}{A_{\rm max}}$, where $A_{\rm max}$ is the maximum amplification of the event. A high magnification event occurs when the source and lens stars are well aligned,

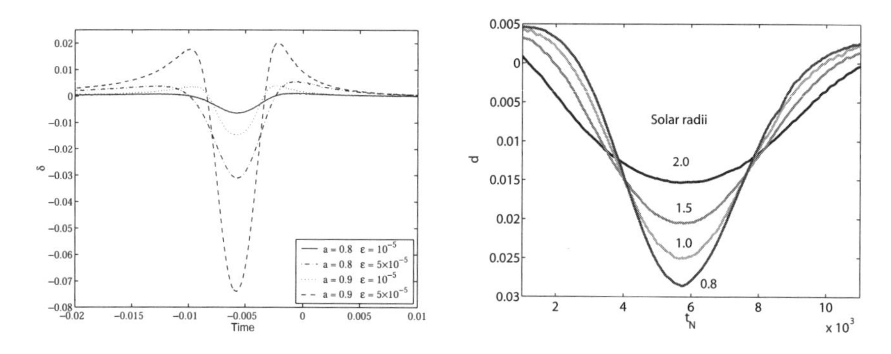

Figure 1. (a) Light curve perturbations for various values of planet mass ratio, ϵ and orbit radii, a. (b) Effect of source star radii on planet perturbation. The units are in $M_\odot$ and $(a, \epsilon) = (0.8R_{\rm E}, 4 \times 10^{-5})$, $M_{\rm L} = 0.3M_\odot$.

i.e. when $A_{\rm max} \simeq \frac{R_{\rm E}}{u_{\rm min}} \gg 1$. Figure 1a shows typical perturbations due to a lens system planet for various mass ratio and orbit radius values. Figure 1b shows the effect on the perturbation shape as a function of source star radius. Simulations and data analysis of high magnification microlensing events are carried out on the University of Auckland cluster computer, *Kalaka* (Rattenbury 2001). The finite source star size is inherent in the inverse ray shooting technique.

2. Critical Observation Time and Detectability Regions

We define the time period around the peak amplification, $t_{\rm FWHM}$, as being the critical time to observe the event in order to detect any perturbations in the peak due to lens system planets. Simulated perturbed light curves were each compared with a single lens light curve. If the difference between the perturbed light curve and the single lens light curve corresponded to a difference in χ^2 of 60, the perturbation was considered detectable by a 1-m class telescope. An example of the detection regions for an Earth-mass planet orbiting a $0.3M_\odot$ lens star is shown in Figure 2a. Simulated observations were only carried out in the time interval $+/-0.5t_{\rm FWHM}$. Figure 2a shows the detection regions extending to either side of the Einstein ring. If observations are made just during the critical time $t_{\rm FWHM}$, then the data has the potential to contain perturbations in about 67% of all possible lens-planet position angles.

3. World-wide Microlensing Event Monitoring

The major advantage of using high amplification microlensing events for the detection of extra-solar planets is the fact that the perturbation due to a lens system planet occurs near the peak of the event. The time of the event peak can be predicted well in advance. Telescope resources can therefore be concentrated during at least the critical time for detecting the planetary perturbations. In order to cover completely the critical time period for most events, a network

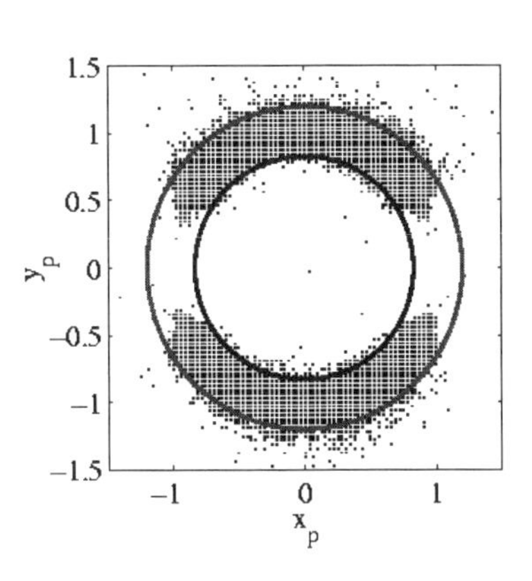

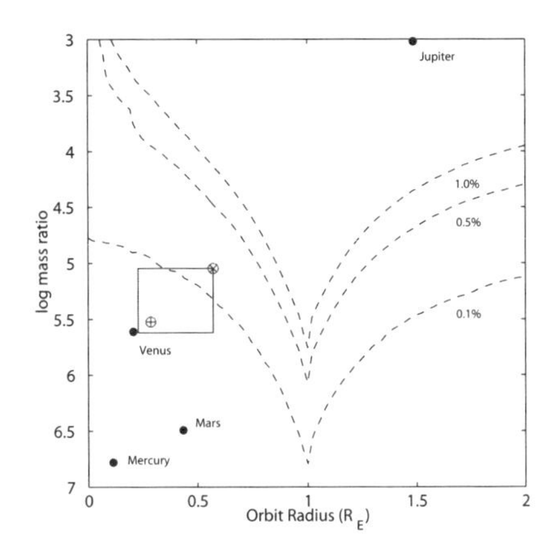

Figure 2. (a) Detection limit maps for an Earth-mass planet orbiting a $0.3M_\oplus$ lens star. The axes are in units of $R_{\rm E}$. The I-band magnitude at maximum is 15, and $A_{\rm max} = 100$. (b) Mass-orbit radius detection limits for the high magnification technique for three values of peak light curve accuracy. The mass ratio and Einstein ring radius are for a solar mass lens star. The simulated light curves were computed over the time interval $[-2t_{\rm FWHM}, 2t_{\rm FWHM}]$, with $A_{\rm max} = 100$.

of telescopes located around the world will be necessary. The Microlensing Observations in Astrophysics (MOA) collaboration issues event alerts via the Internet: http://www.roe.ac.uk/~iab/alert/alert.html and by email. Follow up observations of high magnification events are sought by other telescopes (Bond et al. 2002a).

4. An Example: MACHO 98-BLG-35

Analysis by the MPS and MOA groups found a perturbation in the peak of the light curve for event MACHO 98-BLG-35 that may have been caused by a planet orbiting the lens star (Rhie et al. 2000). Re-analysis of the available data by the MOA collaboration using difference imaging yielded a perturbation that may have been caused by a low mass $(0.4 - 1.5M_\oplus)$ planet (Bond et al. 2002).

5. Capabilities of the High Magnification Technique

The detection regions for Earth, Neptune and Jupiter mass planets for a variety of microlensing situations were computed and are given in Rattenbury et al. (2002). While low mass planets are detectable using this technique, planets in the habitable zone are probably not detectable, see Figure 2b. The planet orbit radius and mass can be determined to within $\pm 5\%$ and $\pm 50\%$, respectively in typical cases, see Figure 3a. Further simulations show that multiple planet systems can be easily modelled, provided that the position angle between planets

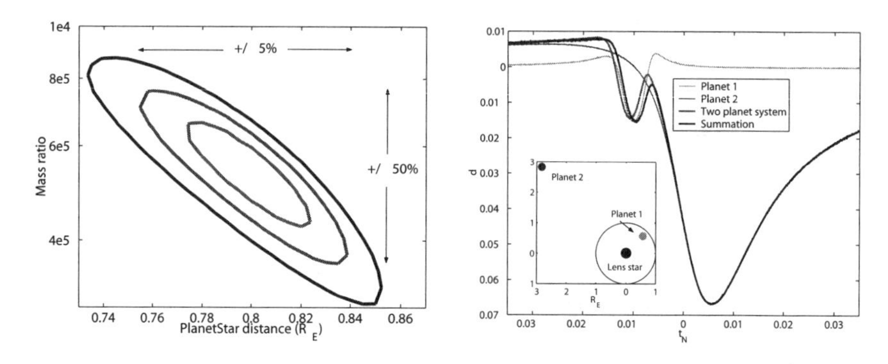

Figure 3. (a) 1-,2- and 3σ contours for a planet at orbit radius and mass of $(a, \epsilon) = (0.8R_{\rm E}, 5 \times 10^{-5})$. (b) Simulated and approximated multi-planet light curves for a two planet system. The planet parameters used correspond to Jupiter and Earth-mass planets at projected orbital radii of 7.6 and 1.7 AU respectively, assuming $M_{\rm L} = 0.3M_{\odot}$.

exceeds about 20°, see Figure. Source star spots are not likely to be reasons for a false planetary perturbation, provided a caustic crossing does not occur.

6. Conclusions

The observation of high magnification microlensing events is an effective method for the detection of extra-solar planets. The necessity of only observing during the critical time around the event peak enables efficient use of telescope scheduling and resources. Terrestrial mass planets are detectable using this technique, in regions bounding the Einstein ring of the lens system. For the most likely lens star mass, this distance is about $\simeq 2$ AU. Probing the distribution of light mass planets at these distances around the host star will complement the current understanding of planet formation and distribution. This technique requires continuous, high accuracy photometry of the event around the peak. A network of 1–2-m class telescopes situated around the world will be able to perform the required observations. Future events are expected to yield initial statistics on the abundance of terrestrial mass planets.

NJR thanks the Graduate Research Fund of the University of Auckland, the LOC and the MOA project for financial assistance.

References

Bond, I. A. et al. 2002a, MNRAS, 331, L19

Bond, I. A. et al. 2002b, MNRAS, 333, 71

Rattenbury, N. J. et al. 2001, in W.-P Chen, C. Lemme, B. Paczyński, eds., ASP Conf. Ser. 246: IAU Colloq. 183, 77

Rattenbury, N. J. et al. 2002, MNRAS, 335, 159

Rhie, S. H. 2000, ApJ, 533, 378

IAU 8th Asia-Pacific Regional Meeting
ASP Conference Series, Vol. 289, *2003*
S. Ikeuchi, J. Hearnshaw & T. Hanawa, eds.

Searches for Extrasolar Planets with the Subaru Telescope: Companions and Free-Floaters

Motohide Tamura

National Astronomical Observatory, Osawa 2-21-1, Mitaka, Tokyo 181-8588, Japan

Yoichi Itoh

Kobe University, Rokkodai 1-1, Nada, Kobe 657-8501, Japan

Yumiko Oasa

NASDA, Harumi 1-8-10, Tokyo 104-6023, Japan

CIAO/CISCO/AO/SUBARU Telescope Teams

Abstract. Infrared searches for very low-mass stars including young brown dwarfs, planetary-mass free-floating objects, and orbiting young planets with the Subaru telescope are introduced. A significant number of candidate planetary-mass objects have been discovered in S106 and other star forming regions. Also explained are an infrared coronagraph imager employing adaptive optics for the Subaru telescope and a survey project of young Jupiter-mass objects around nearby young stars with this instrument, CIAO.

1. Introduction

As a step towards the direct detection of extrasolar planets, we are currently conducting systematic searches for very low-mass stars including young brown dwarfs (YBDs) and young Jupiter-mass objects in star forming regions (SFRs). There are several merits in choosing SFRs for the studies of very low-mass stars and their mass functions: (1) Any objects (stars, BDs, and planets) are much brighter when they are young, therefore they are more easily detected than more aged objects. (2) Contrast ratios of brightness between stars and planets are smaller when they are young, and therefore the contrast problem between the central star and nearby planets is remedied for young objects. (3) The derived mass functions of young objects are almost "initial", being less affected by dynamical changes.

Motivated with this information, we are undertaking near-infrared (NIR, 1-2.5 μm) searches in the first instance. Advantages of NIR searches are as follows: (a) The NIR observations are less affected by the heavy dust extinction in SFRs than the optical observations; (b) the very low-mass objects have intrinsically low temperatures, and therefore are relatively bright at NIR wavelengths; (c) very high spatial resolution can be achieved with adaptive optics at NIR wavelengths; (d) YSO candidates of both low-mass stars (T Tauri stars) and very low-mass stars can be identified and the visual extinction to each object can be estimated from NIR colors (e.g., Tamura et al. 1998). Although individual YSO identification might suffer from some confusion with background stars, we can "statistically" discuss the luminosity functions and mass functions of very low-mass YSOs.

2. Ongoing Observations

Three kinds of near-infrared searches for young very low-mass objects are in progress with the Subaru 8.2-m telescope and other telescopes.

- Deep $JHKs$-band simultaneous surveys of SFRs with the SIRIUS infrared camera on the IRSF 1.4-m telescope and the University of Hawaii 2.2-m telescope. See Nakajima et al. (2002), Sato et al. (2002), Nagashima et al. (1999), and Nagayama et al. (2002) for the details of this SFR survey, the IRSF telescope, and the SIRIUS camera. The aim of the surveys is similar to the next survey described below, but with covering much wider areas and shallower limiting magnitudes.
- Very deep JHK'-band surveys of SFRs with the CISCO infrared camera on the Subaru telescope, seeking for "isolated" very low mass populations in SFRs, especially, YBDs and planetary-mass YSOs (Tamura et al. 1998; Oasa et al. 1999; Lucas and Roche 2000; Zapatero-Osorio et al. 2000). In spite of still various uncertainties, the lowest-luminosity ones among those objects are considered to have a few Jupiter masses. Unlike the Doppler planets, they are not orbiting around stars, and are therefore named as isolated or free-floating planets. We prefer to call them isolated planetary-mass objects. Our survey results for S106 are summarized in section 3.
- High resolution imaging surveys with the CIAO coronagraphic infrared camera on the Subaru telescope, seeking for "companion" YBDs and young planets around T Tauri stars and other low-mass YSOs.

3. Subaru Deep IR Surveys of S106

S106 is a bipolar HII region. The distance to this SFR is determined to be ~600 pc based on optical photometry of field stars (Staude et al. 1982). Only relatively shallow observations at NIR wavelengths were reported so far (K~13.5, Hodapp & Rayner 1991). Our observations were conducted with CISCO (Motohara et al. 1998) with a pixel scale of $0.''116$/pixel, under an excellent seeing condition of $\sim 0.''3$. About $5' \times 5'$ field centered around IRS 4 was imaged at J, H, and K' bands. The 10 σ detection limits as high as J=21.0, H=20.5, and K'=20.0 are achieved where the nebula background is relatively small. Thus our data are amongst the deepest and sharpest images of SFRs so far (Oasa et al. in prep., see http://SubaruTelescope.org/Science/press_release/2001/02/S106.html for a JHK' composite image).

Our deep imaging shows that a significant population of faint sources exist in the S106 SFR. We have identified about 1700 sources that are detected in all JHK' bands with errors in K' band ≤ 0.1. There is a clear concentration of sources around IRS4 (the S106 IR star cluster; Hodapp & Rayner 1991) with some distributed components all over the field of view.

The YSO candidates are selected based on the $J - H/H - K'$ color-color diagram (Oasa et al. 1999) and their number accumulates to $\sim$600 sources. Thanks to the high survey depth, the derived (de-reddened) J-band luminosity functions of the color-selected YSO candidates are complete down to the absolute J magnitude of 9.2–9.5, much deeper than the expected magnitude for the objects of hydrogen burning mass limit at an age of 1 Myr ($J\sim6.0$). The luminosity functions of YSO candidates have no peaks within the T Tauri star luminosity range, and steadily increase down to the completeness limit. This is in contrast to the luminosity functions of the Orion trapezium cluster, where the K band luminosity function has a peak within the T Tauri star luminosity range and the steadily decreases beyond the hydrogen burning mass limit.

If we translate this luminosity function to mass function, assuming a cluster age of 1 Myr, the resulting mass function of YSO candidates shows a monotonous increase from around $\log(M/M_{\odot}) = 0$ to -2.0. As a result, we have found about 100 planetary-mass object candidates in this region.

4. CIAO on the Subaru 8.2-m Telescope: A Tool for Companion Searches

CIAO is a near-infrared stellar coronagraphic imager for the Subaru 8.2-m telescope. The purpose of this instrument is to obtain diffraction limited ($\sim0.''06$ at wavelength of 2 μm) images of faint objects in close vicinity of bright objects. For achieving both high spatial resolution and high dynamic range, the instrument is used with the Subaru Cassegrain adaptive optics and designed to have a cold coronagraphic capability.

CIAO is optimized at NIR wavelengths ($1-5$ μm) where the adaptive optics works most efficiently and the effect of scattering by telescope and instrument optics is smaller. Optical components for coronagraph, occulting masks and Lyot stops, are both cooled and various sizes and shapes of masks and stops are available. CIAO is equipped with the standard broad filters ($zJHKKsL'M'$) as well as a number of narrow band filters. Slit spectroscopy with a grism of several hundred in resolution is also available. CIAO utilizes one 1024×1024 InSb detector array. The instrument achieved its first light on the Subaru telescope in 2000 February and its first combination with the Subaru adaptive optics in 2001 January. It is under an open use from 2001 October.

We have observed GG Tau and its circumbinary disk (ring) with an occulting mask diameter of $0.''8$. The spatial resolution of our data is $0.''09$ and the $0.''3$ separation binary is seen through the central occulting mask with some transmission ($\sim$2 %). The detailed structures of the circumbinary ring such as the brightness gradient from north to south due to the scattering geometry and a gap of the ring in the west are clearer than previous images (Roddier et al. 1996; Silber et al. 2000). More details of these data are discussed in Itoh et al. (2002).

We are currently conducting such AO coronagraphic surveys of nearby known T Tauri stars and our original YSOs (Ogura et al. in prep.) in the Taurus and L1457 clouds. Because of the proximity of these objects and the small mask size, it enables to observe companions at 10-20 AU from a central star. Candidate companions discovered in these surveys will be followed up by checking a common proper motion or NIR spectroscopy.

5. Summary and Future Prospects

There are various approaches for the studies of extrasolar planets on the Subaru 8.2-m telescope in both direct and indirect ways.

- Searches for self-luminous isolated low-mass YSOs down to planetary-mass objects in SFRs are in progress. Many candidate planetary-mass objects have already been detected. Follow-up spectroscopy with Multi Object Spectrometers such as FMOS and MOIRCS (now under construction) for the Subaru telescope and deep thermal imaging with the Japanese ASTRO-F mission will be invaluable for the further identification.
- Searches for self-luminous companion low-mass YSOs down to young several Jupiter-mass objects in nearby SFRs are in progress.
- Also in progress on the Subaru telescope are indirect extrasolar planet searches such as the transit method with the optical mosaic CCDs (Suprime-Cam) making use of the unique wide field of view of the Subaru telescope prime focus, and the Doppler method with the optical spectrometer (HDS).

References

Hodapp, K. W., & Rayner, J. 1991, AJ, 102, 1108

Itoh, Y. et al. 2002, PASJ, submitted

Lucas, P. W., & Roche, P. F. 2000, MNRAS, 314, 858

Motohara, K. et al. 1998, SPIE, 3354, 659

Nagayama, T. et al. 2002, SPIE, 4841, in press

Nagashima, C. et al. 1999, in Star Formation 1999, ed. T. Nakamoto (Nobeyaam: Nobeyama Radio Observatory), 397

Nakajima, Y. et al. 2002, in these proceedings, 177

Oasa, Y., Tamura, M., & Sugitani, K. 1999, ApJ, 526, 336

Roddier, C., Roddier, F., Brahic, A., Graves, J. E., & Jim, K. 1996, ApJ, 463, 326

Sato, S. et al. 2002, in Proc. of the IAU 8th Asian-Pacific Regional Meeting, Vol. I, ed. S. Ikeuchi et al. (Tokyo, Astron. Soc. Japan), 27

Silber, J., Gledhill, T., Duchene, G., & Menard, F. 2000, ApJ, 536, L89

Staude, H. J., Lenzen, R., Dyck, H. M., & Schmidt, G. D. 1982, ApJ, 255, 95

Tamura, M., Itoh, Y., Oasa, Y., & Nakajima, T. 1998, Science, 282, 1095

Zapatero-Osorio, M. R. et al. 2000, Science, 290, 103

IAU 8th Asia-Pacific Regional Meeting
ASP Conference Series, Vol. 289, 2003
S. Ikeuchi, J. Hearnshaw & T. Hanawa, eds.

Low Mass Companions to Stars: Implications for Models of the Formation and Evolution of Binary and Planetary Systems

David C. Black and Tomasz F. Stepinski

Lunar and Planetary Institute, 3600 Bay Area Blvd., Houston, TX 77058, USA

Abstract. Precise radial velocity observations over the past decade have revealed evidence of very low amplitude, periodic spectroscopic variations in some 90 stars. These variations have been interpreted to be due to $\sim$ 100 unseen low mass companions to these stars. The apparent masses of these companions are low enough that they are generally referred to as extrasolar planets. We here review briefly the current model for how planets and planetary systems form, as well as all facets of the data from the radial velocity searches. The radial velocity data provide unambiguous information regarding orbital period and eccentricity, but inferences regarding mass are uncertain for all but two of the companions. We point out that while the apparent masses suggest that the companions are substellar, the orbital data, period and eccentricity, are highly reminiscent of similar parameters for stellar binary systems. There is nothing in the results to date from these observational studies that necessitate a revision in current models of how the solar system and planetary systems in general form. However, the available data do suggest that the bulk of these companions, regardless of their mass, have more in common with stars than planets. Until the true nature of these companions is made clear by future studies we suggest that they be referred to as "U-dwarfs".

1. Introduction

The past decade has witnessed major strides in two areas of observational astronomy. We have been able to detect fluctuations in the cosmic background radiation that presage the formation of the macroscopic structures that dominate the visible universe today. We also have detected, via precise radial velocity observations, evidence of very low amplitude, periodic spectroscopic variations in roughly 90 stars. These variations are generally interpreted as arising from the presence of one or more unresolved companions. Within such an interpretation, the *minimum* masses of the companions can be estimated from the mass function, and the typical values are of the order of 1 M_J. The current list of these stars and inferred properties of $\sim$ 100 companions are listed on the Internet at *cfa-www.harvard.edu/planets/* and *www.exoplanets.org*. While public interest in observations of the earliest phases of the universe is based in large measure on understanding how the universe came into existence and what its fate may be, interest in low-mass companions to stars lies in the possibility that these companions may be Jupiter-like extrasolar planets, perhaps largest members of planetary systems that harbor habitable if not inhabited planets.

However, the true mass of any particular companion remains unknown, subject to determination of its orbital inclination i. Their classification as extrasolar planetary candidates (EPC) is based primarily on an expectation that the distribution of i in the stellar samples chosen to monitor for the presence of low-mass companions is not biased and follows a uniform distribution. The most surprising aspect of EPC is their binary-like orbital properties that must be considered peculiar for objects presumed to be giant planets. The purpose of this communication is to examine the available EPC data with an eye to assessing the extent to which they may inform our models of the process of the formation and evolution of companions to stars in general, and models of planetary system formation in particular. We first provide a brief overview of current models regarding the formation of planets and planetary systems, followed by a summary of key aspects of the radial velocity data. We close with some discussion and a summary.

2. Brief Overview of Current Models for the Formation of Planets and Planetary Systems

Current theoretical constructs regarding the formation of the solar system, and by extension all planetary systems, have at their core the notion of a circumstellar disk of gas and dust formed as a natural byproduct of the formation of the central star. This disk is viewed as the nursery for planet formation. The standard model invokes a "thin" disk, defined by $h(r)/r << 1$, where h is the disk's thickness and r a distance from the central star. Planet formation within such a disk is generally regarded to be a "bottom up" coagulation process, where microscopic dust particles collide and grow into planetesimals that in turn coalesce into solid protoplanets or planetary cores that constitute the backbone of a planetary system (Lissauer 1993). As planets in this context are the end product of a large number of collisions/coalescences, their initial orbital properties are highly regularized. That is, the orbits are approximately circular, and the thin nature of the disk leads to near coplanarity of the orbits of planets. These orbital characteristics reflect the attributes of the solar system, they are also consistent with the orbital characteristics of the multiple, planetary mass companion system found to be revolving around the pulsar PSR 1257+12 (Wolszczan & Frail 1992), but do not, in general, reflect the attributes of EPC.

Within the construct described above a giant planet like Jupiter forms by subsequent accretion of gaseous envelope from the surrounding disk onto a protoplanetary core of sufficient mass to start a rapid capture of gas. This critical mass is thought to be $\sim 10\ M_{\oplus}$. Evidence from spacecraft missions to the outer solar system, specifically determinations of the gravitational moments of the planets and inferences therefrom regarding the structure of those planets, have tended to confirm this basic picture of planet formation. However, the recent estimate for the mass of Jupiter core ranges from 0 to 12 $M_{\oplus}$ (Guillot, Gautier & Hubbard 1997), pointing to the possibility that Jupiter has no core, and thus did not form by an accretionary process. Alternatively, Jupiter could form by gravitational instability (Boss, 2002) in a protoplanetary disk, that effectively isolates $\sim 1\ M_{\rm J}$ on a relatively rapid time scale, several local orbital periods.

There is now firm observational evidence (see review by Beckwith 1994) supporting the notion that pre-main-sequence, solar-mass stars are accompanied by protoplanetary disks with properties much like the ones we envision for the solar nebula. This opens the possibility that some of these stars may develop a planetary system around them. It's tempting to postulate that EPC are indeed the largest members of such systems; however, their orbital properties are not consistent with the formation scenario described above.

3. Review of Key Data

It is important to understand the information that is provided by the radial velocity observations used to detect the EPC. The residual radial velocity data form a periodic (or quasi-periodic) time series that is characterized by three attributes, period, P, semi-amplitude, K, and the functional form of the time series. If the residual motion is attributed to a companion, P is an orbital period and the departure of the time series from purely sinusoidal is measured by an orbital eccentricity e. Thus, within the unresolved companion interpretation of spectroscopic data, orbital periods and eccentricities are directly provided by the data and have no ambiguities associated with their values.

The mass function combines the values of P, e, and K to calculate the value of *projected mass* of the companion, $M \sin i$. However, the value of $M \sin i$ is ambiguous. Two further assumptions are needed to connect known quantities to $M \sin i$ via the mass function. First, one needs to assume that what is being observed is a single source of light, and not a so-called blended binary of two nearly identical stars (Imbert & Prévot 1998). Second, that the companion is not itself an unresolved but massive object like a white dwarf. Only when those assumptions are satisfied, one can relate the amplitude to the minimum mass. Although in any particular case these two assumptions can be substantiated by additional observations, this is not routinely done. Thus, the EPC sample could be in reality a collection of different objects with true masses ranging from that of a star to that of Jupiter. More work is needed to assess relative frequencies of different objects with the EPC spectral signature. An ambiguity remains even if most of the EPC are indeed the low mass companions because the true mass of the companion remains unknown by the factor $1/\sin i$. There are only two EPC where information regarding the true mass is available. One is the companion to HD 209458 where transit observations (Charbonneau et al. 2000) provide a direct measure of the inclination angle and hence a determination of the companion mass of $\sim 0.6\ M_{\rm J}$. The second is ρ Cancri, for which a recent astrometric study (McGrath et al. 2002) using the astrometric capability of the HST has provided a firm upper limit of $\sim 30\ M_{\rm J}$ on the mass of the companion. Thus, at least in these two cases, we are sure that the companion is not massive (i.e., stellar).

Interestingly, the only unambiguous quantities, P and e, reveal that, aside from their (potentially) low masses, the EPC population looks much like the stellar secondaries population, not what we would expect from putative giant planets. To quantify this statement we estimate probability distribution func-

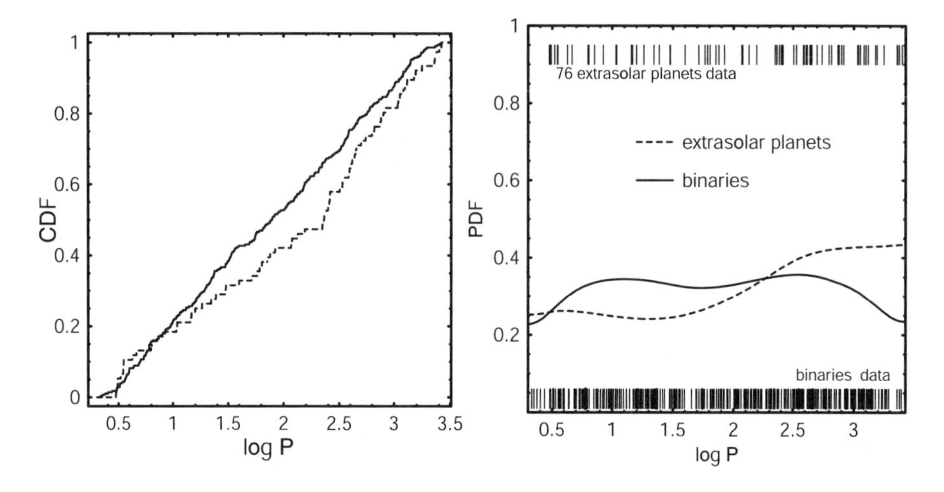

Figure 1. Estimations of orbital period distribution in populations of EPC (dotted lines) and stellar secondaries (solid lines). The left panel shows empirical CDFs of $\log P$ and the right panel shows estimated PDFs of $\log P$. Vertical bars indicated data. Objects with $2 < P < 2650$ days are included, there are 76 such objects in the EPC sample.

tions (PDFs) of P and e for populations of EPC and stellar secondaries using available data. We use all EPC data from web sites cited in Sect. 1. Our estimate is based on 78 EPC objects known as of 2002 April. For stellar secondaries we use a sample of 330 objects found in a survey of spectroscopic binaries in a large sample of G-dwarfs (Udry et al. 1998).

Fig. 1 shows the comparison of distribution of $\log P$ between the two populations. Only those stellar secondaries with a period range covered by the period range defined by the EPC data set are included. Because the values of periods in the samples span 3 orders of magnitude, it is better to estimate a PDF of variable $\log P$, instead of a PDF of P, which can be recovered from a distribution of $\log P$ by an simple transformation. The left panel of Fig. 1 shows empirical CDFs constructed from respective data sets. There is a noticeable difference between two populations. Using the Kolmogorov-Smirnov (K-S) test we have found that there is only 3% chance that the two sets of data came from the same distribution. Thus, it may appear that periods of EPC and stellar secondaries are distributed differently, further boosting an argument for a "planetary" character of the EPC objects. However, the difference is quantitative and not qualitative. This can be seen from the right panel of Fig. 1 which shows the actual PDFs. In both cases the PDFs of $\log P$ are rather flat indicating that PDFs of P are given by the power law functions with indices close to 1. Indeed the direct parametric fit to the power law gives index of -0.99 for stellar secondaries and -0.89 for the EPC. Thus, there is a remarkable qualitative similarity in distribution of orbital periods between EPC and stellar secondaries.

Fig. 2 shows the comparison of distributions of e between the two populations. Because we are interested in companions in uncircularized orbits, only

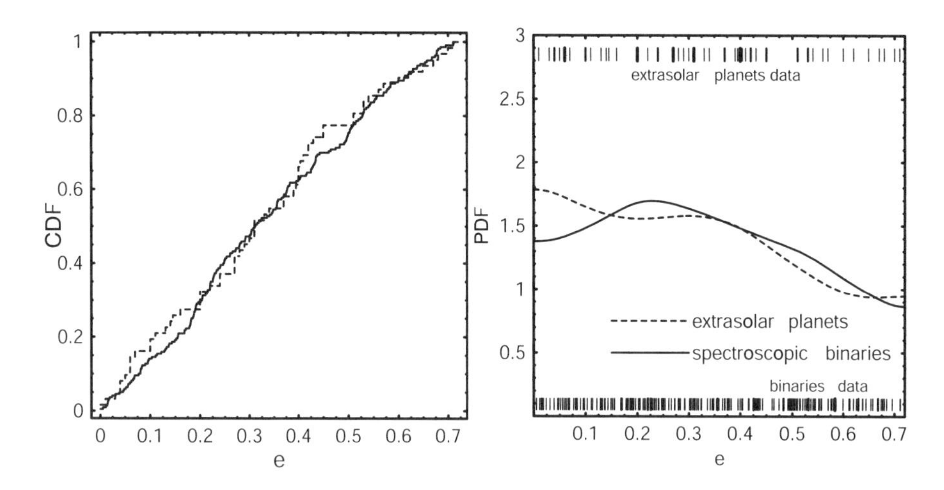

Figure 2. Estimations of orbital eccentricity distribution in populations of EPC (dotted lines) and stellar secondaries (solid lines). The left panel shows empirical CDFs of e and the right panel shows estimated PDFs of e. Vertical bars indicated data. Objects with $10 < P < 2650$ days and $0 < e < 0.72$ are included, there are 62 such objects in the EPC sample.

objects with $P > 10$ days are included. The left panel of Fig. 2 shows a great similarity between empirical CDFs derived from the two data sets. The K-S test gives 90% chance that the two sets of data came from the same population. The actual PDFs, shown in the right panel of Fig. 2 confirm this remarkable similarity.

In addition, variables $\log P$ and e are somewhat correlated (correlation coefficient of about 0.5) to the same degree for both populations. Thus, the EPC have orbital parameters that are similar to those of stellar binaries and very different from what we observe in the solar system.

4. Discussion and Summary

The available data from the high accuracy radial velocity searches for extrasolar planets reveal the following, on the assumption that the observed residual motion is Keplerian in origin:

- The least ambiguous parameters from the observations are the orbital parameters; period and eccentricity, while inferences about companion mass from the data are relatively ambiguous and uncertain.

- The underlying orbital eccentricity distribution of the EPC is statistically indistinguishable from that of normal stellar binary companions with periods in the same range as the EPC sample.

- The underlying orbital period distribution of the EPC is statistically distinguishable from that of normal stellar binary companions, but power law PDFs with very similar indices characterizes both.

- The values of $M \sin i$ strongly suggest that the true masses of the majority of EPC are in the range of a few tens of Jupiter masses down to values near or below a Jupiter mass.

As noted above, inferences regarding the masses of EPC are the least reliable, but it is likely that most are sub-stellar, i.e., have masses less than 80 M_J. Indeed, it is likely, but not proven, that the majority have masses that fall below the limit for burning of deuterium, roughly 12 M_J. This limit is frequently used as the lower mass limit of brown dwarfs. Note that brown dwarfs were originally defined by the upper limit to their masses, as objects that are formed by the same process(es) that form stars (gravitational instability), but which were not massive enough to burn hydrogen. Because the bulk of the EPC appears to have masses below the lower limit of brown dwarfs and more in line with the mass of Jupiter, they are designated as planets. That designation implies the "planetary" mode of their formation, something that seems to be strongly at odds with their orbital data (see Sect. 3).

The theoretical challenge is to come up with a feasible scenario for the formation of the EPC. Currently, all theoretical effort starts with the paradigm that EPC formed by the "bottom up" process (see Sect. 2) and thus at a distance of ~ 5 AU from a central star and on a circular orbit. Subsequent orbital evolution is invoked to account for the observed orbital parameters. Possible mechanism include gravitational interaction between a planet and a disk, gravitational scattering by other massive companions in nearby orbits, and the Kozoi mechanism for those systems that are members of a stellar binary system. However, none of these mechanisms have been shown to work generally, nor to produce distributions of orbital parameters that are similar to what is observed. Also note that in such a paradigm, the solar system appears to be an anomaly.

Alternatively, the EPC may form by means of gravitational instability in the circumstellar disk. Note that such a mode of formation is currently favored for the formation of short-period stellar secondaries (for a review see Bonnell 2001). If indeed stellar secondaries and EPC formed by the same process, the similarity of their orbits would be readily understood. Current computer models have difficulties for many reasons in modelling the formation of low mass companions via the process of disk gravitational instability. However, we might ask: 'Is there any observational evidence that the "star formation process" might operate at masses below 12 M_J?' The recent studies of nearby star formation regions (Lucas & Roche 2000) indicate an existence of free-floating objects with masses at or below 10 M_J. If the estimates of the masses are correct, the masses are not determined directly but inferred from measured luminosity and presumed ages coupled with theoretical models of the luminosity evolution of low-mass objects, the existence of such objects would indicate that the nature can produce "planetary-mass" objects by means of gravitational instability. If such objects are find free-floating, there is no reason why they could not form as companions. In addition, the actual origin of Jupiter may also be in question (see Sect. 2), and it is not inconceivable that it formed via gravitational instability. Therefore, it is premature to consider 12 M_J as a demarcation between two modes of formation. If we wish to retain the lower mass cutoff for "brown dwarfs" at 12 M_J, and the EPC formed by gravitational instability, perhaps we should refer to them as "U-dwarfs", where the "U" could stand for either "ultimate", or "unknown".

The discoveries of these remarkable objects over the past decade challenge us on several fronts. However, the rush to identify them as "extrasolar planets", while understandable, needs to be viewed with extreme caution when one examines all facets of the data. We see nothing at this time that forces a substantial revision of the current notions of how planets and planetary systems form. Indeed, we think that it is entirely possible that these exciting new systems will shed more light on the still unsolved problem of how binary systems form.

Acknowledgments. One of us (DCB) would like to thank the organizers of this IAU Symposium for their support and the opportunity to discuss this exciting area of research with colleagues from Asia. This was conducted at the Lunar and Planetary Institute, which is operated by the Universities Space Research Association under contract No. NASW-4574 with the National Aeronautics and Space Administration. This is Lunar and Planetary Institute Contribution No. 1131.

References

Beckwith, S. V. W. 1994 in NATO ASI Series C Vol. 417, Theory of Accretion Disks - 2, ed. W. J. Duschl, J. Frank, F. Meyer, E. Meyer-Hofmeister, & W. M. Tscharnuter (Kluwer Academic Publishers), 1

Bonnell, I. A. 2001, in IAU Symposium Vol. 200, The Formation of Binary Stars, ed. H. Zinnecker & R. D. Mathieu (San Francisco: ASP), 23

Boss, A. P. 2002, ApJ, 576, 462

Charbonneau, D., Brown, T. M., Latham, D. W., Mayor, M. 2000, ApJ, 529, L45

Guillot, T., Gautier, D., Hubbard, W. B. 1997, Icarus, 130, 532

Imbert, M. & Prévot, L. 1998, A&A, 334, L37

McGrath, M. A., Nelan, E., Black, D.C., Gatewood, G., Noll, K., Schultz, A., Lubow, S., Han, I., Stepinski, T. F., Targett, T. 2002, ApJ, 564, L27

Lissauer, J. J. 1993, ARA&A, 31, 129

Lucas, P. W. & Roche, P. F. 2000, MNRAS, 314, 858

Udry, S., Mayor, M., Latham, D. W., Stefanik, R. P., Torres, G., Mazeh, T., Goldberg, D., Andersen, J., Nordström, B. 1998, in ASP Conf. Ser. 154, Cool Stars, Stellar Systems and the Sun, Tenth Cambridge Workshop, ed. R. A. Donahue & J. A. Bookbinder (San Francisko: ASP), 2148

Wolszczan, A. & Frail, D. A. 1992, Nature, 355, 145

IAU 8th Asia-Pacific Regional Meeting
ASP Conference Series, Vol. 289, 2003
S. Ikeuchi, J. Hearnshaw & T. Hanawa, eds.

Investigation of the Physical Properties of Protoplanetary Disks around T Tauri Stars by a High-resolution Imaging Survey at $\lambda = 2$ mm

Munetake Momose

Institute of Astrophysics and Planetary Sciences, Ibaraki University, Bunkyo 2-1-1, Mito, Ibaraki 310-8512, Japan

Yoshimi Kitamura

Institute of Space and Astronautical Science, Yoshinodai, Sagamihara, Kanagawa 229-8510, Japan

Sozo Yokogawa

Graduate University for Advanced Studies, National Astronomical Observatory of Japan, Osawa, Mitaka, Tokyo 181-8588, Japan

Ryohei Kawabe, Motohide Tamura

National Astronomical Observatory of Japan, Osawa, Mitaka, Tokyo 181-8588, Japan

Shigeru Ida

Tokyo Institute of Technology, Ookayama, Meguro-ku, Tokyo 152-8551, Japan

Abstract. Dust continuum emission at 2 millimeters from the disks around 13 T Tauri stars was imaged with the spatial resolution of $1'' - 2''$. Disk properties are derived from the combination of image-based model fitting and the analysis of the spectral energy distributions. We found a tendency for disk expansion with decreasing accretion activity, which can be interpreted in terms of the evolution of an accretion disk. The derived surface density distribution of the disks may suggest that it is difficult to make planets like the solar system without the redistribution of solids.

1. Introduction

Low-mass pre-main-sequence stars (T Tauri stars) are commonly accompanied by circumstellar disks. Their physical properties have been derived mainly from analysis of spectral energy distributions (SEDs), revealing that the disks contain gas and dust of $(0.1 - 0.001)M_\odot$ within several hundred AU in radius (Beckwith & Sargent 1996). Since such characteristics are reminiscent of the "primordial solar nebula" (e.g., Hayashi, Nakazawa, & Nakagawa 1985), these are believed to be the precursors of planetary systems, or "protoplanetary" disks. Understanding of the internal structure and evolution of the disks in the early T Tauri stage, however, is still limited. It has been revealed that the mass accretion rate from the disk to the central star gets lower as the stellar age increases in the classical T Tauri stage (Calvet, Hartmann & Strom 2000), but it is unclear how this evolutionary trend is related to the internal structure of the disks themselves. We have carried out an imaging survey of protoplanetary disks associated with single T Tauri stars in Taurus Molecular Cloud ($d = 140$ pc) in dust continuum emission at $\lambda = 2$ millimeters with the Nobeyama Millimeter Array (NMA). Physical properties of the disks, including the outer radius and the surface density distribution, have systematically been derived from the combination of SED analysis and image-based model fitting.

2. Observations

Observations were made with the NMA over three winter seasons of 1998 – 2001. In our survey disks around 13 stars were imaged: 11 are T Tauri stars while 2 (HL Tau, Haro 6-5B) are objects embedded in a tenuous envelope. Prior to the high resolution imaging, we measured the total flux density from a disk with a larger beam ($\sim 5''$ beam, or $\sim$ 700AU resolution) of the compact D configuration. We used these total flux densities to check the depth of integration for successive higher-resolution images with the AB ($\sim 1''$ beam) or AB+C ($\sim 2''$ beam) configurations. More details on the sample selection and observations are found in Kitamura et al. (2002).

3. Results and Discussion

Figure 1 shows the obtained high-resolution images. Each emission feature is more extended than the beamsize, indicating that the disk emission is spatially resolved. To estimate the physical properties of the disks, we analyze the disk images and the SEDs on the basis of two disk models; one is a power-law model frequently used in data analysis of disk observations (e.g., Beckwith et al. 1990), and the other is a similarity solution for a viscous accretion disk (Lynden-Bell & Pringle 1974). The disk parameters to be determined are as follows: inner and outer radii ($R_{\rm in}$, $R_{\rm out}$), surface density distribution ($\Sigma(r)$), inclination angle (i), position angle (PA), temperature distribution ($T(r)$), and dust opacity coefficient with the β index. Among these parameters, $R_{\rm out}$, $\Sigma(r)$, i, and PA are determined by the image-based model fitting while $R_{\rm in}$, $T(r)$, and β are by the SED-fitting. Since both the disk models give consistent results with each other, we will only show the case of the power-law disk model in the following (see Kitamura et al. 2002 for more details). Good measures of the disk evolution are necessary when we extract evolutionary trend or diversity in disk physical parameters, i.e., the clock. In this study two clocks are adopted: the age of the central star and the Hα line luminosity (L(Hα)) which is well correlated with the mass accretion rate in the disks (Cabrit et al. 1990). The most intriguing correlation is shown in Figure 2: $R_{\rm out}$ seems to increase with decreasing L(Hα). This trend can be interpreted as radial expansion of an accretion disk due to outward transport of angular momentum (Lynden-Bell & Pringle 1974). The viscosity transports angular momentum from the inner to outer parts of the disk,

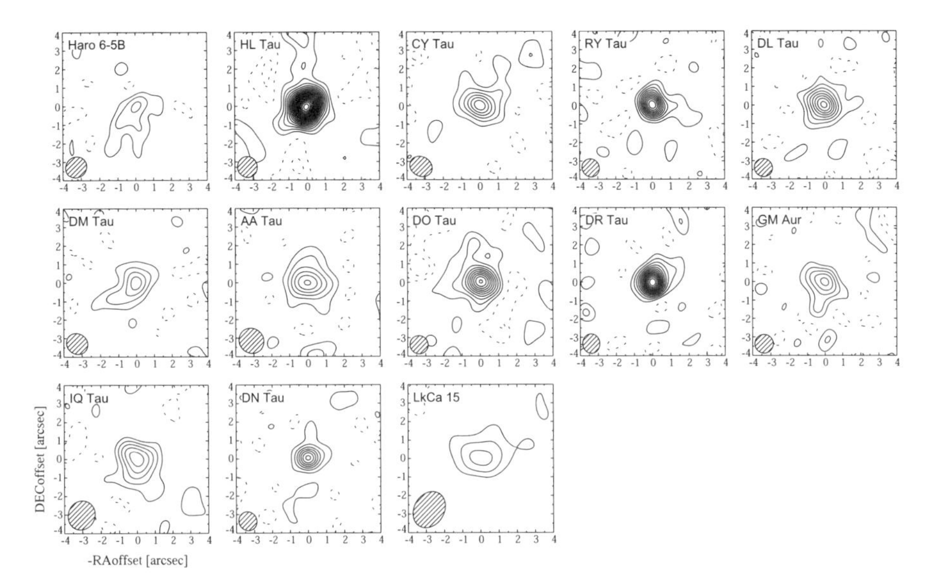

Figure 1. Images of 2 mm dust continuum emission toward 13 T Tauri stars. The contour lines start at 1.5σ and -1.5σ levels with intervals of 1.5σ. The negative levels are written by broken lines. The hatched ellipse in each panel indicates the beamsize (FWHM).

resulting in the accreting motion of the inner part and the expansion of the outer part. If we compare the disk expansion with the similarity solution for viscous accretion disks, the observed increase rate of $R_{\rm out}$ suggests $\alpha \sim 0.01$, which is predicted in theoretical models for the MHD turbulence (Balbus & Hawley 1998). The trend, however, becomes unclear against the stellar age (Figure 2). These results may suggest that the disk evolution does not synchronize well with the stellar evolution. We found no evolutionary trend in surface density distribution. The surface density at 100 AU ranges from 0.1 to 10 g cm^{-2}, which is consistent with the extrapolated value at 100 AU in the Hayashi model, suggesting the outer parts of the disks have enough matter to form small bodies like the Edgeworth-Kuiper Belt Objects. The power-law index p of $\Sigma(r)$ ($\Sigma(r) \propto r^{-p}$), on the other hand, mainly falls into a range of 0 to 1, smaller than the index of 1.5 predicted in the Hayashi model. Such flat distributions of the surface density might prohibit the formation of giant planets in the innermost regions of the disks if the low values of p hold even in the innermost parts, where planets will be formed. One possible solution to this problem is reshuffle of solids within the disks, i.e., inward movement of solids (Stepinski & Valageas 1997). Of course, our observations were insufficient to derive convincing conclusions on this issue. The large scatters of p or $\Sigma_{\rm 100AU}$ might indicate some diversity in the viscosity processes in the disks or initial conditions for planet building. The diversity found in the physical properties of the protoplanetary disks is likely to be one of the main causes that produce the diversity seen in the planetary systems including both our solar system and extra-solar systems being discovered (e.g.,

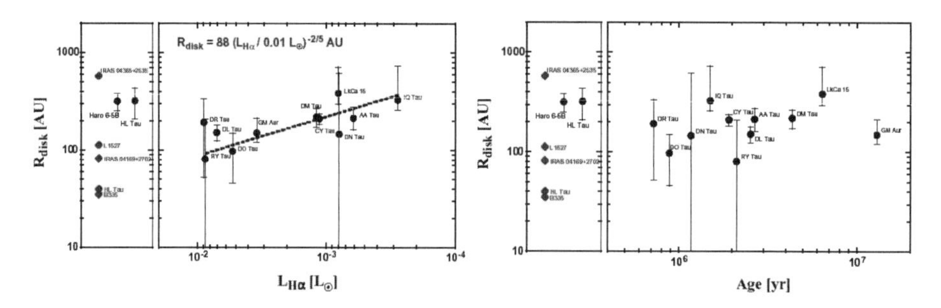

Figure 2. (left) Disk outer radius vs Hα line luminosity. The best-fit curve, shown by a broken line, is $R_{\rm out} = 88(L_{\rm H\alpha}/0.01L_{\odot})^{-2/5}$ AU. The left panel shows the results for the two embedded sources without the abscissa and shows the centrifugal radii derived by Ohashi et al. (1997). (right) Disk radius vs. stellar age.

Marcy & Butler 1998). It has not been well understood as yet, however, which property of the protoplanetary disks produces the difference between the hot Jupiters and our planets. Theoretically it is of great significance to generalize the planet formation processes so as to be applicable to any planetary system. The initial conditions, i.e., the starting point of the generalized models, should be provided by imaging observations with extremely high resolutions and sensitivities. Great advances in understanding of planet formation must be achieved by a next-generation array, the Atacama Large Millimeter and submillimeter Array (ALMA) to be operated in $\sim$ 2010 in Chile.

References

Balbus, S. A. & Hawley, J. F. 1998, Reviews of Modern Physics, 70, 1

Beckwith, S. V. W. & Sargent, A. I. 1996, Nature, 383, 139

Beckwith, S. V. W., Sargent, A. I., Chini, R. S., & Guesten, R. 1990, AJ, 99, 924

Cabrit, S., Edwards, S., Strom, S. E., & Strom, K. M. 1990, ApJ, 354, 687

Calvet, N., Hartmann, L. & Strom, S. E. 2000, in Protostars & Planets IV, ed. V. Mannings, A. P. Boss, & S. S. Russell (Tuscon: Univ. Arizona Press), 377

Hayashi, C., Nakazawa, K., & Nakagawa, Y. 1985, in Protostars & Planets II, ed. D. C. Black & M. S. Matthews (Tuscon: Univ. Arizona Press), 1100

Kitamura, Y., Momose, M., Yokogawa, S., Kawabe, R., Tamura, M., & Ida, S. 2002, ApJ, in press

Marcy, G. W. & Butler, R. P. 1998, ARA&A, 36, 57

Ohashi, N., Hayashi, M., Ho, P. T. P., Momose, M., Tamura, M., Hirano, N., & Sargent, A. I. 1997, ApJ, 488, 317

Stepinski, T. F. & Valageas, P. 1997, A&A, 319, 1007

IAU 8th Asia-Pacific Regional Meeting
ASP Conference Series, Vol. 289, 2003
S. Ikeuchi, J. Hearnshaw & T. Hanawa, eds.

A Dynamical Analysis of the GJ 876 and HD 82943 Systems

Ji Jianghui and Li Guangyu

Purple Mountain Observatory, Chinese Academy of Sciences, Nanjing, 210008, China

Liu Lin

Department of Astronomy, Nanjing University, Nanjing, 210093, China

Abstract. We carry out simulations to investigate the dynamical evolution of the GJ 876 and HD 82943 planetary systems consisting of two resonant Jupiter-like planets respectively, and reveal a possible stabilizing mechanism for maintaining the systems. By simulating and analyzing different coplanar and non-coplanar configurations, we find that all the stable cases are involved in the 2:1 mean motion resonance and that the alignment of the periastron of the two planets also plays an important part in the secular orbital evolution of the multi-planetary systems, indicating that the two kinds of mechanisms could be responsible for the dynamics of the two systems.

1. Introduction

At present, 88 planetary systems[1] , containing 101 giant extrasolar planets, have been discovered in Doppler surveys of solar-type stars (Marcy et al. 2000; Butler et al. 2001), among which there exist 11 multi-planetary systems. In these systems, we concentrate here on the following two systems: GJ 876 (discovered by astronomers from Lick and Keck observatories and UCSB) and HD 82943 (discovered by astronomers from the Geneva Observatory and other research institutes)[2] . These systems both have two Jupiter-like planets, and the outcomes of the observation imply that they are close to the state of 2:1 mean motion resonance (MMR).

Marcy et al. (2001) pointed out that the two planets orbiting the M4 main-sequence star GJ 876 are now apparently locked in the state of 2:1 MMR, with orbital periods 30.1 d and 61.0 d, and semi-major axes 0.13 AU and 0.21 AU. In their paper, they gave GJ 876 an estimated mass of $0.32 \pm 0.05 M_{\odot}$. The masses of two companions are given in Table 1 (Laughlin & Chambers 2001) with respect to an inclination $\sin i = 0.55$. However, HD 82943 is a G0 star with $B - V = 0.623$, and the mass of the parent star is $1.05 M_{\odot}$. Similar to

[1]See http://www.obspm.fr/planets

[2]See http://obswww.unige.ch/~udry/planet/hd82943syst.html

the GJ 876 system (Kinoshita & Nakai 2001; Lee & Peale 2002; Ji, Li & Liu 2002a), the two planets of HD 82943 are also approximately in a 2:1 MMR, with orbital periods 444.6 d and 221.6 d (see Table 1), and semi-major axes 1.16 AU and 0.73 AU. Gozdziewski et al. (2001) used the MEGNO numerical method to investigate the stability of the HD 82943 system and presented a preliminary analysis of its dynamical stability. In the following, we will briefly introduce our work on these systems.

On the basis of N-body codes (Ji, Liu & Li 2002a), our numerical integrations were performed for the two systems by using RKF7(8). We used a time step of one percent of the orbital period of the inner planet when integrating two planets. The integrator was also optimized for close encounters during the orbital evolution. The integration was automatically ceased if either of the planets leave their orbits or move too close to the star. In addition, the numerical errors were effectively controlled over the time span of integration from 1 Myr to 10 Myr. Furthermore, we also adopted symplectic integrators (Feng 1986;Wisdom & Holman 1991) to integrate the same orbit to assure the results for some cases.

We generated the initial six orbital elements of each of the planets of separate systems for the integration. For the coplanar cases of HD 82943, the two planets are considered to locate in the same orbital plane. For all the orbits, we assume that the semi-major axes of the two planets are always unchanged, i.e., 1.16 AU and 0.73 AU respectively. And their initial inclinations are taken as 0.5° for all cases. For the eccentricities, we let the initial eccentricities $e_0 = 0.41$ and 0.54 respectively, be the centers and take the error of the measurement Δe as the radii to randomly produce the eccentricities. We treated the arguments of periastron in the same way. The remaining two angles of nodal longitudes and mean anomalies were randomly chosen between 0° and 360°. The initial orbits of the planets of GJ 876 were similarly produced (Table 1 lists one group of the orbits) and the non-coplanar cases of the mutual highly-inclined orbits were also made for this system.

Table 1. Initial orbital elements of two planets of the GJ 876 and HD 82943 planetary systems

Parameter	GJ 876b	GJ 876c	HD 82943b	HD 82943c
$M\sin i(M_{\rm Jup})$	3.39	1.06	1.63	0.88
Period P(d)	62.092	29.995	444.6	221.6
a (AU)	0.2108	0.1294	1.16	0.73
Eccentricity	0.051	0.314	0.41	0.54
Δe	—	—	0.08	0.05
Inclination(°)	0.5	0.5	0.5	0.5
Ω(°)	20.0	59.2	—	—
ω(°)	40.0	51.8	117.8	138.0
$\Delta\omega$(°)	—	—	3.4	10.2
M(°)	340.0	289.0	—	—

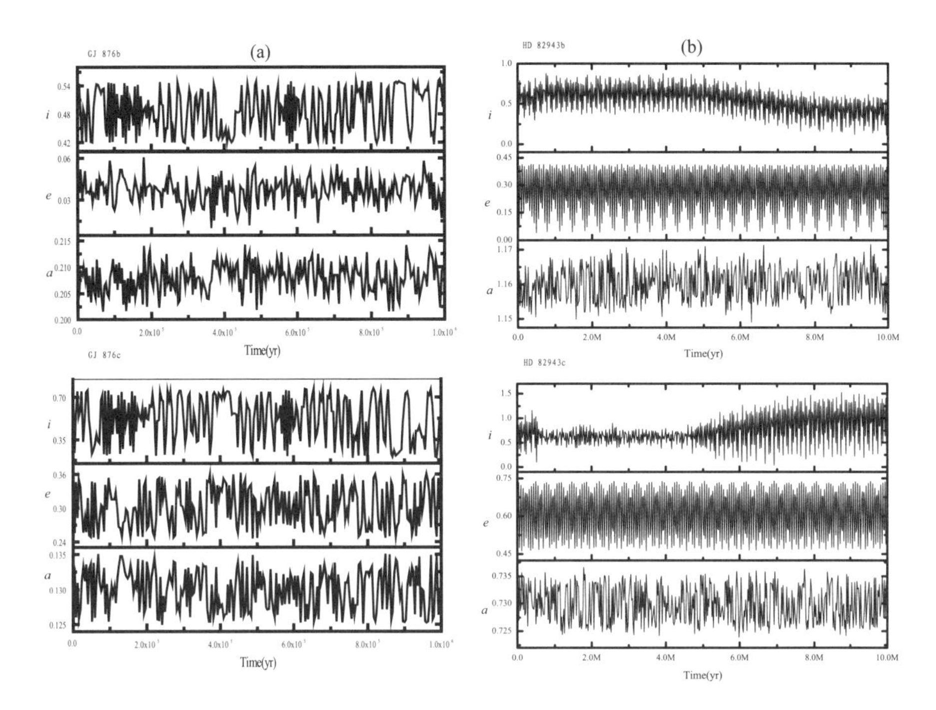

Figure 1. (a) The orbital variations of the two planets of GJ 876 change with time. Note that the semi-major axes, eccentricities and inclinations undergo small oscillations for 1 Myr. (b) The orbital variations of the two planets of HD 82943 system vary for 10 Myr. Notice the small changes of the semi-major axes.

2. Simulation Results and Possible Mechanisms

We here present the chief results of the numerical simulations of the GJ 876 and HD 82943 systems, by exploring different planetary configurations. Next we attempt to introduce two kinds of mechanisms of stabilizing the systems.

In the simulations, we notice that all the stable cases are involved in the 2:1 MMR and easily understand that the stability of a system is sensitive to its initial planetary configuration. The simulation results show that most systems tend to self-destruct in 10^3-10^4 yr, and the lifetime is even shorter for some cases. In greater details, Fig. 1a displays the orbital variations of the two planets of GJ 876 for a stable case, we particularly see that the semi-major axes, eccentricities and inclinations undergo small oscillations for 1 Myr. And Fig. 1b exhibits those of the planets of HD 82943 system for 10 Myr, with the small changes of the semi-major axes. The critical argument θ_1, θ_2 for the 2:1 MMR is

$$\theta_1 = \lambda_1 - 2\lambda_2 + \tilde{\omega}_1, \tag{1}$$

$$\theta_2 = \lambda_1 - 2\lambda_2 + \tilde{\omega}_2, \tag{2}$$

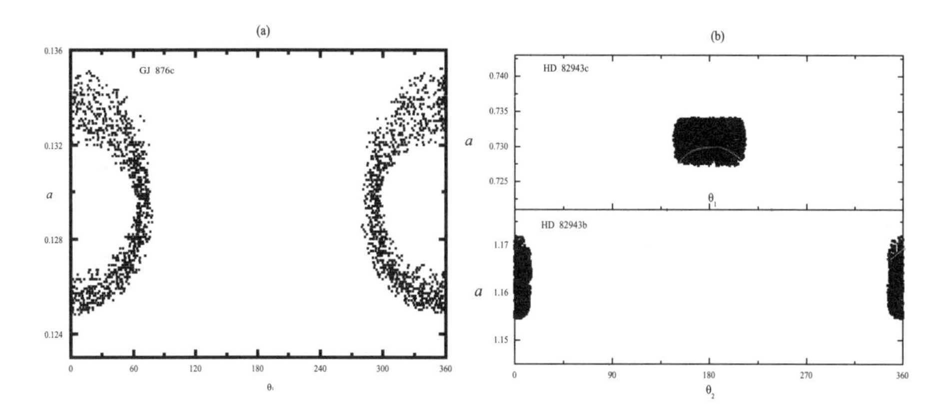

Figure 2. (a)The critical argument θ_1 librates with an amplitude of $\pm70°$ for the inner planet of the GJ 876 system and the equilibrium point appears to be (0.130, 0). (b) The semi-major axes of the inner and outer planets of the HD 82943 system versus the resonant arguments θ_1 and θ_2. Notice that θ_1 librates about 180°, while θ_2 librates about 0°. And the equilibrium points are $(0.730, 180°)$ and $(1.165, 0°)$ respectively.

where λ_1, λ_2 are , respectively,the mean longitudes of the inner and outer planets, and $\tilde{\omega}_1$, $\tilde{\omega}_2$ respectively denote the periastron longitudes of the two planets (hereinafter subscript 1 denotes the inner planet, 2 the outer planet).

According to the simulation results,we plotted the semi-major axis against the critical arguments. Fig. 2a exhibits the critical argument θ_1 librates with an amplitude of $\pm70°$ for the inner planet of the GJ 876 system and the equilibrium point appears to be (0.130, 0). Correspondingly, Fig. 2b displays the semi-major axes of the inner and outer planets of the HD 82943 system versus the resonant arguments θ_1 and θ_2. Still we note that θ_1 librates about 180°,while θ_2 librates about 0°. Notice that the equilibrium points are $(0.730, 180°)$ and $(1.165, 0°)$ respectively. In the same time the semi-major axes undergo small oscillations for the overall time span. Accordingly,these two figures should exhibit that the two planets of each system are indeed close to the 2:1 MMR, further imply that this kind of the resonance may be responsible for the stability of the GJ 876 and HD 82943 systems. Furthermore,we also made theoretical analysis for them(Ji, Li & Liu 2002a; Ji et al. 2002b).

On the other hand, the difference of the apsidal longitudes θ_3 is denoted :

$$\theta_3 = \theta_1 - \theta_2 = \tilde{\omega}_1 - \tilde{\omega}_2 \tag{3}$$

where $\tilde{\omega}_1, \tilde{\omega}_2$ are aforementioned. From Fig.3, we notice that θ_3 librates about 180° with an amplitude of $\pm30°$,which indicates that the alignment of the pericenters (Chiang, Tabachnik & Tremaine 2001; Lee & Peale 2002) for the HD 82943 system. The apsidal alignment corresponds to the libration of the periastron longitudes of the two orbits, such that the two planets have the same averaged rate of apsidal precession. Kinoshita & Nakai (2002) developed a semi-analytical secular method to explain the eccentricities of the two planets of GJ

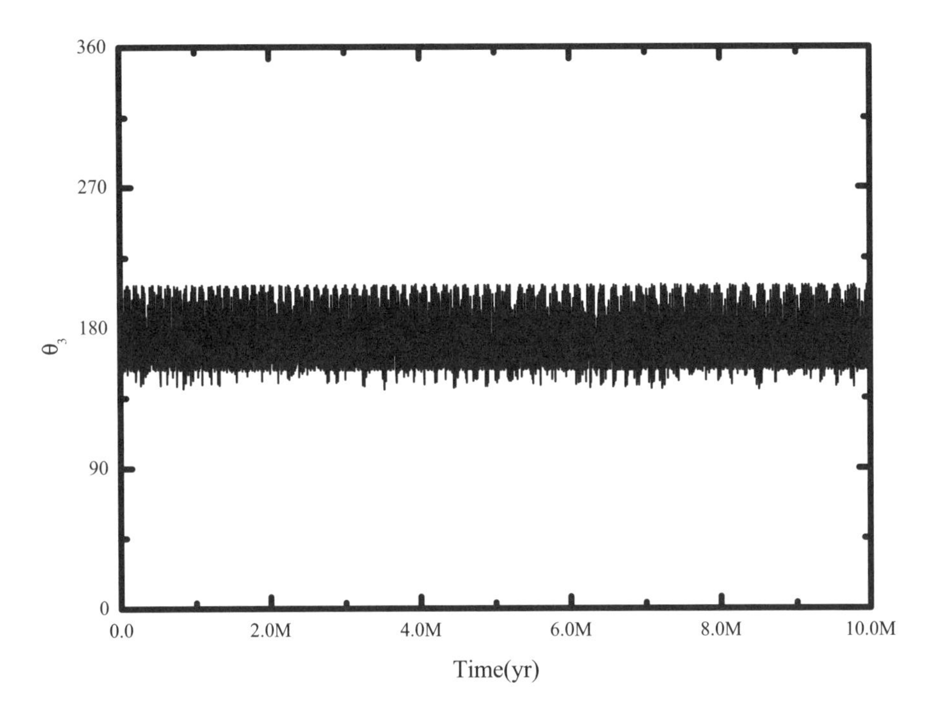

Figure 3. The argument θ_3 librates about $180°$ with an amplitude of $\pm 30°$,which indicates that the alignment of the pericenters for the HD 82943 system.

876 and suggested the stable mechanism of the alignment of the pericenters. Ji et al. (2002b) also applied the method to the cases for HD 82943 system. In this sense, we can understand another stable mechanism of the alignment of the periastron.

3. Conclusions

In this paper, we have mainly explored the dynamical behaviors of the GJ 876 and HD 82943 systems by simulating different planetary configurations in an attempt to discover the possible mechanisms of stabilizing these systems.

In conclusion, for the above-mentioned systems, we found that the configurations in our simulations remain stable for the time span of 1 Myr or 10 Myr. Moreover, the stability of a system is sensitive to its initial planetary configuration. And the numerical results suggested the existence of a 2:1 MMR for the stable systems. We should emphasize that the 2:1 MMR between two planets can act as an effective mechanism for maintaining the stability of the system. In addition, the alignment of the pericenters is playing important part in the secular dynamical evolution of the extrasolar systems and may be acted as one of the mechanisms for the stability of the system.

References

Butler, R. P., Vogt, S. S., Marcy, G. W., Fischer, D. A., Henry, G. W., & Apps, K. 2000, ApJ, 545, 504

Chiang, E. I., Tabachnik, S., & Tremaine, S. 2001, AJ, 122, 1607

Feng K. 1986, J. Comp. Math., 4, 279

Gozdziewski, K. & Maciejewski, A. J. 2001, ApJ, 563, L81

Ji, J. H., Li, G. Y. & Liu, L. 2002a, ApJ, 572, 1041

Ji, J. H., Liu, L., Li, G. Y., Kinoshita H., & Nakai, H. 2002b, submitted to ApJ, preprint(astro-ph/0208025)

Kinoshita, H. & Nakai, H. 2002, in Proc. 34rd Symp. Celest. Mech., in press

Laughlin, G. & Chambers, J. 2001, ApJ, 551, L109

Lee, M. H. & Peale, S. J. 2002, ApJ, 567, 596

Marcy, G. W., Cochran, W. D. & Mayor, M. 2000, in Protostars and Planets IV, ed. V. Mannings, A. P. Boss & S. S. Russell (Tucson: University of Arizona Press), p.1285

Marcy, G. W., Butler, R. P., Fischer, D., Vogt, S. S., Lissauer, J. J., & Rivera, E. J. 2001, ApJ, 556, 296

Wisdom, J., & Holman, M. 1991, AJ, 102, 1528

Large Scale Survey

IAU 8th Asia-Pacific Regional Meeting
ASP Conference Series, Vol. 289, 2003
S. Ikeuchi, J. Hearnshaw & T. Hanawa, eds.

The 6dF Galaxy Survey

Ken-ichi Wakamatsu

Faculty of Engineering, Gifu University, Gifu 501-1192, Japan

Matthew Colless

Mount Stromlo & Siding Spring Observatories, ACT 2611, Australia

Tom Jarrett

IPAC, Caltech, MS 100-22, Pasadena, CA 91125, USA

Quentin Parker

Macquarie University, Sydney 2109, Australia

William Saunders

Anglo-Australian Observatory, Epping NSW 2121, Australia

Fred Watson

Anglo-Australian Observatory, Coonabarabran NSW 2357, Australia

Abstract. The 6dF Galaxy Survey (6dFGS)[1] is a spectroscopic survey of the entire southern sky with $|b| > 10°$, based on the 2MASS near infrared galaxy catalogue. It is conducted with the 6dF multi-fiber spectrograph attached to the 1.2-m UK Schmidt Telescope. The survey will produce redshifts for some 170 000 galaxies, and peculiar velocities for about 15 000 and is expected to be complete by 2005 June.

1. Introduction

In order to reveal large-scale structures at intermediate and large distances, extensive galaxy redshift surveys have been carried out, e.g. the 2dFGRS and SDSS, and the Hubble- and Subaru-deep field surveys. There is now an urgent need to study the large-scale structure of the Local Universe that can be compared with the above deeper surveys. However to do this required hemi-

[1]Member of Science Advisory Group: M. Colless (Chair; ANU, Australia), J. Huchra (CfA, USA), T. Jarrett (IPAC, USA), O. Lahav (Cambridge, UK), J. Lucey (Dahram, UK), G. Mamon (IAP, France), Q. Parker (Maquarie Univ. Australia), D. Proust (Meudon, France), E. Sadler (Univ. Sydney, Australia), W. Saunders (AAO, Australia), K. Wakamatsu (Gifu Univ., Japan), F. Watson (AAO, Australia)

Figure 1. 6dF, an automated fiber positioner, configures magnetic fiber buttons on the curved focus of the field assembly under precise robotic control (5 μm) at the exact co-ordinates of celestial objects.

spheric sky coverage which can only be effectively and efficiently carried out with a dedicated Schmidt telescope with a wide field of view. To this end the AAO implemented a 6dF multi-fiber spectrograph (Fig. 1) for the UK Schmidt Telescope (UKST) and has now commenced a full southern hemisphere galaxy redshift survey of the Local Universe. In this paper an outline of the survey is briefly reported. Further details are given at the following web site: http://www.mso.anu.edu.au/6dFGS/

2. 6dF: A Multi-Fiber Spectrograph on the UKST

6dF is a multi-fiber spectrograph attached to the UKST. It is named after the telescope's field of view which is 6 degrees in diameter, just as 2dF the equivalent 2 degree mulit-fibre system at the 3.9-m Anglo-Australian Telescope. 6dF consists of an automated fiber positioner and a fast $F/0.9$ CCD spectrograph. 6dF is the third generation of multi-fiber spectrograph on the UKST and its early history is described in the Appendix.

Each field assembly has 150 fibers of 100 μm core diameter, which corresponds to 6.7 arcsec on the sky. The 6dF positioner places magnetic fiber-buttons on the curved field plates (mandrel) which matches the telescope's curved focal surface (Fig. 1). The positioner operates off-telescope unlike 2dF. It takes less than 1 hour to accurately place 150 fibers including the defibering process from the previous configuration. There are currently two field plate assemblies, so that one can be configured while the other is on the telescope. Further details of the system are given by Watson et al. (2001).

The 10m optical-fiber cable feeds the existing floor-mounted spectrograph, which has a Marconi 1024 × 1024 CCD detector with 13 μm pixels. Each spectrum is recorded on 3 lines of CCD pixels. The thinned CCD is back illuminated and has broad band coating for enhanced blue sensitivity which is as high as 75% even at 3900 Å, so redshifted H & K lines are detected easily (Fig. 2).

3. Performance of Survey Telescopes

Performance of a survey telescope is simply expressed by how large a volume a telescope can cover in a given observing time. The sky area surveyed in a given exposure is represented by the telescope's field of view Ω in square degree, while an efficiency of the telescope is given by a light-collecting area A, which is proportional to square of an aperture of the telescope. Hence, the survey performance of a telescope can be expressed by the $A\Omega$ product with the larger values indicating increasing survey power. These values for some typical telescopes are given in Table 1.

Table 1. Performance of Survey Telescope

Telescope	Aperture (m)	Field of View (degree)	$A\Omega$ ($\mathrm{m}^2 \cdot \mathrm{degree}^2$)
Gemini	8.1	0.17	1.8
WHT	4.2	0.5	4.4
VLT	8.1	0.4	10
Subaru	8.2	0.5	17
Kiso Schmidt	1.05	5.0	28
Sloan	2.4	2.5	36
UKST	1.2	6.0	52
AAT	3.9	2.0	61

Table 1 shows that the UKST has a very high survey performance due principally to its wide field 6° diameter field of view, while the new generation of large aperture telescopes such as Subaru and the VLT have relatively low survey performance though they can obviously penetrate much deeper. Hence the UKST is ideal for wide-field but shallow surveys.

4. 6dF Galaxy Survey Design

The 6dFGS was designed according to the following strategies:

Differentiation: What does the 6dFGS offer that is not offered by the 2dFGRS, SDSS, or other surveys?

Impact: What survey characteristics are required in order to maximize the science impact?

Timeliness: How quickly must the survey be carried out in order to achieve its goals in a timely and competitive manner?

The 6dFGS has two distinct components: *a redshift survey* and *a peculiar velocity survey*. Target selection is based not on optical galaxy photometric selection like the 2dFGRS and SDSS, but on K-band selection from the recently completed near infrared 2MASS all sky survey (Jarrett et al. 2000b). Our survey area is for the entire southern sky with $|b| > 10°$ and amounts to 17 000 deg^2. For the redshift survey, the surface density of targets must match or exceed the density of 6dF fibres to allow efficient observing. Allowing 10 fibers for sky, this means the sample should have a mean surface density of at least 5 deg^{-2}.

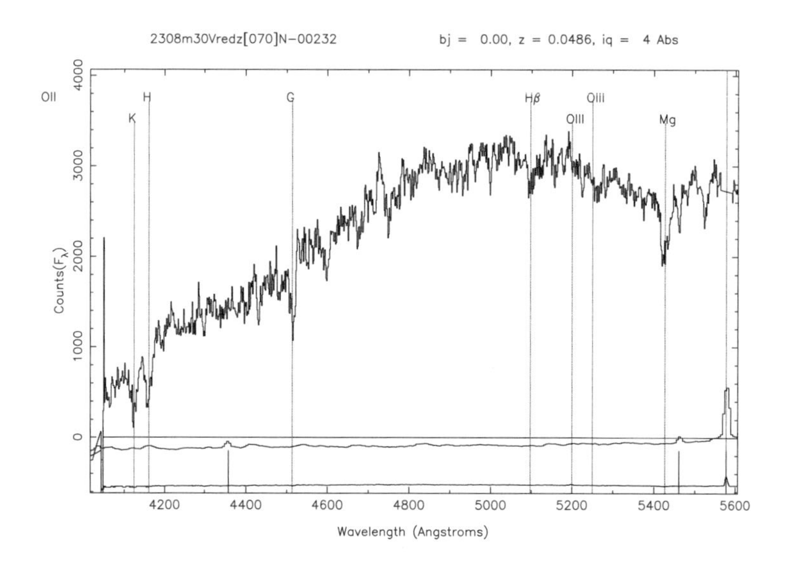

Figure 2. An example of a blue spectrum of a galaxy of $z = 0.0486$. The H and K lines are clearly seen.

Furthermore target galaxies should be bright enough to allow redshifts to be measured in relatively short integration times so that the whole southern sky can be covered in a reasonable amount of time. We set the limiting magnitude at 12.75 in K_s-band[2] , and so finally selected about 120 000 objects from the 2MASS Extended Source Catalog.

Observations for the survey are made with two different gratings for each target field (Table 2). These parameters are likely to change slightly with the imminent commissioning of volume-phase holographic gratings in a new transmissive arrangement which offers enhanced system efficiency.

Table 2. Parameters of Spectroscopic Observations

Spectrum	Grating	Spectral Coverage	FWHM	Exposure
Blue	600V	4000Å – 5600Å	5 – 6 Å	3 × 20 min
Red	316R	5400Å – 8400Å	9 – 12 Å	3 × 10 min

Total exposure times for each survey field is about 1.5 hours. Observing overheads between fields takes a further 30–40 minutes. This permits 5 fields per night in the long winter lunations, reducing to 3 in summer.

To observe target galaxies as efficiently as possible, 6dF field centers have been carefully determined from an adaptive tiling algorithm (Campbell, Saun-

[2]We adopted a corrected total magnitude $K_{\rm tot}$ estimated from the isophotal magnitude K_{20} given in the 2MASS catalogue.

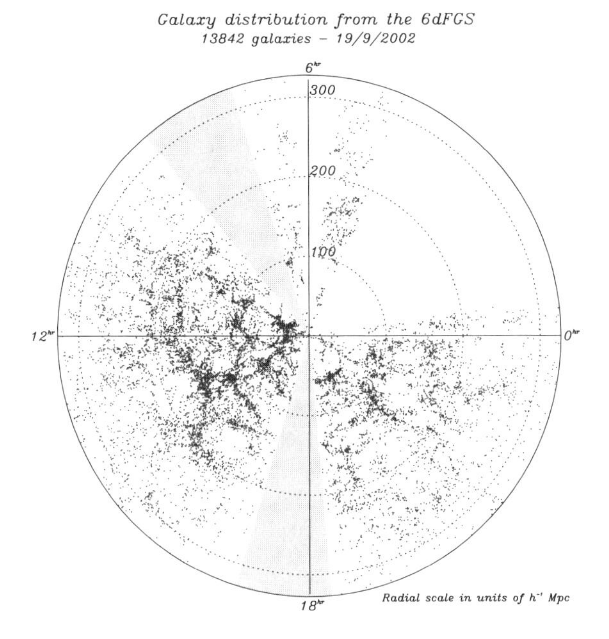

Figure 3. Galaxy distribution from the 6dFGS for a strip $\delta = -30\,^{\circ}$.

ders, & Colless 2002). The total number of field centers is 1360. These sky configurations cover 95% of all the target galaxies, with an efficiency of 87% usage of fibers. In crowded regions like cluster centers, multiple observations with different fiber configurations are required to observe close pairs of galaxies, because fibers cannot be put within a minimum separation 5 arcmin due to the physical proximity constraints imposed by the footprint of the cylindrical magnetic buttons.

Altogether, the survey will produce redshifts for some 170 000 galaxies including objects for several additional target programs (see section 5.4). In addition, peculiar velocities will also be obtained for about 15 000 nearby bright early type galaxies. Both samples will be complete by 2005 June. The first data release of the redshift survey is expected by the end of 2002.

5. Uniqueness of the 6dFGS

The 6dFGS has the following characteristics as compared with 2dFGRS, SDSS, and other redshift surveys.

1. *Near-Infrared Selection: A Clear Window on the Mass Distribution*
   Our survey is based on the new 2MASS near-infrared imaging survey of the whole sky. We use magnitude-limited 2MASS $J, H,$ & $K_{\rm tot}$ galaxy samples, supplemented by complete photometric samples from other optical (SuperCOSMOS B & R) and near-infrared (DENIS) catalogues. Near-infrared (NIR) luminosity, especially in the K_s-band, is not biased by recent star

formation activity, and represents the stellar masses of individual galaxies more accurately than optical magnitudes which can be biased by star formation activity.

Using NIR luminosity also minimizes the effects of internal absorption of galaxies, so that M/L ratios are not affected by orientation, especially for spiral galaxies.

Absorption by dust in our own Galaxy is also much reduced so allowing greater sky coverage and more uniform sample selection over the whole sky (apart from a narrow region around the plane of the Milky Way).

Near-infrared target selection means that the 6dF Galaxy Survey is a very effective means of determining the true mass distribution in the Local Universe.

2. *Peculiar Velocities: Bulk Motions of Galaxies*
As well as measuring redshifts, we also measure galaxies' motions (their 'peculiar velocities'). We concentrate on early-type galaxies, and measure their peculiar velocities using the well-established D_n-sigma relation. 6dF allows us to obtain medium-dispersion, high-S/N spectra from which we can measure the central velocity dispersions of individual galaxies (Fig. 2). By comparing these dispersions with the galaxies' apparent sizes we can determine their distances. Combined with high quality redshift measurements, we can obtain peculiar velocities (the deviations from the Hubble flow) of significant numbers of individual galaxies.

3. *All-Sky Coverage: A Picture of the Local Universe*
The 2dF galaxy survey penetrates deep into space, but is limited to quite a narrow sky region around the south Galactic pole and a section of the celestial equatorial zone, covering 5% of the sky. The Sloan Digital Sky Survey is conducted with a dedicated 2.5-m telescope, but covers less than one-third of the northern sky. The 6dF Galaxy Survey however covers the entire southern sky with galactic latitude greater than 10 degrees. This wide survey area is quite unique among the various current surveys and provides a full hemispheric description of the local universe.

4. *Additional Targets: Wide Windows to the Universe*
In the 6dFGS for fields having insufficient targets to use all 150 available fibers, remaining fibers have been allocated to additional target programs according to scientific merit (such as sampled from the ROSAT All-Sky Survey and NVSS radio survey). These various additional target lists are combined in priority order with the main 6dFGS survey list to provide significant added value to the original survey science

6. Scientific Aims

The 6dFGS will provide a unique snapshot of the Local Universe based on a homogeneous, high-quality database (Fig. 3). These data can be used in a wide range of scientific analyses.

The main scientific goals are:

1. The large-scale structure (density field) of the local universe.

2. The bulk motions (velocity field) of galaxies in the local universe.
3. The estimation of fundamental cosmological parameters, such as the mean mass density and cosmological constant, from the joint analysis of the density and velocity fields.
4. The dependence of the properties of normal galaxies on their local environment and the surrounding large-scale structure.
5. Studies of the properties of rare types of galaxies from additional target samples selected on the basis of their radio, far-infrared, optical or X-ray properties.

7. Survey in Zone of Avoidance

At galactic latitudes below $|b| < 10°$, several important clusters and structures have been discovered, such as the Great Attractor and the Ophiuchus cluster, one of the brightest X-ray cluster in the sky (Wakamatsu et al. 2000). The 2MASS Extended Source Catalog provides target galaxies even in this area despite high extinction and high density of foreground stars (Jarrett et al. 2000a) so we can use 6dF to penetrate as deep as possible into the galactic plane over some limited area to further study the extent and form of these and related important features.

8. Future of Schmidt Telescopes in Spectroscopic Mode

The 6dF spectroscopic survey mode has opened a new era for the UKST. As with all sky imaging surveys, it is also important to extend the 6dF spectroscopic survey into the northern sky. The Kiso Schmidt telescope is one of the best telescopes that could accomplish this if equipped with a 6dF type system. There are already plans for a new innovation at the UKST that will allow more than 2000 fibers to be placed simultaneously on star positions to study the dynamics and chemical evolution of our Galaxy.

The implementation of a much bigger Schmidt telescope for performing muilti-fiber spectroscopy (such as LAMOST in China) is a further extension of this trend.

Appendix: Astronomy with a Glue – Early History of 6dF

There is a long history leading up to the implementation of the powerful new 6dF multi-fiber spectrograph. In the early 1980s when nobody imagined using a Schmidt telescope for spectroscopy, Fred Watson started to put fibers on a curved focal plane with a precision of 20–50 μm.

After many trials, Dr. Watson and his collaborators succeeded in putting fibers in the following manner (Fig. 4): i) a honey-comb mandrel plate holder was manufactured to have room for putting fibers from the backside of the mandrel, ii) a glass plate of a positive contact copy of a sky survey plate is set on the mandrel to use as a template for galaxy positions, iii) inserting fibers into a cut-down syringe needle and housing, and iv) attaching the fiber ferrule on the backside of glass plate with glue. The first spectrograph was made from

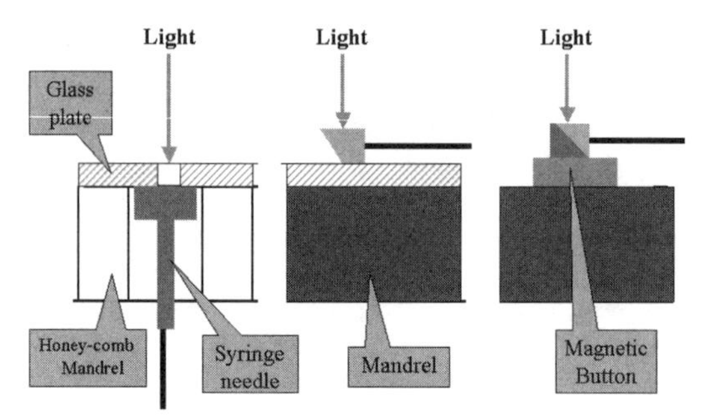

Figure 4. Schematic illustrations of set-up of an optical fiber on the field plate assembly for the original system (left), FLAIR II (middle), and 6dF (right), respectively.

a Pentax camera with hypersensitised Tech-Pan film, and was called FLAIR. System throughput was very poor due to scattering of light on the template.

FLAIR II, the second generation machine, was fabricated under a quite different design: i) a glass plate of a *negative* contact copy was set on a Mandrel, ii) fibers were connect via a modified syringe needle mount or ferrule to a small right-angle prism, iii) with a semi-automated fiber positioner, the prism-ferrule assembly glued at a galaxy position on the glass plate using UV curing cement with a precision of about 20 μm, and iv) 92 fibers running complicatedly on the surface of glass plates were taped and bundled to prevent from blocking of light on the fiber. At this stage, a new spectrograph was fabricated with a fast optics and a CCD detector.

FLAIR II yielded decent throughput, and yielded good performance allowing many useful science projects to be undertaken. However, it had a serious problem; it took 6–7 hours to put 92 fibers to target objects. To overcome this problem, the present day 6dF fully-automated fiber positioner robot was commissioned in 2001 June by Will Saunders and Quentin Parker.

Acknowledgments. We thank all the members of Science Advisory Group. We deeply express our thanks to Mr. M. Hartley and other stuff members at the AAO for their nice observations. KW is supported by a grant-in-aid of Ministry of Education, Culture, Science & Technology of Japan under a No. 13640236.

References

Campbell, L., Saunders, W., & Colless, M. M. 2002, in preparation

Jarrett, T. H., et al. 2000a, AJ, 119, 2498

Jarrett, T. H., et al. 2000b, AJ, 120, 298

Wakamatsu, K., et al. 2000, in ASP Conf. Ser. Vol. 218, Mapping the Hidden Universe: The Universe Behind the Milky Way, eds. R. C. Kraan-Korteweg, P. A. Henning, & H. Andernach (San Francisco: ASP), 187

Watson F. G., et al. 2001, in ASP Conf. Ser. Vol. 232, The New Era in Wide-Field Astronomy', eds., R. Clowes, A. Adamson, & G. Bromage, (San Francisco: ASP), 421

IAU 8th Asia-Pacific Regional Meeting
ASP Conference Series, Vol. 289, 2003
S. Ikeuchi, J. Hearnshaw & T. Hanawa, eds.

Radio Sources in the 2dF Galaxy Redshift Survey: AGN, Starburst Galaxies and Their Cosmic Evolution

Elaine M. Sadler

School of Physics, University of Sydney, NSW 2006, Australia

Carole A. Jackson

Research School of Astronomy and Astrophysics, The Australian National University, Weston Creek, ACT 2611, Australia

Russell D. Cannon

Anglo–Australian Observatory, PO Box 296, Epping, NSW 2121, Australia

Abstract. Radio continuum surveys can detect galaxies over a very wide range in redshift, making them powerful tools for studying the distant universe. Until recently, though, identifying the optical counterparts of faint radio sources and measuring their redshifts was a slow and laborious process which could only be carried out for relatively small samples of galaxies.

Combining data from all-sky radio continuum surveys with optical spectra from the new large-area redshift surveys now makes it possible to measure redshifts for tens of thousands of radio-emitting galaxies, as well as determining unambiguously whether the radio emission in each galaxy arises mainly from an active nucleus (AGN) or from processes related to star formation. Here we present some results from a recent study of radio-source populations in the 2dF Galaxy Redshift Survey, including a new derivation of the local star-formation density, and discuss the prospects for future studies of galaxy evolution using both radio and optical data.

1. Introduction

Most astronomical objects evolve on timescales which are orders of magnitude longer than a human lifetime, so that we effectively view them frozen at a single moment in their life cycle and at a fixed orientation. To piece together the evolutionary history of such objects, we therefore need to observe a sample which is large enough to cover the full range of luminosity, size, age and orientation which exists in nature.

Advances in technology and data processing power mean that surveys of very large numbers of astronomical objects are now becoming feasible. Van den Bergh (2000) recently concluded that "... the astronomy of the twenty-first century will be dominated by computer-based manipulation of huge homogeneous surveys of various types of astronomical objects ...".

The powerful radio emission from some galaxies acts as a beacon which allows them to be seen at very large distances (the median redshift of galaxies detected in radio surveys is typically $z \simeq 1$; Condon 1989). Radio source counts can be used to study the cosmological evolution of active galaxies (e.g. Longair 1966, Jauncey 1975, Wall et al. 1980) but their interpretation is strongly model-dependent. This is especially true at the faint flux densities probed by the new generation of all-sky radio imaging surveys (NVSS, Condon et al. 1998; FIRST, Becker, White & Helfand 1995; WENSS, Rengelink et al. 1997; SUMSS, Bock, Large & Sadler 1999), where AGN and 'normal' star-forming galaxies both contibute significantly (e.g. Condon 1989, 1992).

The scientific return from large radio surveys is enormously increased if the optical counterparts of the radio sources can be identified, their optical spectra classified (as AGN, starburst galaxy, etc.) and their redshift distribution measured. Here, we describe some first results from a study of radio-source populations in the 2dF Galaxy Redshift Survey. Much of this work is discussed in more detail in Sadler et al. (1999, 2002).

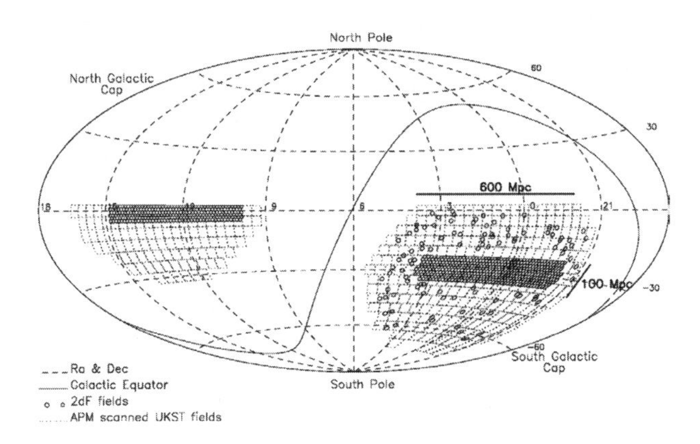

Figure 1. Sky coverage of the 2dF Galaxy Redshift Survey. The main survey regions are the strips around the South Galactic Pole near declination −30° and in the northern Galactic hemisphere along the celestial equator.

2. The 2dF Galaxy Redshift Survey

The 2dF Galaxy Redshift Survey (2dFGRS) is described by Colless et al. (2001), and has also been discussed at this meeting by Peterson. It used two multi-fibre spectrographs at the prime focus of the 4-m Anglo–Australian Telescope (AAT) to measure redshifts for more than 220 000 galaxies in 1700 $\deg^2$ of sky. The main goal of the 2dFGRS team was to study large-scale structure over the redshift range $z = 0$ to 0.2, but the high quality of the spectra in the public 2dFGRS database means that they have also been used for many other studies, such as measuring optical luminosity functions for galaxies of different Hubble types

(Madgwick et al. 2002) and examining the effects of environment on the star formation rate in galaxies (Lewis et al. 2002).

3. Radio Imaging Surveys in the 2dFGRS Area

The 1.4 GHz NRAO VLA Sky Survey (NVSS; Condon et al. 1998) overlaps both the 2dFGRS strips shown in Figure 1, and the southern (dec $< -30°$) part of the SGP strip is also covered by the 843 MHz Sydney University Molonglo Sky Survey (SUMSS, Bock et al. 1999). Part of the northern 2dFGRS strip is overlapped by the higher-resolution 1.4 GHz FIRST survey (Becker et al. 1995), and the properties of the faint 2dFGRS radio sources in this region have been discussed by Magliocchetti et al. (2002).

NVSS and SUMSS have similar resolution ($\sim 45''$) and sensitivity, so in the overlap declination zone $-30°$ to $-40°$ they can be combined to measure a radio spectral index for sources in common (see Figure 2). The NVSS and SUMSS source positions are also accurate enough that we can make unambiguous optical identifications down to at least $B \sim 20$ mag.

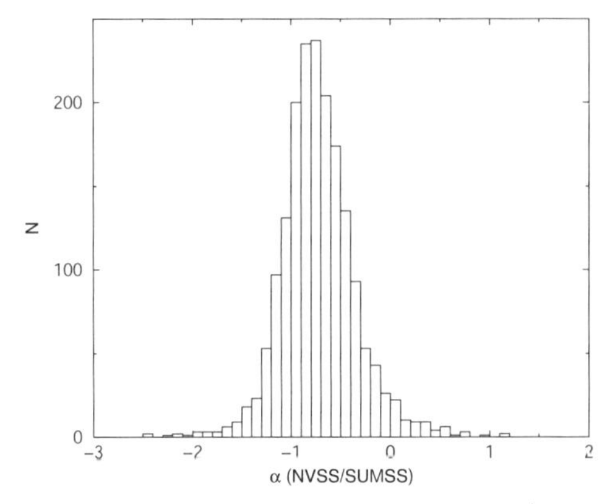

Figure 2. Distribution of radio spectral index α (defined by $S \propto \nu^{\alpha}$ for flux density S at frequency ν, and analogous to an optical colour) for radio sources in the overlap zone which were detected by both NVSS and SUMSS. The mean spectral index between 843 MHz and 1.4 GHz is -0.734 (Mauch et al. 2002).

The NVSS source density is 60 deg^{-2} above 2.5 mJy, and the typical 2dFGRS galaxy density is 180 deg^{-2} above the survey cutoff magnitude of $b_{\rm J} = 19.4$. The intersection of the two surveys is small, with less than 1.5% of the 2dFGRS galaxies detected as radio sources by NVSS (see Table 1), but the size of the 2dFGRS means that this is still the largest and most homogeneous set of optical spectra of radio galaxies so far observed.

4. Radio Source Populations in the 2dFGRS

2dFGRS spectra of each of the radio-source IDs were classified visually as either 'AGN' or 'star-forming' (SF). AGN galaxies have either (i) an absorption-line

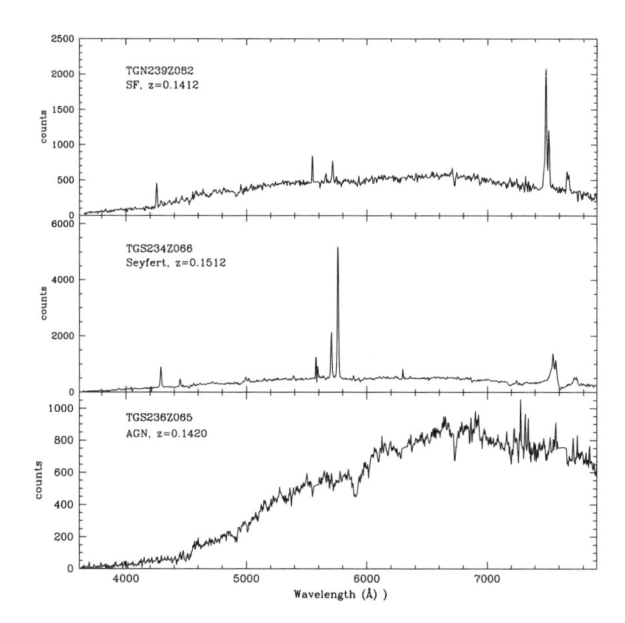

Figure 3. Spectra of three 2dFGRS radio sources with redshift $z \sim 0.15$ (Sadler et al. 1999): (*top*) a star-forming galaxy, with strong, narrow Balmer emission lines of Hα and Hβ and weaker emission lines of [SII] 6716, 6731, [NII] 6583, [OIII] 4959, 5007 and [OII] 3727; (*middle*) an emission-line AGN, with broad Hα emission and [OIII] $>>$Hβ; (*bottom*) a radio galaxy (AGN) with an absorption-line spectrum.

spectrum like that of a giant elliptical galaxy, (ii) an absorption-line spectrum with weak LINER-like emission lines, or (iii) a stellar continuum dominated by nebular emission lines such as [OII] and [OIII] which are strong compared with any Balmer-line emission. In SF galaxies, strong, narrow emission lines of Hα and (usually) Hβ dominate the spectrum.

As can be seen from Table 1, the 2dFGRS radio-source population is a roughly equal mixture of star-forming and active galaxies. The two classes overlap in radio luminosity (though the active galaxies are on average both more distant and more luminous), and in many cases can only be distinguished by examining their optical spectra. Because both populations are well-represented in the 2dFGRS/NVSS data set, we can measure accurate radio luminosity functions for both AGN and star-forming galaxies as a local benchmark for future studies at higher redshift.

4.1. Star-forming Galaxies

The 1.4 GHz radio emission from star-forming galaxies has both a thermal component from individual HII regions and a (dominant) large-scale non-thermal component arising from synchrotron emission in the disk (Condon 1992). Fig-

Table 1. 2dFGRS radio-source statistics

Analysed so far:	
58 454	2dFGRS spectra (25% of total)
757	matched with NVSS radio sources (1.3%)
441	radio galaxies (=AGN)
272	star-forming galaxies (=IRAS galaxies)
44	unclassified spectra (low S/N)

ure 4 shows two star-forming 2dFGRS galaxies which were also detected as NVSS radio sources.

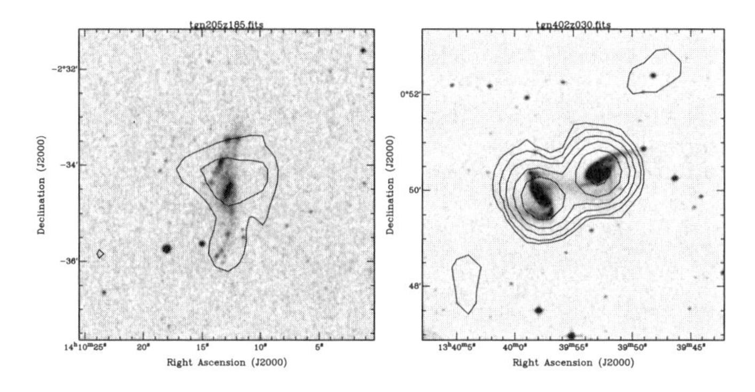

Figure 4. Contours of 1.4 GHz radio flux density overlaid on the optical images of two star-forming 2dFGRS galaxies. The estimated star-formation rates are (*left*) $2 M_{\odot}\,\mathrm{yr}^{-1}$ and (*right*) $120 M_{\odot}\,\mathrm{yr}^{-1}$.

Most of the star-forming 2dFGRS galaxies are also detected as far-infrared (FIR) sources in the IRAS Faint Source Catalogue (FSC: Moshir et al. 1990) and follow the well-known FIR-radio correlation (e.g. de Jong et al. 1985; Helou, Soifer & Rowan-Robinson 1985). The radio luminosity of these galaxies gives an independent estimate of their current star-formation rate (e.g. Sullivan et al. 2001), and by integrating over the radio luminosity function for 2dFGRS/NVSS galaxies we can estimate the star-formation density in the local universe in a way which is unaffected by dust.

The integrated star-formation density of $0.022 \pm 0.004 M_{\odot}\,\mathrm{yr}^{-1}\,\mathrm{Mpc}^{-3}$ derived from the 2dFGRS radio data (Sadler et al. 2002) is slightly higher than the value of 0.013 ± 0.006 derived optically by Gallego et al. (1995). The main reason for the difference is that the 2dF sample has an excess of galaxies with star formation rates above $50 M_{\odot}\,\mathrm{yr}^{-1}$ (see Figure 5). Condon, Cotton & Broderick (2002) suggest that this difference is due to evolution of the galaxies with the highest star-formation rates within the 2dFGRS sample volume. They also note that the data are consistent with luminosity evolution of the form $f(z) \sim (1+z)^3$ for star-forming galaxies.

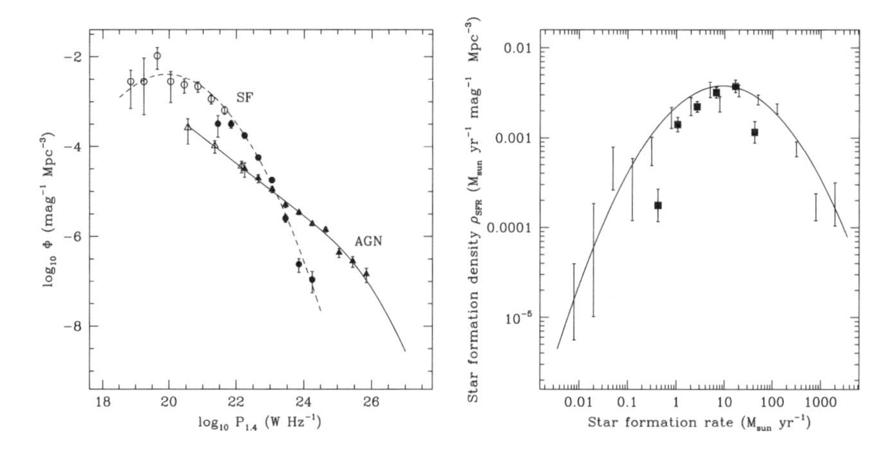

Figure 5. (*Left*) Local radio luminosity functions ($H_0 = 50$ km s^{-1} Mpc^{-3}) for star-forming (SF) galaxies and AGN, from Sadler et al. (2002). Below 10^{22} W Hz^{-1} additional points from nearby bright spiral (Condon 1989) and elliptical (Sadler, Jenkins & Kotanyi 1989) galaxies have been added to the 2dFGRS data. (*Right*) Local star-formation density (in $M_\odot\,\mathrm{yr}^{-1}\,\mathrm{mag}^{-1}\,\mathrm{Mpc}^{-3}$) for galaxies with star-formation rates between 0.01 and $1\,000 M_\odot\,\mathrm{yr}^{-1}$. The solid line is derived from the local radio luminosity function, and the filled squares are values derived by Gallego et al. (1995) from Hα measurements.

4.2. Active Galaxies

The radio luminosity function of 2dFGRS active galaxies (see Figure 5) has a power-law spectrum of the form $\Phi(\mathrm{P}_{1.4}) \propto \mathrm{P}_{1.4}{}^{-0.62\pm0.03}$ over almost five decades in radio luminosity from $10^{20.5}$ to 10^{25} W Hz^{-1}.

Franceschini, Vercellone & Fabian (1998) examined the relation between galaxy luminosity, black hole mass and radio power in nearby active galaxies, and concluded that the radio power of an AGN is both a good indicator of the presence of a supermassive black hole and an estimator of its mass, though Laor (2000) noted that the correlation between radio power and black hole mass has a large scatter. Using a larger data set, Snellen et al. (2002) confirm the relation found by Francescini et al. (1998) but note that it applies only for "relatively passive" elliptical galaxies (i.e. those with radio powers below $\sim 10^{24}$ W Hz^{-1}).

If we apply the relation found by Francescini et al. to the 2dFGRS active galaxies, we derive a value of $\rho_{\mathrm{BH}} = 1.6^{+0.4}_{-0.6} \times 10^5\ M_\odot$ Mpc^{-3} for the total mass density of supermassive ($M_{\mathrm{BH}} > 8 \times 10^7 M_\odot$) black holes in the local universe. This is close to the value of $1.4 - 2.2 \times 10^5$ derived for QSOs by Chokshi & Turner (1992), suggesting that local low-power radio galaxies may be the remnants of most or all of the high-redshift QSOs.

5. PKS 0019–338, A Post-starburst Radio Galaxy

Among the 2dFGRS radio galaxies, we recently discovered the remarkable object PKS 0019–338 at $z = 0.129$. Its optical spectrum (Figure 6) shows both strong emission lines characteristic of an AGN and an underlying blue stellar continuum with strong hydrogen Balmer-line absorption. The strength of the Hδ absorption line implies that this galaxy had a massive starburst which finished only $\sim 1.5 \times 10^8$ years ago (K. Bekki & W.J. Couch, private communication).

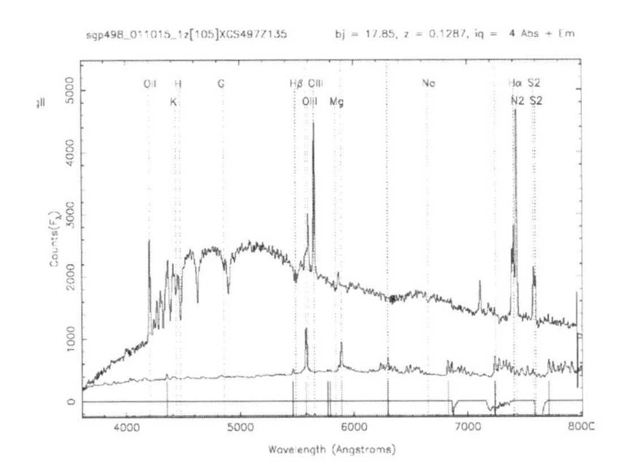

Figure 6. 2dFGRS spectrum of the 'post-starburst' radio galaxy PKS 0019-338.

PKS 0019–338 belongs to the class of 'e+A' galaxies, with an old stellar population overlaid with a substantial younger population created in a massive starburst much less than 1 Gyr ago (e.g. Dressler & Gunn 1983; Zabludoff et al. 1996). The radio source lies within the optical galaxy and has a steep ($\alpha = -1.1$) sepctrum, placing it in the class of compact steep-spectrum (CSS) radio galaxies. This implies an age of less than 10^6 years for the radio source (e.g. de Silva et al. 1999). If the starburst and AGN in this galaxy are causally connected, there is a large time lag ($\sim 10^8$ years) between the peak of the starburst and the onset of radio emission.

6. Future Work

We have so far analysed only 25% of the 2dFGRS radio sample. Adding the rest will give us a large enough sample to measure the evolution of both AGN and star-forming galaxies out to $z \sim 0.3$. The 6dF Galaxy Survey described at this meeting by Wakamatsu (see also Watson et al. 2001) will yield roughly 12,000 more radio-source spectra to $z \sim 0.1$, allowing a detailed investigation of the faint end of the radio luminosity function, and will also have high enough resolution to measure stellar velocity dispersions. By using the velocity dispersion as an independent measure of the mass of the central black hole, we should be able to investigate in more detail the correlation between black hole mass and radio power in nearby elliptical galaxies.

Acknowledgments. EMS thanks the meeting organizers for financial support, and Dr Kenji Bekki and Prof. Warrick Couch for helpful discussions on the ages of 'post-starburst' galaxies.

References

Becker, R. H., White, R. L., Helfand, D. J. 1995, ApJ, 450, 559

Bock, D. C-J., Large, M. I., Sadler, E. M. 1999, AJ, 117, 1593

Chokshi, A., Turner, E. L. 1992, MNRAS, 259, 421

Colless, M. et al. 2001, MNRAS, 328, 1039

Condon, J. J. 1989, ApJ, 338, 13

Condon, J. J. 1992, ARAA, 30, 575

Condon, J. J., Cotton, W. D., Greisen, E. W., Yin, Q. F., Perley, R.A., Taylor, G. B., Broderick, J. J. 1998, AJ, 115, 1693

Condon, J. J, Cotton, W. D., Broderick, J. J. 2002, AJ, 124, 675

de Jong, T., Klein, U., Wielebinski, R., Wunderlich, E. 1985, A&A, 147, L6

de Silva, E., Saunders, R., Baker, J., Hunstead, R. 1999. In *The Hy-Redshift Universe*, ed. A. J. Bunker & W. J. M. van Breugel, ASP Conf. Ser., p. 79

Dressler, A., Gunn, J. E. 1983, ApJ 270, 7

Franceschini, A., Vercellone, S., Fabian, A. C. 1998, MNRAS, 297, 817

Gallego, J., Zamorano, J., Aragon-Salamanca, A., Rego, M. 1995, MNRAS, 455, L1

Helou, G., Soifer, B. T., Rowan-Robinson, M. 1985, ApJ, 298, L7

Jauncey, D. L. 1975, ARAA, 13, 23

Laor, A. 2000, ApJ, 543, L111

Lewis, I. et al. 2002, MNRAS, 334, 673

Longair, M. S. 1966, MNRAS, 133, 421

Madgwick, D. S. et al. 2002, MNRAS, 333, 133

Magliocchetti, M. et al. 2002, MNRAS, 333, 100

Mauch, T. et al. 2002, in preparation

Moshir, M., et al. 1990, IRAS Faint Source Catalogue Version 2.0

Rengelink, R. B., Tang, Y., de Bruyn, A. G., Miley, G. K., Bremer, M. N., Röttgering, H. J. A., Bremer, M. A. R. 1997, A&AS, 124, 259

Sadler, E. M., Jenkins, C. R., Kotanyi, C. G. 1989, MNRAS, 240, 591

Sadler, E. M., McIntyre, V. J., Jackson, C. A., Cannon, R. D. 1999, PASA, 16, 247

Sadler, E. M. et al. 2002, MNRAS, 329, 227

Snellen, I. A. G., Lehnert, M. D., Bremer, M. N., Schilizzi, R. 2002, MNRAS, submitted (astro-ph/0209380)

Sullivan, M., Mobasher, B., Chan, B., Cram, L., Ellis, R., Treyer, M., Hopkins, A. 2001, ApJ, 558, 72

van den Bergh, S. 2000, PASP, 112, 4

Wall, J. V., Pearson, T. J., Longair, M. S. 1980, MNRAS, 193, 683

Watson, F. G. et al. 2001, In *The New Era of Wide Field Astronomy*, ed. R. Clowes, A. Adamson & G. Bromage, ASP Conf. Ser., p. 421

Zabludoff, A. I. et al. 1996, ApJ 466, 104

IAU 8th Asia-Pacific Regional Meeting
ASP Conference Series, Vol. 289, 2003
S. Ikeuchi, J. Hearnshaw & T. Hanawa, eds.

A Large Scale Extragalactic Survey with an Infrared Camera Onboard ASTRO-F

Hideo Matsuhara and Hiroshi Murakami
The Institute of Space and Astronautical Science, Sagamihara, Kanagawa, 229-8510 Japan

Abstract. The scientific capability of deep extragalactic surveys with an infrared camera (IRC) onboard ASTRO-F is presented. ASTRO-F will be launched in early 2004, and $\sim$1 deg^2 ultra-deep as well as $\sim$10 deg^2 medium-deep surveys are planned with the IRC, in order to study the star formation history in galaxies, as well as the structure formation of the universe.

1. Introduction

Recent infrared observations in space have provided us with important information on the birth and evolution of galaxies: the discovery of the far-infrared extragalactic background with COBE (Hauser et al. 1998; Fixen et al. 1998) and its fluctuations (Lagache & Puget 2000; Matsuhara et al. 2000), as well as evidence of the strong evolution seen in the deep number counts at the mid- and far-infrared with ISO (Kawara et al. 1998; Elbaz et al. 1999; Oliver et al. 2002). The subsequent infrared space telescopes such as SIRTF (Deutsch & Bicay 2000) and ASTRO-F (Nakagawa 2001) will expand our understanding of the history of galaxy evolution. ASTRO-F will be launched in early 2004, and will undertake a far infrared all sky survey (Shibai 2002) as well as near and mid-infrared deep surveys of selected areas with an infrared camera (IRC) onboard. Here we present the outline of the extragalactic observations with the IRC and their scientific uniqueness.

2. Outline of Extragalactic Survey with the IRC

ASTRO-F is an ongoing infrared astronomical satellite mission incorporating a 67-cm telescope cooled by liquid helium, and the satellite will be in a sun-synchronous polar orbit with a period of 100 minutes. Two observational modes are considered: one is the survey mode in which the spacecraft spins around the sun-pointed axis once per orbit, in order to perform an all-sky survey, and the other is the pointing mode, in which the spacecraft is fixed in inertial space so that the telescope is pointed at a target in the sky. This pointed observation can be scheduled up to 3 times per orbit and the IRC is mainly used in this mode. The duration of one pointed observation is limited to approximately 10 minutes, because of earthshine avoidance requirements of the telescope's cooled baffle. Before the boil-off of liquid helium (about 500 days after launch) approximately

Figure 1. The whole ASTRO-F satellite assembled during the first end-to-end test (left), and the focal plane instruments under assembly (right). The IRC is in the front.

7000 pointings are expected for the IRC observations. The IRC covers $2-26\,\mu$m in wavelength with nine photometric broad-band filters, as well as several low-resolution spectroscopic elements with a large field-of-view of 10×10 arcmin^2 with a diffraction-limited (at $\lambda \geq 5\,\mu$m) spatial resolution. Details of the latest specifications are described in Wada et al. (2002) and the current estimates of broad-band imaging sensitivities are shown in Table 1. The confusion limits due to faint galaxies are estimated based on the luminosity evolution models of Pearson & Rowan-Robinson (1996) for the 'low evolution' case, and the luminosity plus density evolution models of Pearson (2001) for the 'high evolution' case.

Table 1. Estimated detection limits (5σ) for point sources

Filter	Wavelength (μm)	Detection limits (μJy, 500 s)	Confusion limits low evolution	(μJy) high evolution
N2	1.8-2.7	1.3	0.2	0.2
N3	2.7-3.7	1.2	0.2	0.2
N4	3.7-5.0	1.9	1.0	1.4
S7	5.5-8.5	19	2.7	3.3
S9W	6.0-11.5	19	11.0	17.2
S11	8.5-13.0	51	9.2	14.5
L15	12.5-18.0	101	30	76
L20W	14.0-26.0	92	120	277
L24	22.0-26.0	170	149	363

By using a significant proportion of the pointing mode observation time before the boil-off of liquid helium, $1-2$ deg^2 ultra-deep surveys as well as medium-deep surveys over a wider area (on the order of 10 deg^2) are currently planned. Even after the boil-off of liquid helium, up to 100 deg^2 of survey observations in the near-infrared ($2-5\,\mu$m) are also considered. Because of the limitation of the sky visibility, the survey fields should be chosen from regions close to the ecliptic poles.

3. Scientific Objectives of the IRC Deep Survey

The IRC deep extragalactic survey benefits investigations of the evolution of galaxies and galaxy clusters in two major points: one is the high sensitivity to stellar light from high redshift galaxies, and the other is the capability to detect emissions from hot dust and the unidentified infrared band (UIB) features, which indicate the presence of active star formation in dusty environments.

The stellar mass of a galaxy is a fundamental property of galaxies, and is well traced by the near-infrared ($1-2\ \mu$m) light. Since ground-based telescopes are not sensitive enough to detect the stellar light beyond 2 μm, space-borne telescopes are essentially required for the study of galaxy stellar masses at large redshifts. Near-infrared and shorter mid-infrared filter bands of IRC (see Table 1) continuously cover $2-10\,\mu$m wavelengths, and hence a systematic study of galaxy mass at various redshifts is possible. Using four pointing observations per field, the IRC can detect passively evolving galaxies with $5 \times 10^{10} M_{\odot}$ stellar mass at $z = 3$ (Kodama, T., private communication). With collaborative observations from large ground-based telescopes, such as Subaru at optical and near-infrared wavelengths, we can estimate the redshift of the detected galaxies using spectral energy distribution templates. The large sample of high-z galaxies over a wide area enables us to study the history of the stellar mass evolution in field and cluster environments, and will greatly improve our knowledge of the mass assembling history of the universe.

In the mid-infrared, the IRC is capable of detecting dusty star-forming galaxies located even at $z = 1-2$. The $15\,\mu$m deep galaxy counts, by using ISOCAM onboard ISO, showed an excess below 1 mJy, which can be explained by a large number of star-forming galaxies at $z \sim 1$ (Aussel et al. 1999). The mid-infrared SED of the dusty star-forming galaxies is characterized by the UIB features as well as the hot dust continuum. Hence the six mid-infrared bands of the IRC will be a useful tool to examine the star formation history of the universe at $z = 0-2$ over a much larger volume of the universe than that studied with ISOCAM. An encouraging result from the ISOCAM deep survey of the SSA13 field is shown in Figure 2 (see Sato et al. 2002 for details). In this field three faint submillimeter sources were detected with SCUBA on the JCMT, all of which have counterparts in the deep $7\,\mu$m images. In this way, the IRC deep surveys provide us with a useful and important bridge between the optical and submillimeter deep surveys.

Acknowledgments. The authors would thank all of ASTRO-F/IRC team members as well as the IRC extragalactic science team members in Japan, Korea, and the UK. We especially thank Chris Pearson for providing us with the confusion limit table.

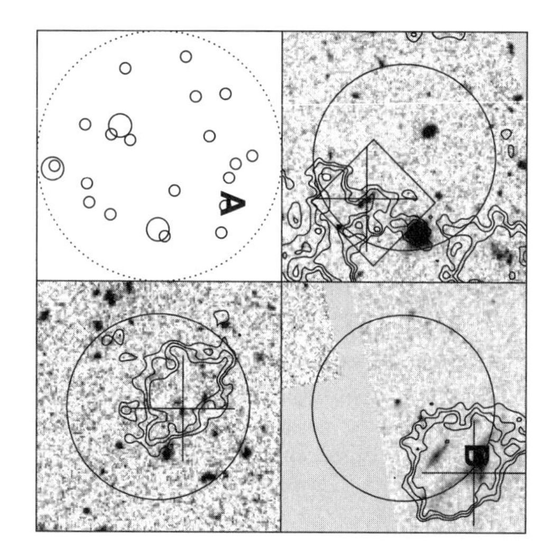

Figure 2. Mid-infrared counterparts of submillimeter sources found in the SSA13 field (Sato et al. 2002). In the upper left panel, SCUBA 850 μm sources and the ISOCAM 7 μm sources are marked with large and small circles. In other panels, magnified views of the submillimeter sources are displayed on the HST/WFPC2 I-band images, where the 7 μm sources are shown with contours.

References

Aussel, H. et al. 1999, A&A, 342, 313

Deutsch, M., & Bicay, M. D. 2000, Ap&SS, 273, 187

Elbaz, D. et al. 1999, A&A, 351, L37

Fixen, D. J. et al. 1998, ApJ, 508, 123

Hauser, M. G. et al. 1998, ApJ, 508, 25

Kawara, K. et al. 1998, A&A, 336, L9

Lagache, G. & Puget, J. L. 2000, A&A, 355, 17

Matsuhara, H. et al. 2000, A&A, 361, 407

Nakagawa, T. 2001, in ESA SP-460, The Promise of the Herschel Space Observatory, ed. G. L. Pilbratt, J. Cernicharo, A. M. Heras, T. Prusti, & R. Harris (The Netherlands: ESA), 67

Oliver, S. et al. 2002, MNRAS, 332, 536

Pearson, C. P. & Rowan-Robinson, M. 1996, MNRAS, 283, 174

Pearson, C. P. 2001, MNRAS, 325, 1511

Sato, Y. et al. 2002, ApJ, 578, L23

Shibai, H. 2002, ASR Conf. Ser., in this volume

Wada T. et al. 2002, Proc. SPIE, 4850, in press

IAU 8th Asia-Pacific Regional Meeting
ASP Conference Series, Vol. 289, 2003
S. Ikeuchi, J. Hearnshaw & T. Hanawa, eds.

Infrared All-Sky Survey with ASTRO-F

Hiroshi Shibai

Nagoya University, Furo-cho, Chikusa-ku, Nagoya 464-8602, Japan

Abstract. ASTRO-F (IRIS) is the first Japanese satellite to be dedicated to infrared astronomy. The primary purpose of this project is to investigate the birth and evolution of galaxies in the early universe by deep and wide-field surveys at wavelengths of 2 to 200 μm. In the far-infrared wavelength band, ASTRO-F will achieve an all-sky survey like the IRAS survey with more than ten times higher sensitivity and several times better spatial resolution. In the near- and mid-infrared regions, wide area sky-surveys will be made for pre-selected fields. In addition to these photometric surveys, low-resolution spectroscopic capabilities are available for all wavelength bands. ASTRO-F will make a fundamental database for future more advanced observatories in the next generation.

1. Introduction

Following successful infrared satellites (IRAS, COBE, and ISO), ASTRO-F (IRIS: the Infrared Imaging Surveyor) project is being developed as the first Japanese satellite to be dedicated to infrared astronomy, as one of the ASTRO-series projects of the Institute of Space and Astronautical Science (ISAS). The ASTRO-F project is managed by ISAS, and is supported by scientists of Nagoya University, the University of Tokyo, and other universities and institutions in Japan, as well as by collaborations with foreign institutions and scientists. Seoul Naional University (Korea), ESA, and Imperial College are participating in the data reduction, analysis, and ground station support. The detailed description of ASTRO-F and pointing observations are written in separate papers of Murakami (2002) and Shibai (2002).

2. Telescope and Instruments

ASTRO-F has a cooled telescope whose effective diameter is 670 mm, which is cooled down to 5.8 K in order to reduce the instrumental thermal emission. ASTRO-F is the first case of a hybrid-type cryostat satellite incorporating a mechanical cooler as well as cryogen. The expected cryogen life is one and a half years. An M-V-6 rocket will launch ASTRO-F into a sun-synchronous circular orbit.

ASTRO-F has two scientific instruments, the Far-Infrared Surveyor (FIS) and the Infrared Camera (IRC). The IRC consists of three independent cameras covering the near- and mid-infrared regions of 1.8–26 μm. The design image

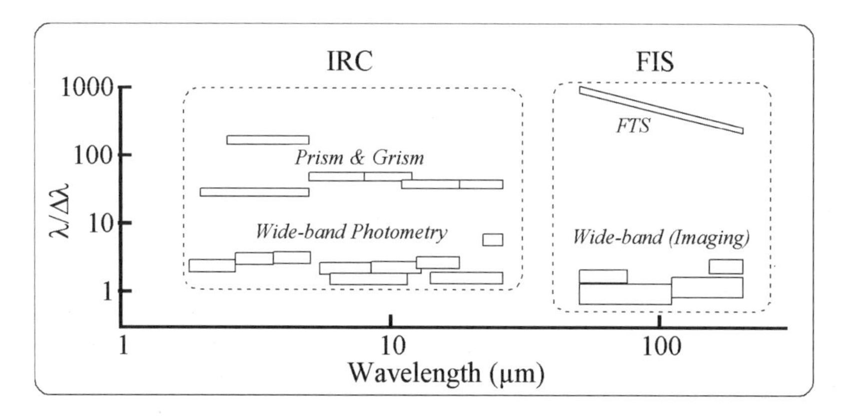

Figure 1. Spectral coverage and resolution

quality is sharper than the size of individual pixels. The large format arrays of the IRC can cover wide fields with better spatial resolution than the present infrared cameras in space, and, by incorporating grisms, achieve highly efficient spectroscopic surveys. The low dispersion spectroscopic capability in the near-infrared is unique and complements SIRTF. The arrays are state-of-the-art, and optimized for the low background conditions in space. The FIS is primarily designed for an IRAS-type all-sky survey in the wavelength region from 50 to 200 μm. Stressed and unstressed Ge:Ga large arrays were recently developed to cover this wavelength region. The former covers the region longer than 100 μm, the latter the spectral range below 100 μm. The pixel sizes were determined to be nearly equal to the diffraction-limited resolution. Each array has one wide-band channel of three rows and one narrow-band channel of two rows. The arrays are tilted by 26.5 degrees from the perpendicular to the scanning direction, which allows us to obtain sampling points by a spacing of a half of the pixel pitch for better resolution in the cross-scanning direction.

3. Scientific Capability

ASTRO-F has wider fields-of-view than SIRTF and has better spatial resolution than IRAS and ISO. This advantage is obtained by adopting state-of-the-art space-optimized array sensors in the near- and mid-infrared and through use of recently developed far-infrared arrays, and these are most important features for wide-sky survey observations. Low-resolution spectroscopy is possible in all bands. SEDs (spectral energy distributions) of starburst galaxies as well as galactic star-forming regions are efficiently obtained by this capability. Figure 1 shows the spectral coverage as well as the resolution. The whole sky is covered in the far-infrared wide-band survey. Up to 100 square degrees will be surveyed with IRC wide-band imaging, and on the order of 1 square degree can be covered with spectroscopy of both instruments. Most of these capabilities are unique.

ASTRO-F is fundamentally a survey-type satellite on a polar orbit. The telescope must always be pointed away from the Earth. In the continuous survey

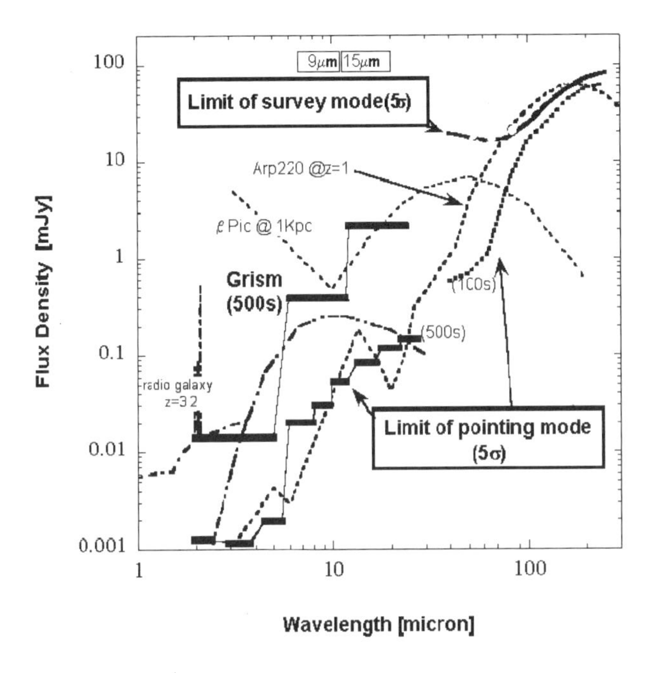

Figure 2. Expected point source sensitivity

mode for the all-sky survey, the telescope scans at a constant rate along the great circle perpendicular to the Sun. This will enable us to completely cover the sky during the first half year of the survey. However, during passages of the South Atlantic Anomaly (SAA), the detectors will be useless, as a result of the impact of high-energy particles and their after effects. In addition to this, the Moon will shine into the telescope when the Moon is within 33 degrees of the telescope axis. Because of these effects, and possibly other miscellaneous failures, malfunctions, calibrations, and maintenance operations, the whole sky cannot be covered within a half year. The parts of the sky which are not covered during this first half-year of the survey will be observed during the next half year and later, while the cryogen lasts.

4. All-Sky and Wide-Area Survey Plan

The expected performance is shown in Figure 2. The primary scientific purpose of the ASTRO-F mission is to investigate the processes leading to the formation and evolution of galaxies in the early universe. In the far-infrared region, the second all-sky survey following that of IRAS will be conducted with remarkably improved performance; more than ten times higher sensitivity for point sources (20–110 mJy), several times better spatial resolution (30–50 arcsec), and an added, longer wavelength band (up to 170 μm). The final point source catalogue is expected to include distant ultra-luminous galaxies, starburst galaxies, proto-planetary disks, and other sources. The expected number of detected point sources, mainly external galaxies, is more than ten million, which should include more than ten thousand distant starbursts at $z > 1$. The data of the all-sky survey will be reduced and analyzed for archiving by a dedicated team, with participation not only by Japanese astronomers but also by Korean, UK, and Dutch astronomers. Several data products, such as the ASTRO-F (IRIS) Flux Catalogue at IRAS Point Sources, Point Source Catalogue at High Latitudes/at Low Latitudes, and All-Sky Images, are planned to be released as soon as possible after a one-year proprietary period. In addition to the all-sky survey, the IRC will conduct a deep wide-sky survey of pre-selected areas of the sky at 2 to 25 μm. IRC has six photometric bands and a capability for low-resolution grism spectroscopy. In particular, we are going to plan two general survey areas, near the Large Magellanic Cloud (LMC) and around the North Ecliptic Pole (NEP). Since ASTRO-F surveys along a great circle, whose axis is directed to the Sun, both the ecliptic pole regions can be surveyed and pointed many times. The LMC is located very near to the south ecliptic pole and can be accessible during two months per half year. By utilizing this benefit, almost 20 square degrees of the LMC will be surveyed with deep pointing observations in addition to the all-sky survey. As a result, we will have complete NIR to FIR maps of the LMC. On the other hand, we also plan a general survey around the NEP region with the pointing observation mode. We will have photometric maps of all bands as well as low-resolution spectroscopic data from NIR to FIR, although the area covered is limited to a few square degrees.

The ASTRO-F all-sky survey catalogue and the selected field images taken by the IRC and by the FIS spectrometer will not only enable us to investigate the history of galaxy evolution in the early universe with statistical means, but also contribute to detailed observations by SIRTF (Gallagher et al. 2002), Herschel (Pilbratt 2002), JWST (Stockman 2002), SOFIA (Meyer & Erickson 2002), SPICA (Matsumoto 2002), and other future observatory-type space missions as well as ground-based telescopes.

References

Gallagher, D. B., Irace, W. R., & Werner, M. W. 2002, in Proc. SPIE Vol. 4850, IR Space Telescopes and Instruments, ed. J. C. Mather, in press

Matsumoto, T. 2002, in Proc. SPIE Vol. 4850, IR Space Telescopes and Instruments, ed. J. C. Mather, in press

Meyer A. W., & Ericksom, E. F. 2002, in Proc. SPIE Vol. 4857, Airborne Telescope System II, ed. R. K. Melugin & H. Roeser, in press

Murakami, H. 2002, in this volume

Pilbratt, G. L. 2002, in Proc. SPIE Vol. 4850, IR Space Telescopes and Instruments, ed. J. C. Mather, in press

Shibai, H. 2002, in Proc. SPIE Vol. 4850, IR Space Telescopes and Instruments, ed. J. C. Mather, in press

Stockman, H. S., Mather, J. C., Smith, E. P., Petro, L., & de Jong, R. 2002, in Proc. SPIE Vol. 4850, IR Space Telescopes and Instruments, ed. J. C. Mather, in press

IAU 8th Asia-Pacific Regional Meeting
ASP Conference Series, Vol. 289, 2003
S. Ikeuchi, J. Hearnshaw & T. Hanawa, eds.

ISOGAL Survey of the Intermediate Galactic Bulge: Mass-losing AGB Stars in Galactic Bulge ISOGAL Fields.

D. K. Ojha

Tata Institute of Fundamental Research, Mumbai - 400 005, India

A. Omont and ISOGAL Team

Institut d'Astrophysique de Paris, F-75014, Paris, France

Abstract. We present a study of ISOGAL sources in the "intermediate" galactic bulge ($|l| < 2°$, $|b| \sim 1° - 4°$), observed by ISOCAM at 7 and 15 μm (Ojha et al. 2002). In combination with near-infrared (I, J, K_s) data of the DENIS survey, we discuss the nature of the ISOGAL sources, their luminosities, the interstellar extinction and the mass-loss rates. A large fraction of the detected sources at 15 μm are AGB stars above the RGB tip, and a number of them show an excess in the ([7]-[15])$_0$ and (K_s-[15])$_0$ colors, characteristic of mass-loss. The latter, especially (K_s-[15])$_0$, allow estimation of the mass-loss rates ($\dot{M}$) and show their distribution in the range 10^{-8} to $10^{-5} M_\odot$/yr.

1. Introduction

ISOGAL is a detailed mid-infrared imaging survey of the inner Galaxy, tracing the Galactic structure and stellar populations (Omont et al. 2002). It combines 7 and 15 μm ISOCAM data with IJK_s DENIS data available for most of the ISOGAL fields. The main goals of the ISOGAL survey are to trace the large-scale inner disk and bulge structure using primarily old red giant (gM) stars; to study the corresponding stellar populations; to trace the age and mass-loss of AGB stars; to determine the number of (dusty) young stars, and to map the star formation regions through diffuse ISM emission and extinction. Multicolor near- and mid-infrared data are essential to analyse these features.

In this paper, we report the study of nine ISOGAL fields in the intermediate galactic bulge with a total area ~ 0.29 deg^2 and more than 2000 detected sources. The 15 μm and 7 μm ISOGAL point sources have been combined with DENIS IJK_s data in the ISOGAL Point Source Catalogue (Schuller et al. 2002) to determine their nature and the interstellar extinction. Analysis at five near- and mid- infrared wavelengths of the ISOGAL sources, together with the relatively low and constant extinction, shows that the majority of the sources are red giants with luminosities just above or close to the RGB tip, and with a large proportion of these with detectable mass-loss.

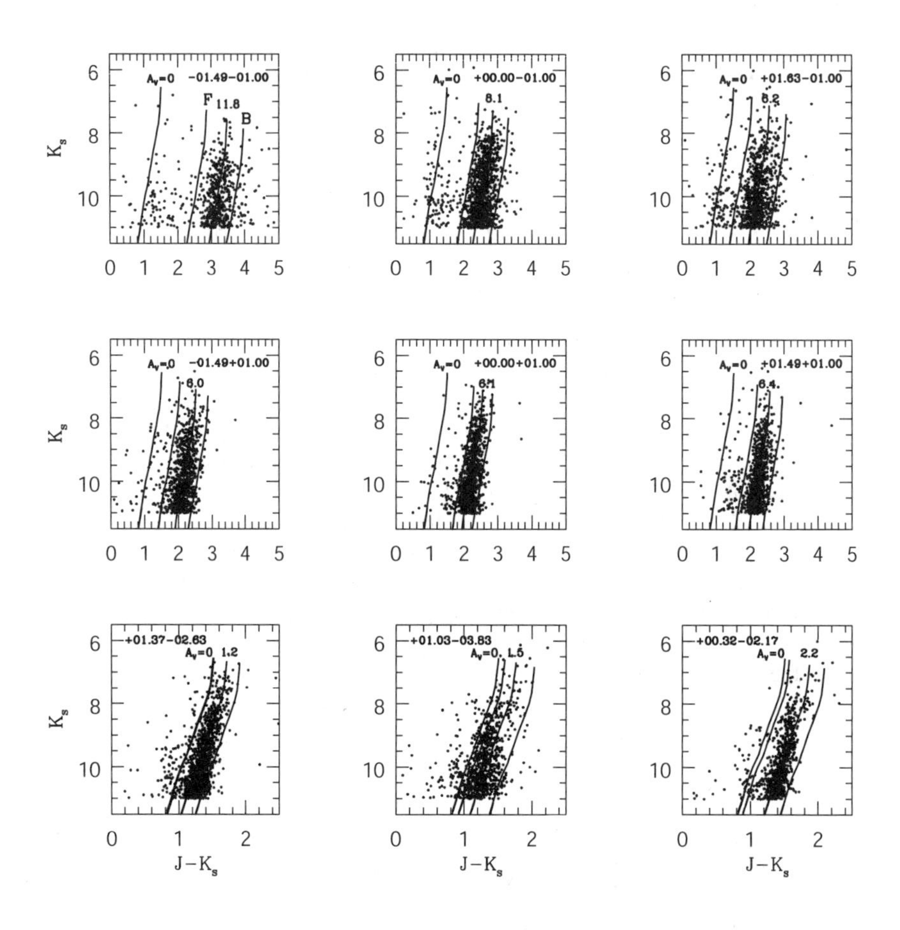

Figure 1. Color-magnitude diagrams ($J - K_s/K_s$) of DENIS sources in the ISOGAL intermediate bulge fields.

2. ISOGAL and DENIS Observations

ISOGAL observations with ISOCAM at $6''$ pixel field of view of nine intermediate bulge fields with good quality of 7 and 15 μm data are used in our study. For all the observed ISOGAL fields, we have used the data of the ISOGAL Point Source Catalogue (PSC, Version 1) described in Schuller et al. (2002). They were derived with an improved data reduction with respect to ISO Archive data, with the use of CIA ISOCAM software and of a special source extraction (Schuller et al. 2002). The near-infrared data used in this paper were acquired from the DENIS survey with special observations of the Galactic bulge in the three bands, I (0.80 μm), J (1.25 μm) and K_s (2.17 μm). Such DENIS observations cover the totality of the nine ISOGAL bulge fields. Fig. 1 shows the $J - K_s/K_s$ color-magnitude diagrams of DENIS sources in the bulge fields.

3. Interstellar Extinction

The $K_s/J - K_s$ magnitude-color diagrams of DENIS sources in the bulge fields show a well-defined red giant sequence shifted by fairly uniform extinction (see Fig. 1), with respect to the reference K_{s0} *vs.* $(J - K_s)_0$ of Bertelli et al. (1994) with $Z = 0.02$ and a distance modulus of 14.5 (distance to the Galactic Center: 8 kpc). The near-infrared colors of this isochrone have been computed with an empirical $T_{\rm eff}$-$(J-K)_0$ color relation built by making a fit through measurements for cool giants. We have assumed that $A_J/A_V = 0.256$; $A_{K_s}/A_V = 0.089$ (Glass 1999).

The average value of the interstellar extinction (A_V) for each bulge field is shown in Fig. 1. The DENIS sources with anomalously low values of A_V are probably foreground sources. They are visible in Fig. 1 in each of the bulge fields (stars around the isochrone with $A_V \sim 0$ mag. and stars clearly left of the bulk of the stars grouped around the isochrone with mean A_V of the field). For each field we define an isochrone "F" (Fig. 1) for which we assume that all the sources left of it are foreground stars which will be no longer considered in the following discussions of "bulge" stars.

To summarize, for the following discussions where dereddening is essential (color–magnitude diagrams, luminosities, mass-loss rates) we have applied the following prescriptions: We have discarded all sources left of the curves "F" in Fig. 1. For the others, with J and K detections, except for a few exceptions (right of the isochrones "B"), we have determined their specific extinction from $J - K_s$ (with Glass (1999) values for A_J/A_V and A_{K_s}/A_V) and the quoted zero-extinction isochrone of M giants. For all the others, especially with no JK associations, we have used the mean extinction A_V of the field. The extinction ratios $A_{[7]}/A_V$ and $A_{[15]}/A_V$ in the ISOGAL bands are still uncertain (Hennebelle et al. 2001). We have used the values $A_{[7]}/A_V = 0.020$; $A_{[15]}/A_V = 0.025$ recommended by Hennebelle et al. for "disk" fields; however, the extinction corrections remain small with the low values of A_V considered.

4. Mass-loss Rate ($\dot{M}$) Determination

It is generally agreed that the color $(K - [12])_0$ (where [12] is the magnitude corresponding to the IRAS 12 μm flux) is a good indicator of the mass-loss rate $\dot{M}$ of AGB stars (see e.g. Le Bertre & Winters 1998, hereinafter LBW98). It is clear that $(K_s - [15])_0$ should be an equally good gauge of $\dot{M}$, with a slightly different calibration. In order to allow nevertheless a rough calibration of the mass-loss rates, to discuss their relative distribution and to try to make a preliminary comparison with the solar neighbourhood, we have applied the results of the previous work of LBW98 relating $\dot{M}$ and $(K - [12])_0$ (Ojha et al. 2002). Fig. 2 shows the $(K_s - [15])_0/[15]_0$ and $(K_s - [15])_0/K_{s0}$ color-magnitude diagrams of ISOGAL sources respectively, as well as the luminosity estimates $(L_\odot)$ and rough indications of mass-loss rates.

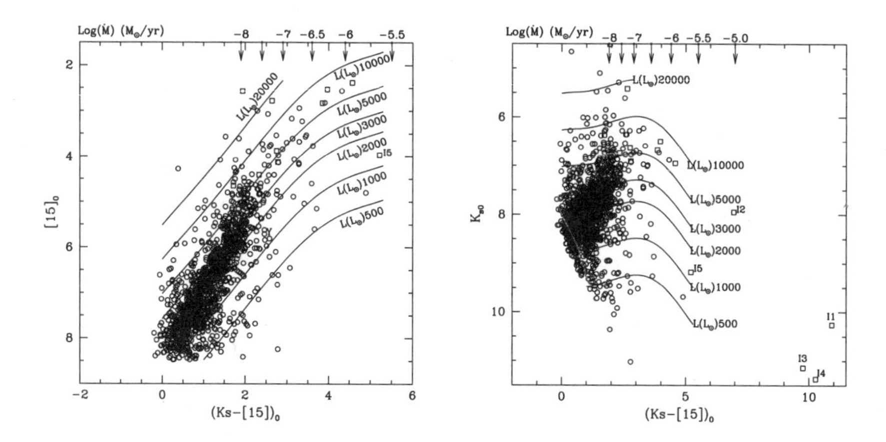

Figure 2. $[15]/K_s$–[15] and K_s/K_s–[15] magnitude-color diagrams of ISOGAL sources with DENIS counterparts. The curves from bottom to top represent the luminosity estimates from 500 to 20 000 $L_{\odot}$, respectively. The approximate scales of mass-loss rates displayed at the top of the figures are derived from LBW98. The IRAS sources are shown by open squares and the red ones as "I" in the figures.

5. Conclusions

We conclude that almost all the ~2000 ISOGAL sources detected both at 7 and 15 μm on the line of sight of the bulge are AGB stars or RGB tip stars (Ojha et al. 2002). A large proportion of these AGB stars have appreciable mass-loss rates, as shown by the excess in $(K_s - [15])_0$ and $([7] - [15])_0$ colors, characteristic of circumstellar dust emission. We have performed a preliminary determination of mass-loss rates from $(K_s - [15])_0$ by extending results of Le Bertre & Winters 1998 for $(K - [12])_0$. However, such estimates of $\dot{M}$ directly depend on the assumed model for circumstellar dust infrared properties and on the value assumed for the dust-to-gas ratio. All mass-loss rates contribute appreciably to the total mass returned to the interstellar medium. Four stars, out of ~2000, have mass-loss rates $\gtrsim 10^{-5} M_{\odot}$/yr. They could dominate the mass return to the interstellar medium if they belong to the bulge, which is still unclear.

References

Bertelli, G., Bressan, A., Chiosi, C. et al. 1994, A&AS, 106, 275
Glass, I. S. 1999, in "Handbook of Infrared Astronomy", Cambridge Univ. Press
Hennebelle, P., Pérault, M., Teyssier, D. et al. 2001, A&A, 365, 598
Le Bertre, T., & Winters, J. M. 1998, A&A, 334, 173 (LBW98)
Ojha, D. K., Omont, A., Schuller, F. et al. 2002, A&A (submitted)
Omont, A., Gilmore, G., Alard, A. et al. 2002, A&A (submitted)
Schuller, F., Ganesh, S., Messineo, M. et al. 2002, A&A (submitted)

IAU 8th Asia-Pacific Regional Meeting
ASP Conference Series, Vol. 289, 2003
S. Ikeuchi, J. Hearnshaw & T. Hanawa, eds.

A Sensitive VLA Image Covering the SIRTF First-Look Survey Area

Q. F. Yin, J. J. Condon, W. D. Cotton

National Radio Astronomy Observatory[1], *520 Edgemont Road, Charlottesville, VA 22903 USA*

Abstract. SIRTF is scheduled for launch in 2003 January and will be orders-of-magnitude more sensitive than earlier infrared telescopes. Its first major science program will be the First-Look Survey (FLS) so that scientists planning to use SIRTF will have a better understanding of the deep infrared sky and can propose observations that will make the best use of SIRTF during its limited operational lifetime. The extragalactic component of the FLS will cover 5 deg^2 near the north ecliptic pole with unprecedented sensitivity at mid- and far-infrared wavelengths. Since most SIRTF FLS sources will be starburst galaxies that obey the far-infrared/radio correlation and since most microJy radio sources are powered by starbursts, a 1.4-GHz VLA survey with a $5\sigma \approx 100\,\mu$Jy detection limit can preselect most of the SIRTF FLS population. The $\theta = 5\farcs0$ resolution of our B-configuration survey minimizes confusion and yields the accurate positions needed for making optical identifications with galaxies as faint as $R \approx 25.5$ mag. The VLA survey of the central 2.5 deg^2 has been completed, and the images and catalogues are available at `http://sirtf.caltech.edu/SSC/T FLS/exgal/vla.html`. The remaining VLA observations will be made during the summer of 2002 and the final data products should be released before the end of the year.

1. Introduction

SIRTF (Space InfraRed Telescope Facility) is the fourth and last of NASA's "Great Observatories." It will be orders-of-magnitude more sensitive than earlier infrared telescopes. The WIRE (Widefield InfraRed Explorer) satellite was supposed to explore the deep infrared sky in 1999, but it failed. Therefore the SIRTF First-Look Survey (FLS) was designed to characterize the mid-infrared sky at depths comparable with the WIRE sensitivity limits and provide the science community with data needed for planning SIRTF observations. The FLS will be the first major science observation of SIRTF about three months after the planned 2003 January launch. To ensure that the FLS can proceed whenever SIRTF is ready, it must be located inside the SIRTF "constant viewing zones"

[1]The National Radio Astronomy Observatory is a facility of the National Science Foundation operated under cooperative agreement by Associated Universities, Inc.

(CVZs) within 10° of the ecliptic poles. The actual area to be covered by the extragalactic component of the FLS is about 5 deg^2 in that part of the northern CVZ having the faintest cirrus foreground $I(100\,\mu\text{m}) \approx 1 - 2$ MJy sr^{-1}. It was chosen in 1999 so that ground-based observations supporting the FLS could be made in advance. They include deep optical and radio imaging.

2. Optical Imaging

Optical identifications are essential for determining the basic nature of the SIRTF sources (e.g., distinguishing between spiral and elliptical galaxies, merging galaxies, and quasars) and obtaining follow-up spectroscopy for redshift distances and clues about the energy-source types (starburst or AGN). The KPNO 4-m telescope MOSAIC CCD camera recently imaged the FLS area to $R \approx 25.5$ mag, the estimated brightness of the faintest galaxies that SIRTF will detect. Thus nearly all of the galaxies hosting SIRTF sources should be visible in those images, but as unrecognizable needles in a haystack of galaxies – we expect a few thousand SIRTF infrared sources among the millions of KPNO optical galaxies.

3. Radio Imaging and Identifications

Only radio images can isolate the needles. The reason is that the microJy radio-source population is nearly the same as the mJy far-infrared (FIR) population, both being dominated by distant starburst galaxies obeying the FIR/radio correlation. The FIR/radio correlation even allows us to estimate the radio sensitivity needed to match the FIR sensitivity of SIRTF, $5\sigma = 3.8$ mJy at $\lambda = 70\,\mu$m and $5\sigma = 33$ mJy at $\lambda = 160\,\mu$m. The corresponding radio limit for distant ($z \sim 0.5$) starburst galaxies is $5\sigma \approx 100\,\mu$Jy at $\nu = 1.4$ GHz. A VLA survey with this sensitivity should detect most of the objects that the SIRTF FLS will detect at 70 and 160 μm. The remarkable similarity of the mJy FIR and microJy radio skies also allows us to exploit technical advantages of the VLA to clarify the SIRTF data.

What does the radio sky look like at the 100 μJy level? What will the SIRTF FLS show? The VLA has already imaged half of the FLS area, so we can begin to answer these questions. A small portion of the VLA image is shown in Figure 1. Its resolution is $\theta = 5''.0$ FWHM. There are about 2000 sources stronger than 100 μJy on the full image, which now covers $\Omega \approx 2.5$ deg^2. Some of the stronger, visibly extended sources are "classical" radio sources powered by AGNs, with cores, jets, and lobes. However, these sources are still so weak that their absolute luminosities are low ($\sim 10^{24}$ W Hz^{-1}) even at cosmological distances. Many of the jets are clearly bent or distorted by their local environment. The majority of the radio sources look like faint points and are probably starburst galaxies, whose angular diameters are only $1'' - 2''$ at moderate redshifts. These are the radio sources that should have FIR counterparts in the SIRTF FLS – the needles in the haystack. The SIRTF FLS images of this area should look very much like the radio image, except with lower angular resolution.

Even after SIRTF has made its part of the FLS, the VLA image will remain essential for interpreting the FIR data. SIRTF is very sensitive but has limited angular resolution because the telescope mirror is only 85 cm in diameter, limited

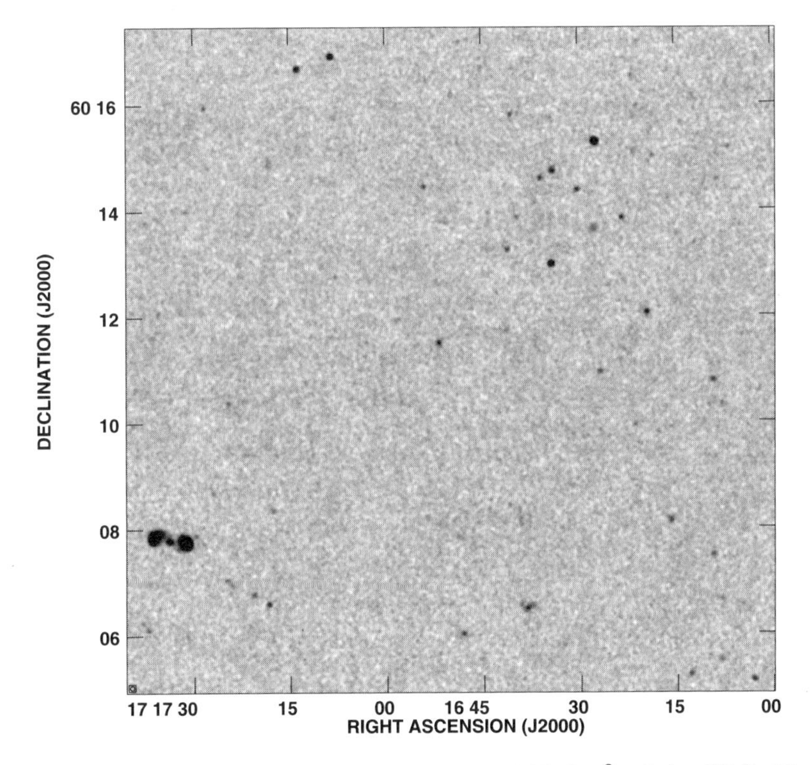

Figure 1. A VLA subimage covering 0.043 deg^2 of the FLS. The angular resolution is $\theta = 5\farcs0$ and the faintest detectable sources have 1.4 GHz flux densities $S \approx 100\,\mu$Jy.

by the size of the rocket used to carry it into orbit. Thus the FWHM of the SIRTF point-source response is $\theta \approx 20''$ at $\lambda = 70\,\mu$m and $\theta \approx 47''$ at $\lambda = 160\,\mu$m. The sky density of detectable galaxies is so high that two or more galaxies will often be blended in one beam area at $\lambda = 160\,\mu$m. The 160μm survey will be strongly confusion limited, and even the 70 μm images will be somewhat confused. Limited angular resolution also reduces position accuracy for faint sources. Most of the sources in an extragalactic survey complete to 5σ have flux densities in the range $5\sigma < S < 10\sigma$. Noise alone will cause rms uncertainties in the source positions $\approx \sigma\theta/(2S)$, or up to $\theta/10$ ($2''$ at $70\,\mu$m, $5''$ at $160\,\mu$m). These uncertainties are too large for making reliable optical identifications with faint galaxies since there is a good chance of finding two or more galaxies inside the SIRTF position error circle—recall that there are millions of optical galaxies in the field, not just thousands. Position errors $< 1''$ are needed to make unique associations on the basis of position coincidence. Since the VLA survey resolution is $\theta = 5\farcs0$, the VLA images can resolve most source blends and the rms position errors of the faintest VLA sources will be only $0\farcs5$, good enough for making reliable optical identifications down to the $R \approx 25.5$ mag. optical limit. Finally, the FIR/radio flux ratio can be used to distinguish between a starburst (obeying the FIR/radio correlation) and a radio-loud AGN as the ultimate energy source for each object.

The status of the FLS is that the individual optical images have been made and are now being mosaicked. The VLA mosaic for the central region is complete and on the web at `http://sirtf.caltech.edu/SSC/T_FLS/exgal/vla.html` The remaining VLA observations will be made during the summer of 2002 and should be released before the end of the year.

IAU 8th Asia-Pacific Regional Meeting
ASP Conference Series, Vol. 289, 2003
S. Ikeuchi, J. Hearnshaw & T. Hanawa, eds.

The Virgo High-Resolution CO-Line Survey

Y. Sofue, J. Koda, H. Nakanishi, S. Onodera, K. Kohno

Inst. Astronomy, Univ. of Tokyo, Mitaka, Tokyo 181-0015, Japan

A. Tomita

Faculty of Education, Wakayama University, Wakayama 640-8510, Japan

S. K. Okumura

NRO, National Astron. Obs., Mitaka, Tokyo 181-8588, Japan

Abstract. We present the results of the Virgo high-resolution CO survey (VCOS) obtained with the Nobeyama Millimeter-wave Array at high-angular and spectral-resolutions in the ^{12}CO ($J = 1 - 0$) line emission for 15 Virgo CO-rich spirals. The central CO distributions can be classified into the "central/single-peak type", which is a majority, and "twin-peaks". We derived exact rotation curves by applying a new iteration method to position-velocity diagrams, which show generally a steep central rise, indicating central massive cores.

1. Introduction

High-resolution CO-line observations play an essential role in studying the kinematics and ISM physics in the central regions of spiral galaxies. The Virgo high-resolution CO survey with the NMA has been performed in order to obtain a high angular- and spectral-resolution database for a large number of CO-bright Virgo Cluster spirals in the ^{12}CO ($J = 1 - 0$) line (Sofue et al. 2003a,b). The major motivation was to investigate the detailed rotation curves from analyses of position-velocity diagrams across the nuclei, which would be effective in detecting central compact massive objects. The data are also useful for investigation of the kinematics and ISM physics of the central molecular disks, and their environmental effect in the cluster circumstance. The advantage of observing the Virgo Cluster galaxies is their identical distance, which has been accurately taken as 16.1 Mpc by Cepheid calibrations (Ferrarese et al. 1996).

2. Observations and Results

The target galaxies in the survey have been selected from the list of Kenney & Young (1988) by the following criteria. The sources were chosen in the order of CO peak antenna temperatures at the optical centers. Inclination angles were limited to be $25° \leq i \leq 75°$ in order to investigate central gas dynamics.

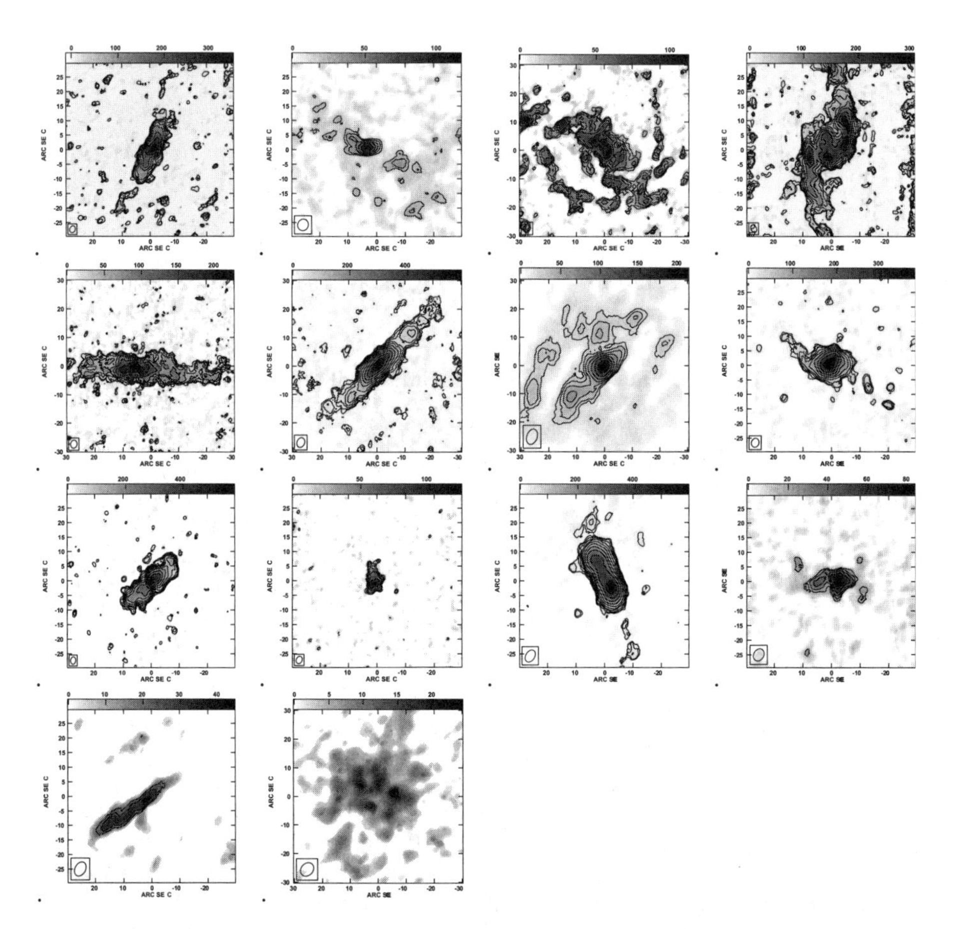

Figure 1. CO atlas of the observed Virgo galaxies. Integrated ^{12}CO ($J = 1-0$)-intensity maps. The image sizes are $1.'0 \times 1.'0$ (4.68×4.68 kpc). The i-th contour corresponds to $[\sqrt{2}]^i$ times 5 K km s^{-1} . Galaxies are from left to right NGC 4192, 4212, 4254, 4303, 4402, 4419, 4501, 4535, 4536, 4548, 4569, 4579, 4654, and 4689.

Galaxies with S0 type were excluded. Interacting galaxies were excluded. Peculiar galaxies in optical images were excluded. Galaxies observed with the NMA since 1994 were excluded. We have thus observed 15 galaxies, among which 14 galaxies were successfully mapped and are shown in Fig. 1, where the typical angular resolutions are $\sim 3''$.

3. Molecular Gas Morphology

The molecular gas distributions show a wealth of variety, although the galaxies have been selected from the brightest CO galaxies from Kenney and Young's (1988) sample without any bias. We can find some characteristic types in the central gas distributions:
Central/Single peak: The majority of the galaxies show a high concentration of molecular gas around the nuclei, which we call the "center/single peak" type. Fig. 3 shows the CO intensity distributions in the central $20'' \times 20''$ regions (1.6 kpc square) for single-peak galaxies. They are not correlated with stellar bars. Some new mechanism other than bar-induced inflow would be responsible for the accretion.
Twin peaks: An example is seen for NGC 4303, which shows two offset arms along the optical bar extending from a molecular ring with two peaks. In so far as the present data set is concerned, this type is not a majority.
Spiral arm type: Some galaxies have prominent spiral arms of molecular gas, mostly along optical dark lanes. NGC 4254 is a typical example.
Bar type: In some galaxies, extended bar-like features are found in the CO distributions. as for CO. NGC 4303. Molecular bar can be also seen in a galaxy without bar in optical bands such as in NGC 4254.
Amorphous type: Some galaxies do not show any regular gas distribution classified in the above types; we classify these galaxies in amorphous type. NGC 4689 is in this type, which has neither particular arms nor central peak. The CO intensities are very weak.

4. Rotation Curves and Dynamical Mass

Rotation curves (RCs) are the basic tool to obtain mass distributions in disk galaxies. We have argued for the advantage to use the CO line for many reasons (Sofue et al. 1999; Sofue and Rubin 2001). In order to solve the central rotation properties, we have developed a new iteration method, which determine the RC so that it reproduces the observed position-velocity diagram (Takamiya and Sofue 2002). We used the obtained position-velocity diagrams along the major axes across the nuclei, and applied this new iteration method.

Fig. 4 shows the PVDs, resulting RCs by the iteration method, and reproduced PVDs by convolution with the intensity distributions. In most cases, the convolved PVDs well mimic the observations.

Fig. 5a shows the calculated SMDs using the RCs obtained in Fig. 4 and combined with those from the literature for outer disks using the method developed by Takamiya and Sofue (2000). The SMDs appear to show a basic, principal structure as the following.
(1) Central massive core: In all cases the mass is highly peaked at the cen-

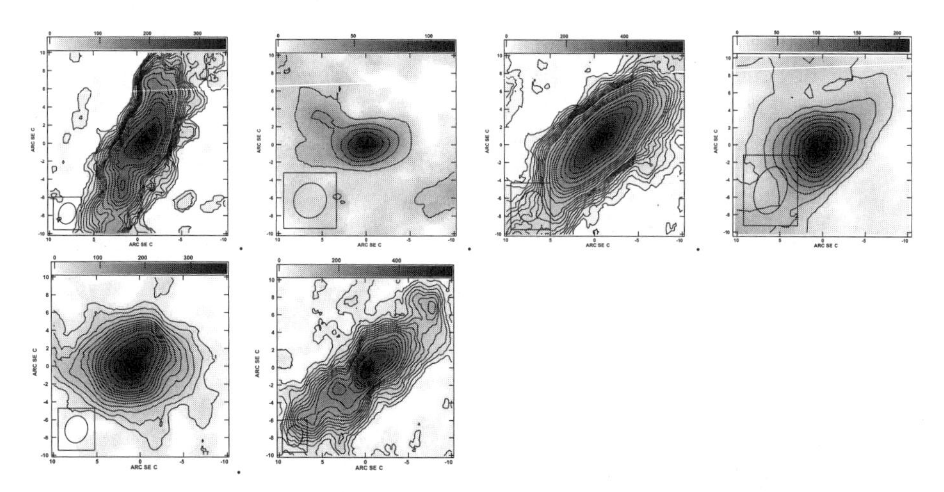

Figure 2. $I_{\rm CO}$ distributions in the central 20″ regions of the "central/single-peak" galaxies. NGC 4192, 4212, 4419, 4501, 4535, 436.

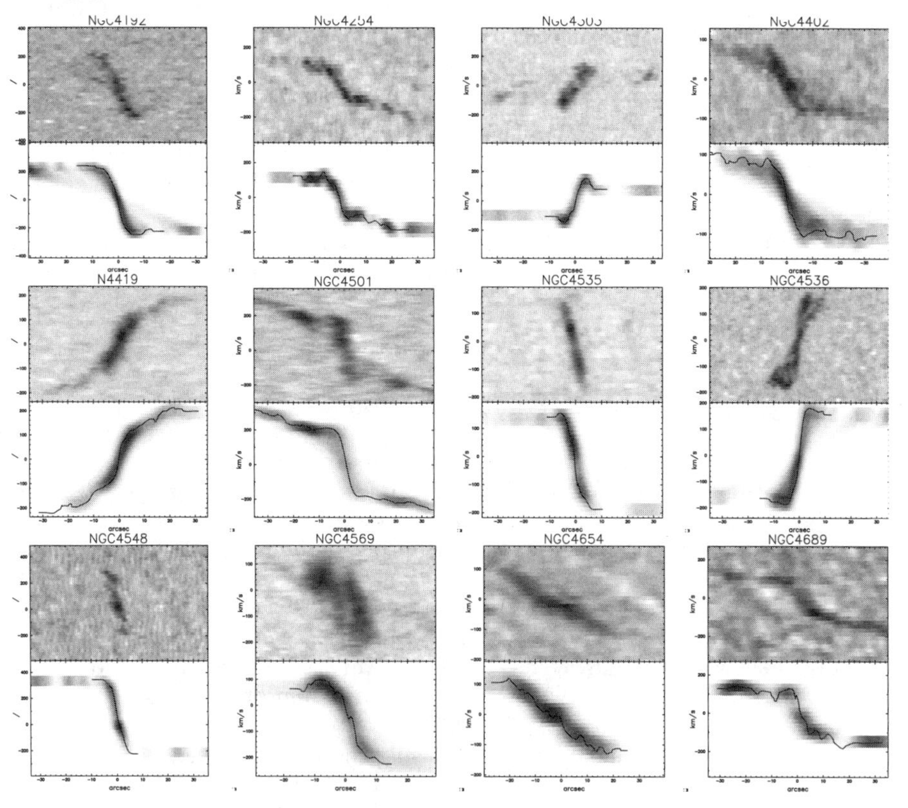

Figure 3. Observed PVDs along the major axes after correcting for the inclination of galaxy disks (upper panels), rotation curves by the iteration method (lower panels), and reproduced PVDs

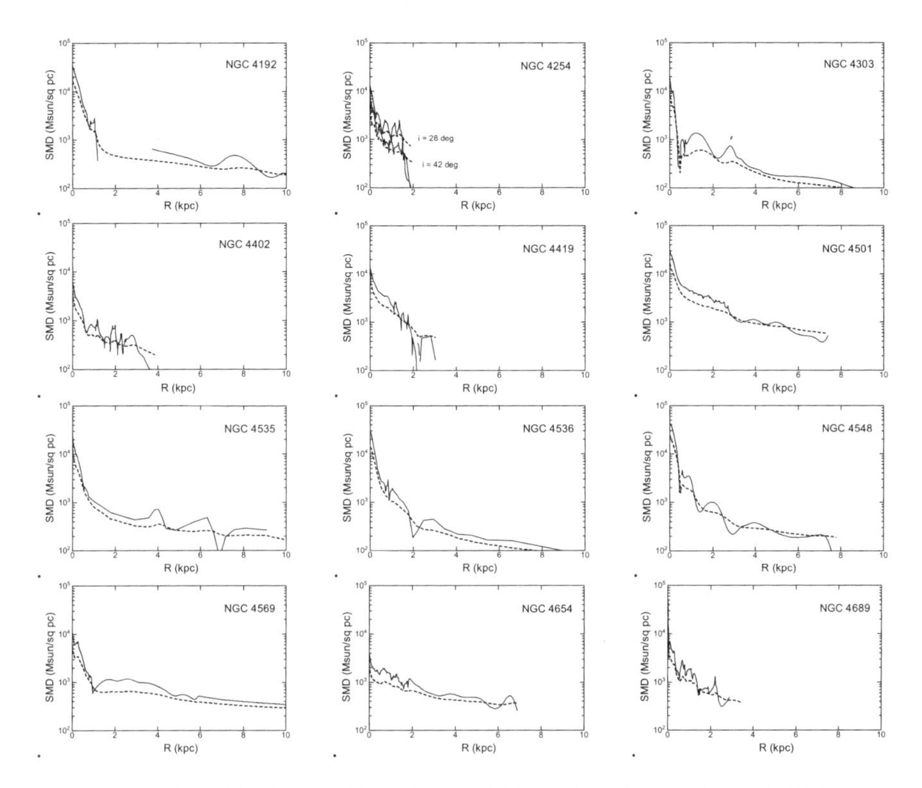

Figure 4. Radial profiles of the SMD of the Virgo spirals. Full lines for spherical symmetry assumption, and thick dashed lines for flat-disk assumption.

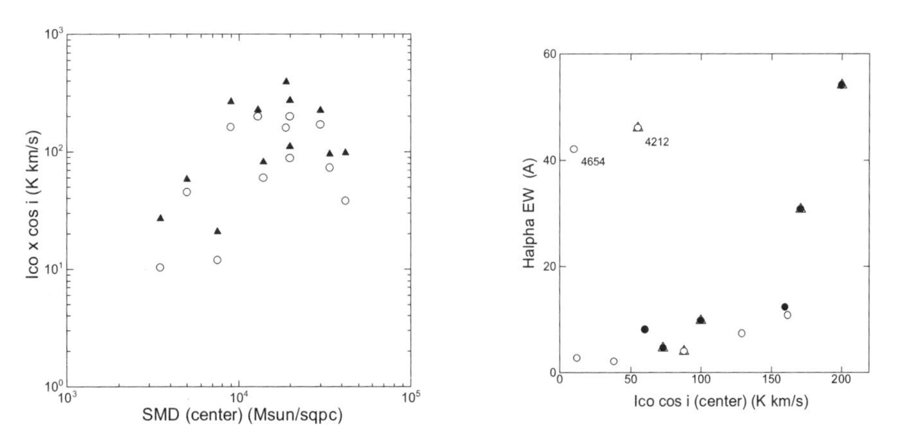

Figure 5. (a) Peak and center I_{CO} vs center SMD (triangles and circles, respectively.) (b) Central Hα activity vs. I_{CO}.

ter, showing a core component with scale radius of about 100 to 200 pc with $\sim 10^4 - 10^5 M_\odot \mathrm{pc}^{-2}$. The dynamical mass of this component within 100 to 200 pc radius is of the order of $\sim 10^9 M_\odot$.
(2) Bulge: The SMD is followed by a more gradually decaying profile at $r = 0.2 - 2$ kpc with scale radius of 500 pc, due to the central bulge.
(3) Disk: The radial profiles at $r \gtrsim 2$ kpc are exponential representing the disk.

5. Correlation among SMD, I_{CO} and Activity

The unambiguous Cepheid distance of 16.1 Mpc enables us to compare the results without suffering from linear scales. We, thus, examined some correlations of the SMD with other parameters. Fig. 5a shows that there is a clear trend that the higher is the SMD, the higher is the peak and central CO intensities. The I_{CO} - SMD correlation indicates simply that *the deeper is the central potential, the stronger is the gas concentration.* This is consistent with our recent results for NGC 3079 (Koda et al. 2002).

Another remarkable correlation is the central activity seen in Hα equivalent widths (Ho et al. 1997) plotted against center I_{CO} in Fig. 5b, where the following activity cycle can be traced. The nuclei gradually get active with increasing I_{CO} from bottom-left corner of the figure, and the activity gets drastically settled when I_{CO} exceeds a threshold of about 150 K km/s to top-right. After exhausting the gas, activity still continues high, with the trajectory returning to top-left.

References

Ferrarese, L., Freedman, W. L., Hill et al. 1996, ApJ, 464, 568
Ho, L. C., Filippenko, A. V. & Sargent W. L. W. ApJS, 112, 315
Kenney, J. D. & Young, J. S. 1988, ApJS, 66, 261
Koda, K., Sofue, Y., Kohno, K. et al. 2002, ApJ, 573, 105
Sofue, Y., Koda, J., Nakanishi, H. et al. 2003a, PASJ, submitted
Sofue, Y., Koda, J., Nakanishi, H., & Onodera, S. 2003b, PASJ, submitted
Sofue, Y. & Rubin, V. C. 2001, ARAA, 39, 137
Sofue, Y., Tutui, Y., Honma, M. et al. 1999, ApJ, 523, 136
Takamiya, T. & Sofue, Y. 2000, ApJ, 534, 670
Takamiya, T. & Sofue, Y. 2002, ApJ, in press

Education and Popularization of Astronomy in Asia

IAU 8th Asia-Pacific Regional Meeting
ASP Conference Series, Vol. 289, 2003
S. Ikeuchi, J. Hearnshaw & T. Hanawa, eds.

Education and Understanding of Astronomy through Total Solar Eclipses

E. Hiei[1,2], N. Takahashi[1], Y. Iizuka[2]

[1]*Meisei University, Hodokubo, Hino, Tokyo 191-8506, Japan*
[2]*National Astronomical Observatory, Mitaka, Tokyo 181-8588, Japan*

Abstract. A total solar eclipse has a great impact on ppeople, and hence there are very old historic records in China, Mesopotamia, etc. The impact still does not change in present times, and inspires wonder in us. The spectacular and magnificent event at a total solar eclipse appeals to our scientific thinking and sensitive feelings. Every one thinks about what nature is. A total solar eclipse is therefore a good and effective opportunity for education and understanding of astronomy.

1. Introduction

Education is indispensable to human life, and IAU Commission 46 has already considered the importance of astronomical education, and the activity of Commission 46 is seen in the publications (IAU Commission 46 Newsletter; Pasachoff and Percy, 1990; Gouguenheim, et al., 1998; Isobe, 1990–).

In these publications, there are a lot of reports and discussions about astronomical education, such as university education, teaching astronomy in the school, planetarium education and training, and the astronomy curriculum, so on.

I am unable to cover astronomical education at all levels, but only describe here how a total solar eclipse has a good effect on both astronomical education and understanding of the nature for students, amateurs and citizens.

2. Eclipse Observations

Astronomy is part of a culture which has a long history and has been supporting humans to survive. Astronomical education should be given not only on up-to-date topics of the universe, but on fundamental/classical astronomy as well. This education should be made both in the class room and also in the field. "Learning by head" and "learning by experience" are practically important for education, and both ways of learning are made efficient use of on the occasion of a total solar eclipse.

One of the essential differences between a physics experiment and an eclipse observation is that in the course of a physics experiment, students are able to try many times for obtaining good data, but an astronomical event occurs only once at a certain time and never again, and therefore should be observed with a mind of "never fail". A sufficient and thorough preparation of fool-proof instruments for an astronomical observation is therefore important and necessary.

During the totality, the surroundings change drastically as day-light becomes dark and the atmospheric temperature decreases by several degrees. One might lose one's normal mind, and be mistaken in operating a telescope during the short period of the totality. In order to avoid such an unhappy accident, it is useful for all participants to join in "a chorus" of counting of time in seconds. Every 30, 20, 10, 5, 3, and 2 minutes before the second contact the time is announced, and then all participants start counting of seconds from one minute before second contact. All students have made preparations for operating a telescope at home, following a planned schedule, and therefore they know what they should do to operate the instruments at each moment in time. The duration of 1 second is rather long when counting, and it is better to insert 0.5 sec.; such as zero five, zero four, zero three, so on.

Students learn not only the procedure of setting-up/operating a telescope, but also study the physics of the solar photosphere, chromosphere, and corona as well. One may also have many questions concerning the eclipse: Can the totality be seen in the far future? What is the corona? Why does the coronal shape change according to the solar activity? Why does a beautiful solar corona move us to tears? – and so on. And they need to carry out data analysis/processing after the eclipse.

3. Eclipse Observations Made by Meisei University

Meisei University has sent students and staff to the observations of the total solar eclipses six times up until now. Table 1 shows the observations made by Meisei University.

The eclipse observation may be disturbed by clouds, and thus carried out at two or three observing sites. It is fortunate to have all coronal data at 6 eclipses, as seen in Table 1, as a result of selecting different observing sites.

Table 2 shows the observations tried at the eclipses, in which the 1st line means the year of the total solar eclipse; the 2nd, an absolute measurement of the K-corona; 3rd line, direct images of 5303 Å corona (green corona), and 6374 Å corona (red corona); 4th line, polarization measurement of K-corona; 5th line, electron temperature derived from smoothness of K-coronal spectrum (Ichimoto et al. 1996); 6th line, polarization of flash spectrum; 7th line, a live program of white light images of the corona, sent at the same time as the eclipse event through the internet relay (Nagai, et al. 2000).

Fig. 1 shows the coronal images taken by the Eclipse Group of Meisei University. The images are reduced by the rotating unsharp masking method (Shiota, 1994). Some results of the eclipse observations are reported (Hiei et al. 1997, 2000; Takahashi et al. 1998, 2000). Fig. 2 shows students preparing the observing instruments for the eclipse.

Students of middle and high school of Meisei Gakuen joined to see the 1994 total solar eclipse in Paraguay. They still vividly remember their first experience of the magnificent and beautiful event of the eclipse, which occurred 8 years ago. They have told their experience to their family and friends, and also are expected to teach their own children in future.

Figure 1. Coronal Images; upper left: 1994 Paraguay, upper right: 1995 India, middle left: 1997 Siberia, middle right: 1998 Guadeloupe, down left: 1999 Turkey, down right: 2001 Zambia, which is taken by Fukushima (2002). The other 5 images are taken by the Eclipse Group of Meisei University. On the image at Paraguay eclipse, inside is soft X-ray image taken with Yohkoh satellite.

Table 1. Eclipse Observation by Meisei University

Eclipse	**Location**	**Weather**	**Stud.**	**Staff**	**Total**
1994 Nov. 3	Vapol Cue, Paraguay	fine	66	15	92
	Chaco, Paraguay	fine	5	1	
	Putre, Chile	thin cloud	3	2	
1995 Oct. 24	Dundlodh, E. India	fine	37	5	42
1997 Mar. 9	Pervomaisky, Siberia	fine	3	3	6
1998 Feb. 26	Venezuela	thin cloud	14	1	21
	Guadeloupe	fine	3	3	
1999 Aug. 11	Elazig, Turkey	fine	7	1	22
	Compiegne, France	cloudy	12	2	
2001 Jun. 21	Zambia	fine	15	2	17

Table 2. Observations carried out by Meisei University

Eclipse	**1994**	**1995**	**1997**	**1998**	**1999**	**2001**
K-corona	•	•	•	•	•	
E-corona	•	•		•		•
Polariz.		•	•			
Electr. Temp.				•		
Flash Spectr.					•	•
Internet Broadcast				•	•	•

4. Eclipse Observations Made by Japanese Amateurs

Japanese amateurs are famous for the activity of finding comets, satellites, and novae, but they also take fine images of the solar corona (Ohgoe and Nakamura, 1994). Table 3 shows the rough number of Japanese amateurs, who go abroad to observe/look at the corona in a group or privately. The numbers of participants seem to depend on the location of easy access at the sight-seeing place. The numbers are roughly several tens in the beginning, and increase by 10–100 times in these ten years. Fig. 3 shows the eclipse images taken by Japanese amateurs.

5. Concluding Remarks

Nature is rich in resources for stimulating our mental and intellectual activity and full of wonder for exciting us with delight. There are many attractive phenomena in nature, but a total solar eclipse is one of the most spectacular and astonishing of phenomena.

A total solar eclipse is an unforgettable event and therefore once one has seen such a magnificent and sublime solar corona as at a total solar eclipse, one

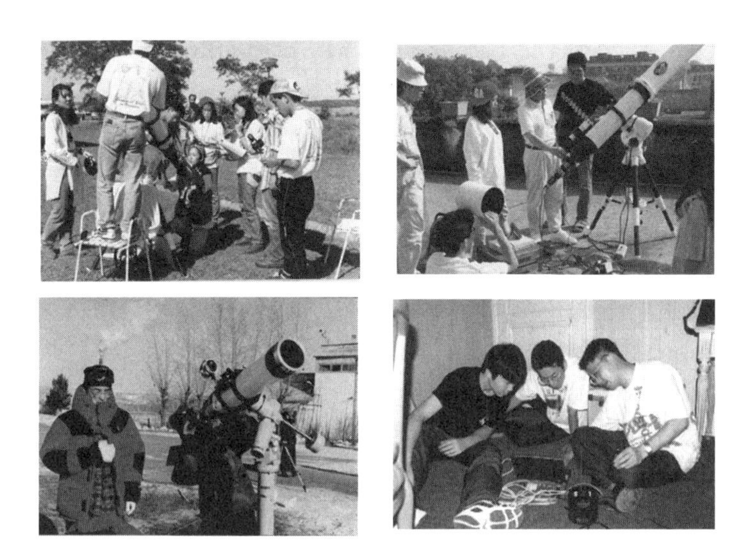

Figure 2. Students, preparing the eclipse observations; upper left: 1994 Paraguay, upper right: 1995 India, down left: 1997 Siberia, down right: 1999 France

is said to become "eclipse-sick" – and would like to visit any place Earth in order to see the corona at a total solar eclipse again.

Some may think of the event of a total solar eclipse philosophically; how a human is small compared with the universe; however, humans are able to predict the exact time of the eclipse occurrence, and not only that, we are able to think about the large space and eternal time of the universe. The total solar eclipse reminds us of the famous words by Blaise Pascal (1640): "L'homme n'est qu'un roseau, le plus faible de la nature, mais c'est un roseau pensant". (Human is a thinking reed).

The eclipse observation calls up our scientific activity of inquiring and gives us an opportunity to think about Nature.

References

Fukushima, H. 2002, "Image Processing of the Solar Corona, wide-field and high -resolution images of the solar corona at the Total Solar Eclipse in Africa on June 21, 2001" Report National Astr. Obs., vol. 5, No.4, 131 (in Japanese)

Gouguenheim, L., Mcnally, D., and Percy, J. R. 1998,"New Trends in Astronomy Teaching", IAU Coll. 162, Cambridge Univ. Press

Hiei, E., Inoue, K., and Takahashi, N. 1997, "Results from the Coronal Observations of 1994 and 1995 Total Solar Eclipses", in Proc. NATO Advanced

Eclipse	Number	Observation Site
1963 July 20	many	Hokkaido
1968 September 22	20	Soviet Union
1970 March 7	40	Mexico
1972 July 10	20	Alaska
1973 June 30	110	Mauritania and Kenya
1974 June 20	20	Australia
1976 October 23	130	Australia
1979 February 26	20	Canada
1980 February 16	20	Kenya and India
1981 July 31	120	Soviet Union
1983 June 11	1000	Indonesia
1984 Nov. 22	20	New Guinea
1988 March 18	2500	Philippines and Ogasawara
1990 July 22	250	Finland
1991 July 11	3500	Hawaii and Mexico
1992 June 30	30	Uruguay
1994 November 3	400	Chile, Bolivia an Paraguay
1995 October 24	500	India, Thailand and Vietnam
1997 March 9	1000	Russia, Mongolia, and China
1998 February 26	200	Venezuela, Guadeloupe
1999 August 11	500	Europe and Turkey
2001 August 21	300	Zambia, Zimbabwe and Madagascar

Table 3. Number of participating amateurs at a total solar eclipse

Research Workshop "Theoretical and Observational Problems Related to Solar Eclipses", Vol. 494, 1

Hiei, E., Takahashi, N., Iizuka, Y., and the Eclipse Group of Meisei University 2000, "Results of the Observations of the Total Solar Eclipses of 1994, 1995, 1997, 1998, and 1999", in the Last Total Solar Eclipse of the Millennium in Turkey, Astr. Soc. Pacific Conf. Ser., 205, 181

Hiei, E., Takahashi, N. 2000, "Ground-based and SOHO observations of polar plumes during eclipse", Advances in Space Research, 25, No.9, 1887

IAU Comm. 46, Newsletter 49-56, 1998-2002

Ichimoto, K., Kumagai, K., Sano, I., Kobiki, T., Sakurai, T. 1996, Publ. Astron. Soc. Japan, 48, 545

Isobe, S. 1990–now, "Teaching of Astronomy in Asian-Pacific Region", Bulletin No.1 – No.17

Nagai, T., Takahashi, N., Okyudo, M., Suginaka, M., and Matsumoto, N. 2000, "Internet Relaying of Total Solar Eclipse on 11 August 1999", in the Last Total Solar Eclipse of the Millennium in Turkey, ASP Conf. Ser. 205, 196

Pasachoff, J. M. and Percy, J. R. 1990, "The Teaching of Astronomy", IAU Coll. 105, Cambridge Univ. Press

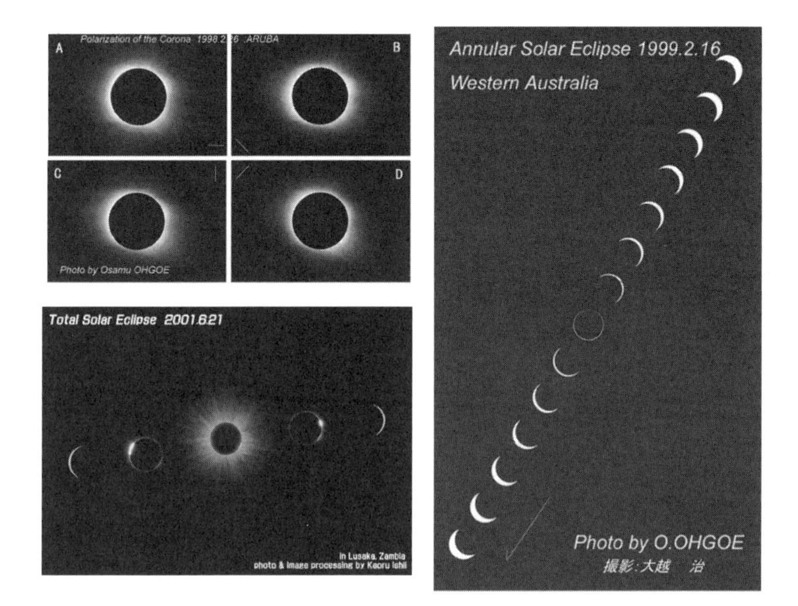

Figure 3. Coronal Images taken by Amateurs: 4 images of upper left: polarization of the corona in 1998 at Aruba, right: 1999 Annular Eclipse at Western Australia, down left: 2001 Zambia

Pascal, B. 1640, in Pensées

Shiota, K. 1994, Sky & Telescope, 88 (5), 19

Takahashi, N., Hiei, E. 1998, "Polar plumes of the 1994, 1995 and 1997 Total Solar Eclipse", in Solar Jets and Coronal Plumes, ESA SP-421, 337

Takahashi, N., Yoneshima, W., Hiei, E., and Ichimoto, K. 2000, "Large-Scale Distribution of Coronal Temperature Observed at the Total Solar Eclipse on 26 February 1998", in the Last Total Solar Eclipse of the Millennium in Turkey, ASP Conf. Ser. 205, 121

IAU 8th Asia-Pacific Regional Meeting
ASP Conference Series, Vol. 289, 2003
S. Ikeuchi, J. Hearnshaw & T. Hanawa, eds.

The Establishment of an Astrophysics Course in the Philippines through the IAU TAD

Cynthia P. Celebre

Chief, Astronomy Research and Development Section PAGASA, Quezon City, Philippines

Abstract. The Japanese Government through its Cultural Grant-aid Program, donated a 45-cm telescope to the Government of the Philippines. It was installed at the PAGASA Astronomical Observatory in May 2000. Its installation had made the officials of PAGASA realize the need to establish an undergraduate astrophysics course in the country. The course will be more economical and practical, compared to training courses and fellowships requested from abroad. It was planned to be established in cooperation with the IAU-TAD and the National Institute of Physics of the University of the Philippines. The activity is discussed in detail in this paper.

1. Introduction

Astronomy has been practiced in the Philippines for more than 100 years now. However, activities in astronomy education and research in the country are limited in scope. Astronomy is taught as a part of the general science subject in elementary schools where it is normally given a three-hour per week period in Grades V and VI classes. It is an elective subject, which is taken in one semester (four months) in the first year at high school level. At college level, there is only one university that offers a subject on astrophysics to students who are enrolled in a course in physics. Hence, at the moment, there is no single university in the country that offers a full course in astronomy (Soriano, et al. 1996). The Philippine Atmospheric, Geophysical and Astronomical Services Administration (PAGASA) is the only government agency in the country that performs astronomical functions. Research activities in the agency are primarily devoted to data collection and publication of astronomical phenomena. It is important to mention that personnel, who do not have any formal education in astronomy, perform the astronomical activities in the agency. The knowledge of science that they possess is obtained through in-service training courses, which are infrequently conducted by the agency due to lack of qualified lecturers. They also read books and publications that are procured, usually from overseas sources. Very few of these personnel gain their knowledge in astronomy by attending workshops, seminars and training courses from abroad, since it is very difficult to gain financial support for such activities locally.

In view of the foregoing, the Atmospheric, Geophysical and Space Sciences Branch (AGSSB), through its Astronomy Research and Development Section

(AsRDS) of the agency, planned the establishment of an undergraduate astrophysics course in the Philippines through the IAU-TAD. This paper will describe the activity.

2. Background Information

In May 2001, a computer-based, 45-cm telescope was installed at the PAGASA Astronomical Observatory, located inside the campus of the University of the Philippines in Diliman, Quezon City. The Japanese Government, through its Cultural Grant-aid Program, donated the equipment. The installation of the highly technical equipment signified a new beginning of astronomy in the country, particularly in the field of research, and greatly supported the new mission of PAGASA to revitalize astronomy in the Philippines.

The arrival of the new telescope has made the officials of PAGASA recognize that its scientific use could only be maximized through a formal course in the field. It was further perceived that such an objective could be accomplished by establishing an astrophysics course in the country inasmuch as it will be more economical and practical, compared to training courses and fellowships requested from abroad. Consequently, the activity was planned to be pursued through the Teaching for Astronomical Development (TAD) of Commission 46 of the International Astronomical Union (IAU).

The activity started when the Chief of AsRDS attended the 24th IAU General Assembly, held in Manchester in August 2000. While undertaking a training in Japan from 2001 March 29 – November 14, she exchanged communication with the Chairman of TAD. The IAU agreed that its representative will visit the Philippines to investigate whether there is really a need for an IAU-TAD in the country. Thus, on 2001 November 19–27, an IAU representative visited the PAGASA, met with AsRDS personnel and discussed the activity with the Director of the National Institute of Physics of the University of the Philippines.

After the visit, a recommendation that an IAU-TAD program could be started in the Philippines was submitted to the Chairman of TAD Program, who prepared a draft of an agreement between PAGASA, UP and IAU. The draft was forwarded to the General Secretary of the IAU, who was willing to support the initiative. At present, negotiations are underway in connection with the stipulations of the Memorandum of Understanding (MOU) prepared in turn by the University of the Philippines.

3. The National Institute of Physics and the New Astronomy Course

The National Institute of Physics (NIP) of UP was established in 1983 with the signing of Executive Order No. 889 by then Pres. Ferdinand E. Marcos. It aims to become the national center of excellence in the education, training, advanced research and development in the area of physics and technology. In 1997, the Philippine Commission for Higher Education accredited the NIP as a Center of Excellence in physics, in recognition of its status as the premiere institute for tertiary physics education in the Philippines.

The Institute offers degree programs in B.S. Physics, B.S. Applied Physics, M.S. and Ph.D. in Physics. In cooperation with other UP colleges and institutes, it also offers M.S. and Ph.D. degrees in Environmental Science, Materials Science and Engineering.

In addition to the high level of instruction in the Institute, the PAGASA officials chose it to become the site where the astrophysics course will be established because the agency's Astronomical Observatory is situated inside the UP campus. It should be noted that the Observatory's 45-cm telescope would be the central equipment that would be used for laboratory exercises of the course.

For the first time since its establishment, the NIP is offering an astronomy course entitled "Physics and Astronomy for Pedestrians" (Physics 10), starting in the school year 2002–2003. The course will serve as an introduction to the different aspects of physics and astronomy, from its emergence up to its current developments. It will be a "walk- through" course for people who enjoy physics and astronomy but want to be spared of the tedious details.

The course aims to introduce concepts from various sub-disciplines of physics and astronomy to students and develop an appreciation of the position of mankind in the universe. It also intends to update the students with the latest developments in physics and astronomy, both local and abroad. It also plans to refine the student's understanding of the role of physics and its sub-disciplines in technological innovations and in the advancement of other fields in the natural and social sciences. Lastly, Physics 10 attempts to enable the students to understand the character and functions of science and technology and develop an appreciation of the key role of science and technology in national development.

The syllabus of the course is divided into four (4) parts, namely: Classical Physics, Post-Classical Physics, Astronomy and Cosmology and Physics and Technological Development. The subjects under each part are as follows:

1. **Classical Physics:** Introduction to the Natural Sciences and Emergence of Physics; Newton's Mechanical Synthesis; The Unification of Electricity, Magnetism, and Light; Thermodynamics

2. **Post-Classical Physics:** The Quantum World of Uncertainties; Probing the Subatomic World

3. **Astronomy and Cosmology:** (a) Einstein's Relativistic Revolution; (b) Connecting Quarks with the Cosmos; (c) Questions and Opportunities

4. **Physics and Technological Development:** (a) Science and Measurement; (b) Physics in a New Era – Macroscopic and Nanoscopic worlds.

The syllabus will be revised later using a developed conceptual approach to space science education by Melek, et al. (2002). The concepts, which are built within the framework of the suggested educational scheme will be used to teach different theories and applications.

4. The Draft Agreement

The Draft Agreement that was prepared by TAD, stipulates that the activity will be a cooperation between the IAU, PAGASA and UP - Diliman. The collaboration is aimed at supporting the long-term development of astronomy and astrophysics in the country. It will be established initially for the period 2002 July - 2006 June, renewable by mutual agreement.

Activities under this program will be planned on an annual basis between the collaborating entities. Activities undertaken, within the approved IAU budget for the TAD program, may include but are not limited to the following:

The IAU will sponsor the international travel for visiting lecturers to establish new courses that are to become part of the regular curriculum at the University and/or raise the astronomical background of the staff at PAGASA. PAGASA and/or the University will provide local support (living costs and participating staff/faculty) to make such visits effective.

The IAU will support international travel as needed to help assure that astronomy can provide an educational science experience to the students, including opportunities for students to use the astronomical observing facilities.

The IAU will provide books and equipment for activities established under the Agreement. These will be under the care of the PAGASA Astronomical Observatory and will be made available on loan without charge for scientific and educational use by faculty and students of the University, PAGASA staff, and qualified teachers.

The University and PAGASA will provide support, including faculty support, with the aim of these activities to become self-sustaining over the long term.

It is estimated that the planned activities will lead to new projects of education or research, which may call for a strengthening of extant academic activities. The University will offer its support, within the framework of its resources, for such long-term developments arising from the present program.

Conclusion

The basic sciences evolved from the study of astronomy. As such, in a developing country such as the Philippines, where space science is considered to be just in its incipient stage, there is a need to emphasize the study of astronomy. The stress should be directed towards educating the youth because it will afford a better understanding of our planet, and thereby encourage them to participate more actively in the conservation and preservation of the environment. More importantly, the youth are the future astronomers of the world. At present, the necessity of addressing the importance of teaching astronomy could be stressed, in the light of the present revision of the elementary school curriculum by the Department of Education where the basic subjects (Mathematics, Science, Filipino, and English) were given focus.

The establishment of an undergraduate course in astrophysics in the Philippines will undoubtedly enhance the capacity for scientific use of the PAGASA telescope donated by the Japanese Government. In addition, the activity will aspire to support the mission of AGSSB/PAGASA to revitalize astronomy in the Philippines by enabling the agency to become self-reliant in the field of astronomy education when graduates of the astrophysics course will be qualified to teach the course in due time. Lastly, the scheme will push the country to establish a name in the international community of astronomers, which will further allow the Philippines to be known as a new emerging developing country in the field of astronomy and basic space science in south-east Asia and the Pacific (Celebre, et al. 2000).

References

Celebre, C. P. and B. M. Soriano, Jr. 2000, "Revitalizing Astronomy in the Philippines" published in the Astronomy for Developing Countries, a Proceedings of a Special Session of the International Astronomical Union, Manchester, United Kingdom, 14-16 August 2000

Soriano, Jr., B. M. and Celebre, C. P. 1996, "Astronomy in the Philippines" to be published in the UN guidebook on "Developing Astronomy and Space Science Worldwide" and presented in the UN/ESA Workshop on Basic Space Science (Bonn, Germany, 9 – 13 September 1996)

Melek, M. and Celebre, C. P. 2002, "Major Dynamical, Physical and Technological Concepts of Space Science: An Educational Approach via Concepts" published in Teaching of Astronomy in Asian Pacific Region, Bulletin No. 18, Tokyo, Japan

IAU 8th Asia-Pacific Regional Meeting
ASP Conference Series, Vol. 289, 2003
S. Ikeuchi, J. Hearnshaw & T. Hanawa, eds.

New Trends in Astronomy Education: A "Mapping" Strategy in Teaching and Learning Astronomy

Sergei Gulyaev

Auckland University of Technology, Private Bag 92006, Auckland 1020, New Zealand

Abstract. The application of a concept of educational "science maps" to astronomy education is discussed. By analogy with geographical maps, scales of educational science maps – scales of integration – are introduced. In astronomy education, scale A represents the level of branches and fields of astronomy and astrophysics, where interconnections between various astronomical disciplines are shown. Scale B represents the level of hypotheses and theories, encompassing a significant segment of a field of astronomy. Scale C represents the level of structures and internal hierarchies, encompassing the "geography" and "anatomy" of the material systems and objects essential for a given astronomical discipline, as well as the principal notions and concepts it uses. Science maps of different scales are illustrated with initial examples exploring the application of this methodology in astronomy and astrophysics.

1. Introduction

If you were to overhear several senior students talking, you might hear one of them say, *our professors have stuffed us full, but what does it all mean?* Many students majoring in astronomy or astrophysics are unable to perceive where all the information they have been given about a particular area of astronomy actually fits into astronomy as a whole. Furthermore, even graduates of astronomy are often unable to conceptualize and understand their field of astronomy and astronomical endeavors. Our students' vision of astronomy is similar to the situation with a metro (underground) passenger. You get out to the surface at one station and see a beautiful square with monuments and buildings around; as you come up to the next station you can see a park and a river. But what is between? What streets, roads or pathways connect these areas? In tertiary astronomy education, each paper or module is like one metro station. Students' knowledge is fragmentary; they lack long-term understanding of astronomy content, much less the ability to apply it. Our students need a map of this city – a map of astronomy.

2. Concept Mapping

Joseph Novak, the founder of concept mapping, writes (1993):

> Knowledge is made of concepts and concept relationships, much like words are made of letters and matter is made of atoms... The principal difference between a so-called genius and an average learner is that the genius has a capacity to use higher-order concepts in obtaining meaning from larger chunks of information... Our task as educators is to empower learners by helping them acquire the ability to organize and generalize knowledge structures.

Systems theory offers a number of practical approaches for conceptualizing knowledge, such as rich picture diagrams, flow charts, influence diagrams, precedence charts, etc. The concept map, developed by Novak and used widely in primary and secondary science education, is one useful learning tool for encouraging systems thinking and "deep" learning approaches with students.

There are examples of successful use of concept maps in tertiary science education, especially as a brainstorming tool in group work. However, an intended "liberty" of concept maps (amorphousness and arbitrariness of structure, mixture of notions, objects and theories in one map), being of value in a brainstorming situation, becomes their drawback when used as a teaching/learning or curriculum design tool.

3. General Systems Theory

General systems theory (GST), as a theoretical tool for representation of the levels and hierarchies associated with science based knowledge, provides needed order, rigour and structure to concept maps (Gulyaev & Stonyer, 2002). GST is designed for specifying systems and defining their interrelationships, and also for pointing to gaps in knowledge. What is really important for us, as tertiary science and astronomy educators, is that GST provides a framework for conceptualizing the subject. It also provides a framework for teaching astronomy based on the continuity of conceptual understanding, rather than the logical yet reductionist structure of astronomy apparent in many texts or curricula. Therefore, GST provides a basis for an integrated approach to astronomy education at the tertiary level.

GST introduces a hierarchy of levels of complexity for basic branches of science. Developing this approach, we use the analogy with geographical maps, and distinguish the three principal "scales" that are necessary for maps of science. In the astronomy context they are:

A. The scale of *branches and fields* of science and astronomy, where interconnections between various disciplines are shown.

B. The scale of *theories and hypotheses*, encompassing a significant segment of astronomy and astrophysics.

C. The scale of *structures and hierarchies*, encompassing the geography and anatomy of the material systems and objects essential for a given discipline, as well as principal methods, notions and concepts it uses.

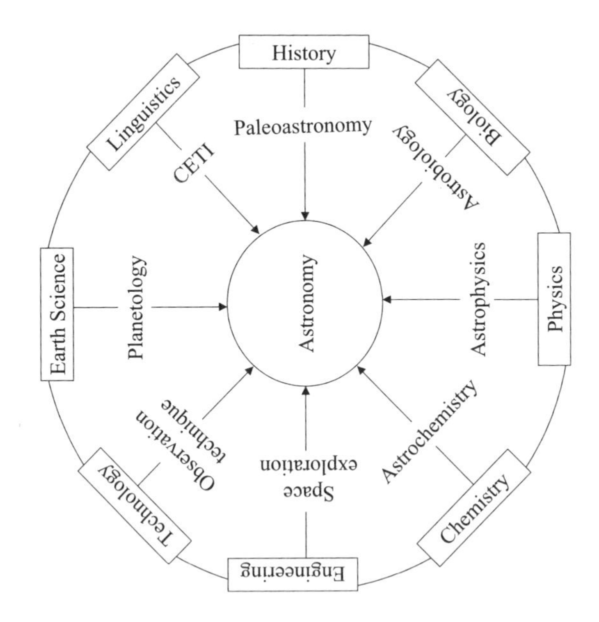

Figure 1. Possible interconnections of astronomy with other fields of science. Corresponding branches of astronomy arising from these interconnections are shown as linking words.

4. Practical Examples of Scales of Astronomy Maps

Scale A: Branches and fields of science and astronomy

Astronomy can be considered both as an independent science with its own tasks, concepts and methods, and as a derivative science that utilizes information and concepts from physics, chemistry, earth science and others. Possible interconnections of astronomy with other fields of science are shown in Figure 1.

Tertiary students who want to specialize in astrophysics should be aware of the route they should follow to become astrophysicists. It is physics and mathematics of course, but still there are astronomical basics of modern astrophysics. One cannot be an astrophysicist without them (see Figure 2).

Astrophysics has many branches and sub-divisions. They can be classified according to the methods of investigation (IR astronomy, neutrino astronomy, etc.) and to the object(s) of study (solar physics, cosmology, stellar astrophysics, etc.).

Scale B: Theories and hypotheses

Each sub-division in the classification of astrophysics presented in Figure 2, in turn, can be sub-divided and classified. Stellar astrophysics, for example, consists of (Figure 3):

(a) Stellar astrophysics as such, which serves to the tasks of establishing of the cosmic distance scale, determination of ages of stars and stellar clus-

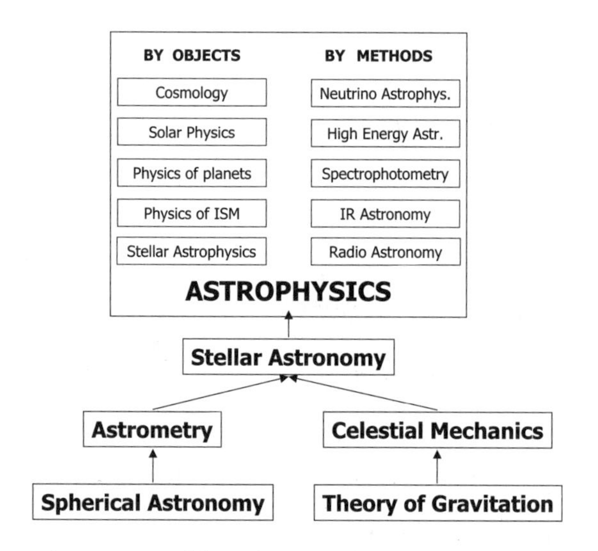

Figure 2. Astronomical foundations of modern astrophysics.

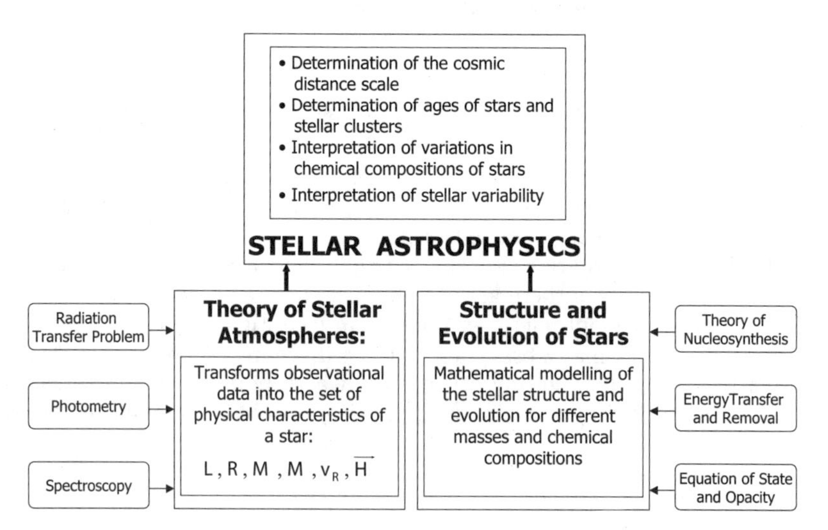

Figure 3. Stellar astrophysics: its tasks, origin and constituents.

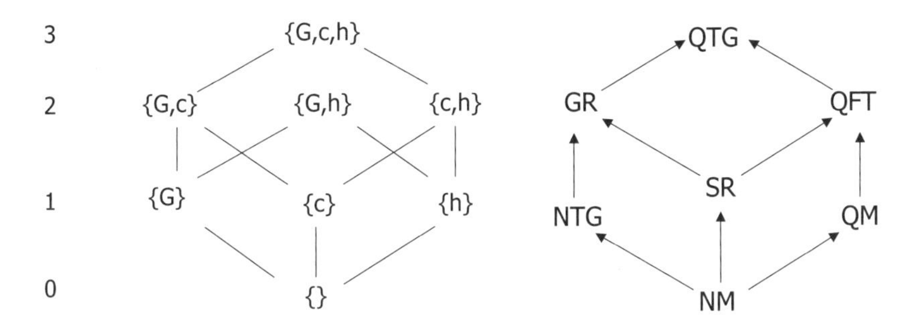

Figure 4. (Left) Eight different subsets that can be created out of the set of three fundamental constants G, c, and h. (Right) Physical theories – physical basis of modern astrophysics presented in the form of the "Cube of theories" (see explanation in the text).

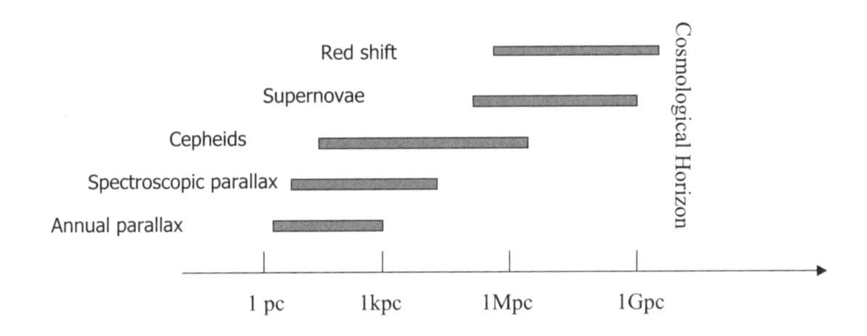

Figure 5. The distance scale ladder (after Rowan-Robinson 1985)

ters, interpretation of variations in chemical compositions of stars, and interpretation of stellar variability;

(b) Theory of stellar atmospheres that transforms observational data into the set of physical characteristics of a star: L, R, M, $\dot{M}$, v_R, $\vec{H}$;

(c) Theory of structure and evolution of stars, which provides stellar astrophysics with mathematical modelling of the stellar structure and evolution for different masses and chemical compositions of stars.

The map in Figure 3 is a map of the next scale – the scale of *theories and hypotheses.* Two basic theories (theory of stellar atmospheres and theory of structure and evolution of stars) are based on a number of physical and astronomical theories and methods such as theory of nucleosynthesis, methods of photometry and spectroscopy, problems of radiation and energy transfer and removal, equation of state, problem of opacity, and others.

Theories that provide *physical* basis for modern astronomy and astrophysics, can be presented in the form of a system shown in Figure 4. Number of fundamental constants incorporated in a given theory is shown in the left column. One can show that eight different subsets can be created out of the set of three constants G, c, and h (Figure 4 left). Algebraically, the empty set $\{\}$, or $\emptyset$, corresponds to Newtonian mechanics (NM), the subset $\{G\}$ is related to Newtonian theory of gravitation (NTG), the subset $\{c\}$ to special relativity (SR), subset $\{h\}$ to quantum mechanics (QM), subset $\{G, c\}$ is associated with general relativity (GR), $\{c,h\}$ with quantum field theory (QFT), and the full set $\{G,c,h\}$ should correspond to quantum theory of gravitation (QTG). A cube of theories (Figure 4 right) is an obvious 3D visualization of these subsets.

Science (astronomy) maps may be very fruitful by themselves. Looking at the cube of theories, one can identify a possible "gap" in knowledge. Within the classroom, the question "What kind of theory should be in the hidden corner of the cube?" can be used as a discussion starter. Students may name this non-existing theory by themselves, for example, as "non-relativistic quantum theory of gravitation".

Scale C: Structures and hierarchies

The distance scale ladder is an example of the next scale of science maps (Figure 5). It is a custom presentation of the hierarchy of methods used for determination of distances of celestial objects.

The structure scale ladder (Gulyaev & Stonyer 2002) represents the hierarchy of material structures in the Universe. It is a system, firstly, because each element (step) of the ladder is a building block for a higher structural element. Another feature of this hierarchical system, which makes it a system, is the mutual connection of the highest elements with the lowest ones: properties of elementary particles determine the structure, properties and evolution of the entire material world (the Universe). A philosophical idea behind this link is the anthropic principle.

5. Conclusion

There is much discussion nowadays about the poor preparation of students and the alarming level of scientific illiteracy within all levels of society. Much of the blame lies with the fragmentary character of education. I believe that the possible way of transferring from fragmentary (reductionist) to the integrated (systems) education passes though the mapping strategy discussed in this report. There is a significant amount of work to be done by experts in specific fields of astronomy and astrophysics, to conceptualize and define how maps of their fields might be drawn, so that science and astronomy educators might utilize them more fully in developing their practice.

References

Gulyaev, S. & Stonyer, H. 2002, IJSE, 24, 753

Novak, J. 1993, The Science Teacher, 60, 50

Rowan-Robinson, M. 1985. The cosmological distance ladder. W. H. Freeman and company

IAU 8th Asia-Pacific Regional Meeting
ASP Conference Series, Vol. 289, 2003
S. Ikeuchi, J. Hearnshaw & T. Hanawa, eds.

Astronomical Education in Azerbaijan: Existing Heritage, Current Status and Perspectives

Elchin S. Babaev

Shamakhy Astrophysical Observatory named after N. Tusi, Azerbaijan National Academy of Sciences, 10, Istiglaliyyat Street, Baku, AZE-370001, Azerbaijan Republic

Abstract. The current situation of astronomical research and education in Azerbaijan is briefly described. Attention is given to the educational process in astronomy and space sciences carried out in higher-secondary schools, lyceums, universities, and academic institutions through lessons, lectures, seminars, conferences, etc. The author's experiences in the popularization of space and astronomical knowledge in the Azerbaijan Republic are discussed. Problems and perspectives, taking into account tendencies and realities, are described.

1. Introduction: Formation of National Astronomy in Azerbaijan

In the Middle Ages, astronomy was in stagnation in Europe, but going through a bloom in the Asian countries. Asian scientists have burned a fire in the candle of ancient science and kept up it for years and even centuries, and later transferred it to Europe in the Renaissance epoch. Astronomers of the Renaissance epoch were educated by works of Asian scientists. The quickest rise of astronomy was mentioned in the Middle East, Middle Asia, the Caucasus, North Africa and Mauritanian Spain.

Azerbaijan, as one of the countries located in the South Caucasus, on the west coast of the Caspian Sea between the East and the West (which is usually called as "the Gate between the East and the West"), has properly contributed in the development of science, particularly so in astronomy. Signs of ancient astronomical knowledge in Azerbaijan are preserved in different historical places, among them thousands of drawings on rocks at Gobustan National Park (VIII-VII B.C., Absheron Peninsula, under UNESCO protection) and Gamigaya National Park (III-II B.C., Ordubad district). One of the interesting and unique historical monuments in the heart of the capital city of Baku, the Maiden Tower ("Azeri Girl's Tower", XII century), is believed to have been a simple observatory. Azerbaijan has got a real astronomical heritage, thanks to the presence of the famous Maragha Observatory in southern Azerbaijan, which was established by the Middle Ages' Azerbaijani astronomer Mohammed Nasiraddin Tusi (XIII century). This astronomical school, which was a prototype of the modern academy of sciences, has influenced the development of astronomy, not only in Eastern countries (Ulugh Beg's Observatory in Samarkand, Beijing Observatory, etc.) as well as in some European countries (e.g. the Danish observatory of Tycho Brahe), and it is indeed difficult to present the significance of Tusi's astronomical school in a few words (Guluyev & Babayev 2002a). Currently there are 11 names in the sky connected with Azerbaijan, and two of them, a minor planet and a crater on the Moon, are named after Tusi.

A remarkable new stage in development of national astronomy in Azerbaijan started in the XX century. The development of national astronomy in the last century could be described through three stages. The first stage (1927–1991) covers the period with activities such as the first astronomical expeditions, establishing of the Shamakhy Astrophysical Observatory (http://astro.aznet.org, www.ab.az) on the eastern-southern slopes of the Great Caucasus Mountain Range, not far from Baku (the official foundation date was 1960 January 13), equipping the observatory with telescopes of German and Russian origin, building up the living settlement with relevant infrastructure for scientists and personnel, preparation of national astronomers in well-known astronomical organizations of Moscow, Leningrad, Kiev, Odessa, etc. Today there is a unique observatory with 11 optical telescopes, including the big "Carl-Zeiss-Jena" 2-meter telescope, horizontal solar telescope, and several astrographs. Up to the end of 1991 Azerbaijan was a part of the former Soviet Union, and as a result, Azerbaijani astronomers could not participate independently in any kind of international cooperation and the Republic could not have enough high-skilled specialists in astronomy and space research.

The second stage (1992–1997) is characterized as a "stagnation period" in the history of national astronomy (because of the collapse of the former Soviet Union and loss of any scientific relations, and economic and political instabilities in the newly independent country in its transition period, and also the Armenian intervention, local conflicts, etc.). A new stage began at the second half of 1997 with repairing, renovation, and reorganization works in the observatory, and in astronomical activity generally (Guluyev & Babayev 2002b).

Astronomical researches in Azerbaijan are conducted mainly in the Shamakhy Astrophysical Observatory (hereinafter ShAO) named after Nasiraddin Tusi and partially in relevant departments of several universities in Baku (e.g. Department of Astrophysics, Baku State University), and in other organizations: Azerbaijan Technical University, Azerbaijan State Pedagogical University and Azerbaijan National Aero-Space Agency (ANASA), some research institutes of the Azerbaijan National Academy of Sciences (e.g. the Institute of Physics). They have made a significant contribution, mainly in extensive ex-Soviet and post-Soviet scientific programs in the fields of astrophysics, fundamental astronomy and space sciences.

Theoretical and experimental investigations in the fields of astronomy, astrophysics and practical astronomy were carried out in ShAO for more than 40 years. There are three main scientific trends at the observatory: the physics of stars and nebulae, investigation of solar system bodies, and solar physics. In the observatory and other institutions there are also carried out to some extent investigations in the following fields: solar-terrestrial relations, space weather effects, the history of astronomy, theoretical astrophysics, cosmology, helioseismology, trans-ionospheric radio wave propagation, radioastronomy, practical astronomy, celestial mechanics, galaxies, interplanetary magnetic field, dynamics of artificial satellites, etc.

2. Current Status of Education and Popularization of Astronomy in Azerbaijan

Azerbaijan has almost all of the attributes required for astronomy. The main contribution comes from the above-mentioned national Astrophysical Observatory, which has headquarters (altitude 1500 m, $48^\circ 35' 04''$ E, $40^\circ 46' 20''$ N) and two high-mountain astronomical stations with favorable geographical locations (altitudes 2100–2200 m, in Nakhchivan AR) and a good astro-climate (averagely 200–250 clear days per year, 40% of which are with a best quality of image).

The second significant fact is the mandatory teaching of astronomy as a separate subject in all higher-secondary schools, lyceums, gymnasiums, as well as the teaching of astronomy and the fundamentals of space science in many university departments. This gives a special status to astronomical education in the State Education Curricula and allows the familiarization of people with astronomical knowledge and involves them in deep astronomical study with later research in this field of science. The above-mentioned research organizations are engaged in astronomical education of and also in space science. The Department of Astrophysics in Baku State University is the main supplier of young astronomers through bachelor's and master's educational levels. Several dedicated Ph.D. study courses, organized permanently in the observatory, in the Azerbaijan National Academy of Sciences, in some universities, as well as Dissertation Councils are the essential institutions for preparing highly skilled national astronomers and in conferring on them the Ph.D. and doctoral degrees.

The main astronomical scientific journals are "the Circular of ShAO" (founded in 1970, 105 issues) and "the Azerbaijani Astronomical Journal" (founded in 2002, which will be published beginning in 2003). Besides, Azerbaijani astronomers publish their papers in Russian, Ukrainian, Turkish, central and western European journals. The Baku City Joint Astronomical Seminar, established in 2001 by joining efforts of seven big research and educational organizations such as ShAO, ANASA, Baku State University, the Research Institute of Physics, etc., plays a great role in astronomical research and educational activities. Serious results of scientific investigations, dissertations, urgent and/or interesting news on astronomy and space science are presented in this Seminar regularly (at least once or twice a month). The role of the permanent "TUSI Majlis" (School of Astronomy named after N. Tusi), which was established in the observatory in 1998, is priceless, especially in the post-stagnation period since 1997. "TUSI Majlis" is very active in organizing conferences and, eight conferences (5 domestic, 2 regional and 1 international) on astronomy and astrophysics have been held from 1998 to 2001. These gave an additional impulse to national astronomy and were recognized as meaningful scientific events. Generally, during the past forty years, more than 50 scientific astronomical events (domestic and international conferences, symposia, workshops) were held in Azerbaijan.

In addition to classical educational activities in schools and universities, the specialized community in Azerbaijan invests a lot of effort in general public outreach (http://europe.unsgac.org/az/). There are regularly broadcasts in the independent (ANS, Space, Lider) and national TV (AzTV-1, programs "Evrika", "Science", Tusi Club of Pupils, etc. with partial involvement in astronomical educational themes) and radio stations (e.g. FM radios ANS-CM and Lider) about astronomical and space activities, daily and/or weekly, dedicated columns in newspapers (e.g. "Elm – Science") and journals (e.g. "Elm ve Heyat – Science and Life"). As an example, the daily space weather forecasts, together with information about interesting astronomical events, are published in the famous newspaper "Zerkalo – Mirror". In order to bring astronomical and space sciences closer to the general public, mobile tutorial telescopes are installed on central squares and streets in Baku, to interest and teach in an active manner a public that normally has limited contact with this science. The observatories are also directly engaged in general public events, such as daily guided and special tours of groups and individuals to telescopes (e.g. during solar or lunar eclipses), and especially school classes, mainly during weekends and summer holidays. Visits to the observatory are usually included in the schedule of official guests and diplomats. In order to stimulate the interest of students and pupils of secondary schools, an Astronomical Tusi Olympiad has been organized, seconded by Pupils' Conferences on the natural sciences and astronomy in the "Zangi" Lyceum. In this Lyceum, astronomers use the Internet, and CD and video facilities in astronomy and space, and hence build up a tutorial on telescopes which enables students to carry out a pilot program of deep study in astronomy. Recognizing the importance of the involvement of the media, special tutorial courses have been organized for selected journalists, in order to avoid mistakes and pseudo-scientific articles on astronomy and space.

For the long-term investment in astronomical education, dedicated books on astronomy and space have been published. Classical books on astronomy from the Soviet period are still in use, but after the gaining of independence, there are some efforts to create national programs and books on astronomical education. R. E. Huseynov has contributed in this program by publishing a classical book "Astronomy" in Azeri for university students. He is preparing an educational astronomical book for higher-secondary schools as well. Another book "Azerbaijani Astronomy in the XX Century", written by A. S. Guluyev (2002), and edited by E. S. Babayev, is one of best works on the history of astronomy, summarizing and analyzing the past and current status of astronomy in Azerbaijan. Publication of some scientific-popular books on astronomy sometimes is conducted with the help of oil companies (e.g. BP-Exxon). Exchange and/or subscription of astronomical scientific literature in different languages and Internet facilities allow the filling of existing gaps in astronomical information.

Mainly professionals, especially astronomers from the observatory, are carrying out the teaching of astronomy. At present, there is a community of astronomers in Azerbaijan, including 60 highly-skilled scientists, 1 academician (full member) and 2 corresponding members in astronomy in the Azerbaijan National Academy of Sciences, 7 doctors and more than 30 people with PhDs in astronomy, who mainly graduated at famous Russian and Ukrainian astronomical schools (in total 12 Doctoral and 83 Ph.D. Theses during 40 years). The number of young astronomers is rising during the last 4–5 years, which is a good sign.

Individual and collective membership of Azerbaijani astronomers in the International Astronomical Union, European Astronomical Society, Euro-Asian Astronomical Society, Joint Organization for Solar Observations, International Council for Scientific Development, International Academy of Science, etc. brings a huge benefit for collaboration in astronomical researches and education. The National Astronomical Society of Azerbaijan will be established very soon.

Young Azerbaijani astronomers are actively involved in the works of some international organizations aiming at enhancing the participation of youth in astronomical and space activities and education: the Space Generation Advisory Council (SGAC) in support of the UN Program on Space Applications, International Forum of Young Scientists, etc. Azerbaijan is the initiator and one of the founders of the internationally recognized organization – Space Association of Turkic States (cf. UN GA Document 2002). On 2002 April 12, the young space generation of Azerbaijan celebrated the World Space Party in Baku, accompanied by a mini-Regional Astro-Space Workshop. Space and astronomy oriented souvenirs and tutorial material provided by NASA, CNES, ESA (books, booklets, etc.), are usually distributed during this kind of event to participants, as well as to higher-secondary schools. It is planned to prepare the UNESCO Space Generation Forum of Turkic States in Azerbaijan to be held in 2003 April.

3. Existing Problems, Perspectives and Ideas

Financing of fundamental science and education is one of main directions of reform and one of the existing problems in Azerbaijan. State financing is the main source, but non-budgetary financing starts to grow. Today Azerbaijan spends a noticeable part of its state budget – up to 23% – for education, but it is not enough: the budget is insignificant, and especially as there are a lot of higher educational institutions. The USA spends on science a huge amount of money – more than 7% of its annual budget, which by 100 times surpasses the state budget of Azerbaijan. In the USA, more than $US200 000 is spent annually for each scientist, and in Japan, up to $US100 000 to $120 000. For comparison, the annual expenditure per scientist in Azerbaijan are less than $US1 000. The boom in the oil industry in Azerbaijan, unfortunately, still bypasses science and education. The expenditure on science makes up just 1.1% of the annual budget. For comparison, the salary of the scientists in Turkey and Iran is 15–20 times higher than the salary of Azerbaijani scientists, while the prices in these neighboring countries are lower than in Azerbaijan.

There is a big list of problems impeding the normal development of astronomy in Azerbaijan which, in turn, affects the level of astronomical education: old astronomical tools and equipment, the need for modern CCDs, PCs, and Internet connections, and tutorial material. In addition one can cite the insufficient number of young specialists, problems in the delivery and subscription of relevant scientific literature, the payment of membership fees in international astronomical organizations, the protection of observatories (light and dust pollution, tourist buildings within a 5-km radius restricted zone around the observatory), electrical power interruptions, local conflicts (two astronomical stations are located near the border with Armenia and are in danger), as well as such kind of specific problems, as astrology and pseudo-scientific popularization of astronomy, the lack of special foundations and/or domestic grants for science, insufficient international and regional contacts and collaborations, etc. (Guluyev 2002).

In the near future, it is planned to publish special scientific-popular journal "Kainat, Yer, Insan – the Universe, the Earth, the Human", to build up several planetariums in Baku and possibly in the other three big regional centers of Azerbaijan, to create a TV-planetarium broadcast, to install educational telescopes in selected secondary high schools, gymnasiums, and an educational observatory in a city, to create a national astronomical outreach program and new departments in universities dealing with astronomy and space sciences, so on.

Besides, there are several ideas to be implemented step-by-step on the basis of international collaboration, and not only in Azerbaijan: establishing of the International Astronomical University (the analogue of the International Space University in Strasbourg), establishing the International and/or Regional Center for "Comet-Asteroid Hazard Problem" and "Tracking of Satellites" on the base of the two high-mountain astronomical stations of ShAO, organizing an Annual Autumn School for Young Astronomers in Azerbaijan, a joint regional education project with the Russian Embassy on preparing/training of 50 young pupils (15–16 years old), in order to send them to prestigious universities in Moscow (the Bauman University, Moscow Physical Technical Institute, etc.), the Silk Road Astro-Space Education Project (from Japan to Europe via Asia and the Caucasus), creating a special web site "Astro-education" under the Euro-Asian Astronomical Society.

Azerbaijani astronomers do hope that these kinds of activities will affect the choice of specialization and the getting of a worldwide outlook of the young generation, who currently prefer juridical, economical and medical specialties more than astronomical ones, under the influence of economical factors and the realities of the modern world.

The IAU, ASJ, NAOJ and SOC/LOC of the IAU 8-th APRM are gratefully acknowledged for financial support. Special thanks go to Prof. N. Kaifu, Dr T. Hasegawa, Dr S. Ikeuchi, Dr M. Doi, Dr K. Akahori, Dr T. Hanawa for their kind attention, help and hospitality. We are grateful to the Japanese Embassy in Azerbaijan for assistance.

References

Guluyev, A. S., & Babayev, E. S. 2002a, in Tusi-800 Majlis Int. Conf. Proc., Nasiraddin Tusi and Modern Astronomy, ed. E.S. Babayev & A.S. Guluyev (Baku: Poliqraf Servis), 5

Guluyev, A. S., & Babayev, E. S. 2002b, in Euro-Asian Astr. Soc. Int. Conf. Abs., International Collaboration in the field of Astronomy: Status and Perspectives (Moscow: Astro), 19

Guluyev, A. S. 2002, Azerbaijani Astronomy in the XX Century, ed. E. S. Babayev (Baku: Elm) UN General Assembly Document A/AC. 105/774, 4 January 2002, 4

Star Formation and ISM

IAU 8th Asia-Pacific Regional Meeting
ASP Conference Series, Vol. 289, 2003
S. Ikeuchi, J. Hearnshaw & T. Hanawa, eds.

The Completion and Release of the AAO/UKST Hα Survey

Q. A. Parker

Department of Physics, Macquarie University, NSW 2109, Australia & Anglo-Australian Observatory, NSW, 1710, Australia

S. Phillipps

Physics Department, University of Bristol, Tyndall avenue, Bristol, UK

and the Hα survey consortium and associates: I. Bond, N. Hambly, H. T. MacGillivray, D. H. Morgan, M. Read, S. B. Tritton, (*IfA*), M. Masheder, R. Morris, M. Pierce, R. Walker (*Bristol*), J. Drew, M. Pozzo (*IC, London*), J. Bland-hawthorn, R. D. Cannon, M. Hartley, D. F. Malin, (*AAO*), M. Cohen (*UC, Berkeley*), A. Green (*Sydney*), S. Mader, V. McIntyre (*ATNF*), A. Walker, W. J. Zealey (*Wollongong*), D. Russeil (*Marseille*), M. Fillipovic (*UWS, Australia*)

Abstract. The AAO UK Schmidt Telescope (UKST) has just completed an Hα survey of the Southern Galactic Plane and Magellanic Clouds. The resultant map represents the last great photographic UKST survey product. A single-element interference filter of exceptional quality was used, the largest of its kind for astronomy. With fine-grained Tech-Pan film as the detector, an atlas with an unequalled combination of resolution, sensitivity and areal coverage has been created, superior to any equivalent optical line emission survey in our Galaxy. The Wide Field Astronomy Unit of the Institute for Astronomy Edinburgh hosts the survey data archive and is responsible for disseminating the data products to the community solely in digital form. As of July 2002, over 150 of the 233 survey fields are accessible on-line and the entire survey should be available early in 2003. A variety of scientific programmes for exploiting the survey are already underway and many more are anticipated now the digital survey data products are available.

1. Introduction

The AAO/UKST has recently finished a special Hα survey of the Southern Galactic Plane and Magellanic Clouds, begun in 1997 July. This effectively brings to a close the UKST photographic survey era (Parker, 2002), though the telescope will continue to operate as a mainly spectroscopic facility thanks to the 6dF multi-object fibre spectroscopy system recently commissioned (Watson et al. 2000, 2001). This latest and final UKST photographic survey was also the first large-scale narrow band survey undertaken on the telescope and is the first where the sole method of dissemination to the community is via access to on-line digital data products (no film copies are produced).

The survey used the world's largest monolithic interference filter in astronomy (Parker and Bland-hawthorn 1998) and high resolution Tech-Pan film based emulsion as detector (Parker and Malin 1999) to produce a map of gaseous galactic emission that is currently unsurpassed in terms of the combination of resolution, coverage and sensitivity. Approximately $4000deg^2$ of the Milky-Way have been covered to $|b| \sim 10-13$ degrees together with a separate contiguous region of 700 deg^2 in and around the Magellanic Clouds. Matching 3-hour Hα and 15-min broad-band R exposures were taken over the 233 fields of the Galactic Plane and 40 fields of the Magellanic Clouds. These were done on non-standard 4-degree centres due to the circular aperture of the Hα interference filter which has a di-electric coating diameter of about 305 mm ($\sim$ 5.7 degrees) deposited on a standard 356 $\times$ 356 mm red glass (RG610) substrate. The overlapping 4-degree field centres enable full, contiguous coverage in Hα despite the circular aperture. The contemporaneous broad band short-red (SR) exposures are well matched to the depth of continuum point sources on the equivalent Hα exposure where the approximate magnitude limit in R is $\sim$ 20.5 (Arrowsmith and Parker, 2001, ROE internal report). Additionally, the use of the same emulsion for both Hα and SR exposures ensures an excellent correspondence of their image point spread functions when film pairs are taken under the same observing conditions. This greatly simplifies the inter-comparability of both types of exposure.

The motivation for undertaking such a survey has been reported previously by Parker & Phillipps (1998a, 1998b) and Parker et al. (1999), whilst a range of refereed papers reporting survey discoveries are now beginning to appear (e.g. Mader et al. 1999, Georgelin et al. 2000, Russeil & Parker 2001, Walker et al. 2001, Morgan, Parker & Russeil 2001). Furthermore, perhaps the most exciting project to arise from the survey to date, is the discovery of a rich source of new Galactic Plane planetary nebulae (e.g. Parker et al. 2001, 2003) which will more than double the number of PNe known (e.g. Acker et al, 1992) and should impact profoundly on nearly all aspects of PNe research.

An in-depth technical description of the survey characteristics and properties will be given by Parker et al. (in preparation). For completeness, a brief summary of the rationale for undertaking the survey is given below.

2. Motivation for the Survey

Although understanding and mapping star formation in our own galaxy is of considerable interest, as is the need to understand the distribution and form of gaseous emission on a wide range of angular scales, surprisingly little optical emission line survey work had been undertaken in a manner that simultaneously combined large angular coverage with decent sensitivity and resolution. A survey with such characteristics was needed to permit proper study of the interaction of the wide range of emission structures seen with their large scale environment on arc second to degree scales. Several recent Hα survey programmes have touched on some of these requirements such as the Gaustad et al. (2001) Southern Hα Survey (SHASSA) undertaken at CTIO and the WHAM Northern and Southern hemisphere survey of Reynolds et al. (2001). Though both have attractive features in respect of full hemispheric coverage and direct sensitivity calibration in rayleighs, neither offer the level of spatial resolution given by the AAO/UKST Hα survey. SHASSA has 48 arc seconds resolution and WHAM has 1 degree resolution, though excellent velocity discrimination along the line of sight. This makes the AAO/UKST Hα survey a particularly powerful tool not only for investigating the detailed morphology of emission features across the widest range of angular scales but also as a means of identifying large numbers of faint point source Hα emitters such as CVs, T-Tauri, Be, symbiotic and Herbig-Haro stars. Most other surveys of this type are largely insensitive to point source emitters as they lack adequate spatial resolution having been set-up primarily to record the faintest levels of resolved and diffuse emission. Hence the AAO/UKST Hα survey fills the need for a high angular resolution optical emission line survey for all classes of object to complement equivalent high resolution studies at other wavebands.

Furthermore, since Hα emission from HII regions is one of the most direct optical tracers of current star formation activity, mapping these features traces the distribution of star formation in the galaxy in particular and the diffuse ionized gas in the ISM in general, revealing the locations of a wide variety of astrophysically interesting Galactic emission phenomena. These range from stellar outflows in regions masked by strong reflection nebulae to shocks from high velocity galactic HI clouds, the optical counterparts of supernova remnants, stellar wind-blown bubbles, shells, sheets and filaments and emission nebulosity close to young stellar sources (e.g. Tenorio-Tagle & Bodenheimer, 1988). The detailed spatial structure of the ionized ISM component traced by the AAO/UKST Hα survey will provide key data for many studies, e.g. mapping of specific areas for detailed spectroscopic follow up to obtain emission line gas kinematics or for dynamical studies of star forming regions, with their implications for the energetics of the central stars. Comparisons with other indicators of star formation activity from other wavebands should also provide essential clues to the mechanisms in operation. The survey also nicely complements the new Galactic Plane radio maps from MOST (Green et al. 1999), the new NIR maps from 2MASS (Jarrett et al. 2000) and the mid-Infrared maps from the MSX satellite (Cohen et al. 2000).

Figure 1 gives a 20 field mosaic of Hα survey fields presented in an l, b plot in and around the Vela Supernova Remnant, courtesy of one of us (M. Read) whilst figure 2 gives a full resolution 0.4×0.52 degree plot from part of the Vela Supernova Remnant shown in the wide-field mosaic of figure 1 and centred on α, δ (J2000): 08h30m, $-44°47'$ (image has north-east corner to top left). These figures demonstrate the power of the AAO/UKST Hα on-line survey: wide field coverage of the widest emission structures present but with the ability to pick out fine detail at the arc-second level.

3. The AAO/UKST Hα Survey On-line

The high speed 'SuperCOSMOS' measuring machine at the Royal Observatory Edinburgh (e.g. Miller et al. 1991) has been used to scan the Hα/SR exposure A-grade pairs at 10 micron (0.67 arcsec) resolution. Strict quality control has been applied to the survey pairs according to well established criteria before they

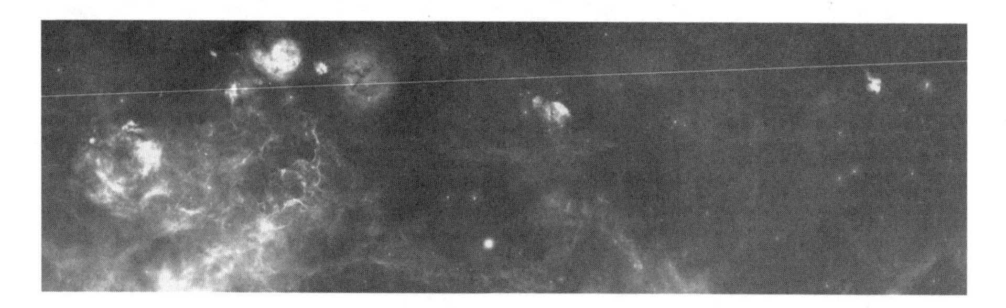

Figure 1. A 30×9 degree galactic l,b plot of 20 fields of the Hα survey mosaicked together in a region around the Vela Supernova Remnant (courtesy of Mike Read). The longitude range is $l = 240 - 270$ degrees and latitude $b = -2.5$ to $+6.25$ degrees

are allowed to be incorporated into the survey atlas. This ensures that the most uniform and homogeneous data product possible is created. The same general scanning and post-processing reduction process is employed as for the directly analogous SuperCOSMOS broad-band surveys of the Southern Sky currently on-line and outlined in detail by Hambly et al. (2001a, 2001b, 2001c). However, due to the special nature of the survey, some additional processing steps and Hα specific options have been added to create the SuperCOSMOS Hα Survey (SHS) described below.

3.1. Basic Characteristics of the SuperCOSMOS Hα Survey

The Wide-Field Astronomy Unit (WFAU) of the Institute for Astronomy Edinburgh is responsible for disseminating the Hα survey data products to the community and the first staged releases of the web-accessed survey product occurred in 2002 April and July. About 150 of the 233 survey fields are now available on-line at:
http://www-wfau.roe.ac.uk/sss/halpha

An example of the web interface for extracting full resolution pixel images from the SHS on-line data is shown in Figure 3 which gives the basic options available. Note that the data products are given as fits files with built-in World Co-ordinate System (WCS). This permits easy incorporation into other software packages such as the STARLINK GAIA and CURSA environments for subsequent visualization, investigation and manipulation. A comprehensive set of web-based documentation has been provided and the entire survey should be accessible via the www by early 2003. The on-line atlas has the following basic characteristics:
- The 10 micron pixel data and associated Image analysis Mode (IAM) parameterized data for both the Hα and SR scanned exposures are stored on-line on a field by field basis
- The SR images have been transformed to exactly match the pixel grid of the master Hα exposures which permits direct image blinking and comparison between the pixel data for each field
- An option to create a difference image of each field following the techniques

Figure 2. A 0.4×0.52 degree high resolution Hα image from part of the Vela Supernova Remnant survey exposure showing the incredible detail available in the on-line data (image oriented so that the North-East corner is to the top left and centred at α, δ (J2000): 08h30m, $-44°47'$)

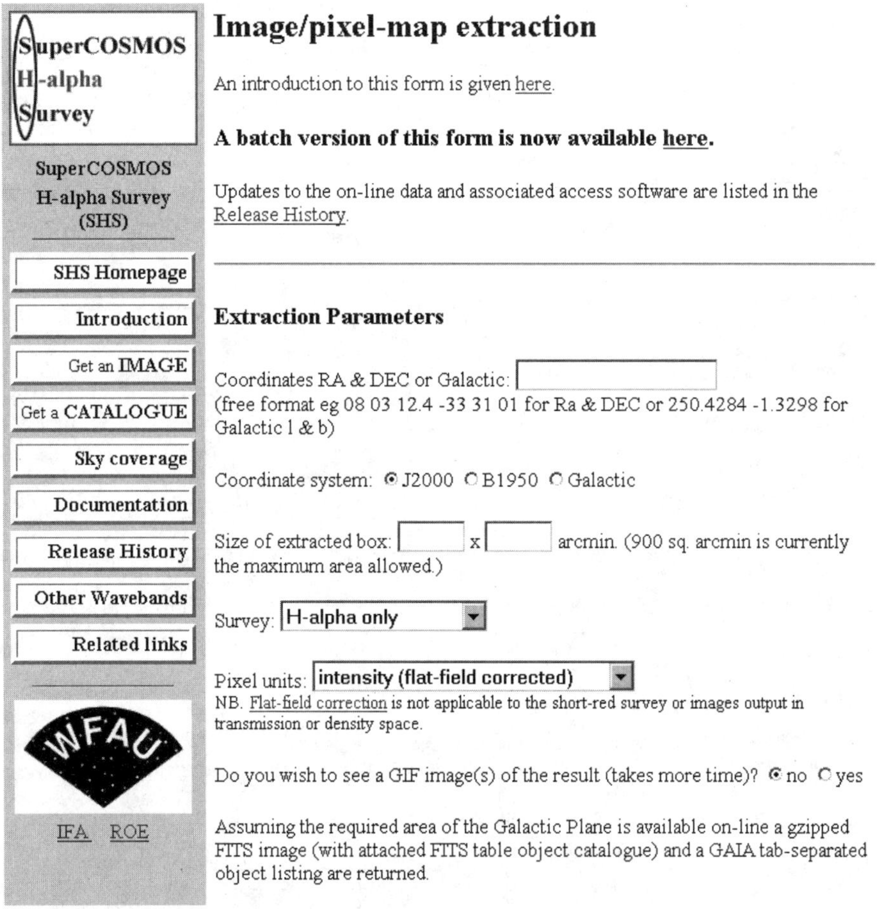

SuperCOSMOS
H-alpha
Survey

SuperCOSMOS
H-alpha Survey
(SHS)

SHS Homepage
Introduction
Get an IMAGE
Get a CATALOGUE
Sky coverage
Documentation
Release History
Other Wavebands
Related links

WFAU

IFA ROE

Image/pixel-map extraction

An introduction to this form is given here.

A batch version of this form is now available here.

Updates to the on-line data and associated access software are listed in the Release History.

Extraction Parameters

Coordinates RA & DEC or Galactic:
(free format eg 08 03 12.4 -33 31 01 for Ra & DEC or 250.4284 -1.3298 for Galactic l & b)

Coordinate system: J2000 B1950 Galactic

Size of extracted box: x arcmin. (900 sq. arcmin is currently the maximum area allowed.)

Survey: H-alpha only

Pixel units: intensity (flat-field corrected)
NB. Flat-field correction is not applicable to the short-red survey or images output in transmission or density space.

Do you wish to see a GIF image(s) of the result (takes more time)? no yes

Assuming the required area of the Galactic Plane is available on-line a gzipped FITS image (with attached FITS table object catalogue) and a GAIA tab-separated object listing are returned.

Figure 3. Web interface to the Hα survey on-line pixel extraction process illustrating the various options available

developed by Bond et al. (2001) permit large-scale resolved emission maps to be straightforwardly created without undue clutter from stars
- The survey data products are accessed via a web interface that has the same look and feel as existing broad-band SuperCOSMOS on-line surveys but with some additional functionality
- A 16× blocked-down version of each field is also available as both a GIF image and as a fits file which has the WCS built in to the fits header. These whole field maps can be studied to select smaller regions of interest for extraction at full 0.67 arcsec resolution
- The full resolution pixel data access limit is currently set at 900 arcmin^2 with regions downloaded as fits files (also with WCS) and both the SR and Hα data for the same region can be downloaded simultaneously
- Areas for extraction can be chosen via equatorial (J2000 or B1950) or galactic (l, b) co-ordinates in an $m \times n$ arcminute rectangular region
- a clickable map of the current fields on-line enables individual field details to be displayed prior to viewing the blocked full field image
- The entire survey data are stored on RAID disks for fast access
- CCD calibration already data exists for each survey field and ultimately this will also be made available
- Parameterized Image Analysis Mode (IAM) 'Catalogue' data has been produced for each field and seamless catalogues can easily be created which cover several adjacent fields
- A batch mode enables large numbers of thumb-nail images to be extracted around objects of interest with the option to return postscript plots of the extracted images
- An option to apply a 'Flat-Field' to the Hα pixel data is included to permit correction of the non-uniformities in the measured exposures arising from the excellent but slightly varying Hα filter transmission profile.

4. Conclusions

The AAO/UKST Hα survey is now complete and on-line. It represents a powerful new tool for the study of the ionized gas content of our galaxy on a range of spatial scales from arc second to tens of degrees. The community is invited to consider use of this valuable new resource when undertaking any study of the Southern Galactic plane. Many exciting discoveries have already been made from preliminary work on the survey by the Hα survey consortium, including the discovery of a rich new vein of Galactic planetary nebulae. This bodes well for general community access.

References

Acker, A. et al. 1992, Strasbourg-ESO Catalogue of Galactic Planetary Nebulae, ISBN 3-923524-41-2

Bond, I. A. et al. 2001, MNRAS, 327, 2001

Cohen, M., Hammersley, P. L. & Egan, M. P. 2000, AJ, 120, 3362

Gaustad, J. E., McCullough, P. R., Rosing, W. & Van Buren, D. 2001, PASP, 113, 1326

Georgelin, Y. M., Russeil, D., Amram, P., Georgelin, Y., Marcelin, M., Parker, Q. A. & Viale, A. 2000, A&A, 357, 308

Green, A., Cram, L. E., Large., M. I., Taisheng, Y. 1999, ApJS, 122, 207

Hambly, N. C et al. 2001, MNRAS, 326, 1279

Hambly, N. C., Irwin, M.J. & MacGillivray, H.T. 2001, 326, 1295

Hambly, N. C., Davenhall, A. C., Irwin, M. J. & MacGillivray, H. T. 2001, 326, 1315

Jarrett, T. H., Chester, T., Cutri, R., Schneider, S., Rosenberg, J., Huchra, J. P. & Mader, J. 2000, AJ, 120, 298

Mader, S. L. et al. 1999, MNRAS, 310, 331

Miller, L., Cormack, W., Paterson, M., Beard, S., & Lawrence, L. 1991, in 'Digitised Optical Sky Surveys', eds. H. T. MacGillivray and E. B. Thomson, Kluwer Academic Publishers, p.133

Morgan, D. H., Parker, Q. A., & Russeil, D. 2001, MNRAS, 322, 877

Parker, Q. A., Bland-Hawthorn, J. 1998, PASA, 15, 33

Parker, Q. A. & Phillipps, S. 1998a, PASA, 15, 28

Parker, Q. A. & Phillipps, S. 1998b, A&G, 39, 10

Parker, Q. A. et al. 1999, ASP. Conf. Ser., 168, 126

Parker, Q. A. & Malin, D. 1999, PASA, 16, 288

Parker, Q. A. & Phillipps, S. 2001, ASP. Conf. Ser., 232, 39

Parker, Q. A. 2002, AAO Newsletter, No. 100, 4

Parker, Q. A., Hartley, M., Russeil, D., Acker, A., Morgan, D., Beaulieu, S., Morris., R., Phillipps, S. & Cohen, M. 2003, in IAU Symposium 209, 'Planetary nebulae', ASP. Conf. Ser. (in press)

Reynolds, R. J., Tufte, S. L., Haffner, L. M., Jaehnig, K. & Percival, J. W. 1998, PASA, 15, 14

Russeil, D., Parker, Q. A. 2001, PASA, 18, 76

Tenorio-Tagle, G. & Bodenheimer, P. 1988, ARA&A, 26, 145

Walker, A., Zealey, W. J. & Parker, Q. A. 2001, PASA, 18, 259

Watson, F. G., Parker, Q. A., Bogatu, G., Farrell, T. J., Hingley, B. E. & Miziarski, S. 2000, SPIE, 4004, 123

Watson, F. G. et al. 2001, AAO Newsletter, No. 97, 14

IAU 8th Asia-Pacific Regional Meeting
ASP Conference Series, Vol. 289, 2003
S. Ikeuchi, J. Hearnshaw & T. Hanawa, eds.

Star Formation in the Southern Hemisphere: A Millimetre-Wave Survey of Dense Southern Cores

Paul A. Jones[1], Maria R. Hunt-Cunningham[2], John B. Whiteoak[1], Graeme L. White[3]

[1]*Australia Telescope National Facility, PO Box 76, Epping NSW 1710, Australia*

[2]*University of New South Wales, NSW 2052, Australia*

[3]*University of Western Sydney, Locked Bag 1797, Penrith South DC NSW 1797, Australia*

Abstract. An extensive survey of molecular emission from dense southern molecular cores has been undertaken with the Mopra telescope and the SEST. Molecular rotational transitions of CO, CS, HCN, HNC, HCO^+, HC_3N, OCS, CH_3OH and SO, and several of their isotopomers have been observed in a sample of 27 molecular clouds with declinations south of −30 degrees. The molecular abundances and physical conditions in the dense cores have been derived using both LTE and LVG models. The observations are discussed here and some results are presented, along with a discussion of planned follow-up observations with the Australia Telescope Compact Array in the 3-mm band.

1. Introduction

The dense regions of molecular clouds contain a wide range of molecules, which are detected most easily through rotational transitions in the millimetre wavelength range. Because the majority of large millimetre-wave radio telescopes, both single dish and interferometers, are sited in the northern hemisphere, the southern part of the Galactic Plane has been relatively poorly studied. We have therefore made single-dish observations in the 3-mm and 2-mm bands, containing a large number of molecular spectral line transitions, with the Australia Telescope National Facility Mopra Telescope and the Swedish-ESO Submillimetre Telesope (SEST), in a sample of the richest southern molecular clouds. This provides not only kinematic information from the line profiles, but also we can estimate molecular excitation temperatures and abundances from modelling the line emission (Hunt-Cunningham et al., in preparation).

A major result of this survey is a database of molecular emission and abundances in southern molecular clouds. This can be used to test the theoretical predictions for molecular abundances in different physical regions of the molecular clouds, and different phases of the star-formation process. Also the Australia Telescope Compact Array (ATCA) is being upgraded to work at the 12-mm and 3-mm bands and is the first large millimetre interferometer in the southern hemisphere, so this survey provides a finding list of the most interesting regions to study in more spatial detail. Similarly, the Atacama Large Millimeter Array (ALMA), and its millimetre/sub-millimetre single-dish precursors in Chile, ASTE and APEX, will benefit from this survey work.

2. Observations

The sample is of dense molecular clouds, most of which are associated with radio continuum from H II regions and IRAS dust emission. The observed sample was selected from previous molecular-line surveys of the southern Galactic Plane with the Parkes 64-m, Tidbinbilla and CSIRO 4-mm telescopes, by Gardner & Whiteoak (1984; Parkes, H_2CO absorption 14.5 GHz), Dickinson et al. (1982; Tidbinbilla, NH_3 23.7 GHz), Batchelor et al. (1981; CSIRO 4 m, HCO^+ 89 GHz), Whiteoak & Gardner (1978; CSIRO 4 m, HCN 88.6 GHz), Gardner & Whiteoak (1978; Parkes, CS 49 GHz) and Whiteoak & Gardner (1974; Parkes, H_2CO absorption 4.83 GHz).

The original list of 50 sources was observed with Mopra in CS, HCN and HCO^+ (lines in the range 88–98 GHz) to select 25 molecular clouds with strong molecular emission, including 5 dark clouds. The well studied sources Orion KL and M 17 SW were added for calibration and comparison. The list of positions is given in table 1 – note that G291.3-0.7 has two pointing centres, and NGC 6334 has four, so the total number of pointing positions is 31.

Lines of 11 different molecules were observed (CO, CH_3OH, C_2H, CS, HCN, HC_3N, HCO^+, HNC, HNCO, OCS and SO), generally in more than one transition, and for some molecules including rarer, isotopically substituted species. The specific rest frequencies were CO (115.271 GHz), ^{13}CO (110.201 GHz), $C^{18}O$ (109.782 GHz), CH_3OH (84.521, 88.940, 96.739–96.756, 107.014, 145.094–145.103 GHz), C_2H (87.284–87.329 GHz), CS (97.981, 146.969 GHz), ^{13}CS (92.494 GHz), HCN (88.630–88.634 GHz), $H^{13}CN$ (86.340 GHz), HC_3N (90.979, 100.076, 109.174, 136.464, 145.561 GHz), HCO^+ (89.189 GHz), $H^{13}CO^+$ (86.754 GHz), $HC^{18}O^+$ (85.162 GHz), HNC (90.9663 GHz), $HN^{13}C$ (87.091 GHz), HNCO (87.898–87.925 GHz), OCS (85.139, 97.301, 109.463, 145.947 GHz) and SO (99.300 GHz).

Observations between 85 and 116 GHz were made with the Mopra telescope, near Coonabarabran Australia, between 1995 Oct. and 1996 Sep., with an effective dish diameter of 15 m (as the outer parts of the 22-m dish were not surfaced for 3-mm observations). Additional Mopra observations were made in 2000 Mar. and Sep., with an effective dish diameter of 22 m, after the dish was resurfaced in 1999. Observations were also made with the SEST, at La Silla Chile, between 84 and 147 GHz, in 1999 Jun., to complement the Mopra observations and observe transitions in the 2-mm band above 115 GHz, that Mopra could not reach.

3. Results

The observed line intensities for different isotopomers give information about the isotope ratios, with some caveats since some lines may be optically thick, notably ^{12}CO. The intensity ratios, mean and standard deviation, are $CO/^{13}CO$

= 3.6 ± 1.9, $CS/^{13}CS$ = 12.0 ± 5.7, $HCN/H^{13}CN$ = 7.8 ± 5.4, $HNC/HN^{13}C$ = 11.9 ± 5.6 and $HCO^+/H^{13}CO^+$ = 6.8 ± 7.6. The intensity ratios of the ^{13}C to ^{18}O isotopomers are $^{13}CO/C^{18}O$ = 6.9±2.6 and $H^{13}CO^+/HC^{18}O^+$ = 15.3±13.7.

Most of the dense cores show some evidence for recent or ongoing star formation, from the detection of thermal CH_3OH and OCS emission, or from line wings suggesting outflow. The molecular clouds G291.3-0.7, G345.5+1.5, NGC 6334 (N), NGC 6334 (N1), G351.6-1.3 and G353.4-0.4 probably contain hot cores.

Table 1. Galactic molecular cloud positions observed.

General Designation	Specific Name	Right Asc. (J2000)	Decl. (J2000)	Right Asc. (B1950)	Decl. (B1950)
Orion MC	Orion KL	05 35 14.5	–05 22 29	05 32 47.0	–05 24 22
HH46 DC		08 25 37.8	–51 04 00	08 24 10.6	–50 54 08
G265.1+1.5	RCW 36	08 59 26.4	–43 45 08	08 57 38.0	–43 33 24
G268.4-0.8		09 01 51.6	–47 44 07	09 00 09.4	–47 32 15
Cham DC		11 06 32.6	–77 23 35	11 05 08.1	–77 07 20
G291.3-0.7	RCW 57E	11 12 04.1	–61 18 29	11 09 56.1	–61 02 09
G291.3-0.7	RCW 57W	11 11 47.4	–61 19 28	11 09 39.6	–61 03 09
Coalsack DC		12 01 27.7	–65 07 54	11 58 53.8	–64 51 11
G301.0+1.2	RCW 65	12 34 57.8	–61 39 56	12 32 06.1	–61 23 24
G305.4+0.2		13 12 33.8	–62 33 49	13 09 21.0	–62 17 54
G311.6+0.3		14 04 56.0	–60 20 21	14 01 20.0	–61 06 00
G322.2+0.6	RCW 92	15 18 38.0	–56 39 00	15 14 47.6	–56 28 05
Lupus DC		15 43 02.2	–34 09 06	15 39 51.2	–33 59 36
G326.7+0.6		15 44 44.8	–54 06 25	15 40 55.0	–53 57 00
G327.3-0.5		15 53 06.1	–54 35 31	15 49 13.0	–54 26 37
G331.5-0.1		16 12 12.3	–51 29 02	16 08 24.2	–51 21 20
G333.0-0.6		16 21 05.5	–50 35 25	16 17 18.3	–50 28 18
G333.4-0.4		16 21 32.9	–50 26 32	16 17 46.0	–50 19 27
G333.6-0.2		16 22 12.1	–50 05 56	16 18 26.0	–49 58 54
G345.5+1.5		16 59 41.9	–40 03 13	16 56 14.1	–39 58 45
G345.5+0.3		17 04 28.5	–40 46 02	17 00 59.0	–40 41 54
NGC 6334	(S)	17 19 55.9	–35 57 53	17 16 34.5	–35 54 51
NGC 6334	(CO)	17 20 23.4	–35 55 00	17 17 02.0	–35 52 00
NGC 6334	(N)	17 20 48.5	–35 46 33	17 17 27.4	–35 43 35
NGC 6334	(N1)	17 20 53.4	–35 45 32	17 17 32.3	–35 42 35
G348.7-1.0	RCW 122	17 20 05.2	–38 57 22	17 16 38.3	–38 54 21
G351.6-1.3		17 29 12.8	–36 40 12	17 25 49.8	–36 37 50
G353.4-0.4		17 30 24.6	–34 41 39	17 27 05.0	–34 39 23
G1.6-0.025		17 49 20.6	–27 34 08	17 46 12.0	–27 33 15
Cor Aust DC	R CrA	19 10 18.2	–37 08 58	19 06 56.0	–37 13 54
M 17	M17 SW	18 20 23.1	–16 11 36	18 17 30.0	–16 12 59

The molecular clouds G265.1+1.5, G291.3-0.7 and G1.6-0.025 have self-absorbed line profiles in many molecules, suggesting a complex cloud structure and kinematics. Further observations show that G1.6-0.025 is probably undergoing a cloud-cloud collision, as is G265.1+1.5 and possibly G291.3-0.7.

The hydrogen column density $N(H_2)$ was obtained from the intensity of the CO 1–0 lines of ^{12}CO, ^{13}CO and $C^{18}O$. The column density of the molecules for which only one transition was observed, was calculated assuming local thermodynamic equilibrium (LTE) and low optical depth ($\tau \ll 1$). Excitation temperatures were estimated these, based on groups of molecules. For the molecules where more than three transitions were observed (HC_3N, OCS and CH_3OH), the rotation diagram analysis (Goldsmith & Langer 1999) was used, giving both the column density and excitation temperature from a fit to the multiple transitions, and testing the assumptions of LTE and low optical depth. For HC_3N, a Large Velocity Gradient (LVG) model was also applied.

4. ATCA 3-mm Observations

The single-dish Mopra and SEST observations considered here have a spatial resolution of between 35 and 59 arcsec (depending on the frequency and telescope used), so that the observed intensities and line profiles, and calculations of abundances and temperatures are averaged over this scale. The scale of physical, and hence chemical, structures involved in star formation is smaller than this, so higher resolution observations with a millimetre-wave interferometer are necessary for the detailed study of individual dense cores. This is now becoming possible for these southern declinations with the upgrading of the Australia Telescope Compact Array (ATCA) to the 3-mm band. We have, for example, ATCA observations scheduled in 2002 Oct. of G291.3-0.7 (RCW 57) for imaging at 3 mm in HCN and HCO^+, to test the cloud-cloud collision scenario. The ATCA will have 5 $\times$ 22-m dishes operating between 85 and 105 GHz over maximum 3-km baseline by mid-2003 (currently, mid-2002 season 3 dishes working over a more restricted frequency range at 3 mm).

References

Batchelor, R. A., McCulloch, M. G., & Whiteoak, J. B. 1981, MNRAS, 194, 911

Dickinson, D. F., Gulkis, S., Klein, M. J., Kuiper, T. B. H., Batty, M., Gardner, F. F., Jauncey, D. L. & Whiteoak, J. B. 1982, AJ, 87, 1202

Gardner, F. F. & Whiteoak, J. B. 1978, MNRAS, 183, 711

Gardner, F. F. & Whiteoak, J. B. 1984, MNRAS, 210, 23

Goldsmith, P. F., & Langer, W. D. 1999, ApJ, 517, 209

Hunt-Cunningham M. R., Whiteoak, J. B., Jones, P. A. & White, G. L., in preparation

Whiteoak, J. B., & Gardner, F. F. 1974, A&A, 37, 389

Whiteoak, J. B., & Gardner, F. F. 1978, MNRAS, 185, 33

IAU 8th Asia-Pacific Regional Meeting
ASP Conference Series, Vol. 289, 2003
S. Ikeuchi, J. Hearnshaw & T. Hanawa, eds.

Star Formation Studies with SIRIUS – JHK Simultaneous Near-infrared Camera

Yasushi Nakajima, Chie Nagashima, Takahiro Nagayama, Daisuke Baba, Daisuke Kato, Mikio Kurita, Tetsuya Nagata, Shuji Sato

Department of Astrophysics, Nagoya University, Furo-cho, Chukusa-ku, Nagoya, 464-8602, Japan

Motohide Tamura, Takahiro Naoi

National Astronomical Observatory of Japan, Osawa, Mitaka, Japan

Hidehiko Nakaya

Subaru Telescope, National Astronomical Observatory of Japan, Hilo, USA

Koji Sugitani

Nagoya City University, Mizuho-ku, Nagoya, Japan

Abstract. Some results on a star formation study from the near infrared camera SIRIUS are presented. SIRIUS is designed for deep and wide JHK$_s$-bands simultaneous surveys, being equipped with three near-infrared (1024 × 1024) arrays. SIRIUS is attached to a dedicated 1.4-m telescope at Sutherland Observatory in South Africa. The field of view is $7.8' \times 7.8'$, the pixel scale is $0.45''$, and the limiting magnitudes are $J = 19.2$, $H = 18.6$, $K_s = 17.3$ (S/N = 10σ and 15 minutes' integration) with the 1.4-m telescope. SIRIUS was also used on the University of Hawaii 2.2-m telescope at Mauna Kea. Having a wide field, high sensitivity, and ability to undertake simultaneous JHK$_s$-color observations, SIRIUS is an ideal camera for investigations of the luminosity function and spatial distribution of YSOs in clusters/clouds, very low-mass stars, and extinction of background stars due to dust in dark clouds. Many star-forming regions have been observed since the first light of SIRIUS in 2000 August. As some of the results, we have obtained the following: discovery of a shining dark cloud in Lupus, which is unaccountably bright in the K-band; star-forming structure in M16; and pre-main sequence cluster formation in the Large Magellanic Cloud.

1. SIRIUS

SIRIUS (Simultaneous-color InfraRed Imager for Unbiased Surveys) is a near-infrared camera (Nagashima et al. 1999) developed by Nagoya University and National Astronomical Observatory of Japan. It is designed for deep and wide

JHK$_s$-bands simultaneous surveys. SIRIUS is equipped with three science-grade HAWAII (1024×1024) arrays, J ($\lambda = 1.25\,\mu$m), H ($1.65\,\mu$m), and K$_s$($2.15\,\mu$m) band filters, and two dichroic mirrors which enable simultaneous observations at the three bands. Simultaneous observations at the three bands provides us precise color information with high time efficiency. More details about SIRIUS are available at

`http://www.z.phys.nagoya-u.ac.jp/~sirius/index_e.html`

SIRIUS is a powerful for star formation studies. Observations at these wavelengths is sensitive to T Tauri stars, as they emit most of their energy in these wavelengths. With these wavelengths, we can see through dusty clouds in which YSOs are often embedded. JHK simultaneous observations provide us with precise color information. JHK colors are useful for selection of YSOs, owing to their color excess, which is due to circum-stellar dust emission.

2. Surveys with SIRIUS

The purpose of SIRIUS is deep and high-resolution surveys of large areas, but not of the whole sky like 2MASS or DENIS. SIRIUS is mainly used with a dedicated 1.4-meter telescope, IRSF (InfraRed Survey Facility), at the South African Astronomical Observatory. The survey of the southern sky began in 2000 November. When it is attached on the IRSF 1.4-meter telescope, the field of view is $7.8'$ by $7.8'$ with the pixel scale of $0.45''$. With a typical integration time of 15 minutes, the 10 σ limiting magnitudes are J = 19.2, H = 18.6, K_s = 17.3 mag. SIRIUS was also used on the University of Hawaii 2.2-meter telescope at Mauna Kea for three times in 2000 August, 2000 October, and 2001 September. When SIRIUS is attached to the UH 2.2-meter telescope, the field of view is $4.9' \times 4.9'$ with a pixel scale of $0.28''$. With a typical integration time of 15 minutes, the 10 σ limiting magnitudes are $J = 20.3$, $H = 19.2$, and $K_s = 18.3$ mag. Surveys of several northern sky areas were done.

We show some survey projects at IRSF and some results obtained with IRSF and the UH 2.2-meter telescope in the following sections.

3. LMC Survey

The main purpose of SIRIUS at IRSF is an unbiased deep survey for 6-degree square area of the Large Magellanic Cloud (LMC). The survey began in 2000 December. The survey area is shown in Figure 1. We completed 25% of the total area so far. The integration time for each field is set to be 5 minutes and the limiting magnitude is $K_s \sim 17$ mag.

Star formation in the LMC is one of the most important researches in this survey. With the limiting magnitude of $K_s \sim 17$ mag., a large fraction of the population of intermediate mass pre-main sequence stars (Herbig Ae/Be stars) can be detected at the distance of the LMC, if the luminosity function in the LMC is same as that of the Galaxy. Covering a 6-degree square area and detecting most of the Herbig Ae/Be stars, the LMC survey of SIRIUS will be a unique one in terms of the survey of star-forming regions in the LMC. We have already detected a number of pre-main sequence star candidates as members of clusters in some HII regions (Nakajima et al. in prep.). By using

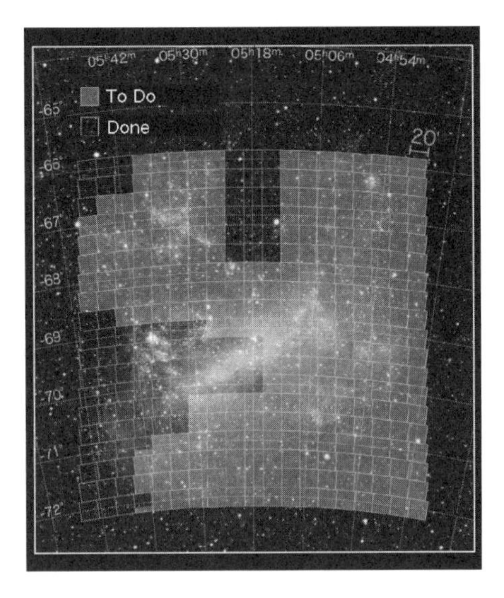

Figure 1. Superposed on an optical image are survey area of LMC current status of the survey. The blank tiles show the area observed by 2002 May while the white-shaded do the area planned to be observed.

JHK_s colors, we can select pre-main sequence candidates among all the point sources. Because pre-main sequence stars are surrounded by hot circumstellar dust, they are intrinsically redder than main sequence stars or giants in the near infrared wavelengths.

4. M16

M16 is a star-forming region of which the distance is about 2 kpc. We observed M16 with SIRIUS and the UH 2.2-m telescope. SIRIUS revealed the following (Sugitani et al. 2002).

1. head-tail structures which is different from the pillars as seen in the optical;
2. a number of previously unreported YSO candidates associated with the head-tail pillars;
3. the youngest ones are located at the tips of the pillars;
4. some of them are associated with NIR jets or cometary reflection nebulae.

5. Lupus 3

Lupus 3 is a nearby dark cloud. We carried out a very deep imaging of this area with SIRIUS and the IRSF 1.4-m telescope. The dark cloud shines in the NIR wavelengths andhave several complex structures as shown in Figure 2 (Nakajima et al. in prep.). Such an image is obtained for the first time for dense clouds of

Figure 2. The K_s band image of the Lupus 3 dark cloud. Most part of the cloud shines in Ks. The field of view is $\sim 8' \times 8'$, which corresponds to 0.4 pc $\times$ 0.4 pc at the distance of the Lupus dark cloud, 150 pc.

which visual extinction exceeds 10 mag. The surface brightness at each band, J, H, and K_s, was measured and the maximum surface brightness was $J = 20.6$, $H = 19.8$, and $K_s = 19.4$ mag. per square arcsecond. The extinction map for the background stars was also obtained. The maximum extinction was $A_v = 47$ mag. The surface brightness and the extinction are well correlated and explained by scattering of background starlight by dust. The results supports the dust size distribution of Weindgartner and Drain (2001). Lager dust is needed than the widely used model of Mathis et al. (1977). Our result also demonstrates that the imaging of NIR diffuse emission can be a new tool to probe the structure of dark clouds.

Acknowledgments. We thank the staff of the UH 2.2-meter telescope, Subaru Telescope, and SAAO for supporting the observations with SIRIUS. This work is financially supported by the Sumitomo Foundation through a Grant-in-Aid for Scientific Researches in Priority Areas (A), and by a Grant-in-Aid for International Scientific Research, from the Ministry of Education, Culture, Sports, Science and Technology. Y. Nakajima, C. Nagashima, and T. Nagayama received Grants-in-Aid for JSPS Fellows.

References

Mathis, J. S., Rumpl, W., and Nordsiek, K. H. 1977, ApJ, 217, 425

Nagashima, C. et al. 1999, in Star formation 1999, ed. T. Nakamoto, (Nobeyama Radio Observatory), 397

Sugitani, K. et al. 2002, ApJ, 565, L25

Weingartner, J. C. and Draine, B. T. 2002, ApJ, 548, 296

IAU 8th Asia-Pacific Regional Meeting
ASP Conference Series, Vol. 289, 2003
S. Ikeuchi, J. Hearnshaw & T. Hanawa, eds.

Evolution of Spatial Structure of Star Clusters

W. P. Chen
Institute of Astronomy and Department of Physics, National Central University, Chung-Li 32054, Taiwan

C. W. Chen
Institute of Astronomy, National Central University, Chung-Li 32054, Taiwan

Abstract. We present the results of a pilot project to study the structure of star clusters with the 2MASS database. While the 2MASS cannot resolve the cores or detect much of the main sequence of globular clusters, the homogeneity and extended angular coverage make the star database suitable to study young star clusters. Even the youngest star clusters which are not yet dynamically relaxed have their stars – regardless of the stellar masses – concentrated progressively towards the cluster centers, a result due more to the cluster formation process than from subsequent gravitational interaction. We show evidence that, as a cluster ages, effects of mass segregation and external disturbances start to dominate the evolution of the spatial distribution of stars in a star cluster.

1. Introduction

The *initial* stellar distribution in a star cluster is dictated by the structure in the parental molecular cloud. As the cluster evolves, the distribution is modified by *internal* gravitational interaction among member stars. Eventually stellar evaporation and *external* disturbances – Galactic tidal force, differential rotation, and collision with molecular clouds – would dissolve the cluster. Star clusters therefore provide a laboratory to study stellar dynamics. The youngest clusters in particular still bear the imprint of their formation history, so their structure, when compared with that of molecular clouds, would shed crucial light on the fragmentation process during cloud collapse.

Stars in a globular clusters are known to concentrate progressively towards the center, more so for massive stars than for low-mass stars. The density distribution is well described by the King model (1962), which is understood as a combination of an isothermal sphere (i.e., dynamically relaxed) in the inner part of the cluster, and tidal truncation by the Milky Way in the outer part.

On the other hand, open clusters appear irregularly shaped, with member stars sparsely distributed. Our perception, or even recognition, of an open cluster is largely biased towards the appearance of the brightest members. Do star clusters form preferentially by sequential star formation, so that low-mass stars would occupy a location different from that for high-mass stars? How do stars of different masses suffer the dynamical interaction and external perturbation during the evolution of a cluster? Are young clusters mass segregated, and if so, to what extent is this due to dynamical relaxation, as opposed to the relics of the structure in the molecular cloud? To answer these questions, it is desirable to study the spatial structure of the youngest star clusters, and see how it evolves as the cluster ages.

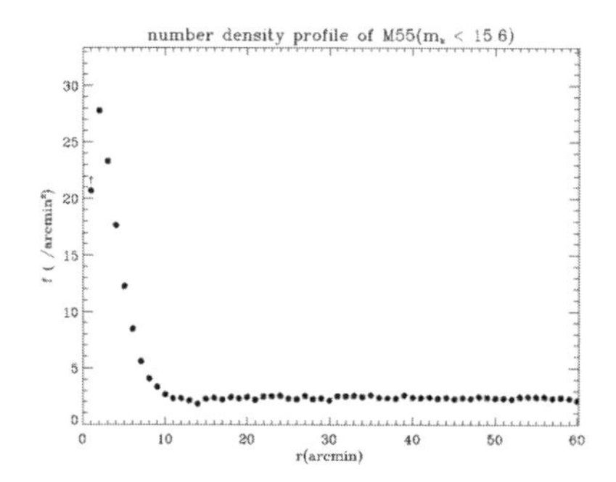

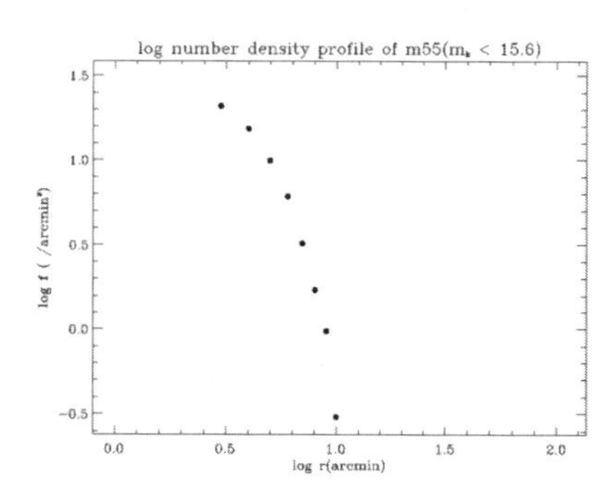

Figure 1. The radial surface density distribution of M 55 by 2MASS. A uniform background (left panel) extends out to large radii. While the core is too crowded to resolve by 2MASS, the outer part (right panel) of the cluster is clearly seen, and follows the King model.

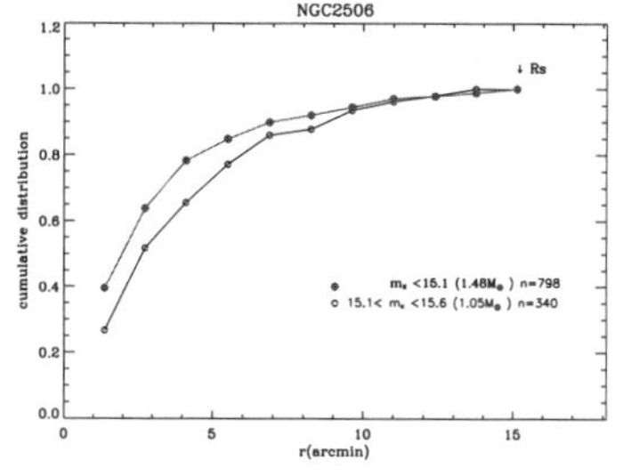

Figure 2. Comparison of the radial distribution of bright and faint stars in NGC 2506 suggests that mass segregation has occurred.

2. Structure of Open Clusters

We analyzed the spatial structure of a star cluster by its stellar density distribution within concentric annuli in the Two-Micron All-Sky Survey (2MASS) star catalogue.

Figure 1 shows the radial density profile of the globular cluster M 55 detected by the 2MASS. A well defined background, important in our star-count study, is seen out to large radii, something leisurely available with a sky survey star caalogue. While the angular resolution and sensitivity ($K \sim 15.6$ mag., 3σ) of the 2MASS cannot resolve the cores or detect much of the main sequence for distant and old globular clusters, the homogeneity and extended angular cover-

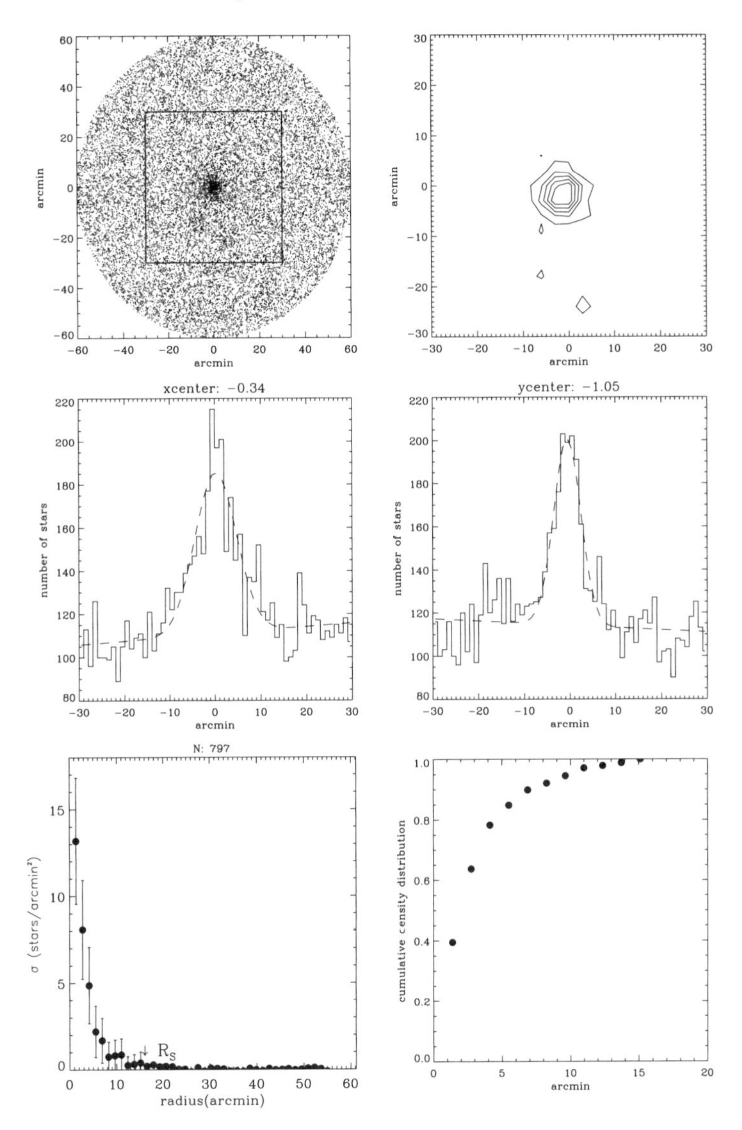

Figure 3. Density distribution of bright stars (K$\lesssim$15.1 mag) in NGC 2506. (*Top*) Each dot on the left panel represents a star entry in the 2MASS catalogue. The right panel plots the the surface number density. (*Center*) The stellar density, projected onto the R.A. (left) and the dec. axis (right), respectively, each with a Gaussian fit. The peak position defines the center of the cluster. The slope of the background is due to the stellar density gradient of the galactic disk. (*Bottom*) The radial (left) and cumulative density profile of the cluster, with the background subtracted.

age of the database make it suitable to study young star clusters (Chen & Chen 2001).

The sample of our pilot study consists of star clusters with a variety of ages and distances. As an example, Figure 2 compares the cumulative density distribution for bright and faint stars in NGC 2506. The age of this cluster (1.9 Gyr, Twarog et al. 1999) is several times the relaxation time (~300 Myr), so NGC 2506 should have been dynamically relaxed already, as indeed seen in Figure 2. Figure 3 shows how the bright stars in NGC 2506 distribute. Even though no membership information for individual stars is available, integration of the density distribution within the cluster boundary (3σ background) allows us to estimate the total number of stars in the cluster. Given the 2MASS sensitivity, we then estimate for each cluster the stellar mass range being detected, from which the relaxation time scale is calculated. Table 1 summarizes the parameters and our results for the seven clusters we have studied.

3. Summary

The 2MASS star catalogue provides a very useful database for open cluster study. As in globular clusters, stars in open clusters, regardless of their masses, are concentrated progressively toward the center. Even the youngest star clusters show evidence of mass segregation. This suggests that spatial structure of a star cluster is governed by the structure of the molecular cloud from which the cluster was formed, and then later modified by the dynamical relaxation and external perturbation.

Table 1. Parameters for open clusters

Cluster	ℓ, b	D (kpc)	τ (Myr)	$N*$	M ($M_\odot$)	R (pc)	τ_{re}	τ/τ_{re}	Segr.
				Young					
NGC1893	174,-02	4.4	4	498	309	8.9	291	0.01	?
IC348	160,-18	0.32	5	322	200	1.6	14	0.2	Y
				Intermediate					
NGC1817	186,-13	2.1	800	236	146	7.9	139	6	N?
NGC2506	231,+10	3.3	1,900	1,038	643	17.3	605	3	Y
NGC2420	198,+20	2.5	2,200	450	279	9.4	223	10	Y
				Old					
NGC6791	070,+11	4.2	8,000	1,095	679	13.2	543	15	?
Be17	176,-04	2.5	9,000	370	229	7.1	142	63	N

References

Chen, J. W., & Chen, W. P. 2001, in ASP Conf. Ser., 246, Small-Telescope Astronomy on Global Scales, ed. W. P. Chen, C. Lemme, & B. Paczynski (San Francisco: ASP), 331

King, I. 1962, AJ, 67, 471

Twarog, B. A., Anthony-Twarog, B. J. & Bricker, A. R. 1999, AJ, 117, 1816

IAU 8th Asia-Pacific Regional Meeting
ASP Conference Series, Vol. 289, 2003
S. Ikeuchi, J. Hearnshaw & T. Hanawa, eds.

Molecular Line Studies of Galactic Young Stellar Objects

Ji Yang, Ruiqing Mao, Zhibo Jiang, Tao Geng and Yi-ping Ao

Purple MountainObservatory, Academia Sinica, Nanjing 210008, China

Abstract. We conducted a large-scale survey for cold infrared sources along the northern galactic plane in CO ($J = 1 - 0$) line. There are 1912 IRAS sources selected on the basis of their color indices over the 12 μm, 25 μm, and 60 μm wavebands and their association with regions of recent star formation. A quick single-point survey was made towards all of the sources, which results in a detection of 1331 sources with significant CO emission above the detection limit of 0.7 K, inferring a CO detection rate of 70% (Yang et al. 2002). The single-point CO survey also revealed 289 sources which exhibit prominent high-velocity wing emission. Detailed studies of the candidate CO outflows were conducted by CO ($J = 2 - 1$) mapping. It shows that most of the high-velocity wing sources are intrinsically high-velocity CO outflows.

1. Introduction

To obtain a quantitative measure of different young stellar populations in the Galactic scale is one of the approaches in the study of star formation over the whole Milky Way. In aid of the sensitive millimeter-wave telescope, we have carried out a large-scale survey of young stellar objects based on the selected sample of cold IRAS point sources. These sources demonstrate the coldest color indices based on their IRAS fluxes with equivalent color temperatures of 162 K for the 12–25 μm bands and 87 K for the 25–60 μm bands (Yang et al. 2002).

Our molecular-line study began with a single-point observation toward the 1912 sources in CO ($J = 1 - 0$) line. Within the average 3σ detection limit of 0.7 K, the detection rate was 70%. These sources provide a uniform database for the study of star formation over the Galactic scale.

Most of the sources are moderately bright. The average antenna temperature of the detected CO sources is 3.9 K. After correcting the beam efficiency of the telescope, 0.44, the average radiation temperature of the sources is 8.9 K.

High-velocity CO wing emission has been detected from 351 CO-emitting sources. Among these sources, identification based on an earlier outflow catalog shows that there are about 290 sources beyond the known catalog and thus are candidates for high-velocity molecular outflows. Detailed study of individual sources provides insight into the intrinsic properties of the sample.

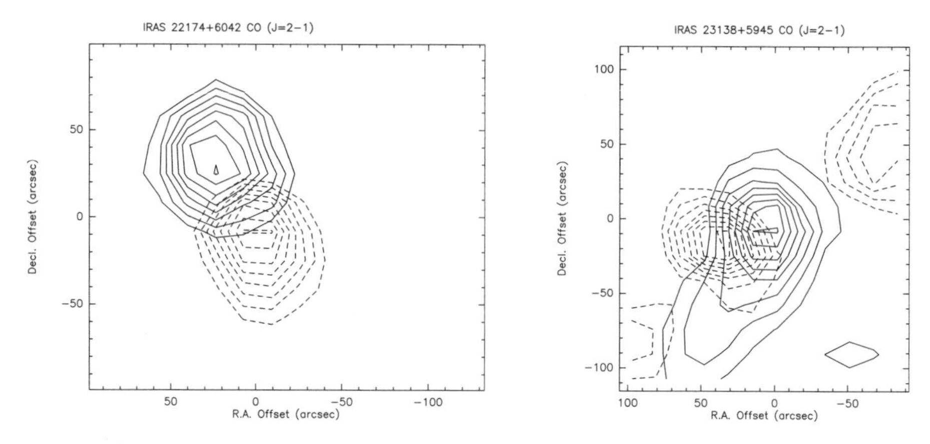

Figure 1. Contour map for the CO ($J = 2 - 1$) high-velocity outflow emission from IRAS 22174+6042 (left) and that for IRAS 23138+5945 (right). The IRAS source is at the (0,0) position. The blue-shifted emission is denoted by the solid lines and the red-shifted emission by dashed lines. The wing components are integrated over -17 to -13 km s^{-1} for the blue lobe and -9 km s^{-1} to -5 km s^{-1} for the red lobe of IRAS 22174+6042. They are integrated over -50 to -45 km s^{-1} for the blue lobe and -40 km s^{-1} to -34 km s^{-1} for the red lobe of IRAS 23138+5945.

2. The Follow-up Studies

In order to understand the outflow nature of the 289 high-velocity wing sources, sub-sample identification and follow-up studies have been performed. There were 48 sources mapped in CO ($J = 2 - 1$) and/or CO ($J = 3 - 2$) using the 10-m HHT submillimeter telescope in Arizona during several runs between 1999 and 2001.

Table 1 lists the identification of these sources. Within this sample, there are 37 sources confirmed to be high-velocity molecular outflows, mostly in a bipolar form.

IRAS 02575+6017 (AFGL 4029) is associated with the HII region IC1848. Snell et al. (1988) observed the source in CO ($J = 1 - 0$) line. There is general agreement between our map with that obtained in CO ($J = 1 - 0$). In addition, our CO ($J = 2 - 1$) map well separated the blue- and red outflow lobes along the E–W direction.

IRAS 04547+4753 (G159.14+3.26) is associated with S217. It is among the high-velocity CO wing candidate sources in the single-point survey by Shepherd & Churchwell (1996) and its far-infrared luminosity is $8.7 \times 10^3 L_\odot$. The outflow gas comes out mainly along the E–W direction but the blue components is apparently stronger than the red component.

IRAS05439+3035, is associated with a cluster of NIR sources with an approximate distance of 3–4 kpc and a luminosity of $10^4 L_\odot$ (Tapia et al. 1997). In our CO ($J = 2 - 1$) map, the high-velocity outflow is separated also in the E–W

Table 1. Summary of CO ($J = 2-1$) mapping.

Num	Source	R.A. (1950.0)	dec. (1950.0)	V_{LSR} (km s^{-1})	Mapping Comments
1	00117+6153	00:11:45.5	+61:53:01	−50.37	multiple outflow
2	00211+6549	00:21:09.6	+65:49:26	−68.66	bipolar outflow
3	00267+6511	00:26:44.9	+65:11:16	−17.37	monopolar outflow
4	01134+6429	01:13:29.0	+64:29:48	−54.09	bipolar outflow
5	02112+5939	02:11:17.6	+59:39:12	−38.98	bipolar outflow
6	02230+5231	02:23:00.2	+52:31:15	−34.89	no outflow
7	02455+6034	02:45:30.1	+60:34:35	−42.33	no outflow
8	02575+6017	02:57:35.6	+60:17:22	−38.17	bipolar outflow
9	03096+5819	03:09:36.5	+58:19:01	−38.57	no outflow
10	03422+3156	03:42:12.3	+31:56:33	−9.70	no outflow
11	04547+4753	04:54:44.7	+47:53:54	−18.36	bipolar outflow
12	05044-0325	05:04:25.8	−03:25:08	7.78	no outflow
13	05155+0707	05:15:35.1	+07:07:54	−1.68	bipolar outflow
14	05357-0710	05:35:42.0	−07:10:12	5.71	monopolar outflow
15	05361+3539	05:36:06.2	+35:39:06	−17.81	bipolar outflow
16	05379+3550	05:37:58.7	+35:50:38	−20.53	no outflow
17	05380+3608	05:38:00.9	+36:08:27	−21.34	multiple outflow
18	05439+3035	05:43:59.7	+30:35:09	−18.19	bipolar outflow
19	06046-0603	06:04:41.8	−06:03:21	11.53	bipolar outflow
20	06073+1249	06:07:23.5	+12:49:24	25.54	bipolar outflow
21	18308-0503	18:30:50.8	−05:03:27	42.89	bipolar outflow
22	19055+0459	19:05:31.4	+04:59:07	53.59	bipolar outflow
23	19242+1944	19:24:13.6	+19:44:39	−38.85	bipolar outflow
24	20020+4135	20:02:04.4	+41:35:01	7.01	no outflow
25	20044+3513	20:04:29.1	+35:13:55	−0.54	monopolar outflow
26	20067+3415	20:06:46.2	+34:15:29	13.39	bipolar outflow
27	20099+3640	20:09:54.6	+36:40:35	−36.67	bipolar outflow
28	20130+3559	20:13:00.7	+35:59:38	2.5	no outflow
29	20143+3634	20:14:18.7	+36:34:06	−1.01	bipolar outflow
30	20149+3913	20:14:54.2	+39:13:55	3.66	bipolar outflow
31	20160+3636	20:16:03.5	+36:36:09	0.25	bipolar outflow
32	20178+4046	20:17:53.0	+40:47:00	0.72	bipolar outflow
33	20220+3728	20:22:03.6	+37:28:25	−2.24	monopolar outflow
34	20265+3830	20:26:35.8	+38:30:47	−28.79	no outflow
35	20326+3757	20:32:40.2	+37:57:39	0.92	no outflow
36	20447+4441	20:44:42.8	+44:41:49	−6.31	bipolar outflow
37	20504+4931	20:50:24.9	+49:31:04	13.30	bipolar outflow
38	21161+6141	21:16:06.8	+61:41:49	0.49	bipolar outflow
39	22157+6127	22:15:42.5	+61:27:47	−10.27	monopolar outflow
40	22174+6042	22:17:27.0	+60:42:44	−10.71	bipolar outflow
41	22475+5939	22:47:30.9	+59:39:03	−49.35	bipolar outflow
42	22539+5758	22:53:56.3	+57:58:44	−54.12	multiple outflow
43	23008+5939	23:00:50.6	+59:39:02	−47.48	no outflow
44	23030+5958	23:03:04.9	+59:58:28	−51.10	bipolar outflow
45	23031+6003	23:03:07.5	+60:03:48	−51.37	multiple outflow
46	23129+6115	23:12:56.4	+61:15:06	−50.09	multiple outflow
47	23133+6050	23:13:21.5	+60:50:47	−56.24	bipolar outflow
48	23138+5945	23:13:53.5	+59:45:37	−42.90	bipolar outflow

direction. Its half-maximum angular size is $\sim 30''$, suggesting that its distance is about 3.4 kpc, provided that the average linear size of an outflow is 0.5 pc for massive molecular outflows (e.g., Shepherd & Churchwell 1996).

IRAS 19242+1944 has no prominent association. Its $V_{\rm LSR}$ equals –39.78 km s^{-1}, suggesting a distance of 12.7 kpc. Its outflow components are aligned along the NW–SE direction.

IRAS 22174+6042 is located inside a high-extinction area without any associated nebulosity. Its ^{12}CO radial velocity, $V_{\rm LSR} = -10.2$ km s^{-1}, suggests a kinematic distance of 0.9 kpc. Figure 1 (lef) shows the morphology of the outflow emission seen in CO ($J = 2 - 1$).

IRAS 22475+5939 is associated with S146 in Cep OB3. High-velocity wing emission was reported by Shepherd & Churchwell (1996). Our CO ($J = 2 - 1$) mapping revealed the spatial distribution of molecular outflows, which shows a slightly separated blue- and red-lobe distribution of the high-velocity gas.

Compared with other sources, IRAS 23138+5945 is a rather extended outflow. In our CO map, it shows three detached red wing emissions along the NW–SE direction, with the driving source centered at the peak position of the blue wing component (Figure 1 right). The survey by Bronfman et al. (1996) detected strong CS emission toward this source. The CS ($J = 2 - 1$) mapping shows that this source is associated with a dense CS core (Ao et al. 2002, in preparation).

3. The Nature of the Cold IRAS Sources

Our sample statistics indicates that, within the present detection limit of CO mapping, about 77% (or 223 sources) of the 289 high-velocity wing sources listed in Table 2 of Yang et al. (2002) are intrinsically molecular outflow sources. Modified by this identification rate, the percentage of the current CO sources in the Galaxy undergoing the outflow phase is 32%.

Acknowledgments. The authors appreciate assistance provided by staff members at Qinghai Millimetre-wave Radio Observatory at Delingha, and at SMTO in Arizona during observations.

References

Bachiller, R. 1996, ARA&A, 34, 111

Bronfman, L., Nyman, L.-A., May, J. 1996, A&AS, 115, 81

Shepherd, D.,& Churchwell, E. 1996, ApJ, 457, 267

Snell, R., Huang, Y., Dickman, R., & Claussen, M. J. 1988, ApJ, 325, 853

Tapia, M., Persi, P., Bohigas, J., & Ferrari-Toniolo, M. 1997, AJ, 113, 1769

Yang, J., Jiang, Z., Wang, M., Ju, B., & Wang, H. 2002, ApJS, 141, 157

IAU 8th Asia-Pacific Regional Meeting
ASP Conference Series, Vol. 289, 2003
S. Ikeuchi, J. Hearnshaw & T. Hanawa, eds.

Observation of Interstellar H_3^+ Using Subaru IRCS: The Galactic Center

Miwa Goto

Institute of Astronomy, University of Hawaii, 640, North A'ohoku Place, Hilo HI 96720, USA

Benjamin J. McCall

Department of Chemistry and Department of Astronomy, University of California, 601 Campbell Hall, Berkeley, CA 94720-3411, USA

Thomas R. Geballe

Gemini Observatory, 670, North A'ohoku Place, Hilo, HI 96720, USA

Tomonori Usuda, Naoto Kobayashi, Hiroshi Terada

Subaru Telescope, 650, North A'ohoku Place, Hilo, HI 96720, USA

Takeshi Oka

Department of Astronomy and Astrophysics and Department of Chemistry, The University of Chicago, Chicago, IL 60637, USA

Abstract. The high resolution infrared absorption spectrum of H_3^+ has been observed towards the luminous Galactic center sources GCS 3-2 and GC IRS 3. With the wide wavelength coverage of Subaru IRCS, six absorption lines of H_3^+ have been detected from 3.5 to 4.0 μm, three of which are new. In particular the $R(3, 3)^l$ transition arising from the metastable (3, 3) level has been detected, indicating the existence of a high temperature cloud. At least four velocity components are found in the H_3^+ absorption profile. The observations have revealed a striking difference between the absorption profiles of H_3^+ and CO, demonstrating the complementary nature of the H_3^+ and CO as astrophysical probes.

1. Introduction

H_3^+ is the third pure hydrogenic species after H and H_2 that can be used for astronomical observations. Since its discovery towards young stellar objects that are deeply embedded in molecular clouds (Geballe and Oka 1996), H_3^+ has been observed in many dense molecular clouds (McCall et al. 1999). Quite unexpectedly, H_3^+ has also been observed in diffuse clouds with similar column densities as in dense clouds (McCall et al. 1998; Geballe et al. 1999; McCall

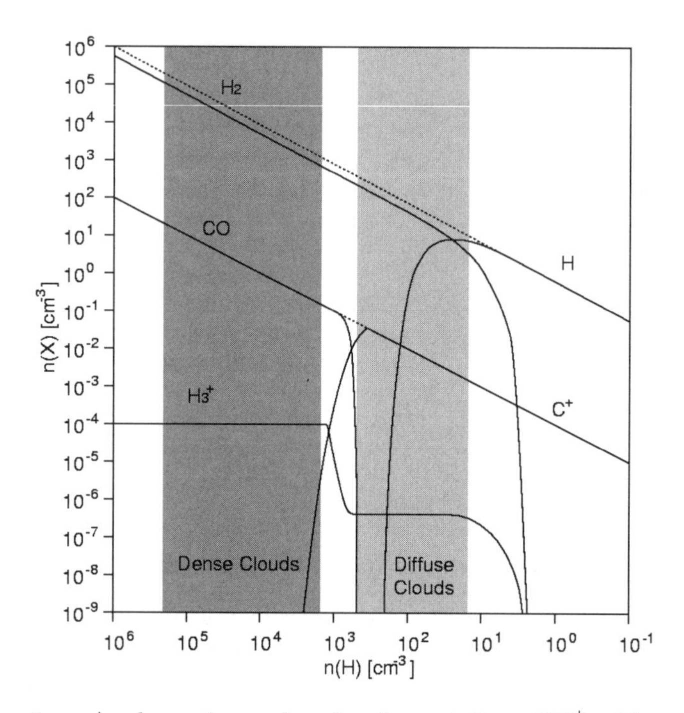

Figure 1. A schematic number density variation of H_3^+ with respect to hydrogen, along with that of major molecules and ions in the interstellar medium.

et al. 2002). This ubiquity and abundance makes H_3^+ a powerful astrophysical probe more generally applicable for the study of interstellar matter than H_2.

H_3^+ is produced by cosmic ray-ionization of H_2 into H_2^+ followed by the very efficient Langevin reaction $H_2^+ + H_2 \rightarrow H_3^+ + H$. It plays the pivotal role in interstellar chemistry as a universal proton donor (acid) and it initiates a network of chemical reactions. The simple chemistry of H_3^+ has the special characteristic that its number density is constant and is independent of the cloud density for typically dense or diffuse clouds (Fig. 1). Thus H_3^+ can be used to measure the dimension of the cloud (Gaballe & Oka 1996; McCall et al. 1998, 1999, 2002). This in turn gives the average number density of the cloud if the column density of H_2 is estimated from extinction. The presence of two rotational levels (J, K) = (1, 0) and (1, 1) separated by 32.87 K, and the rapid equilibration between the two states by the Langevin reaction allows H_3^+ to be used to measure the temperature of the cloud.

These excellent characteristics of H_3^+ as a yardstick, densitometer, and thermometer of clouds are being applied to the survey of the Galactic center whose lines of sight contain diffuse and dense clouds in intervening spiral arms as well as the molecular complex close to the nucleus. Two infrared sources, GCS 3-2 in the Quintuplet cluster and GC IRS 3 near Sgr A have been selected because of their high luminosity in the L-band. We report the initial findings of the study (Goto et al. submitted to PASJ).

2. Observations and Results

The H_3^+ spectra were observed using the Infrared Camera and Spectrograph (IRCS) with the 8.2-m Subaru Telescope on Mauna Kea. The IRCS with an échelle and a cross dispersing grating allowed us to cover a wide wavelength region containing not only five of the six lines arising from the lowest $J = 1$ levels but also many spectral lines from higher J levels with two grating settings. A $0.15'' \times 4.5''$ slit width was used to achieve a spectral resolution of 20 000.

Observed spectra are shown in Fig. 2 for GCS 3-2 (left column) and GC IRS 3 (right column). The atmospheric transmission curves are also shown. All five absorption lines in our coverage starting from the lowest $J = 1$ levels [$R(1, 0)$, $Q(1, 0)$, $R(1, 1)^l$, $R(1, 1)^u$ and $Q(1, 1)$] are detected, while those starting from higher J levels are negative except $R(3, 3)^l$ which arise from the (3, 3) metastable rotational level.

The lines with $J = 1$ show several discrete kinematic components with peaks at LSR velocities of -110 to -140 km s^{-1}, -50 to -60 km s^{-1}, and 0 km s^{-1}. GC IRS 3 shows an additional broad pedestal peak at 20 km s^{-1} and a wing at $+50$ km s^{-1} on the pedestal. A weak absorption is seen at -170 km s^{-1}.

3. Discussions

3.1. Cloud Components

Observed spectral lines of the $R(1, 1)^l$ and the $R(3, 3)^u$ transitions of H_3^+ in GC IRS 3 are compared with spectra of H (Liszt et al. 1985), CO (Geballe et al. 1989) and H_2CO (Güsten & Downes 1981) in Fig. 3. The three sharp lines of H_3^+ at -140 km s^{-1}, -60 km s^{-1} and 0 km s^{-1} have corresponding peaks in H and H_2CO and perhaps in CO. The weakness of the H_2CO and CO spectra suggests that those H_3^+ are in the diffuse interstellar medium in the "expanding molecular ring" (Kaifu, Kato, & Iguchi 1972; Scoville 1972) the"3 kpc arm" and local clouds, respectively.

The strong and saturated infrared CO absorption towards GC IRS 3 and radio H_2CO absorption in Sgr A at the velocity of $+50$ km s^{-1} clearly represent dense clouds. It is remarkable that the H_3^+ lines arising from the $J = 1$ levels are weak at this velocity (Fig. 2 and 3). This indicates the presence of a dense cloud with a short path length in front of the infrared source. The complementary nature of CO and H_3^+, the former representing the total amount of molecules while the latter simply the path length, is clearly seen here. It is surprising that the intensity of the $R(3, 3)$ absorption is the strongest at this velocity. This is also noted for the spectrum of GCS 3-2.

3.2. The (3, 3) Metastable State

The detection of the $R(3, 3)$ line is a breakthrough and it introduces a new dimension into the study of interstellar H_3^+. The detection and our non-detection of other high J transitions is reasonable since the (3, 3) level is metastable, that is, the spontaneous emission from this level to a lower level by centrifugal distortion induced dipole moment (Pan & Oka 1986) is forbidden.

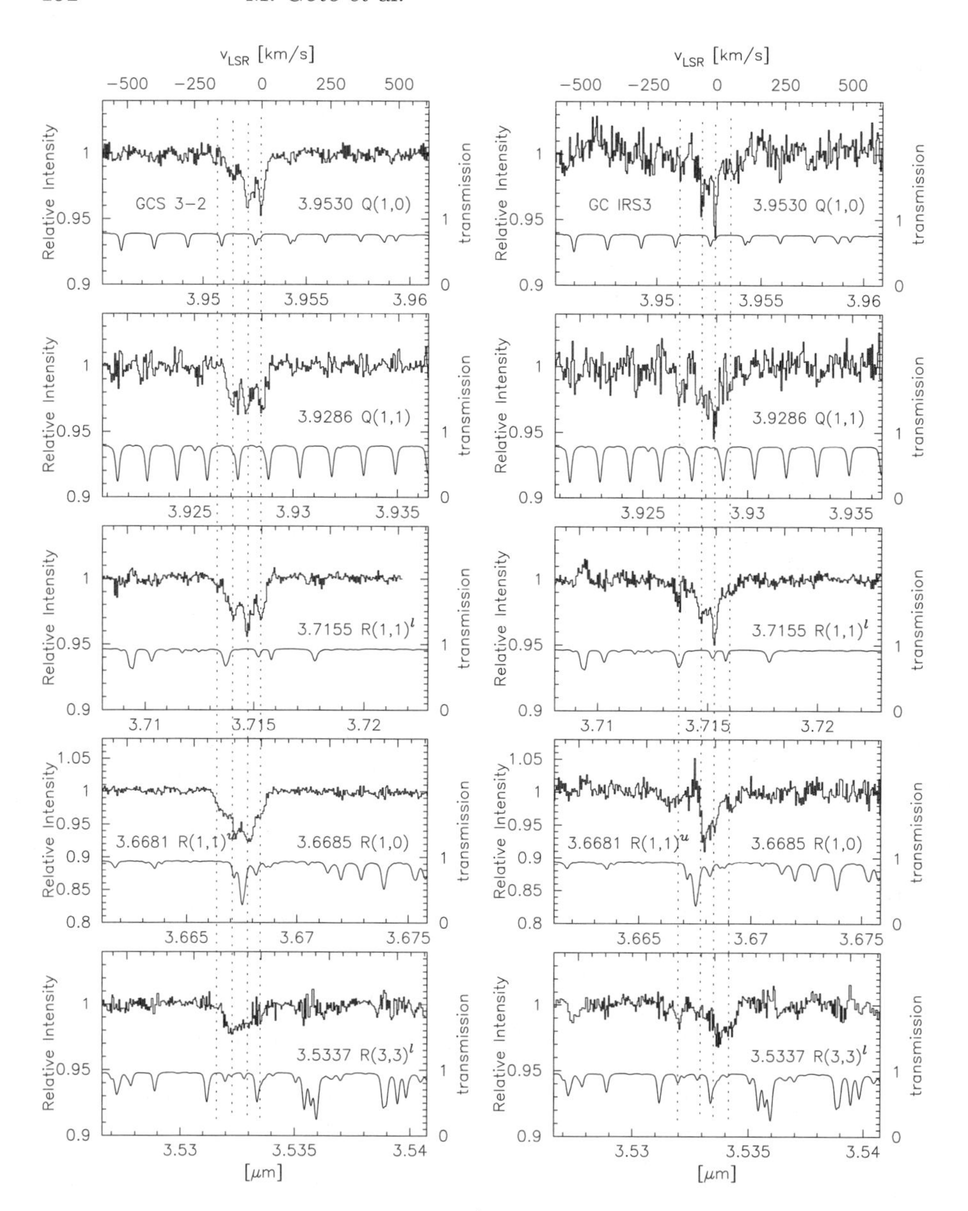

Figure 2. The total detected H_3^+ absorption spectra towards GCS 3-2 (left column) and GC IRS 3 (right column).

References

Geballe, T. R., Baas, F. & Wade, R. 1989, A&A, 208, 255

Geballe, T. R. & Oka, T. 1996, Nature, 384, 334

Geballe, T. R., McCall, B. J., Hinkle, K. H. & Oka, T. 1999, ApJ, 510, 251

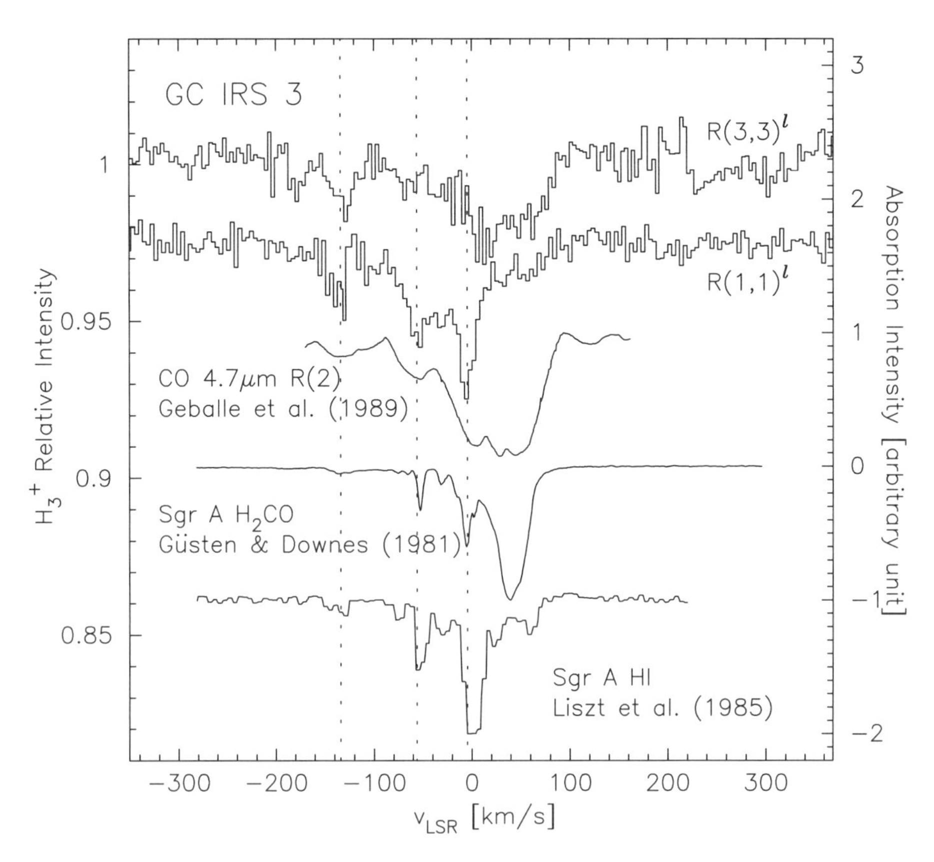

Figure 3. Comparison of the line profiles of H_3^+ $R(1,1)^l$ and $R(3,3)^l$ lines along with those of CO , H_2CO and towards GC IRS 3.

Güsten, R. & Downes, D. 1981, A&A, 99, 27
Kaifu, N., Kato, T. & Iguchi, I. 1972, Nature, 238, 109
Liszt, H. S., Burton, W. B. & van der Hulst, J. M. 1985, A&A, 142, 237
McCall, B. J., Geballe, T. R., Hinkle, K. H. & Oka, T. 1998, Science, 279, 1910
McCall, B. J., Geballe, T. R., Hinkle, K. H. & Oka, T. 1999, ApJ, 522, 338
McCall, B. J. et al. ApJ, 2002, 567, 391
Pan, F.-S. & Oka, T. 1986, ApJ, 305, 518
Scoville, N. Z. 1972, ApJ, 175, L127

IAU 8th Asia-Pacific Regional Meeting
ASP Conference Series, Vol. 289, 2003
S. Ikeuchi, J. Hearnshaw & T. Hanawa, eds.

The Formation of Molecular Clouds: The Origin of Supersonic Turbulence

Shu-ichiro Inutsuka

Department of Physics, Kyoto University, Kyoto, 606-8502, Japan

Hiroshi Koyama

National Astronomical Observatory, Mitaka, Tokyo 181-8588, Japan

Abstract. We have done two- and three-dimensional magnetohydrodynamical simulations of the propagation of a shock wave into the interstellar medium by taking into account radiative heating/cooling, thermal conduction, and physical viscosity. Our results show that the thermal instability in the post-shock gas in the interstellar medium produces high-density molecular cloudlets which have a supersonic velocity dispersion. The dynamical evolution driven by thermal instability in the post-shock layer is an important basic process for the transition from warm gases to cold molecular gases, because the shock waves are frequently generated by supernovae in the Galaxy.

1. Introduction

It is well-known that the physical conditions of the cold interstellar medium (ISM) and molecular clouds are characterized by a clumpy substructure and a turbulent velocity field. Stars are formed in molecular clouds. The maintenance and dissipation processes of the turbulence in molecular clouds are supposed to be important in the theory of star formation. The understanding of the origin of the turbulence and the internal substructure of molecular clouds has fundamental importance for a consistent theory of star formation and the ISM.

We propose that the clumpiness in clouds arises naturally from their formation through thermal instability which acts on time scales that can be much shorter than the duration of the interstellar shocks (e.g., supernova remnants and galactic spiral shocks). Koyama & Inutsuka (2000) have done one-dimensional hydrodynamical calculations for the propagation of a strong shock wave into warm neutral medium (WNM) and cold neutral medium (CNM), including detailed thermal and chemical processes. They have shown that the post-shock region collapses into a cold layer as a result of thermal instability. They expect that this layer will break up into very small cloudlets which have different translational velocities. Subsequent two-dimensional simulations by Koyama & Inutsuka (2002) have confirmed this expectation: the fragmentation of the shock-compressed layer indeed provides turbulent condensations in interstellar clouds. In this short article, we show that the fragmentation process of the shock-compressed layer remains effective in three-dimensional magnetohydro-

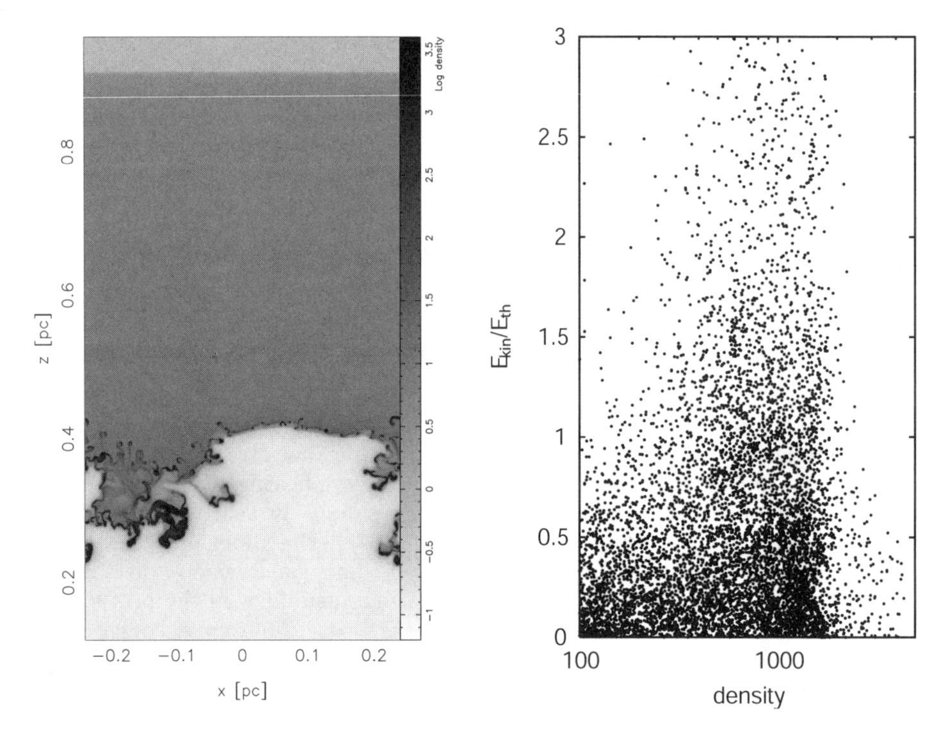

Figure 1. *Left:* Density distribution of the shock-compressed layer. The snapshot is taken at $t = 1.1$ Myr. *Right:* The ratio of the kinetic energy to the thermal energy is plotted as a function of density for every (high-density) grid cell of the simulations shown in the left panel.

dynamic simulations with radiative cooling/heating, thermal conduction, and physical viscosity.

2. Magnetohydrodynamical Simulations

We solve magnetohydrodynamical equations, which include radiative heating, cooling, thermal conduction, and physical viscosity. The detailed descriptions are found in Koyama & Inutsuka (2000, 2002). Thermal conduction could stabilize and even erase a density perturbation whose length scale is smaller than the Field length, $\lambda_{\rm F} = \sqrt{KT/\rho^2\Lambda}$, where K denotes the coefficient of thermal conduction, and Λ is the cooling function (Field 1965). The Field length also determines the width of the intermediate region between the cold clumpy medium and surrounding warm medium, where heating and cooling due to the thermal conduction balance the radiative heating and cooling. A calculation with large grid spacing cannot resolve the conduction front, and the grids on the surface of a cold clump become numerically unstable. Thus a high spatial resolution is required for the calculation of thermal instability. We use 1024×2048 Cartesian

Figure 2. The volume rendering of high-density region of the shock-compressed layer. The arrow denotes the location of the shock front. a) The non-magnetized case. b) The magnetized case. The initial weak magnetic field is in the x-direction and uniform. The magnetic pressure becomes comparable to the gas pressure in the post-shock region.

grid points, covering a $0.48\,\mathrm{pc} \times 1.92\,\mathrm{pc}$ region, so that the spatial resolution is $0.000938\,\mathrm{pc} = 188\,\mathrm{AU}$.

To investigate the shock propagation into WNM we consider a plane-parallel shocked layer of gas. We take the z-axis to be perpendicular to the layer. Uniform flow approaches the layer from the upper side with a velocity $V_z = -26$ km/s and density $n = 0.6$ cm^{-3} and temperature $T = 6000$ K. We assume that the lower side of the shocked layer is occupied by hot, tenuous gas (density $n = 0.12\,\mathrm{cm}^{-3}$) with a high pressure $P/k_\mathrm{B} = 3.5 \times 10^4\,\mathrm{K\,cm}^{-3}$. We set up initial density fluctuations $\delta\rho/\rho \sim 0.05$.

Figure 1a shows density distribution at $t =$ 1.1 Myr. Density discontinuity at $z \sim 0.9\,\mathrm{pc}$ corresponds to the shock front. The shock front itself remains stable in 10^6 years in this problem. Behind the shock front, cooling dominates heating and temperature decreases monotonically. The ISM in the range of 300 – 6000 K is thermally unstable. Thus, the thermally collapsing layer is subject to thermal instability.

Figure 2 shows the snapshots of our three-dimensional calculations with and without a magnetic field. We take the x- and y-axes to be in the layer. Uniform flow approaches the layer from the left-hand side with a velocity $V_z = -13.6$ km/s and density $n = 0.3$ cm^{-3} and pressure $P/k_\mathrm{B} = 2 \times 10^3\,\mathrm{K\,cm}^{-3}$. We assume that the right-hand side of the shocked layer is occupied by hot, tenuous gas (density $n = 0.12\,\mathrm{cm}^{-3}$) with a high pressure $P/k_\mathrm{B} = 8 \times 10^3\,\mathrm{K\,cm}^{-3}$. We set up initial density fluctuations $\delta\rho/\rho \sim 0.05$. We set out-going boundary condition for the z-direction, and periodic boundary condition for the x- and y-directions. The initial uniform magnetic field in the magnetized case corresponds to high plasma β: $P/B_x^2 = 2 \times 10^4$ in WNM, $= 8 \times 10^4$ in the hot medium. In the post-shock layer, however, the magnetic pressure becomes comparable to the gas pressure. This is because the density becomes larger by a factor of 10^2, so that magnetic pressure becomes about 10^4 times the initial one, although thermal pressure does not increase so much owing to the efficient cooling. The snapshots in Figure 2a and 2b are taken at times $t =$ 3.88 Myr and 6.48 Myr. The arrow denotes the location of the shock front. The shock front itself remains stable in multiples of 10^6 years in this plane-parallel problem.

The cold clumps have considerable translational velocity dispersion. In Figure 1b the ratio of the kinetic energy to the thermal energy is plotted as a function of density for every grid cell of the simulations shown in Figure 1a. A substantial fraction (in volume) of high-density gas is supersonic. The typical velocity dispersion is a few km/s. The velocity of the nonlinearly developed perturbation has an upper limit that is essentially determined by the sound speed of the warmer medium ($\approx$ 10 km/s), because the driving force of the instability is the pressure of the less dense warmer medium. This velocity smaller than the sound speed of the WNM is highly supersonic with respect to the sound speed of the cold medium. Thus we can understand why the supersonic velocity dispersion of the cold medium is comparable to (but less than) the sound speed of the WNM.

In the radiative shocked layers, the gases lose thermal energy through radiative cooling. Thus, the initial kinetic energy of the pre-shock gas (in the comoving frame of the post-shock gas) is converted to radiation energy, which escapes from the system. If, however, the post-shock gas becomes dynamically unstable by the thermal instability as shown in this paper, then a considerable fraction of the thermal energy is transformed into the kinetic energy of the translational motions of the cold cloudlets, which does not easily escape from the system. Therefore we can attribute the origin of interstellar turbulence to the conversion of the gas energy via thermal instability. The sizes of the cold clumps are a few orders of magnitude smaller than the Jeans length, $\lambda_J = 1.2\ \mathrm{pc}\sqrt{(T/20\ \mathrm{K})(2000\ \mathrm{cm}^{-3}/n)}$. Thus, each cloud is gravitationally stable.

3. Remarks

The ISM is frequently compressed by supernova remnants (McKee & Ostriker 1977), and thus the real ISM is multiply superposed by the effects considered in this paper. The intersection regions of SNe might produce larger two-phase clouds that consist of cold clouds embedded in the warm medium. The average column density of the two-phase cloud has an upper limit determined by the attenuation of external radiation (Inutsuka & Koyama 1999). Our next work will include the study of the (radiation) hydrodynamical evolution from two-phase clouds to genuinely molecular clouds.

References

Field, G. B. 1965, ApJ, 142, 531

Inutsuka, S. & Koyama, H. 1999, in Star Formation 1999, ed. T. Nakamoto, (Nobeyama Radio Observatory), 112

Inutsuka, S. & Koyama, H. 2002a, Ap & SS, 281, 67

Koyama, H. & Inutsuka, S. 2000, ApJ, 532, 980

Koyama, H. & Inutsuka, S. 2002, ApJ, 564, L97

McKee, C. F. & Ostriker, J. P. 1977, ApJ, 218, 148

IAU 8th Asia-Pacific Regional Meeting
ASP Conference Series, Vol. 289, 2003
S. Ikeuchi, J. Hearnshaw & T. Hanawa, eds.

Shocked Atomic and Molecular Gas in Supernova Remnants

Bon-Chul Koo

SEES, Seoul Natioanl University, Seoul 151-742, Korea

Abstract. HI 21 cm line emission from fast-moving atomic gas has been detected toward 27 SNRs. The HI gas was usually thought to be part of an expanding HI shell, as expected for an old supernova remnant in a uniform medium. But recent studies show that a good ($\gtrsim$ 20–30%) fraction of these remnants are interacting with molecular clouds and that the fast-moving HI gas in some of these remnants is likely to have been part of the molecular cloud at first. I review the results of recent studies on SNRs interacting with molecular clouds and discuss the implications of the scarcity of supernova remnants with expanding HI shells.

1. Introduction

Supernova explosions play a major role in the evolution of the interstellar medium (ISM) and understanding how a supernova remnant (SNR) interacts with the ISM is important. According to theoretical calculations on the evolution of SNRs (e.g., Cioffi et al. 1988), the supernova shock becomes radiative when the SNR is older than a few ten thousand years in a uniform medium with hydrogen density of 1 cm^{-3}, so that we expect to observe a fast expanding HI shell surrounding hot, diffuse X-ray emitting gas for an old SNR. If we observe an old SNR in HI 21 cm line with enough spatial resolution, therefore, we may observe a ring whose size varies with the line-of-sight velocity. On the other hand, if the SNR is not resolved by the telescope beam, we may observe a wide rectangular line because the mass per unit line-of-sight velocity interval is constant.

Most SNRs, however, are located in the Galactic plane where the Galactic foreground/background HI emission is strong, so that what we actually observe is a very faint emission superposed on the strong background emission. But we can still be sure that the emission is associated with the SNR if the velocity is forbidden by the Galactic rotation and if the emission is confined to the SNR. One of the best examples with a resolved expanding HI shell is the SNR W44 (Koo & Heiles 1995). The position-velocity diagram in Figure 1 was obtained by using the Arecibo telescope (FHWM = 3.′3) and clearly shows a portion of an expanding HI shell between $v_{\rm LSR} \simeq$ 130 and 210 km s^{-1}. The velocity is much greater than the maximum LSR velocity ($\simeq$ 90 km s^{-1}) permitted by the Galactic rotation toward this direction ($\ell = 34.^\circ 7$). The HI spectrum toward the SNR HB 21 in Figure 1, on the other hand, was obtained by a telescope with a large beam (FWHM = 36′) and shows an excess emission at $v_{\rm LSR} \gtrsim$ 40 km s^{-1}

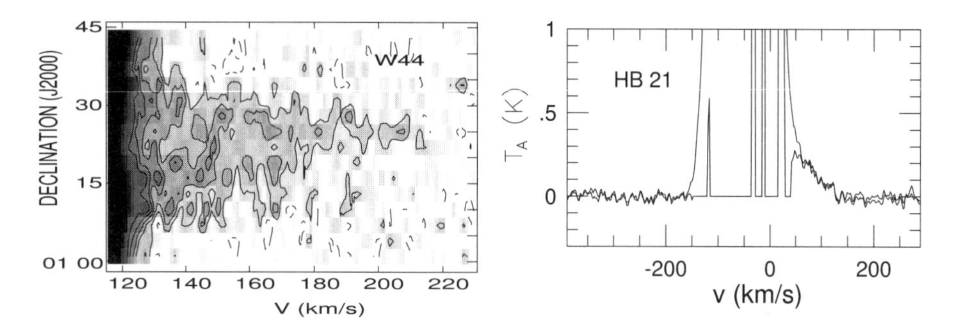

Figure 1. (left) Position-velocity diagram of the expanding HI shell in W44 (Koo & Heiles 1995). (right) HI 21 cm spectrum toward HB 21 (Koo & Heiles 1991). The thick solid line is the total emission, while the thin solid line is the excess emission brighter than the surrounding region.

(Koo & Heiles 1991). The maximum LSR velocity toward the direction of HB 21 ($\ell = 89.^\circ 0$) is only a few km s^{-1}. Koo & Heiles (1991) did a HI 21 cm line survey toward 103 Northern Galactic SNRs and detected such excess emission toward 27 SNRs (including their rank 2 SNRs). High-resolution ($\lesssim 3'$) images have been obtained only for 6 of those SNRs, e.g., W44, W51C, CTB 80, DR 4, VRO 42.05.01, and IC 443 (Koo & Heiles 1995; Koo & Moon 1997a; Koo et al. 1993; Braun & Strom 1986).

In this paper, I want to address the following question: "Is what we observe what we expected?", or "Does the fast-moving HI gas detected toward the SNRs represent the HI shells that we expect in the *standard* SNR evolution model?" where the 'standard' model means a model in a uniform homogeneous medium. The answer may not be yes because a good fraction of the SNRs with fast-moving HI gas (including W44 and HB 21) are found to be interacting with molecular clouds. The HI gas in these (and also possibly in other) SNRs is likely to have been part of molecular clouds at first. In the following, I review the results of recent studies on the SNRs interacting with molecular clouds and discuss the implications of the scarcity of the SNRs with expanding HI shells.

2. SNRs Interacting with Molecular Clouds

Type II and Ib SNRs have massive ($\gtrsim 8\ M_\odot$) stars as their progenitors and may interact with molecular clouds. People have searched for such SNRs from 1970s because we can learn the physics and chemistry of molecular shocks, how the molecular clouds are disrupted by SN shocks, and how the molecular clouds affect the evolution of SNRs. But the early studies are mostly based on circumstantial evidences, such as spatial coincidences or spatial correlations on the sky, e.g., Huang & Thaddeus (1986), Tatematsu et al. (1987, 1990). No direct evidences had been detected for most SNRs except IC 443, for which broad emission lines from shocked CO and OH molecules were detected (DeNoyer 1979a, 1979b).

In 1990s, direct evidences have been discovered in several SNRs. One of the strongest evidences is the broad molecular lines. The broad lines indicate an acceleration by strong shocks. The broad-line emitting molecules cloud be either the molecules reformed behind a dissociative shock or the molecules slowly-accelerated by a non-dissociative shock. The lines usually have high (> 1) ratios of high- to low-transition CO lines. The high ratio indicates that the molecular gas is warmer and denser than general molecular clouds, which is presumably due to the shock heating and shock compression. There are now at least six SNRs with broad molecular lines from the shocked gas: W28 (Arikawa et al. 1999), 3C 391 (Reach & Rho 1999), W44 (Seta et al. 1998), W51C (Koo & Moon 1997b), HB 21 (Koo et al. 2001), and the classical source IC 443. Table 1 summarizes their basic properties where $v_{\rm max}$ is the maximum *line-of-sight* velocity of the CO emission with respect to the systematic velocities of the SNRs. Shocked atomic gas has been detected in all of the SNRs except 3C 391 in Table 1. Conversely, about 20% of the SNRs with fast-moving HI gas has broad molecular lines. Systematic search would increase the fraction. Even if a SNR is interacting with a molecular cloud, the broad lines may be absent if the shock is dissociative and molecules have not reformed.

Another indication of the interaction is the 1720-MHz maser line ($^2\Pi_{3/2}$, $J = \frac{3}{2}$, $F = 2 \rightarrow 1$) of the OH molecule (see the review by Wardle & Yusef-Zadeh 2002 and references therein). This line was first detected toward SNRs as early as 1968, but its association with the shocked material in SNRs was established only within the past decade. About 170 SNRs have been observed, and the maser emission has been detected toward 17 SNRs (Green et al. 1997). All the SNRs in Table 1 except HB 21 have associated OH masers, while 7 ($\sim 30\%$) out of the 27 SNRs with fast-moving HI gas have associated OH masers. According to theoretical calculations (Elitzur 1976; Lockett, et al. 1999), the 1720 MHz OH masers arise only in slow, non-dissociative C-shocks with a large (10^{16}–10^{17} cm^{-2}) OH column density which can be met only when the shock is seen tangentially.

There are other signatures such as the shock-excited infrared emission lines from H_2 and other molecules (e.g., Burton et al. 1988; Oliva et al. 1990; Graham et al. 1991; Reach & Rho 1998, 2000). Also the infrared atomic and ionic forbidden lines could be used to infer the properties of the molecular shocks (e.g., Draine et al. 1983; Hollenbach & McKee 1989).

Table 1. Supernova Remnants with Broad CO Emission Lines

Name	Distance (kpc)	Size (pc)	$v_{\rm max}$ (km s^{-1})	OH maser	Thermal composite	Shocked HI
W28	1.8	25	$\lesssim 30$	y	y	y
3C 391	9	18×13	$\lesssim 30$	y	y	n
W44	3	31×24	$\lesssim 20$	y	y	y
W51C	6	88×66	$\lesssim 50$	y	y	y
HB 21	0.8	28×21	$\lesssim 20$	n	y	y
IC 443	1.5	20	$\lesssim 80$	y	y	y

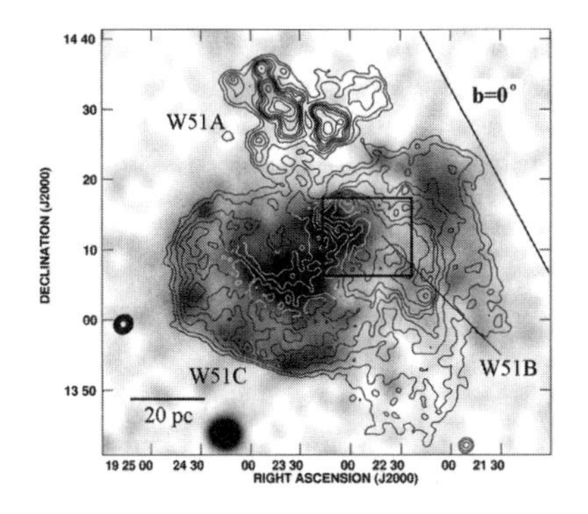

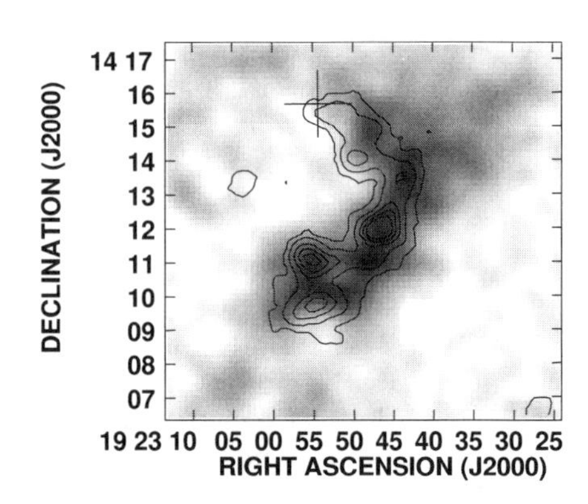

Figure 2. (left) 330 MHz radio contour map of the W51 complex (Subrahmanyan & Goss 1995) overlaid on the *ROSAT* X-ray image. The box marks the area on the right panel, where the shocked gas has been discovered. (right) Map of the shocked CO (contours) superposed on the map of the shocked HI (grey scale) (cf. Koo & Moon 1997b). The cross marks the position of the OH maser.

3. Shocked HI and Shocked Molecules

Shocked atomic gas has been detected in all of the SNRs except 3C 391 in Table 1. Among them, W51C is the only SNR where a detailed comparison between the shocked HI and the shocked CO has been made (Koo & Moon 1997b; see Koo 2002 for a review).

3.1. W51C

W51C appears in radio continuum as an incomplete shell of $\sim 30'$ extent with its upper portion open (Fig. 2). It is part of the radio complex W51, which is composed of two massive star-forming regions, W51A and W51B, and the SNR W51C. In X-rays, the SNR appears centrally-brightened, but it also has shell-like features along the eastern and western boundaries. The X-ray emission is thermal and its temperature $kT \simeq 0.3$ keV. The distance to the SNR is 6 kpc.

Fast-moving atomic and molecular gases have been detected in the western part of the SNR. Their velocities, $v_{\rm LSR} = 80$–150 km s^{-1} for HI and $v_{\rm LSR} = 80$–120 km s^{-1} for CO, are much greater than the maximum velocity (55–60 km s^{-1}) permitted by the Galactic rotation toward this direction, so that the emission is clearly distinguishable from the background emission. The fast-moving HI and CO gas is distributed along an arc structure, which is located at an interface between the X-ray bright central region of the W51C SNR and a large molecular cloud (Fig. 2; see also Fig. 9 of Koo & Moon 1997a). This positional correlation strongly suggests that both the HI and CO gas have been produced by the same SNR shock propagating through the molecular cloud. The large amount of the fast-moving HI gas indicates that the the shock is a fast ($\sim$ 100 km

s^{-1}), dissociative J-shock. Most of the fast-moving CO is considered to be the molecules that had been destroyed and reformed behind the shock front.

The shocked gas generally has the CO/HI abundance ratio less than 4×10^{-5}. The upper limit corresponds to the ratio of CO to hydrogen nuclei in typical molecular clouds. Therefore it implies that almost all of the hydrogen nuclei in the shocked gas are atomic, not molecular. The shock velocity is ~ 100 km s^{-1} and the HI column density perpendicular to the shock front is few times 10^{20} cm^{-2}, which is smaller than the value required for the complete reformation of H_2 molecule (Hollenbach & McKee 1989). Therefore, the HI and CO results suggest that the SNR shock propagating through the molecular cloud in W51C is a fast, dissociative J-shock, and is old enough to reform CO molecules, but not old enough to completely reform H_2. But the detection of an 1720 MHz OH maser in the northern portion of the filamentary structure in Figure 2 suggests that there could be a slow, non-dissociative C-shock too, which may represent an area where the SNR shock is interacting with a dense molecular clump. In any case, the fast-moving HI gas in the SNR W51C is not a shell that we expect in the standard theory, but is most likely to have been part of a molecular cloud at first.

3.2. Other SNRs

No comparison as detailed as in W51C has been made for the other SNRs. IC 443 is the only other SNR where high-resolution ($\lesssim 1'$) HI data is available, but its relationship to the shocked molecular gas has been explored only partly (Burton et al. 1988; Inoue et al. 1993). The fast-moving HI gas in these SNRs could have been produced either by a dissociative molecular shock as in W51C, or due to radiative atomic shocks propagating through the interclump medium of $n_0 \sim 10$ cm^{-3} as have been suggested by Chevalier (1999).

4. Interacting SNRs in X-rays

As in radio, SNRs are basically classified into a shell type or a filled-center based on their X-ray morphology (e.g., Seward 1990). Young SNRs such as Cas A or an old SNR such as the Cygnus Loop are typical shell-type SNRs. The X-ray emission from these shell SNRs is of thermal origin. On the other hand, the filled-center SNRs such as the Crab nebula usually have a pulsar and their X-ray emission is the non-thermal synchrotron emission from pulsar wind nebula. There are also SNRs with irregular X-ray morphology.

The SNRs with broad emission lines, however, appear shell-type in radio but centrally-brightened with *thermal* X-rays. These 'thermal composite' (Jones et al. 1998) or 'mixed-morphology' (Rho & Petre 1998) SNRs draw our attention because, for a SNR expanding in a uniform, homogeneous medium, the Sedov-Taylor solution tells us that the SNR should be limb-brightened, not centrally-brightened, in X-rays. According to Rho & Petre, 15 SNRs, or about $\sim 25\%$ of the X-ray detected SNRs belong to this category. Since all the SNRs in Table 1 are in this category, at least 40% of the thermal-composite SNRs have strong evidences for the interaction with molecular clouds. Among the 27 SNRs with fast-moving HI gas, 15 have been detected in X-ray and 9 (60%) belong to this category.

Several explanations have been proposed for the centrally-brightened thermal X-rays. Probably the most popular one is the so-called evaporation model of White & Long (1991). In this model, the ambient medium is clumpy, so that there are dense clumps inside the SNR which survived the passage of the SN shock. Later in the evolution of the SNR, these clumps deposit extra mass through thermal evaporation and can smooth out the density and temperature gradient. Alternatively, it could be the conduction in hot interior as have been proposed by Cox et al. (1999) and Shelton et al. (1999) for W44. In some old SNRs, the shock speed could be too low to newly generate X-ray emitting hot plasma. The selective absorption of the soft X-rays by the intervening IS gas could be partly responsible too. None of these models, however, explicitly invoke the interaction with molecular clouds

5. Concluding Remark

The number of SNRs where an excess emission from fast-moving HI gas has been detected is 27. A good fraction (20–30%) of them has broad molecular emission lines and/or OH (1720 MHz) masers, which are strong evidences for the interaction with molecular clouds. 15 out of 27 SNRs have been detected in X-ray and 60% of them have centrally-peaked thermal X-rays, which might be another evidence for the interaction. The fast-moving HI gas in the SNRs interacting with molecular clouds is likely to have been part of molecular clouds at first. Therefore, we have a very limited number of SNRs which possibly have an HI shell expanding at 100–200 km/s as in the standard theory. In fact, WE DON'T HAVE A SINGLE OLD SNR THAT WOULD FIT INTO THE STANDARD THEORY. So "Where are all those old SNRs with expanding HI shells?"

Suppose that the SN rate in the Galaxy is $\sim 1/50$ yr and that the period that a SNR has an HI shell expanding at 100–200 km s^{-1} is about 20 000 yr. Then there should be about 400 SNRs with fast-moving HI gas, much greater than what we have discovered. The Galactic background/foreground emission might be responsible partly, but probably not entirely, because the expansion velocity 100–200 km s^{-1} is large, large enough to be discernable from the background emission. One possibility is that there are that many SNRs but we have not detected. This is possible because the HI survey by Koo & Heiles (1991) was made toward the known Galactic SNRs, most of which have been identified in radio. If old SNRs are faint, too faint to be identified in radio, then a survey based on the current SNR catalog might miss most of the SNRs with HI shells. Alternatively, there may not be that many SNRs. According to the three phase model of the ISM by McKee & Ostriker (1977), the SNRs develop dense shells when they are very old ($t = 8 \times 10^5$ yrs), at which time their radius is 180 pc. But this radius is very large, so that it is possible that the SNRs come into equilibrium with the interstellar pressure or overlap with other SNRs before dense shell formation, in which case HI shells are not formed. The SNRs with HI shells may be the ones interacting with molecular clouds. The truth might be between the two possibilities.

Acknowledgments. I wish to thank the organizers of the workshop for their hospitality and for financial support. This work was supported in part by the Korea Research Foundation Grant (KRF-2000-015-DP0446).

References

Arikawa, Y., Tatematsu, K., Sekimoto, Y. & Takahashi, T. 1999, PASJ, 51, L7

Burton, M. G., Geballe, T. R., Brand, P. W. J. L. & Webster, A. S. 1988, MNRAS, 231, 617

Chevalier, R. A. 1999, ApJ, 511, 798

Cox, D. P. et al. 1999, ApJ, 524, 179

DeNoyer L. K. 1978, MNRAS, 183, 187

DeNoyer, L. K. 1979a, ApJ, 228, L41

DeNoyer, L. K. 1979b, ApJ, 232, L165

Draine, B. T., Roberge, W. G. & Dalgarno, A. 1983, ApJ, 264, 485

Elitzur, M. 1976, ApJ, 203, 124

Giovanelli, R. & Haynes, M. P. 1979, ApJ, 230, 404

Graham, J. R., Wright, G. S., Hester, J. J., & Longmore, A. J. 1991, AJ, 101, 175

Green, A. J., Frail, D. A., Goss, W. M. & Otrupcek, R. 1997, AJ, 114, 2058

Hollenbach, D. & McKee, C. F. 1989, ApJ, 342, 306

Huang, Y.-L. & Thaddeus, P. 1986, ApJ, 309, 804

Inoue, M. Y. et al. 1993, PASJ, 45, 539

Jones, T. W. et al. 1998, PASP, 110, 125

Koo, B.-C. 2002, in the HI 50 Conference, Seeing through the Dust (in press)

Koo, B.-C. & Heiles, C. 1991, ApJ, 382, 204

Koo, B.-C. & Heiles, C. 1995, ApJ, 443, 679

Koo, B.-C. Lee, J.-J., & Seward, F. D. 2002, AJ, 123, 1629

Koo, B.-C. & Moon, D.-S. 1997a, ApJ, 475, 194

Koo, B.-C. & Moon, D.-S. 1997b, ApJ, 485, 263

Koo, B.-C. Rho, J., Reach, W. T., Jung, J. H., & Mangum, J. G. 2001, ApJ, 552, 175

Koo, B.-C., Yun, M., Ho, P. T. P. & Lee, Y. 1993, ApJ, 417, 196

Lockett, P., Gautheir, E. & Elitzur, M. 1999, ApJ, 511, 235

McKee, C. F. & Ostriker, J. P. 1977, ApJ, 218, 148

Oliva, E., Moorwood, A. F. M. & Danziger, I. J. 1990, A&A, 240, 453

Reach, W. T. & Rho, J. 1998, ApJ, 507, L93

—— 1999, ApJ, 511, 836

—— 2000, ApJ, 544, 843

Rho, J., & Petre, R. 1998, ApJ, 503, L167

Seta, M. et al. 1998, ApJ, 505, 286

Seward, F. D. 1990, ApJS, 73, 781

Shelton, R. L. et al. 1999, ApJ, 524, 192
Subrahmanyan, R. & Goss, W. M. 1995, MNRAS, 275, 755
Tatematsu, K., Fukui, Y., Landecker, T. L. & Roger, R. S. 1990, A&A, 237, 189
Tatematsu, K., Fukui, Y., Nakano, M., Kogure, T., Ogawa, H. & Kawabata, K. 1987, A&A, 184, 279
Wardle, M. & Yusef-Zadeh, F. 2002, Science, 296, 2350
White, R. L & Long, K. S. 1991, ApJ, 373, 543

IAU 8th Asia-Pacific Regional Meeting
ASP Conference Series, Vol. 289, 2003
S. Ikeuchi, J. Hearnshaw & T. Hanawa, eds.

Discovery of Overionized Plasma in the Mixed-Morphology SNR IC 443

Masahiro T. Kawasaki

The Institute of Space and Astronautical Science, 3-1-1, Yoshinodai, Sagamihara, Kanagawa, Japan 229-8510

Masanobu Ozaki, Fumiaki Nagase

The Institute of Space and Astronautical Science, 3-1-1, Yoshinodai, Sagamihara, Kanagawa, Japan 229-8510

Abstract. We discovered overionized plasma in IC 443, which is one of the "Mixed-Morphology" supernova remnants (MM SNRs). Using *ASCA* data, we found that IC 443 has a plasma structure of an inner-hot (1.0 keV) and an outer-cool (0.2 keV) region. In addition, we found that the silicon and sulfur H-like line emission is too strong, considering the gas temperature $\simeq$1.0 keV suggested from the continuum shape of the spectrum. The intensity ratio of H-like Kα to He-like Kα requires a temperature $\simeq$1.5 keV, which is more highly ionized than what would be expected if the plasma were in 1.0 keV collisional equilibrium. We therefore conclude that the 1.0 keV plasma is overionized, which is the first such detection in SNRs. To make the plasma overionized, the gas should cool faster than the ions recombine to their equilibrium level, and thermal conduction from the hot interior to the cool outer shell could cause the interior plasma to become overionized. Among other MM SNRs, W44 shows no evidence of overionization while Kes 27 shows an implication of it, which suggests that not all of MM SNRs have overionized plasma.

1. Introduction

Supernova remnants (SNRs) have been traditionally classified into three categories: shell-like, Crab-like (Plerion), and composite (shell-like containing a Plerion). These SNRs usually have similar X-ray and radio morphologies. However, there is a recently established class characterized by a center-filled morphology and thermal emission in the X-ray band, contrasting with a shell-like morphologies in the radio and optical bands. Rho & Petre (1998) termed them "Mixed-Morphology" SNRs (MM SNRs).

IC 443 (G189.1+3.0) is one of the prototypical MM SNRs, which is located at the distance of 1.5 kpc. The H I shell covering the eastern part of the remnant (Giovanelli & Haynes 1979) suggests that at least part of IC 443 is in the radiative stage.

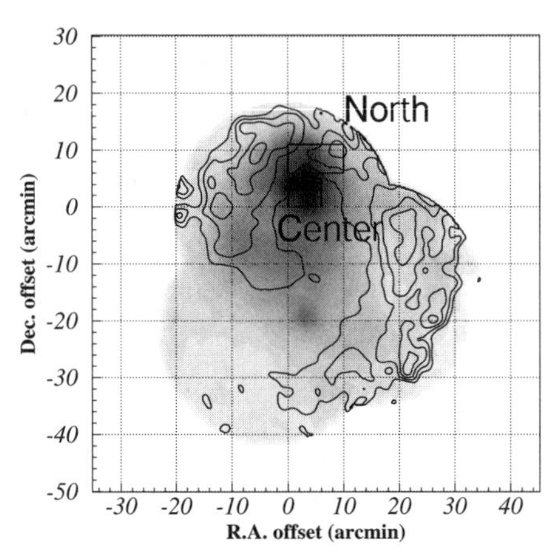

Figure 1. GIS softness ratio ($F_{0.7-1.5\ \mathrm{keV}}/F_{1.5-10.0\ \mathrm{keV}}$) contours over GIS 0.7–10.0 keV band image. The offset center of the coordinates is R.A.=$06^{\mathrm{h}}17^{\mathrm{m}}20^{\mathrm{s}}$, Dec.=$22^{\circ}40'52''$.

2. *ASCA* Observations and Results

ASCA observed IC 443 three times (1993 Apr., 1994 Apr. and 1998 May), and we extracted necessary data from the *ASCA* public archive.

In every energy band, the composite GIS image shows IC 443 to have a centrally-filled morphology. The remnant is larger in images below 1.5 keV than in higher energy bands. Thus, the softness-ratio ($F_{0.7-1.5\ \mathrm{keV}}/F_{1.5-10.0\ \mathrm{keV}}$) map reveals a shell-like structure, contrasting with the center-filled appearance in the X-ray surface brightness map (see Figure 1).

We then extracted the SIS spectra from two distinct regions named "North" and "Center" in Figure 1. Both spectra are well reproduced with two thermal plasma; 0.2 keV non-equilibrium ionized (NEI) one and 1.0 keV collisional ionization equilibrium (CIE) one with 7×10^{21} cm^{-2} interstellar absorption shown in Figure 2. The only difference between two spectra is the intensity ratio of the 0.2 keV plasma to the 1.0 keV one; the former is stronger in North than in Center. Taking the shell-like structure of the softness-ratio map into consideration, these results suggest that IC 443 has an inner-hot and outer-cool plasma structure.

In addition, the H-like lines of Ne, Si, and S are so strong that it is necessary to introduce additional Gaussian components to represent them. In order to investigate the anomalous S line feature, we fit the 2.2–6.0 keV spectra with a model consisting of a CIE plasma and three narrow Gaussian components (He-like Kα, H-like Kα, and He-like Kβ of S). (We can ignore the contribution of the 0.2 keV plasma component at the energies around the S lines.) The intensity ratio of H-like Kα to He-like Kα (hereafter $R_{\mathrm{H/He}}$) for S requires an ionization temperature (hereafter T_z) of about 1.5 keV in each region, which is higher than the continuum temperature (hereafter T_{e}) of 1.0 keV as shown in Figure 3. The same result can be concluded for Si. As the origin of these highly ionized component, another higher temperature plasma component such

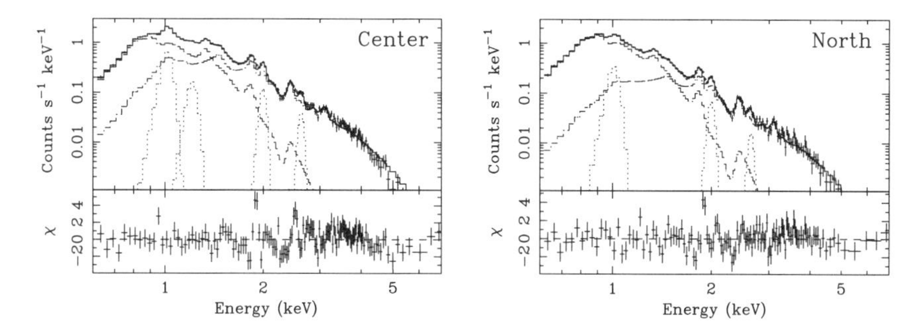

Figure 2. SIS spectra of the Center (left panel) and the North (right panel) regions with best-fit two plasma and Gaussian models. Dashed, dash-dotted, and dotted lines represent the 0.2 keV plasma, 1.0 keV plasma, and additional Gaussians respectively. The lower part of each panel shows the residuals of the fit.

as the ejecta is implausible because it requires extraordinary large abundance. Hence, we conclude that the 1.0 keV plasma itself is overionized, which is the first detection in SNRs.

3. Discussion

3.1. Possible Mechanisms of Overionization

In SNRs, we are accustomed to consider the plasma underionized; as in a collisionally-ionized low-density plasma, the ionization timescale is typically much larger than the elapsed time since the bulk of the gas was shock heated. Nevertheless, conditions can (and apparently do) exist in which such a plasma can become overionized.

For the gas to become overionized in the absence of a photo-ionizing flux, it must cool faster than the ions recombine. There are three cooling mechanisms in the gas: radiation, expansion, and thermal conduction. The first two have much longer timescales than the recombination timescale. On the other hand, IC 443 has an inner-hot and outer-cool plasma structure, and the thermal conduction timescale ($t_{\rm cond}$) to the outer plasma is

$$t_{\rm cond} \simeq 2 \times 10^{11} \left(\frac{n}{1\ {\rm cm}^{-3}}\right) \left(\frac{l_{\rm T}}{10^{19}\ {\rm cm}}\right)^2 \left(\frac{kT}{1.0\ {\rm keV}}\right)^{-5/2} \left(\frac{\ln \lambda}{32.2}\right)\ {\rm s},$$

which can be shorter than the recombination timescale of $\simeq 9{\times}10^{11}(n/1\ {\rm cm}^{-3})^{-1}$ s. (n and kT are the density and the electron temperature of the inner-hot plasma, respectively. $l_{\rm T}$ is the temperature gradient scale length, and $\ln \Lambda$ is the Coulomb logarithm.) Therefore, we suggest that strong gas cooling via thermal conduction can make the interior plasma to become overionized.

3.2. Comparison with Other SNRs

The process of thermal conduction should arise in all SNRs. Thus, we expect that observations of other evolved SNRs (such as mixed-morphology types)

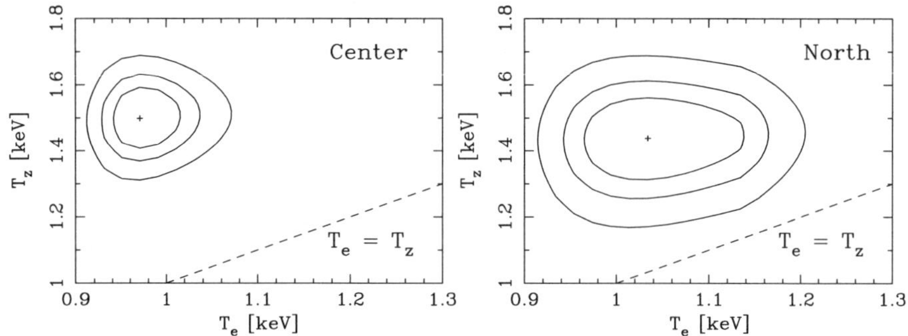

Figure 3. Confidence contours of the ionization temperature (T_z) to the continuum temperature ($T_{\rm e}$) of the SIS spectra in Center (left panel) and North (right panel). The confidence levels are 99%, 90%, and 67%. The contours are above the $T_{\rm e} = T_z$ line which implies that the ionization temperature is higher than the continuum temperature in each region.

should detect further evidence of overionized plasma, and searched this in several *ASCA* observations.

W44 is also one of the prototypical MM SNRs, having similar properties to IC 443. However, the SIS spectrum of the center region shows single and lower electron temperature of $\simeq$0.65 keV and no strong emission from the H-like Si or S. The $R_{\rm H/He}$ of Si indicates that this plasma is in the collisional ionization equilibrium.

Kes27, another sample, has a temperature gradient from the inner region ($\simeq$0.84 keV) to the outer region ($\simeq$0.59 keV) like IC 443 (Enoguchi et al. 2002). The SIS spectrum of the entire region shows a little excess of the H-like S from the fitted CIE model. However, due to the poor statistics, it is hard to confirm that it is overionized.

4. Summary

We have found that IC 443 has a plasma structure consisting of a central hot region surrounded by an envelope of a cool plasma. In addition, we discovered that the inner plasma is overionized, and thermal conduction to the cooler plasma can produce it. However, W44 that has similar properties to IC 443 shows no evidence of overionization. Therefore, we indicate that not all of the MM SNRs has an overionized plasma.

References

Enoguchi, H., Tsunemi, H., Miyata, E. & Yoshita K. 2002, PASJ, 54, 229
Giovanelli, R. & Haynes, M. P. 1979, ApJ, 230, 404
Rho, J. -H. & Petre, R. 1998, ApJ, 503, L167

IAU 8th Asia-Pacific Regional Meeting
ASP Conference Series, Vol. 289, *2003*
S. Ikeuchi, J. Hearnshaw & T. Hanawa, eds.

On the Relation between Abundances in Planetary Nebulae and Their Central Star Evolution

Mezak A. Ratag

National Institute of Aeronautics and Space (LAPAN), Jl. Junjunan 133 Bandung 40173, Indonesia

Abstract. Using recent model-based abundance determinations, which simultaneously determined also the central star temperatures and luminosities, the relation between nebular abundances in galactic bulge planetary nebulae and their central star evolution is studied. The currently existing (hydrogen burning) post-AGB evolutionary scenarios lead to the expectation that there is a strong correlation between the luminosity during the 'horizontal' phase and the enrichment in He and N or the abundances of elements such as O, Ne, Ar and S which are unaffected in the previous stages of evolution. Contrary to that expectation, this work clearly shows that such correlations are difficult to observe.

1. Introduction

Planetary nebula (PN) represents the short-lived evolutionary stage linking the tip of the asymptotic giant branch (AGB) and the white dwarf phase in the late evolutionary history of stars with initial masses between 0.8 and ~5–8 $M_\odot$. The upper limit depends on model details. This initial mass range reflects also a significantly large range in formation time. A ~1 $M_\odot$ star needs about 13 Gyr to reach the PN stage, while a 6 $M_\odot$ star reaches that phase in only about 0.1 Gyr. As the abundances in the interstellar medium also evolve, one would expect to see this evolution by examining abundances in a large sample of PNe with various ages: objects with lower heavy element abundances have lower mass progenitors. Stellar evolution theory predicts also significant changes in surface abundances of He, C, N, and *s*-process elements during the course of evolution from main sequence to PN stage, due to various dredge-up events. The changes are expected to be strongly dependent upon initial mass [see e.g. the review by Iben & Renzini (1983), and recent results by Marigo (1998)].

In this paper we address the problem of initial–final mass relation by correlating PN nebular abundances and their central star luminosities. Theoretically, we should expect that nebular abundances vary as a function of central star position across the HR diagram. Particularly one would expect a strong correlation between the luminosity during the 'horizontal' phase and heavy element abundances or enrichment in He and N. The 'horizontal' phase is a post-AGB stage in which the energy production is dominated by the envelope nuclear burning; the stars evolve with approximately constant luminosities.

2. Nebular Abundances and the Central Star Parameters

More detailed description on the nebular abundance and the central star parameter determinations is given by Ratag et al. (1997). The planetary nebulae in the sample are presumably located in the galactic bulge. The abundances are derived by employing nebular models as interpolation devices in establishing the ionization correction factors. For heavy elements, the accuracies are generally between 25 and 50%. The element abundances are expressed in logarithmic form such that for element X the logarithmic abundance $[X] = \log[N(X)/N(H)] + 12$. The central star luminosities and temperatures are determined by employing a variant of the energy balance method in combination with detailed nebular modeling. The accuracies of the derived luminosities are likely better than 40%.

In this study, oxygen and argon has been chosen as standard elements. Many investigators (see e.g. Wheeler et al. (1989)) have argued that, compared to iron, the abundances of elements such as O and Ar should be a better chronometer.

The scatter in the luminosity vs abundance diagrams (Fig. 1a-c) is partly due to observational uncertainties, but most of it is undoubtedly real. It is clear that there is no obvious relation between nebular abundances and the central star luminosity. This is also true for the abundance ratio N/O, but this is possibly caused by the fact that the ratio does not necessarily correlate with progenitor mass, as has traditionally put forward by several investigators (see e.g. Peimbert, 1978). The analyses of the N/O ratios in a large sample of bulge PNe lead Ratag et al. (1992) to the following important remarks: (i) the on average higher N/O ratios in the bulge PNe compared to the disk nebulae are in contradiction with theoretical expectation that significantly large nitrogen enhancement can occur only in objects with a progenitor more massive than ~ 2 $M_{\odot}$, (ii) very large N/O ratios are found not only limited to the case of high metal abundances (which presumably corresponds to the upper limit of initial mass), but also in the whole range of abundances. Thus, there is strong evidence that the N/O ratio does not necessarily indicate the progenitor mass.

3. Discussion

The results of Terndrup (1988), van der Veen & Habing (1989), and several other investigators (see Ratag, 1991, for a review) suggest that the stars in the galactic bulge were formed in a 'starburst-like' mode. The duration of star formation, however, is not yet known. What can be ruled out is the possibility for the bulge region to have a large scale star formation during the last ~ 8 Gyr. In this study we consider the two possible cases introduced by Ratag (1991), namely the 'short-burst' (in less than 1 Gyr; Case 1), and the 'long-burst' (in about 6 Gyr; Case 2), with an average age of about 11 Gyr.

The first case implies a very small initial mass range between 1.05 and-1.1 $M_{\odot}$ (Maeder & Meynet, 1988). In this case, the fact that the age differences among the objects are so small, may be argued as an explanation for the lack of correlation between nebular abundances and central position in the HR diagram. However, Case 1 scenario suggests that the very large variation as large as an order of magnitude in luminosity (Fig.1), and the possible ranges of final masses

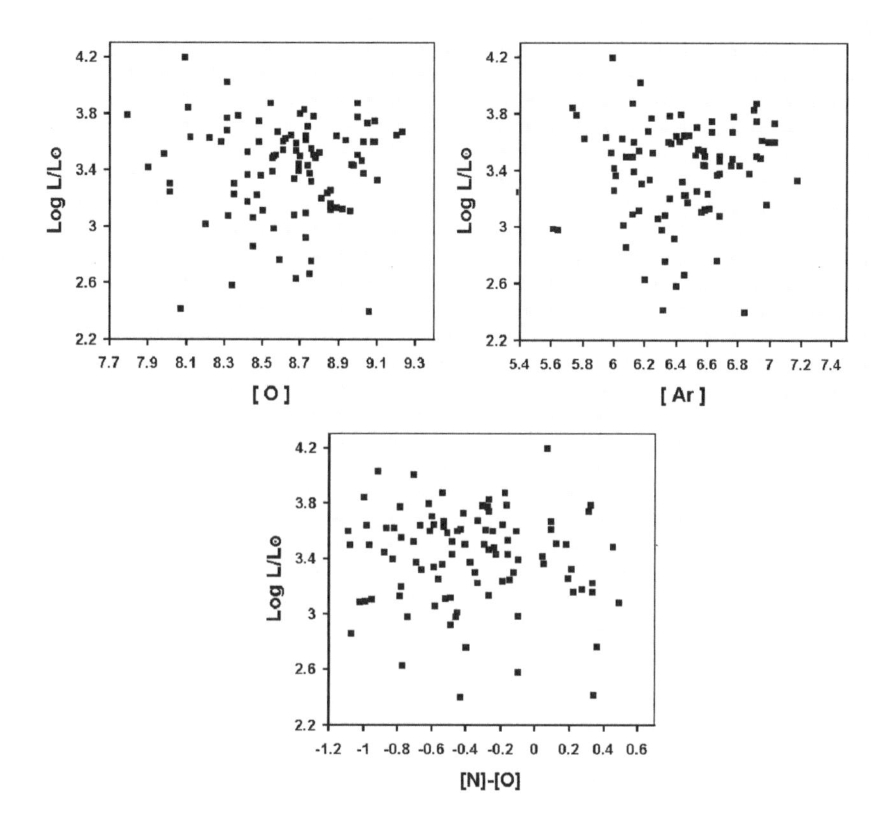

Figure 1. (a) Oxygen abundance as a function of 'horizontal' luminosity. For objects already in their cooling tracks the 'horizontal' luminosities have been estimated with the help of the evolutionary tracks. The distribution pattern is not much affected if we exclude them. No correlation is seen here. (b) same as (a) but for argon. (c) same as (a) but for the abundance ratio N/O

of larger than 0.2 $M_\odot$ (Ratag, 1991), have originated from a very narrow initial mass range with mass differences of less than 0.05 $M_\odot$.

Case 2 assumes a range of initial masses between 0.9 and 1.3 $M_\odot$ with a peak at about the initial mass range of case 1. This scenario can explain the large variation in luminosity but fails in explaining the lack of correlation between nebular abundances and the central star 'horizontal' luminosities (or positions in the HR diagram). The abundance distributions shown in Fig. 1a-c are best explained in terms of chemical enrichment in the bulge during the 6 Gyr star formation burst. This means that the lower end of the distributions should correspond to the lower end of the initial mass range ($\sim$0.9 $M_\odot$), and the upper end is related to the higher mass end of the initial mass range ($\sim$1.3 $M_\odot$). If there is an initial–final mass relation, one would expect the presence of strong correlation between the 'horizontal' luminosity (or the central star position in the HR diagram) and the nebular abundances. The fact that we do not find such a correlation argues against the presence of a unique initial–final mass relation in the bulge PN sample.

It is also possible that we have been using an inappropriate core mass-luminosity relation since we cannot determine precisely at which phase of helium flash a star has left the AGB. The luminosity during the PN stage depends on the helium flash phase at which the central star has left the AGB. In particular, there is a possibility that a certain group of stars would tend to leave the AGB at the helium shell flash. In this case the stars follow a core mass-luminosity relation significantly different from that for the hydrogen burning case.

As suggested by Ratag (1991), the chance is large that many of the galactic bulge PNe are illuminated by helium-burning post-AGB stars. Thus, the galactic bulge PN nuclei studied here are likely to be a mixture of hydrogen- and helium-burning objects. Additional problem arises not only due to the different core mass-luminosity relations, but also because their evolutionary tracks are significantly different from each other. The hydrogen-burning stars tend to have a constant luminosity track during the shell-burning dominated stage, while the helium-burning ones show a declining track (Wood & Faulkner, 1986; Vassiliadis & Wood, 1993). Their calculations indicate that in the temperature range between 3×10^4 and $\sim10^5$ K, the 0.6 $M_\odot$ helium-burning post-AGB evolutionary track crosses the tracks of hydrogen-burning stars with various masses.

4. Summary and Concluding Remarks

The relation between nebular abundances and central star evolution is studied using a large sample of galactic bulge planetary nebulae. No correlation is found between nebular abundances and the central star 'horizontal' luminosities. This fact contradicts the theoretical expectation. The possibility that nebular ejection at the tip of the AGB may take place at various (unspecifiable) phases during the helium flash cycle, creates additional complications due to the possible range of core mass-luminosity relations resulting from it.

For the star formation scenario for the bulge region, we have distinguished two possible cases, namely 0.1 Gyr burst (case 1) and 6 Gyr burst (case 2). Assuming either of these two cases, the presence of a unique initial–final mass relation contradicts the existing observational data.

References

Iben, I. Jr., Renzini, A. 1983, ARA&A, 21, 271
Maeder, A., Meynet, G. 1988, A&AS, 76, 411
Marigo, P. 1998, Ph.D. Thesis, University of Padova, Italy
Peimbert, M. 1978, IAU Symp. No. 76, ed. Y. Terzian, Reidel, Dordrecht, p.215
Ratag, M. A. 1991, Ph.D. Thesis, University of Groningen, Holland
Ratag, M. A., Pottasch, S. R., Dennefeld, M., Menzies, J. 1992, A&A, 255, 255
Ratag, M. A., Pottasch, S. R., Dennefeld, M., Menzies, J. 1997, A&AS, 126, 297
Terndrup, D. M. 1988, AJ, 96, 884
van der Veen, W. E. C. J., Habing, H. J. 1990, A&A, 231, 404
Vassiliadis, E., Wood, P. R. 1993, ApJ, 413, 641
Wheeler, J. C., Sneden, C., Truran, Jr., J. W. 1989, ARA&A, 27, 279
Wood, P. R., Faulkner, D. J. 1986, ApJ, 307, 659

Cosmology, Galaxy Formation and Evolution

IAU 8th Asia-Pacific Regional Meeting
ASP Conference Series, Vol. nnn, 2003
S. Ikeuchi, J. Hearnshaw & T. Hanawa, eds.

The Two Degree Field Galaxy Redshift Survey (2dFGRS): The Instrument, Data and Science

Bruce A. Peterson [1]

Mt. Stromlo Observatory, Research School of Astronomy and Astrophysics, The Australian National University, ACT 2611, Australia

Abstract. The 2dFGRS contains over two hundred twenty thousand galaxy redshifts obtained with the two degree field fiber spectrograph facility at the prime focus of the 4-m Anglo-Australian Telescope. The photometric catalogue, and all spectra are publicly available on-line, and as a CD-ROM/DVD-ROM set. The main cosmological results are: the measurements of the correlation function and of the power spectrum of the galaxy distribution, the determination of the mass density of the universe, $\Omega_m = 0.29 \pm 0.05$, the determination of the baryon fraction through the effects of baryon-photon coupling, $\Omega_b/\Omega_m = 0.15 \pm 0.07$, a direct measurement of the absolute value of the bias parameter, $b = 1.0 \pm 0.1$, a limit on the neutrino mass, $m_\nu < 1.8$ eV, and by combining the 2dFGRS results with CMB anisotropy measurements, a determination of the value of the cosmological constant, $\lambda = 0.7 \pm 0.1$, and of the Hubble constant, $h = 0.72 \pm 0.07$. In addition, a wide range of results on the properties of the galaxy population have been obtained.

1. Introduction

The 2dF Galaxy Redshift Survey has obtained a quarter of a million spectra using the two degree field fiber spectrograph facility at the prime focus of the 4-m Anglo-Australian Telescope, located at Siding Spring Observatory, near Coonabarabran, New South Wales, Australia.

The source catalogue is derived from Automatic Plate Measuring machine (APM) scans of UK Schmidt Telescope plates (Maddoox et al. 1990a, 1990b, 1990c). The Kodak IIIaJ plates and GG395 filter define a b_j magnitude that is

---

[1] *On behalf of the 2dFGRS team and collaborators:* Ivan K. Baldry (JHU), Carlton M. Baugh (Durham), Joss Bland-Hawthorn (AAO), Sara L. Bridle (IoA), Terry J. Bridges (AAO), Russell D. Cannon (AAO), Shaun Cole (Durham), Matthew M. Colless (ANU), Chris A. Collins (LJMU), Warrick J. Couch (UNSW), Nicholas Cross (St Andrews), Gavin B. Dalton (Oxford), Roberto De Propris (ANU), Simon P. Driver (ANU), George P. Efstathiou (IoA), Oysten Elgaroy (IoA), Richard S. Ellis (Caltech), Carlos S. Frenk (Durham), Karl Glazebrook (JHU), Edward Hawkins (Nottingham), Alan Heavens (ROE), Carole A. Jackson (ANU), Bryn Jones (Nottingham) Ofer Lahav (IoA), Ian Lewis (Oxford), Stuart Lumsden (Leeds), Steve J. Maddox (Nottingham), Darren S. Madgwick (IoA), Manuela Maglocchetti (SISSA), Stephen Moody (IoA), Peder Norberg (Durham), John A. Peacock (ROE), Will J. Percival (ROE), Bruce A. Peterson (ANU), Elane M. Sadler (Sydney), Will J. Sutherland (ROE), Keith Taylor (Caltech), Licia Verde (Rutgers).

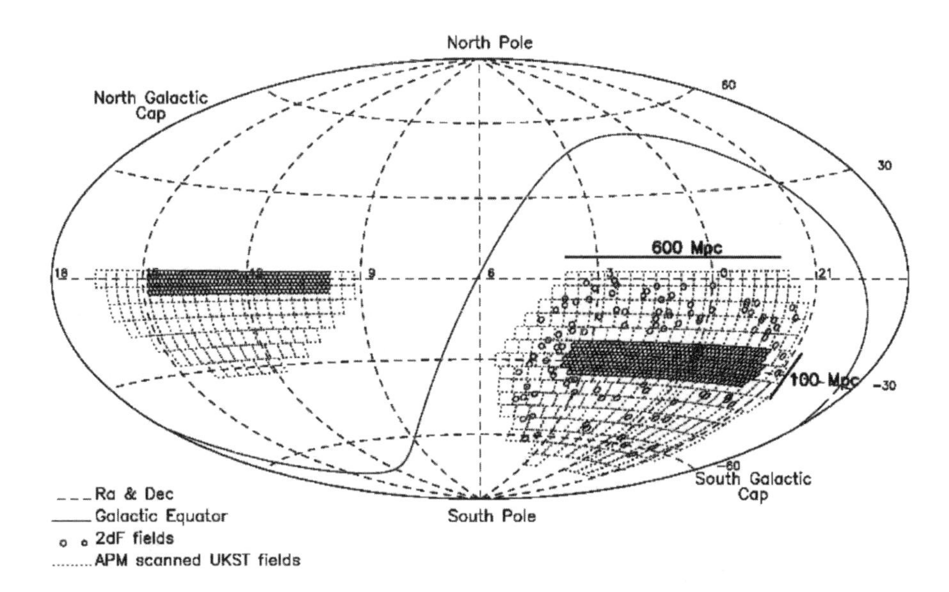

Figure 1. The 2dFGRS survey area. The small circles represent the individual 2° survey fields. The squares represent the scanned plates in the APM catalogue.

related to the Johnson-Cousins B magnitude by $b_j = B - 0.28(B - V)$. The limiting magnitude of the survey is an extinction corrected $b_j = 19.45$.

The area of the galaxy redshift survey covers two contiguous strips, and 100 random 2° fields spread uniformly over the 7000 deg^2 of the southern APM catalogue, as shown in Figure 1. The $75° \times 7.5°$ strip along the celestial equator in the northern Galactic hemisphere contains 70 000 target galaxies. The $75° \times 15°$ strip through the South Galactic Pole contains 140 000 target galaxies. The random fields in the southern Galactic Cap contain another 40 000 target galaxies. At the end of the survey (2002 May) 271 229 non-unique, 246 677 unique redshifts had been obtained. Of the unique redshifts, 14 049 are stars ($z < 0.002$), 180 are QSOs, and 221 283 are galaxies with good quality ($Q \geq 3$) spectra. The overall completeness is 93% in the 2° fields that were observed.

2. The 2dF Instrument

The Anglo-Australian Telescope is a 4-m class telescope with three top-ends that can be changed to provide an f/3.3 prime focus (3 correctors with fields of 60 arcmin, 25 arcmin, and 10 arcmin), an f/8 Ritchey-Chrétien Cassegrain secondary (30 arcmin field), and two f/15-f/36 Cassegrain-coudé secondaries that can be interchanged within the single top-end (small on axis fields).

A new top-end was constructed by the Anglo-Australian Observatory for their 2dF fiber spectrograph facility. The 2dF corrector consists of 5 elements. The first two elements are counter-rotating prismatic doublets 0.87 m and 0.84 m in diameter that act as an atmospheric dispersion compensator for zenith

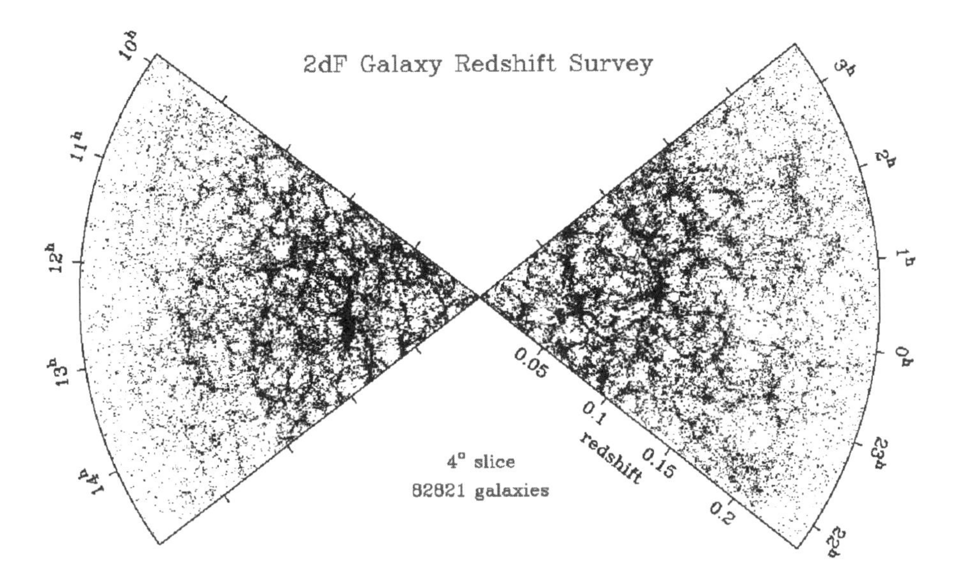

Figure 2. The distribution of galaxy redshifts in the northern and southern 2dFGRS strips, in a slice 4° thick in declination (Peacock 2002). The galaxy distribution has a frothy nature. All structures are significantly smaller than volume of the universe surveyed.

distances of up to 67°. These doublets are followed by two single element lenses. This corrector lens system provides an f/3.5 corrected field that is 2.1 degrees in diameter, at a scale of 67 μm/arcsec.

There are two spectrographs mounted on the top-end ring. Each spectrograph records the spectra produced by 200 fibers, so that 400 spectra are obtained simultaneously. There are two interchangeable field plates, each with its own set of fibers, so that fibers can be positioned on one plate, while observing continues using the fibers on the other plate. The time needed to interchange the two plates and fiber sets under remote control is similar to the time required to slew the telescope to a new field.

The spectrographs use a 150-mm beam, f/3.15 off-axis Maksutov collimators and f/1.0 (spectral), f/1.2 (spatial) modified Schmidt cameras with a CCD detector at the camera prime focus. The dispersion is 4.8 Å per pixel, and the spectra are separated on the detector by 5 pixels. Total system throughput is about 5 percent, including the atmosphere, telescope, corrector, fibers, grating, and spectrograph. The spectrum of a typical $B = 19$ galaxy has a signal/noise ratio of > 10 per pixel in less than one hour.

Each fiber is 8 m in length, 140 μm (2.1 arcsec) in diameter, and is held in the focal plane by a magnetic button positioned on a steel plate. The fibers run out across the focal plane under tension from retractors around the circumference of the field. A microprism on the end of each fiber bends the incident light through 90° into the fiber which lies parallel to the focal plane. The magnetic buttons are positioned in the focal plane by a gripper on an XY gantry above the focal plane. The XY gantry is moved by computer control of linear AC servo motors. Each axis has optical encoders with 1.25 μm resolution. In order to avoid any telescope shake, motions of the XY gantry are balanced by counter weights that move in the opposite direction. The fibers are back illuminated during positioning so that a CCD camera in the gripper can 'see' the fiber core and ensure that the final position of the fiber is correct. The RMS positioning precision is 20 μm (0.3 arcsec). Because fibers can cross on top of other fibers, a typical re-configuration of 400 fibers requires about 600 moves to untangle an old configuration and setup a new configuration. At 6 s per move, this takes about 1 hour.

Guiding during the exposure is done by a digital camera that views the light from four guide bundles. Each guide bundle has 7 fibers in a hexagonal arrangement around one central fiber.

A typical observing sequence consists of calibration exposures of a incandescent lamp and an emission line arc, followed by three integrations on the galaxy field. A pipeline data reduction system assigns wavelengths to the spectra, subtracts a weighted sky spectrum derived from a few fibers placed on blank sky, cross-correlates the galaxy spectrum with several templates, and displays the spectrum and best cross-correlation for the observer to assess. The spectrum is displayed with common galaxy emission and absorption lines marked on the display at the cross-correlation redshift, and the redshift is marked on the cross-correlation display. The observer may intervene to manually determine a redshift, or may simply assign a quality to the redshift. Every spectrum has been through the data reduction pipeline and been viewed and assessed before the start of the next night of observations.

The Anglo-Australian Observatory's 2dF facility is described in more detail by Lewis et al. (2002).

3. The 2dFGRS Database

The 2dFGRS database has two main parts: a collection of FITS files and an SQL query engine with HTML pages for submitting queries and displaying results.

There is one multi-extension FITS file for each target in the survey. The first extension contains a DSS image of the target, and APM photometric parameters in keywords. If one or more spectra were obtained, subsequent extensions contain the sky-subtracted, wavelength calibrated, spectrum. Parameters derived from the spectrum, such as the redshift, are in keywords in the extension containing the spectrum to which they apply. The spectrum is stored as a two dimensional array, with the spectrum in row 1, and the variance array, and sky spectrum in the other rows.

The SQL query engine allows the selection of targets by various criteria, such as magnitude, redshift, position on the sky, etc. Results can be returned as the FITS files, or as a text or HTML data table for the targets that meet the SQL query criteria. If the HTML table is requested, then the DSS image and the spectrum of each target in the table can be selected for display with the web browser.

Users can also search for survey targets around specified positions on the sky, and submit lists of objects to be matched with survey targets.

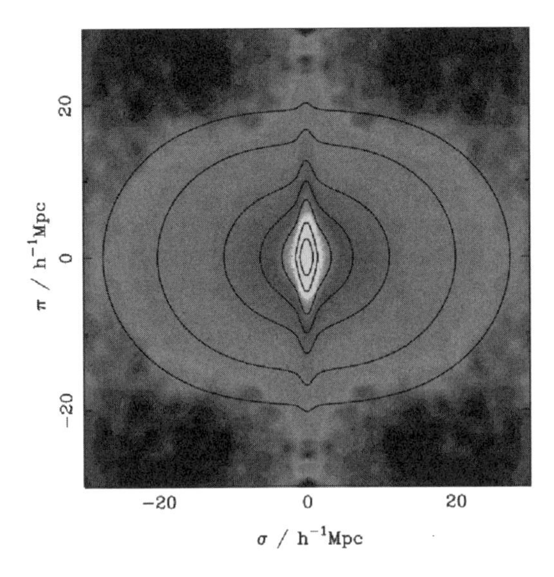

Figure 3. The galaxy correlation function, $\xi(\sigma, \pi)$, for pair separations along the line of sight, π, and in the plane of the sky, σ (Peacock 2002). The amplitude is represented by the grey scale image. The contours for a model with $\beta = 0.4$ and a pairwise dispersion velocity of 400 km s^{-1} are shown. At small angular pair separations, large peculiar velocities along the line of sight in non-linear structures, such as in galaxy groups and clusters, elongate the correlation function. At larger transverse scales, coherent in-fall causes a flattening of the correlation function. The degree of flattening is determined by total mass density and the biasing of the galaxy distribution $\beta = \Omega_m^{0.6}/b = 0.43 \pm 0.07$ (Peacock et al. 2001)

The first public release of about half of the data was made in June 2001 on a set of 4 CD-ROMs. This release contains all the HTML pages and the SQL and associated software needed to setup a database server on a Linux PC. All of the data will soon be available on a CD-ROM/DVD-ROM set, and on-line. The order form for the ROMs and on-line access to the database can be found at `http://www.mso.anu.edu.au/2dFGRS`

More details of the database, including the FITS keywords and SQL parameters, are given in Colless et al. (2001), and can be found on-line at the URL cited above.

4. Science from the 2dFGRS

The science results that have been obtained from the survey are listed in summary form along with references to the papers describing these results in more detail. The galaxy two-point correlation function is described in Figure 3, and the galaxy power spectrum is described in Figure 4.

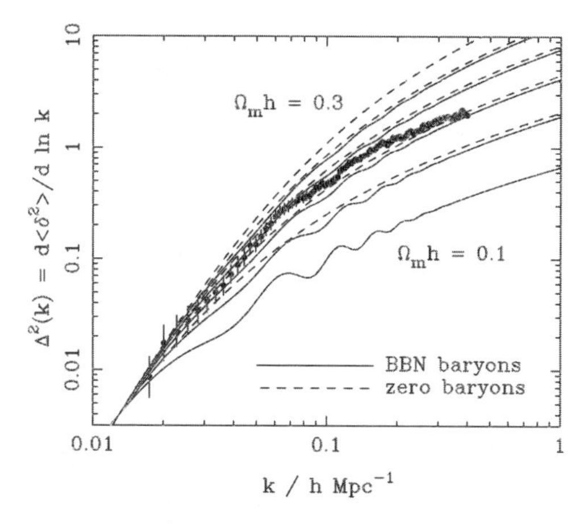

Figure 4. The redshift-space dimensionless power spectrum (Peacock 2002). The points with error bars represent the 2dFGRS results. The lines represent $n = 1$ CDM models for $\Omega_m h = 0.1 - 0.3$. The BBN models assume the big-bang nucleosynthesis value of $\Omega_b h^2 = 0.02$ and $h = 0.7$ A good fit is obtained with the BBN model for $\Omega_m h \approx 0.2$, consistent with the conclusion that the galaxies are unbiased tracers of the mass on large scales.

4.1. Properties of the Galaxy Distribution

- A precise determination of the galaxy power spectrum, quantifying the large-scale structure of the the galaxy distribution on scales up to 300 h^{-1} Mpc (Percival et al. 2001).

- The unambiguous detection of redshift-space distortions due to coherent collapse on large scales, confirming that structures grow through gravitational instability (Peacock et al. 2001).

- Measurements of Ω_m (the mean mass density of the universe) from both the power spectrum and the redshift-space distortions: $\Omega_m = 0.29 \pm 0.05$ (Peacock et al. 2001; Percival et al. 2001)

- The detection of acoustic oscillations in the galaxy power spectrum, resulting from baryon-photon coupling in the early universe (Percival et al. 2001)

- An estimate of the Hubble constant, H_0, and the cosmological constant, λ, from a comparison of the galaxy and CMB power spectra: $H_0 = 72 \pm 7$ km s^{-1} Mpc^{-1} and $\lambda = 0.7 \pm 0.1$ (Efstathiou et al. 2002)

- Measurements of the baryon fraction from the acoustic oscillations in the power spectrum and from a comparison with the CMB power spectrum: $\Omega_b/\Omega_m = 0.15 \pm 0.07$ (Percival et al. 2001; Efstathiou et al. 2002)

- An improved upper limit on the neutrino fraction, $\Omega_\nu/\Omega_m < 0.13$, implying a limit on the total mass of all neutrino species of $m_\nu < 1.8$ eV (Elgaroy et al. 2002).
- The first measurements of the galaxy bias parameter, from the bispectrum of the galaxy distribution alone, and from a comparison of the galaxy and CMB power spectra: $b = 1.00 \pm 0.09$ (Verde et al. 2002; Lahav et al. 2002)

4.2. Properties of the Galaxy Population

- The best determinations to date of the optical and near-IR galaxy luminosity functions (Cole et al. 2001; Norberg et al. 2002)
- The detailed characterization of the variations in the luminosity function with spectral type (Folkes et al. 1999; Madgwick et al. 2002).
- A preliminary determination of the bivariate distribution of galaxies over luminosity and surface brightness (Cross et al. 2001).
- The first precise measurement of the dependence of galaxy clustering on both luminosity and spectral type (Norberg et al. 2001, 2002).
- A constraint on the space density of rich clusters from the velocity dispersion distribution for identified clusters (De Propris et al. 2001)
- Separate radio luminosity functions for AGN and star-forming Galaxies (Sadler et al. 2001; Magliocchetti et al. 2002)
- Constraints on the star-formation history of galaxies from the mean spectrum of galaxies in the local universe (Baldry et al. 2002)
- A measurement of the environmental dependence of star-formation rates of galaxies in clusters (Lewis et al. 2002).

5. Conclusions

The 2dFGRS results support the conclusion that the universe is flat and lambda dominated, as was first indicated by the galaxy counts (Peterson 1993; Yoshii and Peterson, 1995). Together with the CMB measurements, the 2dFGRS results determine the values of the cosmological parameters, $(\Omega_b, \Omega_m, \lambda, h) = (0.04, 0.29, 0.71, 0.72)$, to an accuracy of 10%.

Acknowledgments. The 2dFGRS team acknowledges and thanks the many people at the AAO whose extraordinary efforts have made the survey not only possible but successful.

References

Baldry, I. et al. 2002, ApJ, 569, 582 [astro-ph/0110676]

Cole, S. et al. 2001, MNRAS, 326, 255 [astro-ph/0012429]

Colless, M. M. et al. 2001, MNRAS, 328, 1039 [astro-ph/0106498]

Cross, N. et al. 2001, MNRAS, 324 ,825 [astro-ph/0012165]
De Propris, R. et al. 2002, MNRAS, 329, 87 [astro-ph/0109167]
Efstathiou, G. et al. 2002, MNRAS, 330, 29 [astro-ph/0109152]
Elgaroy, O. et al. 2002, Phys.Rev.Lett, (in press) [astro-ph/0204152]
Folkes, S., et al. 1999, MNRAS, 308, 459 [astro-ph/9903456]
Lahav, O., et al. 2002, MNRAS, (in press) [astro-ph/0112162]
Lewis, I. J. et al. 2002a, MNRAS, (in press) [astro-ph/0202175]
Lewis, I. J. et al. 2002b, MNRAS, (in press) [astro-ph/0203336]
Maddox et al. 1990a, MNRAS, 242, 43P
Maddox et al. 1990b, MNRAS, 243, 692
Maddox et al. 1990c, MNRAS, 246, 433
Madgwick, D. S. et al. 2002, MNRAS, 333, 133 [astro-ph/0107197]
Magliocchetii, M. 2002, MNRAS, 333, 100 [astro-ph/0106430]
Norberg, P. et al. 2001, MNRAS, 328, 64 [astro-ph/0105500]
Norberg, P. et al. 2002a, MNRAS, 332, 827 [astro-ph/0112043]
Norberg, P. et al. 2002b, MNRAS, (in press) [astro-ph/0111011]
Peacock, J. A. et al. 2001, Nature, 410, 169 [astro-ph/0103143]
Peacock, J. A. 2002, in ASP Conf. Ser., A New Era in Cosmology, ed. T. Shanks & N. Metcalfe (San Francisco: ASP) [astro-ph/0204239]
Percival J. A. et al. 2001, MNRAS, 327, 1297 [astro-ph/0105252]
Percival J. A. et al. 2002, MNRAS, (submitted) [astro-ph/0206256]
Peterson, B. A. 1993, in Relativistic Cosmology, ed. M. Sasaki (Tokyo: Universal Academy), 3
Sadler, E. M. et al. 2002, MNRAS, 329, 227 [astro-ph/0106173]
Verde, L. et al. 2002, MNRAS, (in press) [astro-ph/0112161]
Yoshii, Y. and Peterson, B. A. 1995, ApJ, 444, 15

IAU 8th Asia-Pacific Regional Meeting
ASP Conference Series, Vol. 289, 2003
S. Ikeuchi, J. Hearnshaw & T. Hanawa, eds.

The Deep Near-Infrared Universe Seen in the Subaru Deep Field

Tomonori Totani

Princeton University Observatory, Peyton Hall, NJ08544, USA, and Theory Division, National Astronomical Observatory, Mitaka, Tokyo 181-8588, Japan

Abstract. The Subaru Deep Field provides the currently deepest K-selected sample of high-z galaxies ($K' \sim 23.5$ at 5σ). The SDF counts, colors, and size distributions in the near-infrared bands are carefully compared with pure-luminosity-evolution (PLE) as well as CDM-based hierarchical merging (HM) models. The very flat faint-end slope of the SDF K count indicates that the bulk (more than 90%) of cosmic background radiation (CBR) in this band is resolved, even if we take into account every known source of incompleteness. The integrated flux from the counts is only about a third of the reported flux of the diffuse CBR in the same band, suggesting that a new distinct source of this missing light may be required. We discovered unusually red objects with colors of ($J - K \gtrsim$ 3–4), which are even redder than the known population of EROs, and difficult to explain by passively evolving elliptical galaxies. A plausible interpretation, which is the only viable one among those we examined, is that these are dusty starbursts at high-z ($z \sim 3$), whose number density is comparable with that of present-day ellipticals or spheroidal galaxies, as well as with that of faint submillimeter sources. The photometric redshift distribution obtained by $BVRIz'JK'$ photometries is also compared with the data, and the HM model is found to predict too few high-z objects at $K' \lesssim 22$ and $z \lesssim 2$; the PLE model with a reasonable amount of absorption by dust looks more consistent with the data. This result is apparently in contradiction with some previous ones for shallower observations, and we discuss the origin of this. These results raise a question for the HM models: how to form massive objects with starbursts at such high redshifts, which presumably evolve into present-day elliptical galaxies or bulges?

1. Introduction

The Subaru Deep Survey is a systematic project of the 8.2-m Subaru telescope to study the deep extragalactic universe. The Subaru Deep Field was selected near the north Galactic pole, avoiding large Galactic extinction and nearby galaxy clusters, and the airmass of this field is smaller than the Hubble Deep Field at Mauna Kea (Maihara et al. 2000). The wide field near-infrared (NIR) camera CISCO took a very deep NIR $2' \times 2'$ image in J and K' bands, with 5σ magnitude limits of 25.1 and 23.5. This is the deepest image in the K band taken so far, providing a unique K-selected sample of galaxies which should be useful for study of faint, high-z galaxies. The field was also deeply followed-up by optical instruments of FOCAS and Suprime-Cam. Here we review some interesting implications obtained by these data set, focusing on NIR galaxy counts, colors, and photometric-redshift distribution, compared with some theoretical models of galaxy formation and evolution. Although omitted here, some interesting results for the clustering of Lyman break galaxies and Lyman alpha emitters at $z \sim 4$ have been obtained in the SDF and another project of the Subaru/XMM-Newton deep survey, thanks to the very wide field of the Suprime-Cam. See Ouchi et al. (2001, 2002) for these.

2. NIR Galaxy Counts and Contribution to CBR

Figure 1 shows K band SDF galaxy counts, compared with those estimated by other observations (Totani et al. 2001a). Here, we plot counts multiplied by flux, rather than count itself, to show the contribution to the cosmic background radiation (CBR) per magnitude. Both the raw and corrected counts assuming point sources are showing very flat faint-end slope, with rapidly decreasing contribution to CBR beyond $K \gtrsim 18$. Therefore the extrapolation of the galaxy counts into fainter magnitudes does not significantly increase EBL but converges to a finite EBL flux, and this means that the bulk of EBL from galactic light has already been resolved into discrete galaxies. These results require that the diffuse EBL in NIR bands should not be different from the count integrations, provided that the ordinary galactic light is the dominant source of the EBL in these bands, as generally believed. However, a few recently reported detections of diffuse EBL in these bands suggest that the diffuse CBR flux in K bands is consistently higher than the count integrations by a factor of ~ 3: $\nu I_\nu = 27.8 \pm 6.7$ nW m^{-2}sr^{-1} (Cambrésy et al. 2001), 29.3 ± 5.4 (Matsumoto 2000), and 20.2 ± 6.3 (Wright 2001), which should be compared with the integration of K counts (~ 8 nW m^{-2}sr^{-1}).

If the discrepancy between the diffuse EBL and count integration is real, it might suggest the existence of very diffuse component which is different from normal galaxies. Before deriving this extraordinary conclusion, however, all possible systematic uncertainties in the above estimates must extensively be checked. One of such systematics is the contribution to EBL by the galaxies missed in deep galaxy surveys. Since galaxies are extended sources, the detectability near the detection limit is not as simple as point sources. Furthermore, the well-known effect of the cosmological dimming of surface brightness $[S \propto (1+z)^{-4}]$ should make high-z galaxies very difficult to detect, while such objects may have a significant contribution to EBL. The photometry scheme could also be a problem, because there is considerable uncertainty in the estimate of the magnitude of faint galaxies because of 'growing' the photometry beyond the outer detection isophotes of galaxies.

Therefore we estimated the contribution to CBR by galaxies missed in SDF, based on a realistic theoretical model of galaxy counts which includes all known physical or observational selection effects and incompleteness, based on the method presented in Yoshii (1993) and Totani & Yoshii (2000). First, we construct a model of galaxy counts which best fits to the observed raw (i.e.,

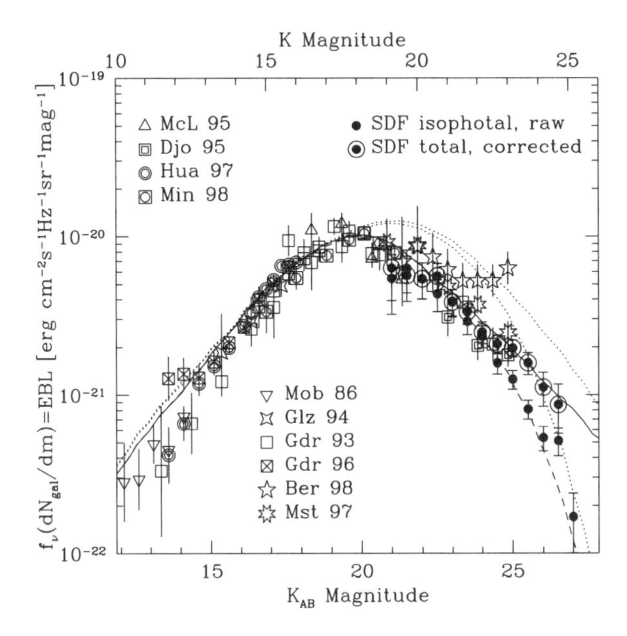

Figure 1. The contribution to EBL by galaxies in the K band. The filled circles are the raw SDF counts in isophotal magnitude, while the symbols ⊙ are the counts in total magnitude which are corrected for incompleteness assuming point sources (Maihara et al. 2000). The dashed line is the prediction by a PLE model for which the selection effects under the observational conditions of SDF are taken into account, fitting to the raw counts. The solid line is the same prediction, but the selection effects are not included. For detail, see Totani et al. (2001a). The two dotted lines are model predictions with and without selection effects, using the same PLE model but including a simple number evolution of $\eta = 1$.

uncorrected) counts, taking into account all the above selection effects. Then we can calculate the true galaxy counts and EBL flux using the same model without selection effects, and comparison between the true counts and observed counts gives an estimate of contribution by missing galaxies. The results are shown in Fig. 1, where the dashed line is the best-fit model to raw SDF counts taking into account the selection effects in theoretical calculation, while the solid line is the same model but without the selection effects. By this procedure, we found that the correction by the incompleteness would increase the CBR from galaxies by at most 10%, which is too small to reconcile the count integrations and diffuse CBR measurements (Totani et al. 2001a). Therefore we conclude that there must be a new source of CBR in the NIR band, which must be very different and distinct populations from known galaxies, unless some unknown systematics have affected the diffuse CBR measurements significantly.

The model shown in the dashed and solid lines is a so-called pure-luminosity-evolution (PLE) model without number evolution, including five types of galaxies (E/S0, Sab, Sbc, Scd, and Sdm), whose detail is given in Totani & Yoshii (2000). In fact, all the SDF NIR data of counts, $(J-K)$ colors, and size distributions are well described by this rather simple model in the popular Λ-dominated flat universe, without any indication of number evolution. This is somewhat in contrast to the result of the same model against HDF galaxies, where a modest number evolution of $\eta \sim 1$ is required to fit the data [$\phi^* \propto (1+z)^\eta$, $L^* \propto (1+z)^{-\eta}$] (Totani & Yoshii 2000). This is the most naturally explained by different merging histories for different galaxy types; longer wavelengths are dominated by earlier types. Therefore, these results indicate that elliptical galaxies which are dominant in the K band are evolving without significant change of number density from $z \sim 2$ to the present (Totani et al. 2001c). A number evolution of $\eta \sim 1$ for elliptical galaxies is already inconsistent with the data, as shown by dotted lines in Fig. 1.

On the other hand, a study of SDF K counts by a hierarchical merging (HM) model based on CDM-based structure formation is presented in Nagashima et al. (2002), where the selection effects are carefully taken into account in a similar way. The HM also fits to the SDF counts in low density cosmological models, but the PLE and HM models are giving different predictions in the redshift distribution, which will be discussed later.

2.1. Unusually Red Objects

An interesting discovery by SDF is the existence of unusually red objects in NIR colors, with $J-K \gtrsim$3–4. The brightest four objects of them are presented in Maihara et al. (2000), and a more careful analysis has been performed to estimate the number fraction of such objects as a function of K magnitude (Totani et al. 2001b). They found that the number fraction of such objects sharply rises with increasing magnitude from $K \sim 20$, reaching a few percent at the faintest magnitudes (see Fig. 2). It should be noted that such NIR color is even redder than the known population of the extremely red objects (EROs), which are defined by red optical-NIR colors such as $R-K > 5$; typical EROs have $J-K$ of at most 2. Furthermore, such red NIR color cannot be explained by passively evolving elliptical galaxies at any reasonable redshift without extinction by dust (Totani et al. 2001b), while recent studies of EROs indicate that the majority of them are passively evolving ellipticals (e.g., Daddi et al. 2000).

Then two interpretations remain for these hyper extremely red objects (HEROs). One is ultra-high z objects at $z \sim 10$, the red color being due to the Lyman-break between J and K bands. Since HEROs are too faint to measure redshifts, it is difficult to verify this observationally. However, theoretically it is very unlikely; the rest-frame UV luminosity indicates star formation rate of more than $100 M_\odot$/yr, but there should be very few objects which are massive enough to allow such high SFR at $z \sim 10$, according to the widely believed CDM-based structure formation theory. The other interpretation is that they are dusty starbursts, which often show very red colors. In fact, Totani et al. (2001b) has shown that the colors and counts of HEROs are well reproduced by a simple model if present-day elliptical galaxies have formed by starbursts with a reasonable amount of dust (i.e., inferred from model metallicity) at $z \sim 3$.

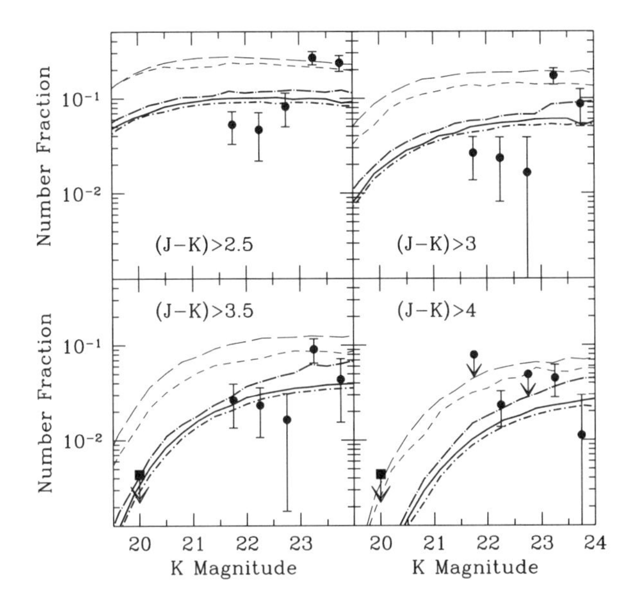

Figure 2. Number fraction of galaxies redder than several threshold $J - K$ colors (indicated in each panel), as a function of K magnitude. Filled circles are the data of the SDF. The error bars are 1σ, while the upper limits shown by arrows are at the 95% confidence level. The upper limit at $K = 20$ is from Scodeggio & Silva (2000, filled square). The solid line is the model prediction with the formation redshift $z_F = 3$ and our standard dust-extinction normalization. See Totani et al. (2001b) for the detail for other curves.

This redshift is similar to those estimated for the faintest submillimeter sources in recent years. Interestingly, the counts of HEROs are roughly the same with the faintest SCUBA sources, and sub-mm flux expected by the dusty-starburst model of HEROs is in fact close to the SCUBA sensitivity limit. A detailed modelling of FIR–sub-mm counts by Totani & Takeuchi (2002) has indeed shown that SCUBA counts are nicely explained by the dusty starbursts of forming elliptical galaxies which are also responsible for HEROs. These results suggest an interesting possibility that HEROs and SCUBA sources are the same population, which will evolve into present-day elliptical galaxies or bulges. It is very important to examine the correlation between these two. Ultimate confirmation of this hypothesis would be brought in the NGST and ALMA era.

2.2. Photometric Redshift Distributions: PLE vs. HM

Both the PLE and HM models fit to the SDF K galaxy counts. It may be because the galaxy counts do not have enough power to discriminate these two, or may be because galaxies dominant in the K band in the HM model have only small or negligible number evolution. This degeneracy can be broken by

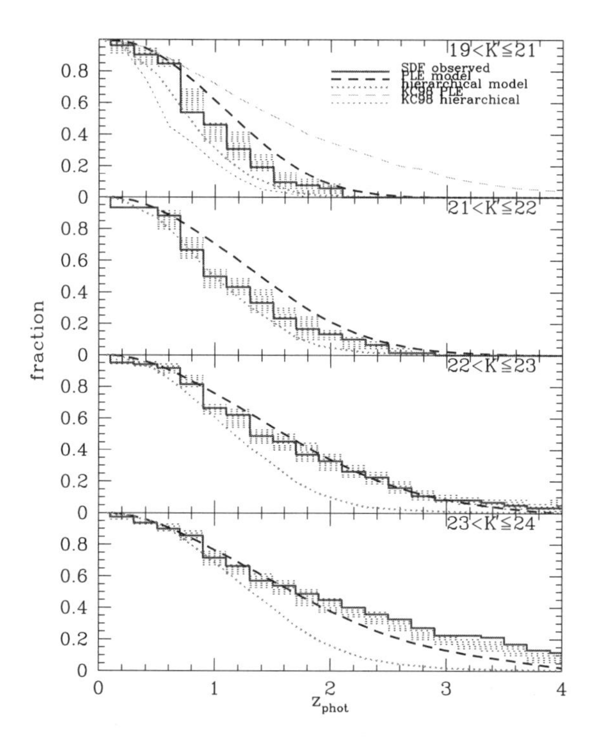

Figure 3. The normalized cumulative redshift distribution of SDF sample separated by K' magnitude. The thick solid histograms show the result of the SDF photometric redshifts. The shaded regions show $\pm 3\sigma$ deviated counts estimated by Monte Carlo realizations when photometric redshift errors are taken into account. (Poisson fluctuation of number of galaxies is not included.) The thick dashed lines denote the prediction of the PLE model in Totani et al. (2001), while the thick dotted lines for the HM model in Nagashima et al. (2002). The predictions of dust-free PLE and HM in Kauffman & Charlot (1998) are also shown in the top panel by thin dotted and dashed lines, respectively.

redshift distribution; especially a K-selected sample has a strong power for the discrimination, since K-band light traces the stellar mass which has formed so far, rather than star formation rate at that time. Though spectroscopic redshifts are unavailable for the faintest SDF galaxies, a reasonably reliable test is possible by the photometric redshift technique, as is done by Kashikawa et al. (2002). The result is shown in Fig. 3, in a form of cumulative z-distribution separated by K' magnitude intervals.

In bright magnitude ranges of $K' < 22$, the observed distribution is somewhat between the two models of PLE and HM. Although in the range of $21 \leq K' \leq 22$ the HM model looks better than the PLE, statistical fluctuation is large because of the small number of galaxies. The Kolmogorov-Smirnov (KS) test gives a chance probability of 2 and 1% of getting the observed distribution from the PLE model. Considering model uncertainties which are not taken into account in the KS test, it is impossible to reject this model. On the other hand, the deficit of high-z galaxies in the HM model at $K' > 22$ is statistically much more significant (chance probabilities of 6×10^{-6} and 5×10^{-15} at $22 \leq K' \leq 23$ and $23 \leq K' \leq 24$, respectively). It should also be noted that the HM model used here gives z-distribution more weighted to high redshifts compared with another HM model of Kauffman & Charlot (1998) (see the top panel). We emphasize that this is a "blind" test, i.e., the models shown here are those fitting best only to the SDF counts as described in Totani et al. (2001c) and Nagashima et al. (2002) without further tuning of parameters to the new photo-z data of Kashikawa et al. (2002).

These results seem to be controversial when compared with several previous papers doing a similar test (but in shallower magnitudes). Fontana et al. (1999) and Rudnick et al. (2001) claimed that the photometric redshift distribution at $K < 21$ is consistent with HM models, but not with PLE models. Firth et al. (2002) reached a similar conclusion using a sample of $H < 20$. It should be noted that PLE models used by these groups are either dust-free or assuming only constant extinction at the level of Galactic extinction. Such PLE models are inconsistent with the observed z-distribution since they predict too many high-z galaxies which are visible because of strong starbursts assumed in the formation of elliptical galaxies (see top panel of Fig. 3 for the dust-free PLE prediction by Kauffmann & Charlot 1998). However, dust-free model is obviously unrealistic, especially for initial starbursts expected for elliptical galaxies. We know that starbursting populations of galaxies quite often show strong extinction and reddening. A chemical evolution model of elliptical galaxies also suggests that the amount of metal produced in the initial starburst phase is huge, and hence strong extinction seems quite plausible (Totani & Yoshii 2000). The PLE model used here (dashed line in Fig. 3) is taking into account the extinction by a reasonable amount of dust inferred from chemical evolution. In addition, observational selection effects discussed in §2 may also have affected previous results. (The theoretical predictions by our PLE and HM models appropriately included all known selection effects under the SDF condition).

Cimatti et al. (2002) compared a spectroscopic redshift distribution at $K < 20$ obtained by the K20 survey with the latest PLE and HM models. In fact, they found that, if dust extinction is taken into account (another option is using the Scalo IMF rather than the Salpeter), the difference between PLE and HM models becomes much smaller than previously claimed. They found that such PLE models are in reasonable agreement with the data, but on the other hand, HM models predict too many low redshift galaxies.

3. Concluding Remarks

From these results, we conclude that the deepest K-selected sample of the SDF is in overall agreement with the simple picture of PLE for early type or elliptical galaxies, but the present version of HM models has a problem in the redshift distribution of the faintest galaxies. Considering the overall success of the CDM structure formation theory against various tests *not* based on galaxy luminosity and star formation activities (e.g., clustering properties or abundance of galaxy clusters), it is reasonable to think that the problem identified for HM models is related with the treatment of star formation activity. It must incorporate a population of massive and dusty starbursts at high redshift ($z \gtrsim 3$), which presumably evolved into present-day elliptical galaxies or bulges without significant number evolution.

The author would like to thank all the collaborators of the SDF project, and the Subaru Telescope staffs who made this project possible. He has been financially supported in part by the JSPS Fellowship for Research Abroad.

References

Cambrésy, L., Reach, W. T., Beichman, C.A., & Jarrett, T.H. 2001, ApJ, 555, 563

Cimatti, A. 2002, A&A in press, astro-ph/0207191

Daddi, E. et al. 2000, A&A, 361, 535

Firth, A.E. et al. 2002, MNRAS, 332, 617

Fontana, A. et al. 1999, MNRAS, 310, L27

Kauffmann, G. & Charlot, S. 1998, MNRAS, 297, L23

Kashikawa, N. et al. 2002, AJ, submitted.

Maihara, T. et al. 2001, PASJ, 53, 25

Matsumoto, T. 2000, in the proceedings of The Extragalactic Infrared Background and its Cosmological Implications, International Astronomical Union. Symposium no. 204. Manchester, England, August 2000.

Ouchi, M. et al. 2001, ApJ, 558, L83

Ouchi, M. et al. 2002, ApJ in press, astro-ph/0202204

Rudnick, G. et al. 2001, AJ, 122, 2205

Scodeggio, M. & Silva, D. R. 2000, A&A, 359, 953

Totani, T. & Takeuchi, T. T. 2002, ApJ, 570, 470

Totani, T. & Yoshii, Y. 2000, ApJ, 540, 81

Totani, T., Yoshii, Y., Iwamuro, F., Maihara, T. & Motohara, K. 2001a, ApJ, 550, L137

Totani, T., Yoshii, Y., Iwamuro, F., Maihara, T. & Motohara, K. 2001b, ApJ, 558, L87

Totani, T., Yoshii, Y., Iwamuro, F., Maihara, T. & Motohara, K. 2001c, ApJ, 559, 592

Yoshii, Y. 1993, ApJ, 403, 552

Wright, E. L. 2001, ApJ, 553, 538

IAU 8th Asia-Pacific Regional Meeting
ASP Conference Series, Vol. 289, 2003
S. Ikeuchi, J. Hearnshaw & T. Hanawa, eds.

The Cluster 'Sphere of Influence': Tracking Star Formation with Environment via Hα Surveys

Warrick J. Couch

School of Physics, The University of New South Wales, Sydney 2052, NSW, Australia

Michael Balogh, Richard Bower

Department of Physics, University of Durham, UK

Ian Lewis

Astrophysics, Nuclear & Astrophysics Laboratory, Oxford, UK

Abstract. Two major observational programs are described which trace the galactic star formation as a function of environment in the vicinity of rich clusters at both low and intermediate redshifts. Both involve the use of the Hα emission line to provide a direct measure of the instantaneous star formation rate. Both of these surveys provide strong evidence that, relative to the field, star formation is suppressed in these environments out to very large clustercentric radii ($R \gtrsim 3R_v$) or, equivalently, where the local galaxy density has dropped to that typical of the group environment. We also find further evidence that significant amounts of the star formation in distant clusters is obscured when viewed at optical rest wavelengths.

1. Introduction

The relationship between galactic star formation and galactic environment is fundamental to understanding the basic properties of galaxies. Furthermore, the connection between the two has important implications for galaxy evolution, given that a galaxy's environment is thought to become increasingly clustered with cosmic time if hierarchical models of structure growth are correct. Indeed the significant decline in the global star formation rate (SFR) observed between $0 < z < 1$ (Lilly et al. 1996, Madau et al. 1996, Cowie et al. 1999) raises the interesting question as to whether this is environment driven, or simply reflects mechanisms internal to galaxies which operate on a cosmic time-scale (e.g. gas exhaustion; Larson, Tinsley & Caldwell 1980).

These two effects can only be disentangled by carefully quantifying star formation activity in galaxies over their full range of environments and at different epochs. To date, studies have generally concentrated on the extremes of galaxy environment: the low-density field and the dense cores of rich clusters. For the latter, attention has been very much drawn by the discovery of Butcher & Oemler (1978) that such systems harboured many more star-forming galaxies in the past. But galaxies in cluster cores comprise only a small fraction of the stellar content of the universe and may be subject to environmental effects that are peculiar to these very high density environments (e.g., ram-pressure stripping; Gunn & Gott 1972). Much more pertinent to the evolution of the general galaxy population is the environment *between* cluster cores and the field, spanning three orders of magnitude in galaxy density and which remains largely unchartered in terms of tracking star formation.

In this paper, we describe two new studies which probe this 'intermediate environment' regime in an unprecedented way at low and intermediate redshifts. The first involves the 2dF Galaxy Redshift Survey (2dFGRS; Colless et al. 2001), where the great wealth of data has been exploited to track the SFR continuously from the centres of clusters out to arbitrarily large radii at $z \sim 0.05$–0.1. The second involves the use of a novel "nod-and-shuffle" technique to survey spectroscopically hundreds of members in clusters at $z \sim 0.2$–0.3 out to at least twice the virial radius (R_v). A key feature of both of these studies is the use of the Hα emission line to *directly* measure the instantaneous SFR. Complete details of these studies can be found in Lewis et al. (2002) and Balogh et al. (2002). Throughout this paper we adopt a cosmology with $\Omega_\Lambda = 0.7$, $\Omega_{\rm m} = 0.3$, $H_0 = 70\,{\rm km\,s^{-1}\,Mpc^{-1}}$.

2. SFR versus Environment at Low Redshift with 2dFGRS

2dFGRS has obtained over 220 000 spectra of galaxies located in two contiguous strips (one in each of the northern and southern galactic hemispheres) plus 99 randomly located fields. For this study, 17 of the known rich clusters located within the strips (De Propris et al. 2002) were selected, 10 of which had 'high' velocity dispersions ($\sigma_v > 800\,{\rm km\,s^{-1}}$), and the remaining with 'low' velocity dispersions ($400 \leq \sigma_v \leq 800\,{\rm km\,s^{-1}}$). All the 2dFGRS galaxies within a projected distance of 20 Mpc from the centres of these clusters and within the range $0.06 < z < 0.10\,{\rm km\,s^{-1}}$ were selected for analysis. The restriction in redshift is to limit the effects of aperture bias and poor sky-subtraction. A luminosity cutoff of $M_b = -19$ was then applied to the sample, this being the limit to which 2dFGRS is complete over our adopted redshift range. Finally, all galaxies where there were continuum/line fitting problems at Hα or whose EW([NII] λ6583/EW(Hα) ratios showed evidence of a non-thermal component (values >0.55), were excluded from the analysis. This resulted in a final sample of 11 006 galaxies.

The equivalent width of the Hα line (be it in absorption or emission) was measured using an automated Gaussian profile fitting algorithm, which simultaneously fitted and deblended the Hα line from the neighbouring [NII] λ6548 Å and [NII]λ6583 Å lines. Since only equivalent width rather than line flux measurements are possible with the 2dFGRS spectra, we can only infer the star formation rate, μ, per unit luminosity: $\mu/L_{\rm cont} = 7.9 \times 10^{-42}\,{\rm EW(H\alpha)}$, using Kennicutt's (1983) SFR–EW(Hα) relation. We then normalized this to a characteristic luminosity L^*: $\mu^* = \mu/(L_{\rm cont}/L^*) = 0.087\,{\rm EW(H\alpha)}$, where L^* is taken to be the knee in the luminosity function in the r'-band (near rest-frame Hα), as determined by Blanton et al. (2001).

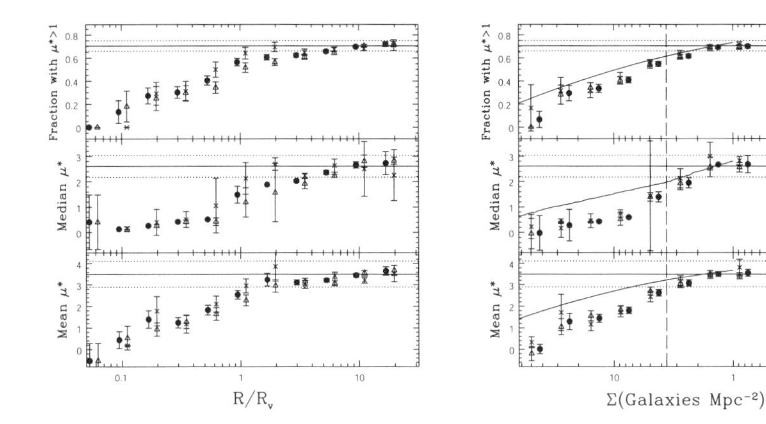

Figure 1. *Left:* Star formation rate as a function of cluster-centric radius, derived from our 2dFGRS analysis. *Filled* points represent the full galaxy sample, while the *triangles* and *crosses* represent the 'high'and 'low' velocity dispersion clusters (see text for details). The *solid horizontal* lines represent the value of each statistic for our field galaxy sample. *Right:* as for the left-hand panel, but plotted as a function of local projected galaxy density, Σ. The *dashed* vertical line represents the mean value of Σ within R_v. The *curved* lines indicate the variation expected as a result of the varying morphological mix.

In the left panel of Figure 1, the mean and median value of μ^* is plotted as a function of the projected cluster-centric radius. Here the latter has been normalized by the cluster virial radius, R_v, which has been calculated for each individual cluster and ranges from 1.4–2.4 Mpc. Also plotted in the top panel of Figure 1 is the fraction of galaxies with $\mu^* > 1\,M_\odot\,\mathrm{yr}^{-1}$, which represents the tail of the distribution, comprised of galaxies that are currently forming stars at a high rate relative to their luminosities. The solid horizontal line represents the values derived for the 'field' galaxies within our sample; that is, galaxies that lie within the 20 Mpc selection radius, but which from their redshifts are identified as non-members. The bracketing dashed horizontal lines represent the 1σ standard deviation from field to field, giving some estimate of the cosmological variance in the field value.

Irrespective of which statistic is used, it is very clear that within the clusters, μ^* falls significantly below the value of the field, the difference being at a maximum at the cluster centre and then monotonically decreasing with increasing cluster-centric radius. *Importantly, convergence does not occur until $R \gtrsim 3R_v$!* Hence, compared to the field, cluster galaxies differ in their mean star formation properties as far out as ~ 6 Mpc from their centres. Also of note is that these radial trends in μ^* appear to be insensitive to cluster mass, with there being no discernible difference between the 'high' and 'low' velocity dispersion clusters.

Such a radial analysis, however, is likely to be sub-optimum, since many of the clusters in our sample are clearly not spherically symmetric, and show sub-structure. We have therefore analysed μ^* as a function of the local projected galaxy density, Σ, based on the distance to the tenth nearest neighbour (Dressler 1980). The relationship between μ^* and Σ is shown in the right-hand panel of Figure 1, using the same three statistics that we used in the radial analysis. The vertical line shows the mean value of Σ within R_v, and once again the horizontal lines show μ^* and its 1σ variance for the field.

We see that cluster galaxy star formation is suppressed relative to the field, the difference being greatest at the highest local densities, and decreasing monotonically with decreasing density until the two converge at $\Sigma \sim 1.5\,\mathrm{gals\,Mpc}^{-2}$, a factor of ~ 2.5 times lower than the mean projected density of the cluster virialized region. Yet again the trend is no different in the 'high' and 'low' velocity dispersion clusters, indicating that star formation rates depend only on the local density, regardless of the large-scale structure in which they are embedded. Further evidence that local density is the key variant is also provided by a plot based on just those galaxies with $R > 2R_v$, where the same trend observed for the full sample is seen. This would suggest that a more general view of star formation suppression be taken: that it will be low relative to the global average in *any* region where the local density exceeds a value of $\sim 1\,\mathrm{gal\ Mpc}^{-2}$ ($M_b \leq 19$).

It is well known that galaxy morphology is very strongly correlated with local galaxy density (Dressler 1980), and hence consideration needs to be given to what underlying contribution this has to the $[\mu^*, \Sigma]$–relationship seen in Figure 1. Using a simple model based on Dressler's morphology-density relation, we have calculated the expected variation in μ^*; this is represented by the *solid curves* in Figure 1. This appears to be shallower than the observed relation, suggesting that the morphology-density relation is distinct from the SFR-density relation.

3. Hα Surveys at Higher Redshift

The successful implementation of the "nod-and-shuffle" technique (Glazebrook & Bland-Hawthorn 2001) on high-throughput multi-slit spectrographs at the 3.9m Anglo-Australian Telescope, has made feasible highly multiplexed faint surveys, even at wavelengths dominated by bright and highly variable airglow emission ($\lambda > 7000$ Å). We have exploited this capability to survey $z \sim 0.2-0.3$ clusters for Hα-emitting galaxies, and thus begin to establish the SFR-environment relation at these earlier epochs. Using the upgraded Low Dispersion Survey Spectrograph (LDSS++), we can detect redshifted Hα to a limit of $f(\mathrm{H}\alpha) \sim 1 \times 10^{-17}\,\mathrm{erg\,s^{-1}\,cm^{-2}}$, which at the redshift of the clusters corresponds to a SFR of $\sim 0.1\,M_{\odot}\,\mathrm{yr}^{-1}$. Hence our survey is the most sensitive that has been conducted at these redshifts.

In addition, we employ a special narrow-band filter to isolate a ~ 400 Å–wide region centred on redshifted Hα. This restricted wavelength coverage together with the small 'micro-slits' that can be used in nod-and-shuffle mode, enables 'ultraplexed' spectroscopy, with as many as ~ 900 objects being observed simultaneously with a single mask configuration over the $8.7' \times 8.7'$ field of LDSS++. Hence it is feasible to survey almost the entire galaxy population down to $I = 22.5$ ($M_R = -17.3$) within the field, providing a complete inventory of star-forming galaxies out to a radius of $\sim 2R_v$.

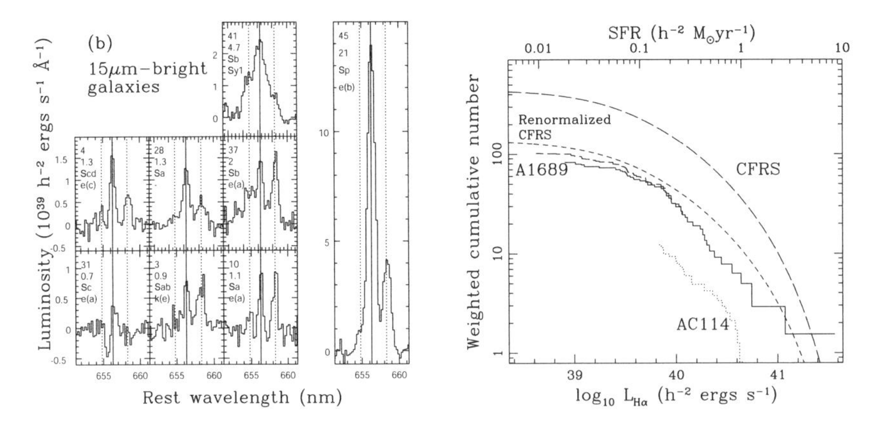

Figure 2. *Left:* Our LDSS++ spectra in the vicinity of Hα for the 15μm ISO sources observed in A1689. *Right:* Hα luminosity functions for the clusters A1689 ($z = 0.18$) & AC114 ($z = 0.32$), and the field at $z \sim 0.2$ (CFRS).

The first cluster that we targeted was AC114 at $z = 0.32$, and the results of that study have already been reported in Couch et al. (2001). More recently, we have observed the Abell cluster, A1689, at $z = 0.18$, and it is the results from this study that we shall concentrate on here. Apart from the interest in the 'demographics' of A1689's star-forming population, this cluster is of additional interest because of its excess (relative to local clusters) of MIR (15-μm) *ISO* sources, many of which are thought to be dusty starburst galaxies (Fadda et al. 2000, hereafter F2000). Hence we have an opportunity directly to compare Hα– and MIR–based SFRs for these galaxies and gauge just how much of their star formation is obscured by dust.

A total of 522 galaxies were observed in the field of A1689, of which 60 were detected in Hα (spectra for 8 of these, the sample of 15-μm ISO sources, are shown in the left panel of Figure 2). This corresponds to a $24 \pm 4\%$ incidence amongst cluster members, with no evidence of this varying with radius out to ~ 1 Mpc. In the right panel of Figure 2, we show the Hα luminosity function (HαLF) for A1689 and, for comparison, that of the field at $z \sim 0.2$ and that for AC114 (at $z = 0.32$). The field HαLF comes from CFRS (Tresse & Maddox 1998); we show both the uncorrected version (*long-dashed* line) and one which has been renormalised to account for the different morphological mix in the cluster and field samples (*short-dashed* line). The CFRS data have not been corrected for aperture effects or reddening, this not being necessary for a fair comparison with our data.

It can be seen that the A1689's HαLF is $\sim 50\%$ below that of the renormalized CFRS function in terms of normalization, indicating that the total amount of ongoing star formation in A1689 is this much lower than that in field galaxies at the same redshift. Surprisingly, perhaps, we see that the HαLF for AC114 is a factor of ~ 2 below that of A1689, even though this cluster is at a higher redshift. The reason for this is not completely understood; both clusters have a $\sim$ 20% fraction of blue galaxies (Couch & Sharples 1987; Duc et al. 2002, hereafter D02), and any effects of global evolution would lead to more star formation in AC114, contrary to what is observed. The most likely explanation for the difference is the dynamical state of the clusters. AC114 appears to be reasonably relaxed, whereas A1689 appears to have a substructured galaxy distribution and to be undergoing a dynamical merger (D02). A clearer picture of the link between cluster dynamics and star formation activity will be obtained when we complete our study of what are four quite dynamically diverse clusters.

Of the 522 galaxies targeted in A1689, 29 were known MIR sources from the F2000 *ISOCAM* observations. Previous spectroscopy had confirmed 19 of these were cluster members, and our observations detected Hα at the cluster redshift in 2 more of these sources. Confining our attention to these 21 members, all but 2 of them were detected at 6.7 μm, while only eight were detected at 15 μm. Half of the 15-μm sources, however, are unusually bright at 15 μm ($f_{15\,\mu\mathrm{m}}/f_{6.7\,\mu\mathrm{m}} > 4$), relative to normal spiral galaxies. Their spectra are likely to be dominated by a hot dust component, although it is not clear whether the dust is heated by star formation or central nuclear activity. Revisiting the spectra of the eight 15-μm sources in Figure 2, each of which is labelled with the object's Hubble type and spectral class (D02), we see that they are *all* spiral galaxies with securely detected Hα emission. It is also evident that *many* of these galaxies have atypical spectra in the vicinity of Hα (putting ISO#41 aside, since it is a known Seyfert 1 galaxy), with particularly strong [NII] λ6583 emission.

A more quantitative judgement of the significance of this strong [NII] λ6583 emission can be gathered from the left panel of Figure 3, where the ratio of [NII] λ6583 to Hα is plotted as a function of Hα luminosity, for all galaxies with Hα fluxes above our detection limit. The MIR sources are plotted with error bars. Veilleux & Osterbrock (1987) have shown that galaxies with [NII] λ6583/H$\alpha > 0.55$ are almost always associated with non-thermal emission, and we see that four of the MIR sources lie in this regime. However, the Hα fluxes are not corrected for underlying stellar absorption, and thus the plotted [NII] λ6583/Hα ratios are overestimates. Indeed, galaxies #31, 37, and 10 are known from their full optical spectra to have strong Balmer absorption lines (D02). If these three galaxies were to have a EW$\sim$ 3 Å absorption correction applied to their Hα line (as suggested by the strength of the absorption seen in their other Balmer lines), all would fall below the "non-thermal" line. Hence with the exception of ISO#41, we have no clear evidence of a non-thermal contribution to the Hα emission in these galaxies.

On this basis, we can now examine the claims of F2000 and D02 that much of the star formation in A1689 is obscured by dust, and is therefore only observable in the MIR. In the right panel of Figure 3, we compare the MIR SFRs measured from the 15μm luminosities by D02 with our values based on Hα. Taken at face value, the plot indicates that Hα underestimates the star formation rate in the 15-μm sources by a factor of 2–20 with respect to the MIR. If we correct our observed Hα fluxes for say 5 Å of stellar absorption, this reduces the discrepancy to a factor of 1.7–10. As indicated by the dashed horizontal lines in the figure, this corresponds to as much as 3 mags of dust extinction at Hα. Subject to the caveat that non-thermal sources are *not* contributing to the Hα and MIR emission – a possibility that still must be taken seriously

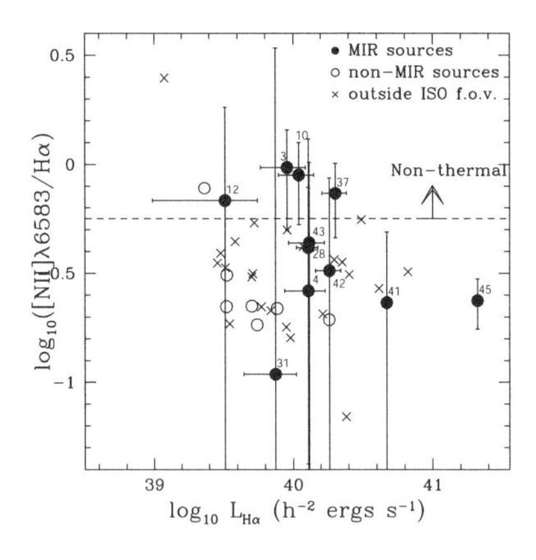

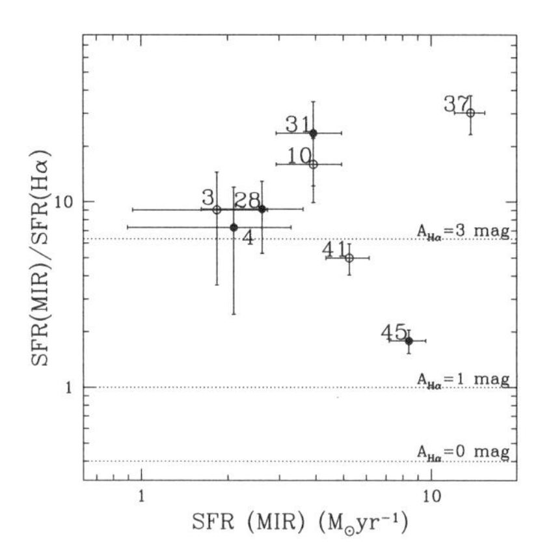

Figure 3. *Left:* The [NII] λ6583/Hα ratio plotted as a function of Hα luminosity. All members of A1689 with Hα fluxes above our detection limit are plotted; different sub-samples are indicated in the legend. *Right:* The ratio of the SFR measured in the MIR to that measured using Hα, plotted as a function of the former. The amount of dust extinction required to explain this ratio is indicated by the *dotted* horizontal lines.

and investigated further – our study of A1689 would suggest, at least, that significant amounts of star formation in distant clusters remains unseen at rest optical wavelengths, supporting the conclusion drawn previously by Smail et al. (1999) on the basis of a radio versus optical comparison.

4. Summary

The salient results from our two studies can be summarized as follows:

- Star formation in cluster galaxies is *suppressed* relative to the field out to very large radii: *the cluster 'sphere of influence' extends out to* $R \sim 6$ Mpc ($\sim 3R_v$).

- The SFR is also strongly correlated with the local projected galaxy density, and this correlation persists to densities as low as ~ 1 gal Mpc^{-2}, which are typical of those found in the group environment.

- The physical mechanisms responsible for this suppression cannot therefore be those confined to rich cluster cores; less severe processes such as halo gas stripping are highly favoured.

- The strong evolution in the abundance of galaxy groups over recent epochs makes it likely that hierarchical growth of structure plays a significant role in decreasing the global SFR.

- Subject to the absence of non-thermal emission, further evidence is found for star formation being significantly obscured in distant clusters, with the

SFR measures based on MIR fluxes being 2–10 times larger than those based on our Hα measurements in A1689.

References

Balogh, M. L., Morris, S. L., Yee, H. K. C., Carlberg, R. G. & Ellingson, E. 1997, ApJ, 488, L75

Balogh, M. L., Schade, D., Morris, S. L., Yee, H. K. C., Carlberg, R.G. & Ellingson, E. 1997, ApJ, 504, L75

Balogh, M. L., Couch, W. J., Smail, I., Bower, R. G. & Glazebrook, K. 2002, MNRAS, 335, 10

Blanton, M. R., Dalcanton, J., Eisenstein, D., Loveday, J., Strauss, M. A., SubbaRao, M., Weinberg, D. H. & the Sloan collaboration 2001, AJ, 121, 2358

Butcher, H. & Oemler, A. 1978, ApJ, 219, 18

Colless, M., Dalton, G., Maddox, S., Sutherland, W., Norberg, P., Cole, S., , & the 2dFGRS team 2001, MNRAS, 328, 1039

Couch, W. J., Balogh, M. L., Bower, R. G., Smail, I., Glazebrook, K. & Taylor, M. 2001, ApJ, 549, 820

Couch, W. J. & Sharples, R. M. 1987, MNRAS, 229, 423

Cowie, L. L., Songaila, A. & Barger, A. J. 1999, AJ, 118, 603

De Propris, R., Couch, W. J., Colless, M., Dalton, G. B., Collins, C., Baugh, C. M. & the 2dFGRS team 2002, MNRAS, 329, 87

Dressler, A. 1980, ApJ, 236, 351

Duc, P-A., Poggianti, B. M., Fadda, D., Elbaz, D., Flores, H., Chanial, P., Franceschini, A., Moorwood, A. F. M. & Cesarsky, C. J. 2002, A&A, 382, 60 (D02)

Fadda, D., Elbaz, D., Duc, P-A., Flores, H., Franceschini, A., Cesarsky, C. J. & Moorwood, A. F. M. 2000, A&A, 361, 827 (F2000)

Glazebrook, K. & Bland-Hawthorn, J. 2001, PASP, 113, 197

Gunn, J. E. & Gott, J. R. 1972, ApJ, 176, 1

Kennicutt, R. C. 1983, ApJ, 272, 54

Larson, R. B., Tinsley, B. M. & Calwell, C. N. 1980, ApJ, 237, 692

Lewis, I., Balogh, M., De Propris, R., Couch, W. J., Bower, R., Offer, A. & the 2dFGRS team 2002, MNRAS, 334, 673

Lilly, S. J., Le Fevre, O., Hammer, F. & Crampton, D. 1996, ApJL, 460, L1

Madau, P., Ferguson, H. C., Dickinson, M. E., Giavalisco, M., Steidel, C. & Fruchter, A. 1996, MNRAS, 283, 1388

Smail, I., Morrison, G., Gray, M. E., Owen, F. N., Ivison, R. J., Kneib, J. P. & Ellis, R. S. 1999, ApJ, 525, 609

Tresse, L. & Maddox, S. J. 1998, ApJ, 495, 691

IAU 8th Asia-Pacific Regional Meeting
ASP Conference Series, Vol. 289, 2003
S. Ikeuchi, J. Hearnshaw & T. Hanawa, eds.

A Search for Galaxy Clustering at z=4 Using an SDSS Quasar Pair[1]

Osamu Nakamura, Masataka Fukugita

Institute for Cosmic Ray Research, University of Tokyo, Kashiwa, 2778582 Japan

Mamoru Doi

Institute for Astronomy, University of Tokyo, Mitaka, 1818588 Japan

Nobunari Kashikawa

National Astronomical Observatory, Mitaka, 1818588 Japan

Donald P. Schneider

Department of Astronomy, Pennsylvania State University, PA 16802, U. S. A.

Abstract. We have carried out deep V, R and I Subaru/FOCAS imaging of the field centred on a pair of $z = 4.25$ quasars to search for an associated protocluster. No significant enhancement in the galaxy surface density or any recognizable structure is found within an arcminute of the centre of the pair.

1. Introduction

Quasars are rare phenomena in the universe. When the comoving density of luminous quasars reaches a maximum at $z \approx 2$, it is about 10^{-7} times the density of galaxies. Taking an advantage of wide-field multicolour CCD imaging of the Sloan Digital Sky Survey (SDSS; York et al. 2000). Schneider et al. (2000) discovered a pair of $z = 4.25$ luminous quasars separated by $33''$ (SDSSp J1439−0034). From spectroscopic features, this pair is most likely to be a physical association, not a gravitational lens. The projected distance is $0.16h^{-1}$ Mpc with a non-zero lambda ($\Omega_0 = 0.3$, $\lambda = 0.7$) cosmology. This system's separation is among the smallest known at redshift above four.

In the widely-accepted standard model of galaxy formation based on the cold dark matter (CDM) dominance, quasars are usually ascribed to a phenomenon associated with very rare, high peaks of Gaussian fluctuations (Efstathiou & Rees 1988). This suggests that a close pair of bright quasars may be

[1]Based on data collected at the Subaru Telescope, which is operated by the National Astronomical Observatory of Japan.

embedded in a very rich environment, such as a protocluster of galaxies (e.g., Djorgovski 1998; 1999; see also Martini & Weinberg 2001; Haiman & Hui 2001). On the other hand, the structure formation in the CDM model with fluctuations normalized by the cosmic microwave background predicts that clusters or even groups in excess of $10^{13} M_{\odot}$ cannot have formed at $z \approx 4$.

To investigate this situation, we have carried out deep V, R and I imaging of the field around the SDSS quasar pair using the Subaru Telescope at Mauna Kea. Our prime purpose is to examine the spatial distribution of galaxies that show Lyman break features (LBG) in the V passband, which indicate $z \gtrsim 4$.

Table 1. Properties of the SDSS Quasar Pair

Quasar	RA(J2000)	DEC(J2000)	z	r'
A	14:39:52.58	−00:33:59.2	4.255	20.97
B	14:39:51.60	−00:34:29.2	4.258	21.79

2. Observations and Data Reduction

Table 1 shows the pair of quasars given in Schneider et al. (2000). We carried out imaging of this field in the V, R and I passbands using FOCAS instrument (Kashikawa et al. 2000) on 2001 June 19 and 20. The condition was photometric. We also carried out imaging of off-quasar field which we call a 'blank' field. The total exposure times are 6000 s, 2700 s, 2700 s for V, R and I for the quasar field, and 4800 s, 3000 s, 1800 s for the blank field, respectively. Data were processed using a standard IRAF photometry package, and objects were detected in the R band employing Sextractor version 2.2.1 imposing a 3σ threshold after 5×5 pixel smoothing. This detection threshold corresponds to $R \approx 26$ mag., or roughly $I \approx 25.5$ mag. This is compared with the characteristic magnitude of $z \approx 4$ Lyman break galaxies, $I_* \simeq 24.5$ derived by Steidel et al. (1999). Our depth with FOCAS is slightly shallower than the sample of Ouchi et al. (2001), who also studied $z \approx 4$ galaxies at prime focus of the Subaru Telescope, by about 0.5 mag. Our depth is sufficient to detect significant numbers of L_* galaxies at $z \approx 4$. Sextractor was also applied to V and I images, and cross-identification is made for the objects found in the R band.

Galaxy candidates with $z \approx 4.25$ are identified with the V drop-out technique (Steidel et al. 1999), using the following colour selection criteria:

$$V - R > 0.90, \tag{1}$$
$$V - R > 0.48(R - I) + 0.67, \tag{2}$$
$$V - R > 5.79(R - I) - 3.38, \tag{3}$$

which are obtained from the galaxy track expected in population synthesis models of Bruzual & Charlot (1995; 1993) and Kodama & Arimoto (1997). The loci of the galaxy tracks in this colour plane depend little on the models used and the assumed epoch of galaxy formation.

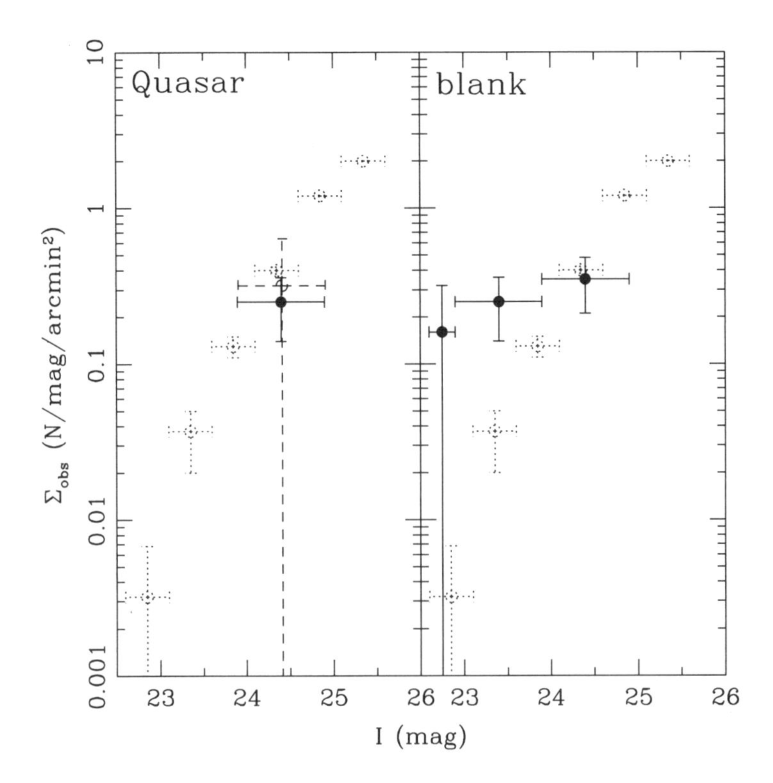

Figure 1. Surface number density of galaxies with $z = 4.25$. The data denoted by solid circles show the density averaged over the FoV of FOCAS (5′ radius) for the quasar filed (the left panel) and the blank filed (the right panel). Dashed symbol indicates the density of galaxies within 1 arcminute from the centre of the quasar pair. The dotted data show the results of Ouchi et al. (2001) for comparison. Incompleteness corrections are not applied.

3. Discussion

We detect 5 LBG candidates in the full FOCAS field (5 arcminutes diameter) that includes the quasar pairs. We find no recognizable structure in their positions. Only one candidate is located within 1 arcmin ($0.28h^{-1}$ Mpc at $z = 4.25$) from the centre of the pair quasars. For the blank filed we detected 13 LBG candidates in the same FoV.

We present in Figure 1 surface number densities of the $z \approx 4$ LBG candidates for the two fields. We have not corrected the numbers of detections for incompleteness regarding the colour selection, but we only argue for relative numbers detected in each field. The left panel concerns the FOCAS field including quasar pairs, and the thick dashed data point represents the surface density within 1 arcmin from the centre of the quasar pairs. Our results may be compared with the thin dotted data points, which indicate the surface number density obtained by Ouchi et al. (2001) from i' band imaging for significantly wider field of view (without corrections for incompleteness), after transforming to our I band. Our data, although statistics are low, match well those of Ouchi et al. They also agree with Steidel et al.'s (1999) surface density. We note in particular that the surface density around the quasar pair (dashed circle) is about the field value (solid circle). The conclusion is unchanged even if we count two quasars as candidate galaxies.

We conclude that there is no significant enhancement of $z \sim 4$ galaxies in the field of the SDSS quasar pair; this is surprising given that the presence of two luminous quasars suggests a particularly rich protocluster environment. Although this is but one example (though the pair of quasars make it a particularly attractive one), our results suggest that luminous quasars are not always signposts for high-density regions; perhaps this is an example of a case where two galaxies happened by chance to display the quasar phenomenon simultaneously. Perhaps the scarcity of luminous, high redshift quasars is not solely (or even primarily) due to the requirement that they reside in the rare, high density perturbations; a relatively short lifetime of the quasar phenomenon at high redshift could play an important role.

Acknowledgments. We would like to thank Dr Youichi Ohyama for his very efficient help for our observations at the Subaru Telescope.

References

Bruzual A., G. & Charlot, S. 1993, ApJ, 405, 538

Djorgovski, S. G. 1998, in Fundamental Parameters in Cosmology, XXXIII Rencontre de Moriond, eds Y. Giraud-Héraud et al. (Editions Frontières, Gif sur Yvette)

Djorgovski, S. G. 1999, in ASP Conf. Ser. 193: The Hy-Redshift universe: Galaxy Formation and Evolution at High Redshift, 397

Efstathiou, G. and Rees, M. J. 1988, MNRAS, 230, P5

Haiman, Z. & Hui, L. 2001, ApJ, 547, 27

Kashikawa, N. et al. 2000, in Proc. SPIE, 4008: Optical and IR Telescope Instrumentation and Detectors, 104

Kodama, T. & Arimoto, N. 1997, A&A, 320, 41

Martini, P. & Weinberg, D. H. 2001, ApJ, 547, 12

Ouchi, M. et al. 2001, ApJ, 558, L83

Schneider, D. P. et al. 2000, AJ, 120, 2183

Steidel, C. C. et al. 1999, ApJ, 519, 1

York, D. G. et al. 2000, AJ, 120, 1579

IAU 8th Asia-Pacific Regional Meeting
ASP Conference Series, Vol. 289, 2003
S. Ikeuchi, J. Hearnshaw & T. Hanawa, eds.

The Discovery of Submillimeter Galaxies

A. W. Blain

Department of Astronomy, Caltech 105-24, Pasadena CA91125, USA

Abstract. I briefly describe some results about luminous distant dusty galaxies obtained in the 5 years since sensitive two-dimensional bolometer array cameras became available. The key requirements for making additional progress in understanding the properties of these galaxies are discussed, especially the potential role of photometric redshifts based on radio, submillimeter (sub-mm) and far-infrared(IR) continuum observations

1. Introduction

In 1997 the SCUBA camera was commissioned at the 15-m JCMT in Hawaii, providing 2.5-arcmin-wide images at wavelengths of 450 and 850 μm with resolutions of 7 and 15″respectively. The MAMBO 1.25-mm camera at the 30-m IRAM telescope provides a similar capability (see Dannerbauer et al. 2002), while the 350-μm SHARC-II and 1.1-mm BOLOCAM cameras being commissioned at the 10-m Caltech Submillimeter Observatory (CSO) should provide significantly enhanced performance within a year. In the future, BOLOCAM is expected to observe from the 50-m mm-wave GTM/LMT telescope under construction in Mexico, a larger-format 8×8-arcmin^2 camera SCUBA-II is being designed in the UK, and developments of MAMBO are planned for APEX – a new 12-m telescope at the 5000-m ALMA site in Chile. APEX will join the 10-m-class Japanese ASTE sub-mm telescope at the ALMA site for which a large bolometer camera is being designed.

SCUBA was the first sub-mm camera able to survey fields large enough to detect the redshifted thermal dust emission from previously unknown galaxies (Smail, Ivison & Blain 1997). The peak of the emission from galaxies, typically at 60–200 μm in their rest-frame, corresponding to a range of dust temperatures between about 60 and 20 K, is redshifted into SCUBA's observing bands. The steep sub-mm Rayleigh–Jeans slope of the dust emission ensures that distant galaxies with similar bolometric luminosities and spectral energy distributions (SEDs) would produce similar flux densities at all redshifts from $z \simeq 0.5$ to $z \sim 5$, assuming the same spectral energy distribution (SED). The detectability of galaxies does depend on the details of the SED, in general being greater for cooler temperatures at a fixed luminosity and redshift (see Blain et al. 2002a).

More than 300 high-z sub-mm galaxies (SMGs) have now been detected by SCUBA and MAMBO, while BOLOCAM (Glenn et al. 1998) could detect new examples at a rate of order 1 per hour. Various types of surveys have been made: in narrow fields to exploit the gravitational lensing effect of clusters of galaxies (Smail et al. 2002; Chapman et al. 2002a; Cowie, Barger & Kneib 2002); in the Hubble Deep Field (Hughes et al. 1998); and in wider shallower surveys (Eales et al. 1999; Borys et al. 2002; Scott et al. 2002) covering a total of about $0.25\,\mathrm{deg}^2$.

These SMGs are responsible for a significant fraction of the star-formation/AGN activity at $z > 1$ (Blain et al. 1999a). It is vital to understand their individual properties if we are to understand the formation of galaxies as a whole. Their inferred space density is similar to that of giant elliptical galaxies in the local universe, and it has been suggested that they are high-z formation events of these rare galaxies (Eales et al. 1999), presumably in the most massive dark-matter halos with the lowest specific angular momenta. It is more likely that they reflect a more common, short-lived phase involving the formation of galactic bulges in perhaps episodic mergers (Blain et al. 1999b). The test of these ideas is to measure the redshifts, clustering properties and mass distribution of a representative sample of SMGs.

2. Finding Redshifts and Studying Astrophysics

The coarse ($\sim$15$''$) resolution of sub-mm images ensures that there are many possible faint optical counterparts to the detected galaxy. Hence, while some SMGs have bright optical counterparts (Ivison et al. 1998, 2000), most remain unidentified at optical wavelengths. Final confirmation of a correct identification in both position and redshift is provided by detecting (sub)mm-wave CO molecular line emission with a suspected redshift in the narrow spectral window of existing (sub)mm-wave spectrographs (Frayer et al. 1998, 1999).

Mm-wave continuum interferometer images of the fields have reduced positional uncertainties to $\sim$ 1 arcsec for several SMGs (Downes et al. 1999; Frayer et al. 2000; Dannerbauer et al. 2002), but are very expensive in observing time. Deep near-IR imaging to $K > 22$ tends to reveal plausible red counterparts for many SMGs (Smail et al. 1999; Frayer et al. 2003) by their $J - K$ and $I - K$ colors. Deep optical/near-IR spectroscopy of these galaxies can then be attempted to find redshifts. The most efficient technique for determining redshifts, however, appears to be to exploit very deep radio images. For reasonable SEDs, SMGs with 850-μm flux densities of about 5–10 mJy (with luminosities $\sim 10^{13}\,\mathrm{L}_\odot$) should be detectable in $\sim 10\,\mu$Jy RMS 1.4-GHz VLA images out to $z \simeq 3$, if they lie on the far-IR–radio correlation observed for local galaxies (Condon 1992): (see Barger, Cowie & Richards (2000) and Chapman et al. (2001, 2002b). The wide field ($0.25\,\mathrm{deg}^2$) and accurate sub-arcsec astrometry of VLA images, coupled with the low surface density of the faintest radio sources as compared with optically-selected galaxies yield accurate positions for a large fraction ($\sim$ 70%) of the SMGs brighter than 5 mJy at 850 μm. This provides an opportunity for efficient multi-object optical spectroscopy. In 2002 March, about 25 spectra were obtained using the Keck-LRIS spectrograph (Chapman et al. 2002c). These will be subject to CO molecular line spectroscopy to confirm the identifications and to study both their gas masses (via velocity dispersions) and excitation conditions (via line–line and line–continuum ratios), using mm-wave interferometers in the 2002–2003 northern winter. Only a handful of reliable redshifts were available for SMGs in 2001: now it is likely that a luminosity function of these radio-selected SMGs should be available in 2003.

3. Photometric Redshifts

The key target of investigating the SMGs is now to find their physical properties, especially their masses. However, just obtaining a reasonably complete redshift distribution is important for fixing their form of evolution and ensuring their fractional contribution to the energy emission of all galaxies is correctly accounted for (Eales et al. 1999; Blain et al. 1999a).

As redshift surveys were generally unsuccessful until (Chapman et al. 2002c), photometric techniques have been proposed to provide redshift information (Eales et al. 1999; Hughes et al. 1998, 2002). The key information available is radio (Carilli & Yun 1999), sub-mm and mid-/far-infrared photometry, typically from VLA, SCUBA and *ISO* respectively. The *IRAS* survey is not sufficiently deep to detect SMGs; *ISO* data is deeper, but covers only a small area and is only useful for the closest (Soucail et al. 1999) or brightest examples (Ivison et al. 1998). The *SIRTF* space telescope will be ideal for finding far-IR SEDs.

Fitting a thermal spectrum to a galaxy at uncertain redshift leads to an unavoidable degeneracy between the inferred dust temperature T and redshift z. The peak wavelength of the SED is determined in the observers frame, but this is shifted in exactly the same way by either a fractional increase in T or a corresponding fractional decrease in $(1 + z)$. This makes any far-IR/sub-mm-based photometric redshift only as accurate as the knowledge of the temperature (Blain, Barnard & Chapman 2002b), and not to $\Delta z \simeq 0.5$ as claimed by Hughes et al. (2002). Despite non-thermal radio emission being due to an entirely different process, the sub-mm–radio properties of galaxies on the far-IR–radio correlation conspire to produce a similar T–$(1 + z)$ degeneracy if $T < 60\,\mathrm{K}$ (Blain 1999). If a reliable link exists between dust temperature and luminosity, then it is possible to break this degeneracy; however, the accuracy of the result is then determined by the scatter in the LT relation, which is likely to be at least 30%, implying at least this great an uncertainty in redshift (Blain et al. 2002b).

The addition of K-band near-IR data (Dannerbauer et al. 2002) could also help, but first the intrinsic scatter in the ratio between the K-band and far-IR luminosity of the galaxies must be known. Based on observations of fairly complete samples of SMGs (Ivison et al. 1998; Frayer et al. 2003), the K-band magnitudes are certainly scattered by as much as $\Delta K \sim 2\,\mathrm{mag}$. Spectroscopic redshifts remain essential for accurate study of SMGs.

4. Spectroscopic Mm-wave Redshifts

Correct identifications of SMGs via either samples of faint radio galaxies that feed targets to multi-object spectrographs or deep near-IR imaging and spectroscopy, must be confirmed and verified using mm-wave interferometers or single-antenna telescopes to detect CO molecular rotational line emission at integer multiples of 115 GHz in the galaxy's rest frame.

There are other possibilities for obtaining spectroscopic redshifts, especially the direct detection of molecular or atomic fine-structure spectral lines at mid-/far-IR and sub-mm wavelengths. The key is to obtain wide-band spectral coverage at these wavelengths, in order to allow searches for redshifts. At present, mm-wave spectrographs cover a total frequency range of only $\Delta\nu/\nu \simeq 0.05$, and so a redshift must be known to well within 5% before confirmation can be made.

Powerful correlators for the 100-m GBT will allow searches for CO(1-0) lines (rest frequency 115 GHz) from high-redshift galaxies in the 22 & 44-GHz radio bands. At shorter sub-mm wavelengths new very stable 230/345-GHz spectral line receivers at the CSO will have $\sim$ 10 GHz bandwidths (Rice et al. in prep.). Dispersive techniques may allow very-wideband, low-resolution spectrographs at millimeter or far-infrared wavelengths. These include the ZSPEC and WaFIRS waveguide/grating concepts for space-borne and ground-based applications (Bradford et al. in prep).

Acknowledgments. I thank Vicki Barnard, Matt Bradford, Scott Chapman, Dave Frayer, Naveen Reddy, Frank Rice and Chip Sumner.

References

Barger, A. J., Cowie, L. L. & Richards E. A. 2000, AJ, 119, 2092

Blain, A. W. 1999, MNRAS, 309, 955

Blain, A. W., Barnard, V. E. & Chapman, S. C. 2002b, MNRAS, submitted

Blain, A. W., Smail, I., Ivison, R. J. & Kneib, J.-P. 1999a, MNRAS, 302, 632

Blain, A. W. et al. 1999b, MNRAS, 209, 715

Blain, A. W., Smail, I., Ivison, R. J., Kneib, J.-P. & Frayer, D. T. 2002a, Physics Reports, in press (astro-ph/0202228)

Borys, C., Chapman, S. C., Halpern, M. & Scott, D. 2002, MNRAS, 330, 63

Carilli, C. L. & Yun, M.S. 1999, ApJ, 513, L13

Chapman, S. C., et al. 2001, ApJ, 548, L147

Chapman, S. C., Scott, D., Borys, C. & Fahlman, G. G. 2002a, MNRAS, 330, 92

Chapman, S. C. et al. 2002b, ApJ, in press

Chapman, S. C. et al. 2002c, Nature, submitted

Condon, J. J. et al. 1992, ARA&A, 30, 575

Cowie, L. L., Barger, A. J., & Kneib, J.-P. 2002, AJ, 123, 2197

Dannerbauer, H., et al. 2002, ApJ, in press (astro-ph/0201104)

Downes, D. et al. 1999, A&A, 347, 809

Eales, S. et al. 1999, ApJ, 515, 518

Frayer, D. T. et al. 1998, ApJ, 506, L7

Frayer, D. T. et al. 1999, ApJ, 514, L13

Frayer, D. T. Smail, I., Ivison, R. J., & Scoville, N. Z. 2000, AJ, 120, 1668

Frayer, D. T. et al. 2003, AJ, submitted

Glenn, J., et al. 1998, Proc. SPIE, 3357, 326

Hughes, D. H. et al. 1998, Nature, 394, 241

Hughes, D. H. et al. 2002, MNRAS, in press (astro-ph/0111547)

Ivison, R. J. et al. 1998, MNRAS, 298, 583

Ivison, R. J. et al. 2000, MNRAS, 315, 209

Scott, S. E. et al. 2002, MNRAS, 331, 817

Smail, I. Ivison, R. J., & Blain, A. W. 1997, ApJ, 490, L5

Smail, I. et al. 1999, MNRAS, 308, 1061

Smail, I. et al. 2002, MNRAS, 331, 495

Soucail, G. et al. 1999, A&A, 343, L70

IAU 8th Asia-Pacific Regional Meeting
ASP Conference Series, Vol. 289, 2003
S. Ikeuchi, J. Hearnshaw & T. Hanawa, eds.

Prospects for Detecting Neutral Hydrogen Using 21-cm Radiation from Large Scale Structure at High Redshifts

J. S. Bagla

Harish-Chandra Research Institute, Chhatnag Road, Jhunsi, Allahabad 211019, India

Martin White

Department of Physics and Astronomy, University of California, Berkeley, CA 94720, USA

Abstract. We estimate the signal from large scale structure at high redshifts in redshifted 21-cm line. We focus on $z \simeq 3$ and the Λ-CDM cosmology. We assume that neutral hydrogen is to be found only in galaxies, and normalize the total content to the density parameter of neutral hydrogen in damped Lymanα absorption systems (DLAS). We find that the *rms* fluctuations due to the large scale distribution of galaxies is very small and cannot be observed at angular scales probed by present day telescopes. We have used the sensitivity of the Giant Meter-wave Radio Telescope (GMRT) for comparison. We find that observations stretching over 10^3 hours will be required for a 3σ detection of such objects.

1. Introduction

Observations of galaxies and absorption systems at high redshifts have shown that the assembly of present day galaxies started around $6 \geq z \geq 1$. At higher redshifts, the fraction of matter that had collapsed to form stars was small. By the end of the redshift range mentioned here, the rate of star formation was starting to decline and assembly of larger structures like the clusters of galaxies is the dominant process. In order to understand the process of galaxy formation completely, it is important to study galaxies in this redshift range in as many wavelengths as possible.

In this paper we present results for the large scale distribution of HI at $z \simeq 3$. More detailed results for the entire range of redshifts will be presented elsewhere. References for earlier theoretical work as well as related observations can be found in Bagla (1999a) and Bharadwaj (2001).

Observations of galaxies and quasars at high redshifts provide convincing proof that the inter-galactic medium is ionized out to $z \sim 6.2$ (Pentericci et al. 2002). Thus most of the neutral hydrogen at $z \simeq 3$ is to be found in galaxies. We have an estimate of the total neutral hydrogen content from observations of DLAS (Storrie-Lombardi, McMahon and Irwin, 1996). There is no observational evidence to support the hypothesis that DLAS and Lyman break galaxies (LBG) have different properties (Fynbo et al. 2002). Thus we can safely assume that neutral hydrogen is distributed uniformly amongst galaxies at $z \simeq 3$, and that the total amount of neutral hydrogen adds up to give us the density parameter contributed by neutral hydrogen in DLAS.

2. From Haloes to Radio Maps

In this section we outline the method used for estimating the signal from neutral hydrogen at high redshifts.

We ran N-body simulations of the Λ-CDM model. The size of the simulation box in physical units is $50h^{-1}$ Mpc (comoving) and the number of particles used in this gravity only simulation was 256^3. We used the TreePM method (Bagla, 1999b) for these simulations.

The density around each particle was estimated by measuring the distance to the nth neighbour, where neighbours were sorted by distance. The results presented here used $n = 16$ but the numbers do not change much for $n = 32$ or for a different choice of estimator for density.

Particles in regions with over-density higher than a threshold were then selected from the simulation. Each of these particles was assigned an equal amount of neutral hydrogen such that the density parameter of neutral hydrogen was 0.002. Results can be scaled trivially if one prefers a different value of Ω_{HI}.

The dependence of emission on the local spin temperature is weak enough to be ignored in physical situations of interest. The optical depth in 21-cm is sufficiently small for us to ignore any absorption. Thus the problem of making radio maps essentially reduces to that of assigning frequency and angular position to each particle in high density regions and adding up the signal in relevant frequency channel and pixel. The conversion from neutral hydrogen mass to signal is

$$S_\nu = 75\ \mu\text{Jy} \left(\frac{M_{\text{HI}}}{4 \times 10^{13} M_\odot} \right) \left(\frac{1\ \text{MHz}}{\Delta\nu} \right). \tag{1}$$

The simulated radio maps can then be used to look for optimum frequency window and angular scales at which a search for signal should be carried out. In such an exercise, parameters of present day telescopes need to be considered and we have used sensitivity levels of the GMRT (Swarup et al. 1991) for this. Figure 1 shows the expected signal from the largest objects for a frequency channel of 1 MHz as a function of angular scale. We scanned the radio maps with a very fine resolution in order to locate the maximum signal. The expected signal is compared with the sensitivity of the GMRT at 327 MHz for a 10^3-hour observation. This figure suggests that a 3σ detection of the largest structures is possible at angular scales $3' - 6'$ in such an observation. Signal expected from typical structures is much smaller.

The expected observation time is very large and hence it is important to ask whether such extreme structures are likely to be there in the volume sampled by the GMRT beam or not. The volume sampled by a GMRT beam at 327 MHz is much larger than the volume of the simulation used here. It is also much larger than the volume of the fields in which spikes in the redshift distribution of LBGs have been observed (Steidel et al. 1998). As the rate at which spikes occur in the redshift distribution of LBGs is close to one per field, we expect that the GMRT beam will contain at least one extreme object.

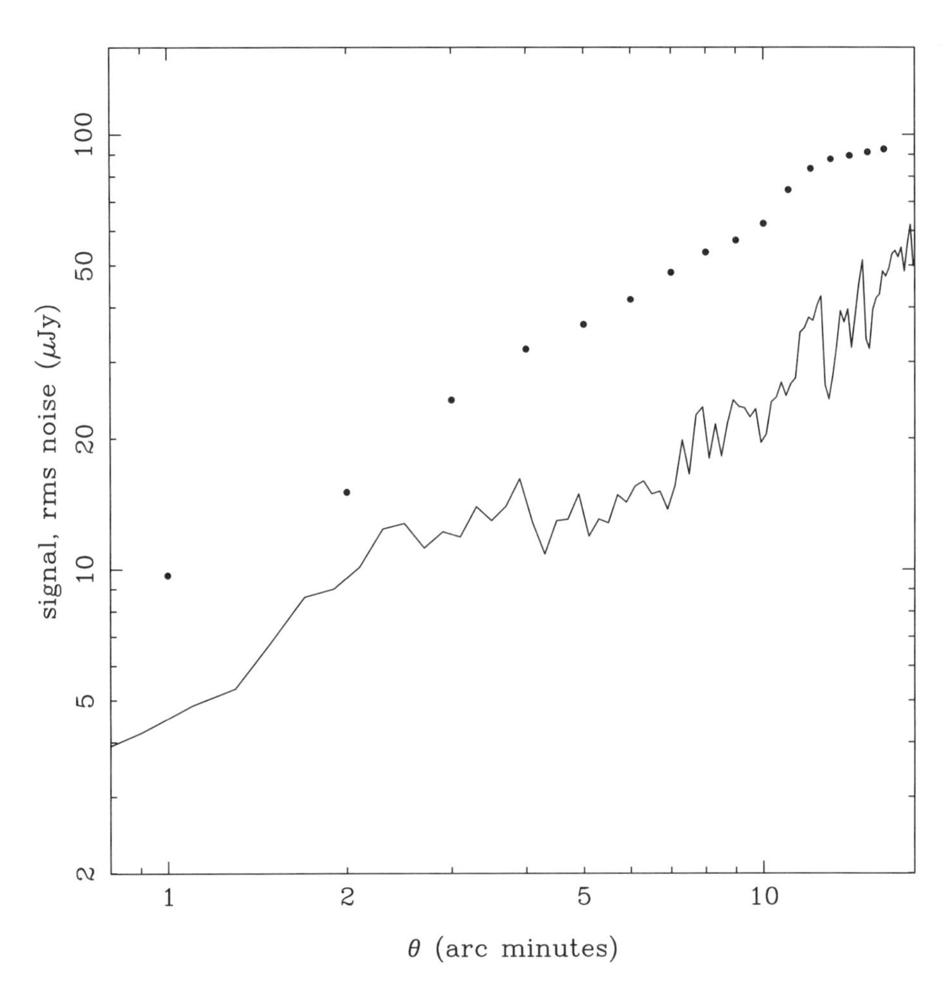

Figure 1. The expected signal (dots) from largest structures at $z \simeq 3$ as a function of angular scale and frequency width of 1 MHz. Estimated noise level (line) for the GMRT is shown as a function of scale for the same bandwidth and 10^3 hours of integration. We used the sensitivity of GMRT at 327 MHz, which corresponds to a redshift slightly higher than 3. A 3σ detection is possible at angular scales around $5'$.

3. Discussion

We have used N-body simulations to estimate the signal in redshifted 21-cm line from large scale structure at $z \simeq 3$. We find that the present day telescopes can detect extreme objects in the large scale structure at these redshifts.

In our estimation, we made use of gravity only simulations and ignored any effects coming from gas physics. These effects play a very important role in the distribution and state of gases at small scales. However, we are interested only in gross properties and that too at large scales and there is no reason to suspect that we will get these wrong in the method that we have used as long as we restrain our estimates to scales larger than 1 Mpc (comoving). The weakest point in our method is that we assumed that the amount of neutral hydrogen assigned to an N-body particle did not depend on the mass of the collapsed structure that contained this particle. We expect larger structures ($M > 10^{13} M_{\odot}$) to be like groups of galaxies and have less neutral hydrogen by fraction. Also, we expect smaller structures ($M < 10^{10} M_{\odot}$) to have less neutral fraction as these will be influenced strongly by the photo-ionizing background. However, the fraction of mass in haloes at these extremes is negligible and hence we do not expect these effects to invalidate our results at these redshifts. Such effects will be very important at lower redshifts where very massive haloes are more common, or at high redshifts where the low mass haloes contain most of the mass in collapsed structures.

Acknowledgments. A part of the work presented here was done using the Beowulf cluster at the Harish-Chandra Research Institute (http://cluster.mri.ernet.in).

References

Bagla, J. S. 1999a, *Highly Redshifted Radio Lines*, Ed. C. L. Carilli, S. J. E. Radford, K. M. Menten and G. I. Langston, ASP Conf. Ser., 156, 9

Bagla, J. S. 1999b, astro-ph/9911025

Bharadwaj, S. and Sethi, S. K. 2001, J. Ap. A., 22, 293

Fynbo, J. P. U., Ledoux, C., Møller, P., Thomsen, B., Burud, I. and Leibundgut, B. 2002, astro-ph/0205234

Pentericci, L. et al. 2002, AJ, 123, 2151

Steidel, C. C., Adelberger, K. L., Dickinson, M., Giavalisco, M., Pettini, M. and Kellogg, M. 1998, ApJ 492, 428

Storrie-Lombardi, L. J., McMahon, R. G. and Irwin, M. J. 1996, MNRAS, 283, 79

Swarup, G., Ananthakrishnan, S., Kapahi, V. K., Rao, A. P., Subrahmanya, C. R. and Kulkarni, V. K. 1991, Cu. Sc., 60, 95

IAU 8th Asia-Pacific Regional Meeting
ASP Conference Series, Vol. 289, 2003
S. Ikeuchi, J. Hearnshaw & T. Hanawa, eds.

Cosmological Constant with Sunyaev-Zel'dovich Effect towards Distant Galaxy Clusters

Masato Tsuboi

Institute of Astrophysics and Planetary Science, Ibaraki University, Ibaraki, 310-8512, Japan

Takashi Kasuga

Department of System and Control Engineering, Hosei University, Tokyo, 184-8584, Japan

Atsushi Miyazaki, Nario Kuno, Akihiro Sakamoto, and Hiroshi Matsuo

Nobeyama Radio Observatory, Nagano, 384-1305, Japan

Abstract. We have observed the Sunyaev-Zel′dovich (S-Z) effect towards four distant galaxy clusters from $z = 0.18 - 0.55$ at 43 GHz using a newly developed 6-beam receiver installed in the Nobeyama 45-m telescope. The S-Z effects at the cluster center at the Rayleigh-Jeans limit are -1.38 ± 0.07 mK for CL0016+16, -0.95 ± 0.20 mK for MS1358+62, -0.63 ± 0.16 mK for MS1008-12, and -0.73 ± 0.10 mK for A2218, respectively. The S-Z effect for MS1008-12 is a new detection. Assuming that the dimensionless cosmological constant is $\Omega_\Lambda = 0.7$, the Hubble constant is estimated to be $H_0 = 67 \pm 13$ km s^{-1} Mpc^{-1}. This is consistent with the value in the nearby universe obtained by the Hubble Space Telescope (HST). On the contrary, assuming that the derived value is equal to that by HST, the results are in agreement with the dimensionless cosmological constant of $\Omega_\Lambda \sim 0.7$.

1. Introduction

The cosmological parameters, $H_0, \Omega_m, \Omega_\Lambda$ etc. decide the structure and the evolution of the universe. But, the measurements of these parameters are very difficult issues in astronomy. For example, the determination of the Hubble constant in 10% accuracy requires great efforts over 70 years. With recent dramatic improvements by the Hubble Space Telescope (Freedman et al. 2000), Hubble constant is presumably $H_0 = 71 \pm 4$ km s^{-1} Mpc^{-1}. This value is, however, measured in nearby universe, $z \leq 0.1$. This does not depend on Ω_Λ. On the other hand, determinations of other cosmological parameters were more difficult. The first acoustic peak on the angular spectrum of CBR is observed by Boomerang and MAXIMA (de Bernardis et al. 2000, Balbi et al. 2000). The result of $l \sim 200$ suggests a flat universe; $\Omega_m + \Omega_\Lambda = 1$. By the way, is the cosmological constant itself well known? We should answer NO! Of course, there are many good observations about it. For example, a type Ia SN at high z suggests that $\Omega_\Lambda \sim 0.7$ (Riess et al. 2001) etc. But this should not be taken as an established result.

The S-Z effect is the phenomenon that the photons of CMB are scattered through inverse Compton process with the electrons in the hot gas of a galaxy cluster (Sunyaev & Zel'dovich, 1972). The S-Z effect makes a silhouette, $\Delta T_0 \leq 1$ mK, of the hot gas towards the CMB under $\nu \approx 220$ GHz. If the isothermal spherical $\beta-$model of hot gas in a galaxy cluster and flatness of the universe are assumed, the angular diameter distance of a galaxy cluster can be estimated by the combination of X-ray surface brightness and the S-Z effect;

$$d_A = \frac{\Delta T_0^2 \Lambda_e m_e c^2 \Gamma(3\beta - 1/2)\Gamma(3\beta/2)^2}{16\pi^{1/2} b_{x0}^2 k T_{e0} \sigma_T^2 T_{CBR}^2 \theta_c \Gamma(3\beta)\Gamma(3\beta/2 - 1/2)^2} \tag{1}$$

And Hubble constant can be given as a function of Ω_Λ and z;

$$H_0(z, \Omega_\Lambda) = \frac{c \int_0^z [(1+z')\{1 + (1-\Omega_\Lambda)z'\} + z'(2+z')\Omega_\Lambda] dz'}{d_A(1+z')} \tag{2}$$

(e.g. Inagaki, et al. 1995). Then, we can estimate the cosmological constant from the comparison between the Hubble constants derived by S-Z/X-ray observations and HST.

2. Instruments and Observations

The observation of the S-Z effect had been one of the most difficult observations in radio astronomy because the effect is only on the order of 100 μK and it is extended over several arcminutes. The drastic improvement of the sensitivity and the reliability of radio telescopes during the last decades have, however, changed such a situation (e.g. Reese, et al. 2000). A small beam size is important to avoid the beam-dilution of the S-Z effect. A low side lobe level is also important. In these points, the Nobeyama 45-m radio telescope is one of most suitable telescopes for the observation. It is important to perform the observation at mm-wavelength because the background point sources decreases steeply in mm-wave. These have motivated us to observe the S-Z effect at 43 GHz with the Nobeyama 45-m telescope. Because it is clear that an array receiver system is superior in sensitivity to a single beam receiver system, we developed an array receiver at 40 GHz for the Nobeyama 45-m telescope. This receiver has 2×3 grid beams in $80''$ interval on the celestial sphere. A typical receiver noise temperature is about 50 K. The detail of the receiver will be shown in our upcoming paper (Tsuboi et al. 2002).

We observed S-Z effect towards 4 distant clusters from $z = 0.18 - 0.55$ (CL0016+16 , MS1358+62, MS1008-12, and A2218) at 43 GHz using the newly developed array receiver installed in the Nobeyama 45-m telescope. They have round shape X-ray images and have no strong central radio source.

3. Results and Discussion

Figure 1 shows the result of our observation. The filled circles in the figure show data points. The curves are best-fit curves for the isothermal spherical $\beta-$model.

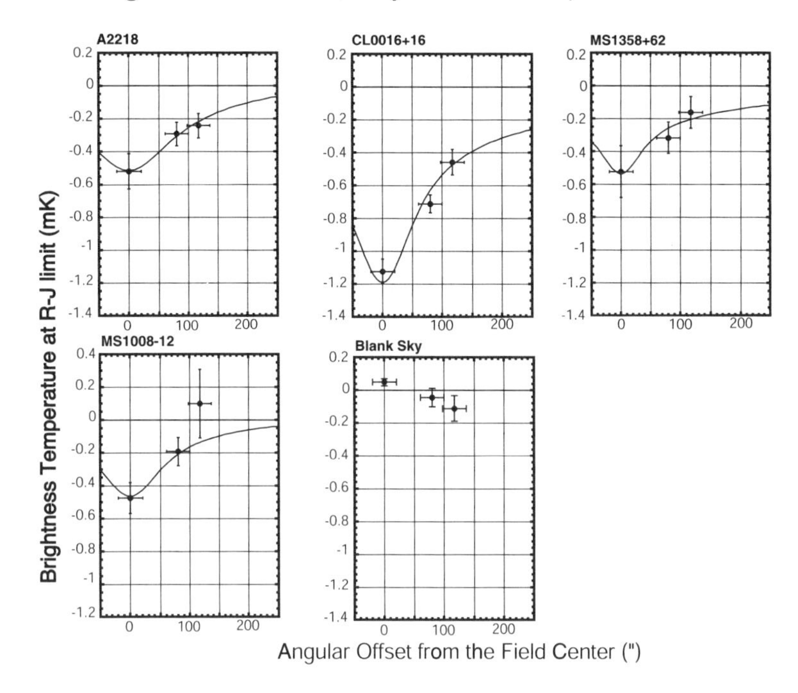

Figure 1. The radial distributions of the S-Z effect. The curves are best-fit curves for the isothermal spherical β−model.

The β−model well reproduces our data. From these data, the S-Z effects at the cluster center at Rayleigh-Jeans limit are estimated to be -1.38 ± 0.07 mK for CL0016+16, -0.95 ± 0.20 mK for MS1358+62, -0.63 ± 0.16 mK for MS1008-12, and -0.73 ± 0.10 mK for A2218, respectively. The S-Z effect for MS1008-12 is a new detection. The values for A2218 and CL0016 are consistent with those of the previous observations (e.g. Birkinshaw & Hughes 1994).

We compared these S-Z effects with X-ray data of ROSAT and ASCA (Ohta 2001) and derived Hubble constants by the formulae (1) and (2). Figure 2 shows Hubble constants as a function of cosmological constant. These curves are derived from our observations. The thick curve shows weighted average of our data. The average Hubble constants are $H_0 = 67 \pm 13$ km s^{-1} Mpc^{-1} for $\Omega_\Lambda = 0.7$ and $H_0 = 55 \pm 11$ km s^{-1} Mpc^{-1} for $\Omega_\Lambda = 0.0$, respectively. The Hubble constant for $\Omega_\Lambda = 0.7$ is consistent with the value from HST Key project.

On the contrary, we estimate the cosmological constant with S-Z effect. This horizontal thick dot line in the figure is the Hubble constant from HST key project. As mentioned previously, this value is not affected by the cosmological constant. If the Friedmann–Remeter model is accepted, our average line and the horizontal line by HST cross at the cosmological constant of the universe. Then, this crossing point in the figure suggests the cosmological constant is about $\Omega_\Lambda \sim 0.7 - 0.9$. Although there is large scatter, this agrees with the value derived by type Ia supernova observation independently (Riess et al. 2001). This supports the existence of the low Ω_m universe.

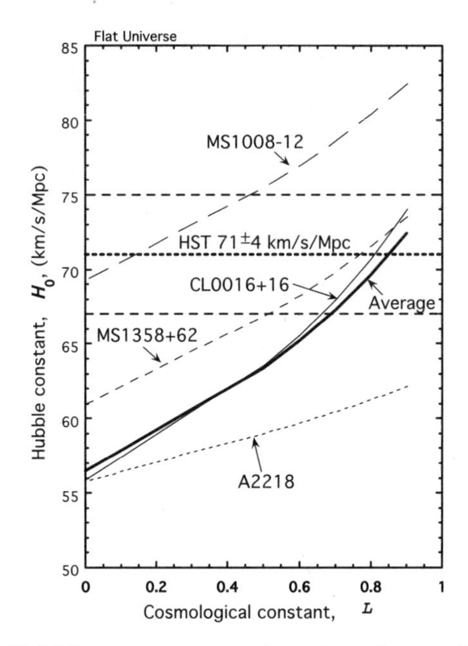

Figure 2. Hubble constants as a function of cosmological constant

References

Balbi, A., Ade, P., Bock, J., et al. 2000, ApJ, 545, L1

Birkinshaw, M.& Hughes, J.P. 1994, ApJ, 420, 33

de Bernardis, P., Ade, P. A. R., Bock, J. J., et al. 2000, Nature 404, 955

Freedman, W.L., Madore, B.F., Gibson, B. K., et al. 2001, ApJ, 553,47

Inagaki Y., Suginohara T. & Suto Y. 1995, PASJ, 47, 411

Ohta, N. 2001, PhD thesis, the University of Tokyo

Reese, E. D., Mohr, J. J., Carlstrom, J. E., et al. 2000, ApJ, 533, 38

Riess, A. G., Nugent, P. E., Gilliland, R. L., et al. 2001, ApJ, 560, 49

Sunyaev, R. A. & Zel'dovich, Ya. B. 1972, Comm. Astrophys. Space Phys., 4, 173

IAU 8th Asia-Pacific Regional Meeting
ASP Conference Series, Vol. 289, 2003
S. Ikeuchi, J. Hearnshaw & T. Hanawa, eds.

An Ellipsoid Model for the Halo Density Profile

Y. P. Jing

Shanghai Astronomical Observatory, the Partner Group of MPI für Astrophysik, Nandan Road 80, Shanghai 200030, China

Abstract. We present for the first time a detailed non-spherical modelling of dark matter halos on the basis of a careful analysis of our state-of-the art N-body simulations. The fitting formulae presented here form a complete and accurate description of the triaxial density profiles of halos in Cold Dark Matter models. The current description of the dark halos is very useful for interpreting many observations related to the non-sphericity effects of halos, including the weak and strong lens statistics, the orbital evolution of galactic satellites and triaxiality of galactic halos, the non-linear clustering of dark matter, and X-ray and Sunyaev-Zel'dovich morphologies of rich clusters.

1. Introduction

The density profiles of dark matter halos have attracted a lot of attention recently since Navarro et al. (1996, hereinafter NFW) discovered the unexpected scaling behavior in their simulated halos. Subsequent independent higher-resolution simulations confirmed the validity of the NFW modelling, in particular the presence of the central cusp, although the inner slope of the cusp seems somewhat steeper than they originally claimed. Those previous models, however, have been based on the spherical average of the density profiles. Actually it is also surprising that the fairly accurate scaling relation applies after the spherical average despite the fact that the departure from the spherical symmetry is quite visible in almost all simulated halos (e.g. Fig. 1 of Jing & Suto 2000).

A more realistic modelling of dark matter halos beyond the spherical approximation is important in understanding various observed properties of galaxy clusters and non-linear clustering (especially the high-order clustering statistics) of dark matter in general. In particular, the non-sphericity of dark halos is supposed to play a central role in the X-ray morphologies of clusters (e.g. Jing et al. 1995), in the cosmological parameter determination via the Sunyaev-Zel'dovich effect (e.g. Yoshikawa et al. 1998) and in the prediction of the cluster weak lensing and the gravitational arc statistics (e.g. Bartelmann et al. 1998). Nevertheless useful analytical modelling of the non-sphericity is almost impossible, and numerical simulations are the only practical means to provide statistical information.

While the non-sphericity of the dark matter halos is a poorly studied topic, some seminal studies do exist which attempt to detect and characterize the non-spherical signature (e.g. Dubinski 1994). Nevertheless there is no systematic and

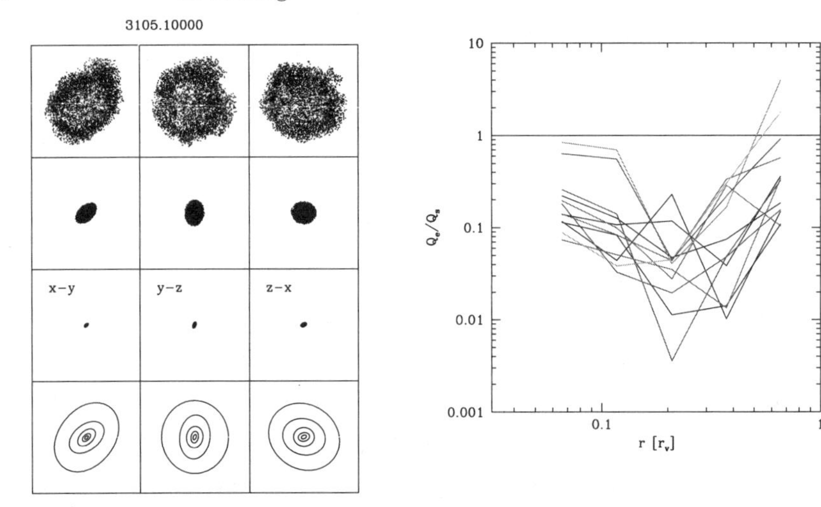

Figure 1. *Left panel:* A typical example of a projected particle distribution within a halo. The bottom panels show the triaxial fits to five isodensity surfaces projected on those planes. *Right panel:* Ratio of the quadrupole moments defined in the triaxial model (Q_e) and in the spherical model (Q_s) for five shells at radii from $0.05r_{vir}$ to $0.65r_{vir}$. The figure clearly shows that the triaxial model describes more accurately the density profiles of halos than the conventional spherical model. From Jing & Suto (2002).

statistical study to model and characterize the density profiles of simulated halos. This is exactly what we will present in the rest of the paper. In particular, much higher mass and spatial resolutions of our current N-body simulations enable us to characterize the statistics of the halo non-sphericity with an unprecedented precision.

2. Modeling the Non-spherical Density Profiles of Dark Matter Halos

We use two sets of state-of-the art simulations for the current purpose. The first is our new set of cosmological N-body simulations with $N = 512^3$ particles in a $100\,h^{-1}$ Mpc box (Jing & Suto 2002), and the other is a set of high-resolution halo simulation runs (Jing & Suto 2000).

To model the shape of dark matter halos, we first find the iso-density surfaces. This begins with the computation of a local density at each particle's position. We adopt the smoothing kernel widely employed in the Smoothed Particle Hydrodynamics (SPH) method (Hernquist & Katz 1989). We use 32 nearest neighbor particles to compute the local density. The left panel of Figure 1 plots a typical example of the projected particle distributions within the isodensity surfaces for four different halos. This plot clearly suggests that the isodensity surfaces are typically approximated as triaxial ellipsoids.

The isodensity ellipsoids at different radii are approximately aligned, and the axial ratios of the ellipsoids are nearly constant. These facts suggest the possibility that the internal density distribution within a halo can be approximated by a sequence of the concentric ellipsoids of a constant axis ratio. To show this to be an improved description over the conventional spherical description, we compute the quadrupole of the particle distribution within a spherical shell (Q_s) or an ellipsoidal shell (Q_e). If the spherical (triaxial) model is exact, Q_s (Q_e) vanishes. The ratio of these two quantities is plotted in the right panel of Figure 1. It clearly shows that the triaxial model works much better than the conventional spherical model.

We found that the density of the isodensity surfaces changes with the major axis R in a way similar to the spherical model. The density profile can be approximately described by the NFW-like profiles, though the concentration parameter is different from the spherical model. In Jing & Suto (2002), we have given recipes, based our analysis of the simulations, for calculating the density profile for a halo of given mass and given axial ratios in a general CDM cosmogony.

It is known that the mass distribution of CDM halos is well described by the Press-Schechter formula (or its modified form). Thus, in order to predict the density profiles for a sample of CDM halos, we need to know the distribution function $p(a/c, b/c)$ of the axial ratios a/c and b/c of the halos at a given mass. The function can be written as

$$\begin{aligned} p(a/c, b/c)d(a/c)d(b/c) &= p(a/c)d(a/c)\, p(b/c|a/c)d(b/c) \\ &= p(a/c)d(a/c)\, p(a/b|a/c)d(a/b)\,. \end{aligned} \quad (1)$$

The probability function $p(a/c)$ and the conditional probability $p(a/b|a/c)$ have been measured for halos in our SCDM and LCDM simulations at different epochs. An example is given in Figure 2. In Jing & Suto (2002), we have proposed accurate universal functions to describe these two probability functions.

3. Conclusions

We have presented a triaxial modelling of the dark matter halo density profiles extensively on the basis of the combined analysis of the high-resolution halo simulations (12 halos with $N \sim 10^6$ particles within their virial radius) and the large cosmological simulations (five realizations with $N = 512^3$ particles in a $100\,h^{-1}$ Mpc box size). In particular, we found that the universal density profile discovered by NFW in the spherical model can be also generalized to our triaxial model description. Our triaxial density profile is specified by the concentration parameter c_e and the scaling radius R_0 (or the *virial* radius R_e in the triaxial modelling) as well as the axis ratios a/c and a/b.

We have obtained accurate fitting formulae for those parameters which are of practical importance in exploring the theoretical and observational consequences of our triaxial model. Because the page limit of the proceedings, we could not have typed in these formulae. We refer the interested reader to the journal paper of Jing & Suto (2002) for the fitting formulae as well as for many details of the work.

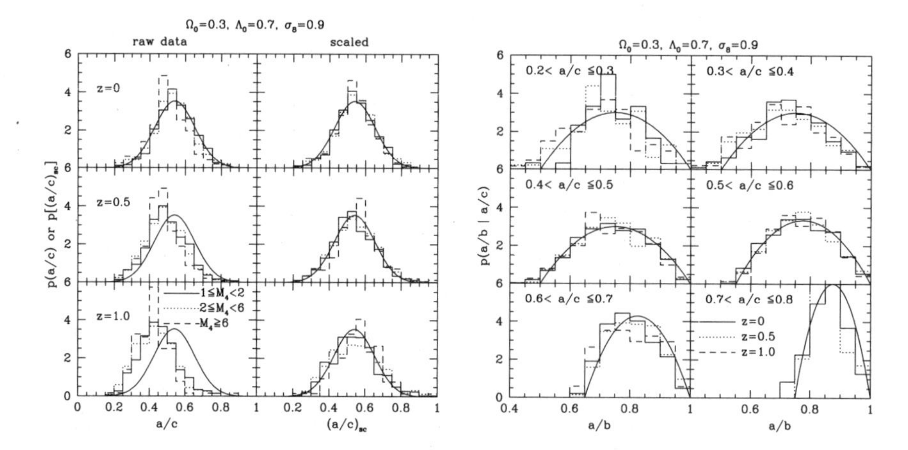

Figure 2. **Left figure:** The distribution of the axis ratio a/c of the halos in the cosmological simulations of the LCDM model before (*left panels*) and after (*right panels*) the scaling described in Jing & Suto (2002). Solid, dotted and dashed histograms indicate the results for halos that have the number of particles of $M_4 \equiv (N_{\rm halo}/10^4)$ within the virial radius. **Right figure:** The conditional distribution of the axis ratio a/b of the halos in the cosmological simulations of the LCDM model for a given range of a/c. The smooth solid curves in all the panels represent our fitting formula. From Jing & Suto (2002).

Acknowledgments. I would like to thank my collaborator, Yasushi Suto, for his important contribution to this work. Numerical simulations presented in this paper were carried out at ADAC (the Astronomical Data Analysis Center) of the National Astronomical Observatory, Japan, and at KEK (High Energy Accelerator Research Organization, Japan). Y.P.J. was supported in part by by NKBRSF (G19990754) and by NSFC (No. 10125314).

References

Bartelmann, M., Huss, A., Colberg, J. M., Jenkins, A., & Pearce, F. R. 1998, A&A, 330, 1

Dubinski, J. 1994, ApJ, 431, 617

Hernquist, L. & Katz, N. 1989, ApJS,70,419

Jing, Y. P., Mo, H. J., Borner, G., & Fang, L. Z. 1995, MNRAS, 276, 417

Jing, Y. P., & Suto, Y. 2000, ApJ, 529, L69

Jing, Y. P., & Suto, Y. 2002, ApJ, 574, 538

Navarro, J. F., Frenk, C. S., & White, S. D. M. 1996, ApJ, 462, 563

Yoshikawa, K., Itoh, M. & Suto, Y. 1998, PASJ, 50, 203

IAU 8th Asia-Pacific Regional Meeting
ASP Conference Series, Vol. 289, 2003
S. Ikeuchi, J. Hearnshaw & T. Hanawa, eds.

Star Formation History in Galaxies Inferred from Stellar Elemental Abundance Patterns

T. Shigeyama

Research Center for the Early Universe, Graduated School of Science, University of Tokyo, Bunkyo-ku, Tokyo 113-0033, Japan

T. Tsujimoto

National Astronomical Observatory, Mitaka-shi, Tokyo 181-8588, Japan

Abstract. Star formation history in the Milky Way and local dwarf spheroidal galaxies is discussed in the context of an inhomogeneous chemical evolution model developed by the authors in which supernovae induce star formation.

1. Introduction

Since the middle of 1990s, abundance patterns of heavy elements have been obtained for individual metal-poor stars in the halo of the Milky Way (McWilliam et al. 1995; Ryan, Norris, & Beers 1996) and local dwarf spheroidal galaxies (dSphs) (Shetrone, Côté, & Sargent 2001). The observed fact that stars with similar metallicities $[Fe/H] \equiv \log(N(Fe)/N(H)) - \log(N_\odot(Fe)/N_\odot(H))$ have quite different quantities of other elements such as barium (Ba) and europium strongly suggests that these stars were formed from matter affected by only a few supernovae (SNe) (Ryan et al. 1996; Audouze & Silk 1995) . Comparison of the observed abundance patterns of α-elements with those in the theoretical SN models led Shigeyama & Tsujimoto (1998) to argue that metal-poor stars were formed from the matter swept up by individual SNe. Assuming that stars are formed from the matter swept up by individual SNe, Tsujimoto, Shigeyama, & Yoshii (1999) formulated a chemical evolution model and succeeded in reproducing the abundance distribution function (ADF) of halo field stars constructed by Ryan & Norris (1991). Thus star formation in a molecular cloud is triggered in the gas shock-compressed by individual SNe until SNe cannot accumulate enough gas to form stars of the next generation. These last SNe insert enough energy into gas to dispel the cloud. Applying the same model to globular clusters (GCs) in the Milky Way and local dSphs for both of which abundance data are available, pictures of star formation history in these objects could be obtained and will be reported in the following sections. In the next section, globular cluster formation will be studied in the framework of cloud-cloud collisions. Conditions to reproduce the observed ADF are briefly discussed. In section 3, the star formation history of local dSphs discussed in Tsujimoto & Shigeyama (2002a) will be summarized. .

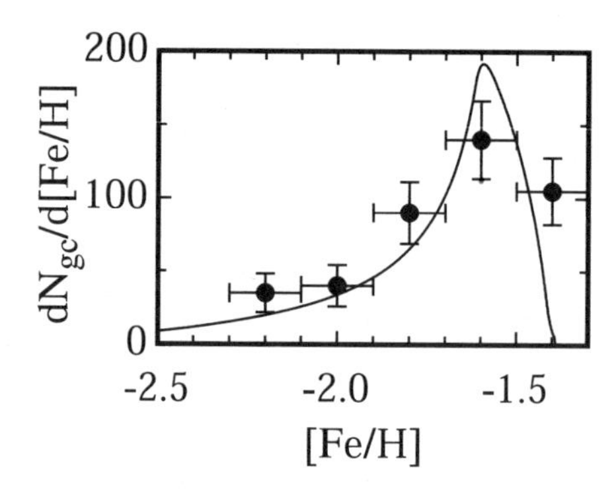

Figure 1. The abundance distribution functions. The solid line represents the calculated ADF. The filled circles are the ADF of the Galactic GCs constructed from the catalogue of Harris (1996).

2. Globular Cluster Formation through Cloud-cloud Collisions

No extremely metal-poor stars with [Fe/H]< -2.5 in GCs suggests that the proto-cluster clouds already possessed a certain amount of heavy elements. A mechanism to form such clusters is cloud-cloud collisions in which each cloud undergoes star formation (Murray & Lin 1993). Cloud-cloud collisions can leave shocked gas with some heavy elements and without either stars or dark-matter. This merging scenario must be reconciled with the observed ADF for the Galactic GCs. Lee, Schramm, & Mathews (1995) tried to reproduce the ADF with a merger model and succeeded in reproducing a certain feature. However, their age-metallicity relation is different from ours that reproduces the observed ADF for the halo field stars (Tsujimoto, et al. 1999). Thus it is not clear if the model of Lee et al. (1995) can explain the ADFs for the field stars and GCs at the same time. We have performed some calculations similar to Lee, et al. (1995) to reproduce ADFs for both field stars and GCs. Suppose that identical star forming clouds move in the halo under its gravitational field. Then the formation rate of GCs $dN_{\rm gc}/dt$ through cloud-cloud collisions will be written as

$$\frac{dN_{\rm gc}}{dt} = f\sigma n_{\rm cl}^2 vV, \tag{1}$$

where $f\sigma$ denotes the effective cross section for GC formation through cloud-cloud collisions. Each cloud has a geometrical cross section of σ and moves at speed v in the halo with the volume of V. The radius of each cloud is assumed to be 20 pc. Clouds are uniformly distributed in the halo with the number density $n_{\rm cl}$. The evolution of $n_{\rm cl}(t)$ is described by the differential equation

$$\frac{1}{V}\frac{dn_{\rm cl}V}{dt} = -2\sigma n_{\rm cl}^2 v - \frac{n_{\rm cl}}{t_{\rm SN}}\theta(t - t_{\rm SF} - t_{\rm SF,0}), \tag{2}$$

where the second term denotes the destruction of a cloud due to SNe at the end of star formation. Here two time scales $t_{\rm SF}$ and $t_{\rm SN}$ have been introduced

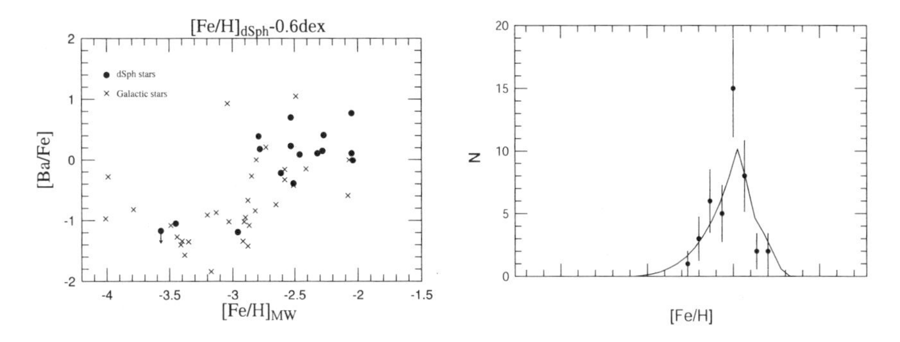

Figure 2. Left panel (a): Correlations of [Ba/Fe] with [Fe/H] for dSph stars (filled circles; Shetrone et al. 2001) and Galactic halo field stars (crosses; McWilliam 1998). Right panel (b): The frequency distribution of dSph stars as a function of the iron abundance, compared with the observation for Sextans (Suntzeff et al. 1993).

to describe the duration of star formation in each cloud and the time scale of SN contribution, respectively. The duration of star formation is specified by our chemical evolution model for the halo field stars as $t_{\rm SF} = 0.3$ Gyr. The time scale of SN contribution is determined by the life time of massive stars as $t_{\rm SN} = 10$ Myr. The time when the star formation starts is denoted by $t_{\rm SF,0}$. As discussed in Lee, et al. (1995), the volume V of the galaxy also evolves according to the self-gravity as

$$\frac{dV}{dt} = 4\pi r^2 \frac{dr}{dt} = -4\pi r^2 \sqrt{\frac{2GM_{\rm MW}}{R_{\rm max}} \left(\frac{R_{\rm max}}{r} - 1 \right)}. \tag{3}$$

Here G denotes the gravitational constant, $M_{\rm MW}$ the mass of the Milky Way, and $R_{\rm max}$ is the maximum radius of the halo. This equation describes the Milky Way shrinking from its maximum size. Thus the time is measured from when the halo has its maximum size ($R_{\rm max} = 270$ kpc is used as a fiducial value). The cloud velocity is given by the virial velocity as $v = \sqrt{0.4GM_{\rm MW}/r}$. The resultant ADF is shown in Figure 1 compared with the observations. The free parameters of $f = 0.09$ and $t_{\rm SF,0} = 1.98$ give the best fit to the halo component of the observed ADF. The small f implies that the collision to produce a GC is nearly head-on with an impact parameter significantly smaller than the cloud size. The shape of the ADF is found to be sensitive to the value of $t_{\rm SF,0}$. Thus it is expected that different galaxies have quite different ADFs of metal-poor GCs. The collapse needs to halt when $r \sim 90$ kpc to reproduce the metal-poor end of the ADF irrespective of $R_{\rm max}$. Otherwise, there would be too few GCs with [Fe/H]< -1.8. The subsequent star formation needs to finish soon after the first SNe explode in order to account for no dispersion in the metallicity of the cluster stars (see Tsujimoto & Shigeyama 2002b and Shigeyama & Tsujimoto 2002).

3. History of Local Dwarf Spheroidal Galaxies

Stellar abundance pattern of n-capture elements such as Ba in dSphs do not show any sign of the s-process. This suggests that the star formation lasted for a few hundred Myr at most to prevent the stars from being contaminated by s-process products from AGB stars. It is also found that the abundance correlation of Ba with Fe in stars belonging to dSph galaxies orbiting the Milky Way, i.e., Draco, Sextans, and Ursa Minor have a feature similar to that in Galactic metal-poor stars (Fig. 2a). The common feature of these two correlations can be realized by our inhomogeneous chemical evolution model based on the SN-driven star formation scenario if dSph stars formed from gas with a velocity dispersion of $\sim$ 26 km s^{-1} larger than the value $\sim$ 10 km s^{-1} used to account for the abundance pattern of Galactic metal-poor stars. The larger velocity dispersion the interstellar matter have, the less amount of matter can a supernova remnant sweep up. Thus the metallicity of the swept up gas and that of stars formed from the gas are enhanced. This velocity dispersion together with the stellar luminosities strongly suggest that dark matter dominated dSph galaxies. The tidal force of the Milky Way links this velocity dispersion with the currently observed value $\lesssim$ 10 km s^{-1} by stripping the dark matter in dSph galaxies. As a result, the total mass of each dSph galaxy is found to have been originally $\sim$ 25 times larger than at present. Our inhomogeneous chemical evolution model succeeds in reproducing the stellar [Fe/H] distribution function observed in Sextans (Fig. 2b). In this model, SNe immediately after the end of the star formation can expel the remaining gas from the gravitational potential of the dSph galaxy.

References

Audouze, J. & Silk, J. 1995, ApJ, 451, L49

Harris, W. E. 1996, AJ, 12, 1487

Lee, S., Schramm, D. N. & Mathews, G. J. 1995, ApJ, 449, 616

McWilliam, A., Preston, G. W., Sneden, C. & Searle, L. 1995, AJ, 109, 2757

McWilliam, A. 1998, AJ, 115, 1640

Murray, S. D. & Lin, D. N. C. 1993, in The Globular Cluster-Galaxy Connection, eds. G. H. Smith, & J. P. Brodie (ASP Conf. Proc. 48), 738

Oh, K. S., Lin, D. N. C. & Aarseth, S. J. 1995, ApJ, 442, 142

Ryan, S. G., & Norris, J. E. 1991, AJ, 101, 1865

Ryan, S. G., Norris, J. E. & Beers, T. C. 1996, ApJ, 471, 254

Shetrone, M. D., Côté, P. & Sargent, W. L. 2001, ApJ, 548, 592

Shigeyama, T. & Tsujimoto, T. 1998, ApJ, 507, L135

Shigeyama, T. & Tsujimoto, T. 2002, in preparation

Suntzeff, N. B., Mateo, M., Terndrup, D. M., Olszewski, E. W., Geisler, D. & Weller, W. 1993, ApJ, 418, 208

Tsujimoto, T., Shigeyama, T. & Yoshii, Y. 1999, ApJ, 519, L64

Tsujimoto, T. & Shigeyama, T. 2002a, ApJ, 571, L93

Tsujimoto, T. & Shigeyama, T. 2002b, ApJ, submitted

IAU 8th Asia-Pacific Regional Meeting
ASP Conference Series, Vol. nnn, 2003
S. Ikeuchi, J. Hearnshaw & T. Hanawa, eds.

Cosmic Star Formation History Associated with QSO Activity: An Approach by the Black Hole to Bulge Mass Correlation

Y. P. Wang

Purple Mountain Observatory, Academia Sinica, China; National Astronomical Observatories of China

T. Yamada

National Astronomical Observatory of Japan, Mitaka, Japan

Y. Taniguchi

Tohoku University, Aramaki, Aoba, Sendai 980-8578, Japan

Abstract. Based on our previous work about a joint evolution and the consequential black hole to bulge mass correlation, we use the observed X-ray luminosity function of AGNs and their evolution to estimate the star formation history which is associated with the black hole growth. We show that the total amount of star formation associated with the massive black hole growth is almost the same as that detected by the current optical deep surveys. Meanwhile, the far infrared emission from the dust heated by star formation on-going during the black hole growth could sufficiently account for the observed SCUBA counts, and would be the good candidates of the SCUBA population.

1. Introduction

The main purpose of this paper is to fairly reconstruct the global star formation history mostly associated with AGN accretion and to give a proper estimation of the co-moving star formation rate vs. redshift accounting for the active phase of those massive spheroids in formation. We adopt the joint evolutionary scheme and the tight correlation between the masses of central black holes and their host spheroids in nearby galaxies and AGNs. We also assume the X-ray luminosity L_{X} of AGN is powered by accretion onto a central massive black hole, especially the luminous one. Deep X-ray surveys provide a direct probe of the history of AGN accretion history, the black hole growth as well as the joint star formation, based on the assumption that the luminous AGNs reflect the stage of major black hole growth and spheroid formation. In this paper, we trace the AGN evolution with X-ray (0.5–2 keV) local luminosity function and the evolution rate by Miyaji et al. (2000), which gives an excellent fit to the ROSAT All-sky surveys, including the number counts and the soft X-ray background level.

The present day black hole density and the SCUBA number counts are used as two important model constraints in the calculation, with the set of cosmological parameters $(\Omega_{\rm m}, \Omega_{\Lambda}) = (0.3, 0.7)$ and $H_0 = 50\,{\rm km/s/Mpc}$.

2. Model Calculation and Discussion

We estimate the black hole mass from the X-ray luminosity by assuming an Eddington ratio $\epsilon = \frac{\eta \dot{m} c^2}{L_{\rm Edd}}$. With the bolometric correction β ($L_{\rm bol} = \beta\, L_{\rm x}$), we could get the black hole mass $M_{\rm bh} = \frac{\beta\, L_{\rm x}}{0.013\,\epsilon}$. Where $L_{\rm x}$ is the AGN $0.5-2\,{\rm keV}$ luminosity in units of $10^{40}\,{\rm erg\ s^{-1}}$ and $M_{\rm bh}$ in units of $M_{\odot}$. The black hole (BH) mass function of different redshift is converted from the X-ray luminosity function $\Phi(L_{\rm x}, z)$ by assuming a reasonable value of the Eddington ratio ϵ and the "duty cycle $f_{\rm on}$" of the AGN active phase. Following the various AGN observations, ϵ depends weakly on the luminosity, where most luminous QSOs radiate at about the Eddington limit and low luminosity AGNs ($L < 10^{44}\,{\rm erg/s}$) show $\epsilon \sim 0.1 - 0.05$. We thus set $\epsilon = 10^{\gamma\,(\log L - 49)}$ to simulate such an reality (Padovani 1989, Wandel 1999). The "duty cycle $f_{\rm on}$" is the fraction of black holes that are active at a given time of redshift z. In this work, we simply assume that each AGN shines for a constant time scale $t_{\rm Q}$, $f_{\rm on} = t_{\rm Q}/t_{\rm Hub}(z)$, where $t_{\rm Hub}(z)$ is the redshift dependent Hubble time (Haiman & Menou 2000). The QSO life time $t_{\rm Q} = 5 \times 10^8\,{\rm yr}$ is adopted in the calculation, which is close to the e-folding time $t_e = M_{\rm BH}/\dot{M} = 4 \times 10^8\,\eta/\epsilon$ shown in theoretical models with the radiation efficiency $\eta \sim 0.1$, and consistent with the new results of Chandra for the QSO accretion duration (Wang & Biermann 1998, Barger et al. 2001).

We consider all AGNs in ROSAT sample as unobscured type 1 since non type 1 AGNs are only a small fraction of the ROSAT sample. $\beta = 20$ is adopted in our calculation based on the mean type 1 AGN spectral energy distribution from Elvis et al. (1994). So far, the black hole mass function $\frac{{\rm d}\,\Phi(z, M_{\rm bh})}{{\rm d}\,M_{\rm bh}}$ can be derived from the observed X-ray LF by eq. 1. We thus follow the LDDE cosmological evolution suggested by ROSAT survey, and trace the black hole mass density accreted during luminous QSO phase with lookback time.

$$\frac{{\rm d}\,\Phi(z, M_{\rm bh})}{{\rm d}\,M_{\rm bh}} = \frac{0.013\,\epsilon}{\beta\, f_{\rm on}} \frac{{\rm d}\,\Phi(z, L_{\rm x})}{{\rm d}\,L_{\rm x}} \,. \tag{1}$$

There are several lines of arguments which suspect that optical and soft X-ray surveys may miss a large fraction of type 2 QSOs at high redshift. A ratio between type 2 to unobscured type 1 AGNs is usually in the range of 4–10, especially from the synthesis models of the X-ray background (Fabian & Iwasawa 1999, Gilli et al. 2001). Estimating roughly the amount of star formation related to the obscured QSOs, we simplify the abundance ratio R_{2-1} of the type 2 to type 1 AGNs as $\Phi_2(z, L_{\rm x}) = \alpha\,(1+z)^p\,\Phi_1(z, L_{\rm x})$ based on the unification scheme. Considering the present day black hole density $\rho_{\rm BH} \sim 3-5 \times 10^5\,M_{\odot}/{\rm Mpc}^3$ and the upper limit of the co-moving star formation rate from submillimeter deep surveys, we found the ratio R_{2-1} has an upper limit of about 2. There is not much room left for a ratio far beyond this number. This is actually consistent with the results of recent Chandra deep surveys by Alexander et al. (2001) who suggests this ratio is unlikely to be more than 8 and is probably considerably lower.

The spheroid mass function may have a similar form as the black hole mass function according to the joint evolution scheme of the black hole growth and the spheroid formation.

The spheroid mass distribution $\frac{\mathrm{d}\,\Phi(z, M_{\mathrm{sph}})}{\mathrm{d}\,M_{\mathrm{sph}}}$ could be derived by:

$$\frac{\mathrm{d}\,\Phi(z, M_{\mathrm{sph}})}{\mathrm{d}\,M_{\mathrm{sph}}} = \frac{\mathrm{d}\,\Phi(z, M_{\mathrm{bh}})}{\mathrm{d}\,M_{\mathrm{bh}}} / R(M_{\mathrm{bh}}/M_{\mathrm{sph}}) \; . \tag{2}$$

$R(M_{\mathrm{bh}}/M_{\mathrm{sph}})$ is the black hole to bulge mass ratio. We adopt here the mean value $M_{\mathrm{bh}}/M_{\mathrm{sph}} \sim 0.002$ as a first approximation (Kormendy & Richstone 1995, Magorrian et al. 1998, Merritt & Ferrarese 2001).

The star formation time scale $\tau_{\mathrm{sf}} \sim t_{\mathrm{Q}} \sim 5 \times 10^8$ yr is adopted for the starburst coupled with the black hole growth. The star formation rate during such a phase would be $SFR = M_{\mathrm{sph}}/\tau_{\mathrm{sf}}$. Following the cosmological evolution from X-ray deep surveys, we get the redshift dependent black hole mass distribution by eq. 1, the spheroid mass function with redshift by eq. 2. The co-moving star formation rate vs. redshift from our calculation is shown in Fig. 1a.

The SCUBA number counts are used as a model constraint for the amount of star formation, especially at $z > 1$. However, it is not so easy to convert the star formation rate to far-infrared emission for individual sources because of the high internal extinction. A number of authors have discussed how the star formation rate in a galaxy can be inferred from its optical, UV or far-infrared luminosity (Scoville & Young 1983, Leitherer & Heckman 1995, Rowan-Robinson et al. 1997). The transformation factor of different set of star formation parameters could vary more than one order of magnitude. We adopt a mean value $L_{\mathrm{FIR}}/L_{\odot} = 3.8 \times 10^9\,\mathrm{SFR}(M_{\odot}\,\mathrm{yr}^{-1})$, and a mean color ratio $R_{\mathrm{c}} = L_{\mathrm{FIR}}/L_{850} \sim 5 \times 10^3$ by Chary & Elbaz (2001) from IRAS, ISO and SCUBA surveys.

Similar to the case in eq. 1, we can derive the 850 μm luminosity function at different redshift as below:

$$\frac{\mathrm{d}\,\Phi(z, L_{850})}{\mathrm{d}\,L_{850}} = f_{\acute{o}n} \frac{\mathrm{d}\,\Phi(z, M_{\mathrm{sph}})}{\mathrm{d}\,M_{\mathrm{sph}}} \frac{\mathrm{d}\,M_{\mathrm{sph}}}{\mathrm{d}\,M_{\mathrm{L}_{850}}} \; . \tag{3}$$

Here $f_{\acute{o}n} = t_{\mathrm{sf}}/t_{\mathrm{Hub}}$ reflects the fraction of galaxies which are in the active stage with intensive star formation ongoing. $\frac{\mathrm{d}\,M_{\mathrm{sph}}}{\mathrm{d}\,L_{850}} = \frac{R_{\mathrm{c}} \times \nu \times t_{\mathrm{sf}}}{3.8 \times 10^9}$, R_{c} and t_{sf} are the parameters adopted in our model and discussed above. $\nu = 3.5 \times 10^{11}$ Hz is the frequency at 850 μm. The 850 μm number count fitting is shown in Fig. 1b. From our results, we found that the amount of star formation in the early stage of QSO evolution could sufficiently heat the dust, and account for the far-infrared emission in most of the SCUBA deep surveys.

Acknowledgments. YPW acknowledges the COE Fellowship of Japan and the NSFC 10173025 of China.

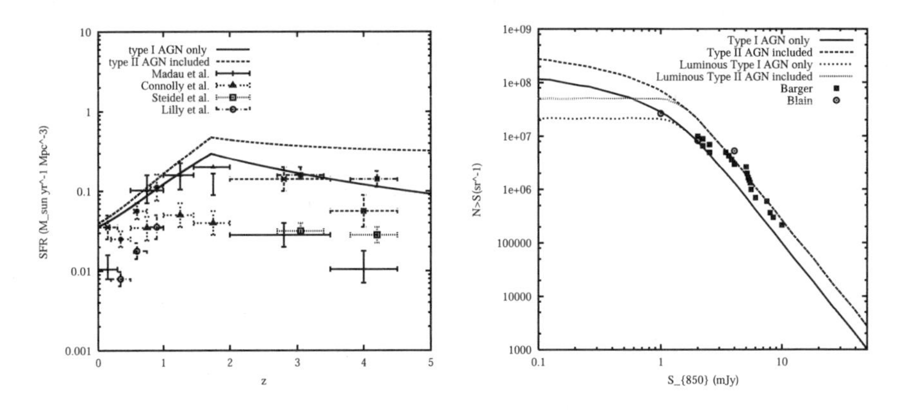

Figure 1. a: Co-moving star formation rate vs. redshift with the fraction of type 2 AGNs included; b: 850 μm number count fitting from the model calculation in cases of type 1 AGNs only and type 2 AGNs included. The data are from Barger et al. (1999) and Blain et al. (1999).

References

Alexander, D. M., Brandt, W. N., Hornschemeir, A. E. et al. 2001, ApJ, 122, 2156

Barger, A. J., Cowie, L. L., Bautz, M. W. et al. 2001, AJ, 122, 2177

Chary, R., Elbaz, D. 2001, ApJ, 556, 562

Elvis, M., et al. 1994, ApJS, 95, 1

Fabian, A. C., Iwasawa, K. 1999, MNRAS, 303, L34

Gilli, R., Salvati, M., Hasinger, G. 2001, A&A, 366, 407

Haiman, Z., Menou, K. 2000, ApJ, 531, 42

Kormendy, J., Richstone, D. 1995, ARA&A, 33, 581

Leitherer, C., Heckman, T. M. 1995, ApJS, 96, L9

Magorrian, J., Tremaine S., Richstone D. et al. 1998, AJ, 115, 2285

Merritt, D., Ferrarese, L. 2001, MNRAS, 320, L30

Miyaji, T., Hasinger, G., Schmidt M. 2000, A&A, 353, 25

Padovani, P. 1989, A&A, 209, 27

Rowan-Robinson, M., Mann, R. G., Oliver, S. J. et al. 1997, MNRAS, 289, 490

Scoville, N.Z., Young, J.S. 1983, ApJ, 265, 148

Wandel, A. 1999, ApJ, 519, L39

Wang, Y. P., Biermann, P. L. 1998, A&A, 334, 87 (WB98)

Compact Objects and High Energy Astrophysics

IAU 8th Asia-Pacific Regional Meeting
ASP Conference Series, Vol. 289, 2003
S. Ikeuchi, J. Hearnshaw & T. Hanawa, eds.

Evolution of Low- and Intermediate-Mass X-Ray Binaries

Xiang-Dong Li

Department of Astronomy, Nanjing University, Nanjing, China; email: lixd@nju.edu.cn

Abstract. In the standard scenario, low-mass binary pulsars (LMBPs) are the descendants of low-mass X-ray binaries (LMXBs). More recent investigations indicate that a large fraction of the observed LMXBs may actually have started their lives as intermediate-mass X-ray binaries (IMXBs). In this paper we summarize the existing problems and progress in the evolutionary studies of LMXBs and IMXBs with respect to LMBP formation.

1. Introduction

It has been suggested (Joss & Rappaport 1983; Savonije 1983; Paczynski 1983) that binary pulsars with low-mass white dwarf companion (low-mass binary pulsars or LMBPs) are the descendants of low-mass X-ray binaries (LMXBs). Formation of LMBPs can be described by the "recycled" scenario as follows (Bhattacharya & van den Heuvel 1991). A high-field ($\sim 10^{12} - 10^{13}$ G), rapidly rotating neutron star born in a binary with a low-mass ($\lesssim 1 M_{\odot}$) main-sequence companion star, spins down under magnetic dipole radiation, and switches off its radio emission within $\sim 10^6 - 10^7$ yr after passing by the so-called "death line" in the magnetic field - spin period ($B - P$) diagram. When the companion evolves to overflow its Roche-lobe, mass transfer occurs by way of an accretion disk. The mechanisms that drive mass transfer in LMXBs depend on the initial separations of the binary components. In narrow systems with $P_{\rm orb} < 1 - 2$ d, mass transfer is driven by loss of orbital angular momentum via gravitational radiation and/or magnetic braking. The evolution of the binary orbit and the mass transfer rate is not well constrained, however, because of poor understanding of the mechanism of magnetic braking. Mass transfer in wide LMXBs with $P_{\rm orb} \gtrsim 1 - 2$ d is driven by the nuclear expansion of the secondary. Such kind of systems form a quite homogeneous group whose evolutionary history seems well understood (Webbink, Rappaport & Savonije 1983; Taam 1983). The binary orbits widen during the evolution, and there exists a positive correlation between the mass transfer rate $\dot{M}$ from the secondary and $P_{\rm orb}$ (Webbink et al. 1983).

Mass accretion onto the neutron star gives rise to X-ray emission, induces magnetic field decay, and spins the star up to short period (the mechanisms for the field-decay induced by accretion are, however, not well understood). When mass transfer ceases, the end point of the evolution is a circular binary containing a neutron star visible as a low-field, millisecond radio pulsar, and a low-mass ($\sim 0.2 - 0.4 M_{\odot}$) white dwarf, the remaining helium core of the companion.

The above picture is strongly supported by recent discovery of millisecond X-ray pulsars in low-mass binary systems, i.e., SAX J1808.4-3658 (Wijnands & van der Klis 1998; Chakrabarty & Morgan 1998), XTE J1751-305 (Markwardt et al. 2002), and XTE J0929-314 (Galloway et al. 2002), as well as the rapid quasi-periodic oscillations observed in LMXBs (see van der Klis 2000 for a review).

However, there exist two big problems in the standard scenario about the evolutionary link between LMXBs and LMBPs.

(1) The birth rate problem. Statistical analyses have questioned whether the known LMXB population could produce the observed LMBPs (e.g., Kulkarni & Narayan 1988; Grindlay & Bailyn 1988). The LMBP lifetime is $\sim$ a few $10^9 - 10^{10}$ yr. For the estimated total number of LMBPs in the Galaxy, $N_{\rm LMBP} \sim 10^5$, the observed LMBPs require a birth rate of $\sim N_{\rm LMBP}/\tau_{\rm LMBP} \sim 10^{-5}$. On the other hand, the spin-up of a slowly rotating neutron star to the spin period of an LMBP would require accreted mass $\sim 0.1\,M_\odot$. For LMXBs with luminosities $L \sim 10^{36} - 10^{38}$ erg s^{-1}, the typical spin-up time scale $\tau_{\rm LMXB} \sim 10^9$ yr. For the total number of LMXBs $N_{\rm LMXB} \sim 10^2$, the estimated birth rate is $\lesssim N_{\rm LMXB}/\tau_{\rm LMXB} \sim 10^{-7}$ yr^{-1}. This rate falls short of the LMBP formation rate roughly by at least 1–2 orders of magnitude.

(2) The pulsar mass problem. The companions of LMBPs have masses of $\sim 0.1 - 0.4 M_\odot$. Their progenitors should be more massive than $\sim 0.9 M_\odot$ so that their main-sequence life time is shorter than the age of the Galaxy. In most cases of LMXB evolution, the mass transfer rates are sub-Eddington, almost all of the mass transferred from the secondary is accreted by the neutron star. Hence LMBPs should be massive ($\sim 2.0 M_\odot$) neutron stars. However, the analyses by Thorsett & Chakrabarty (1999) reveal that radio pulsar masses lie in a narrow range of $1.35 \pm 0.04 M_\odot$, with no evidence that extensive mass accretion ($> 0.1 M_\odot$) has occurred in these systems. Although the sample was contaminated by binary neutron star systems, there is strong evidence that some pulsars with low-mass white dwarf companions are not massive.

These two problems have strong implications for LMXB evolution: the lifetime of short period LMXBs may be much shorter than previously thought, and there must be large mass losses during the binary evolution. In the following sections we will discuss the relevant aspects in recent investigations of binary evolution.

2. Mass Loss in LMXB Evolution

Disk accretion in cataclysmic variables (CVs) is believed to be thermally and viscously unstable, leading to outbursts in dwarf novae (DNe), if the rate of mass transfer from the secondary is lower than a critical value, at which the temperature at the outer edge of the disk becomes below the hydrogen ionization temperature ~ 6500 K (see Cannizzo 1993 for a review). Such a thermal-viscous instability may also occur in soft X-ray transients (SXTs; Mineshige & Wheeler 1989; Cannizzo, Chen & Livio 1995), which are LMXBs characterized by episodic X-ray outbursts during which a soft X-ray spectrum is observed (Tanaka & Shibazaki 1996). The critical mass transfer rate for the disk instability depends

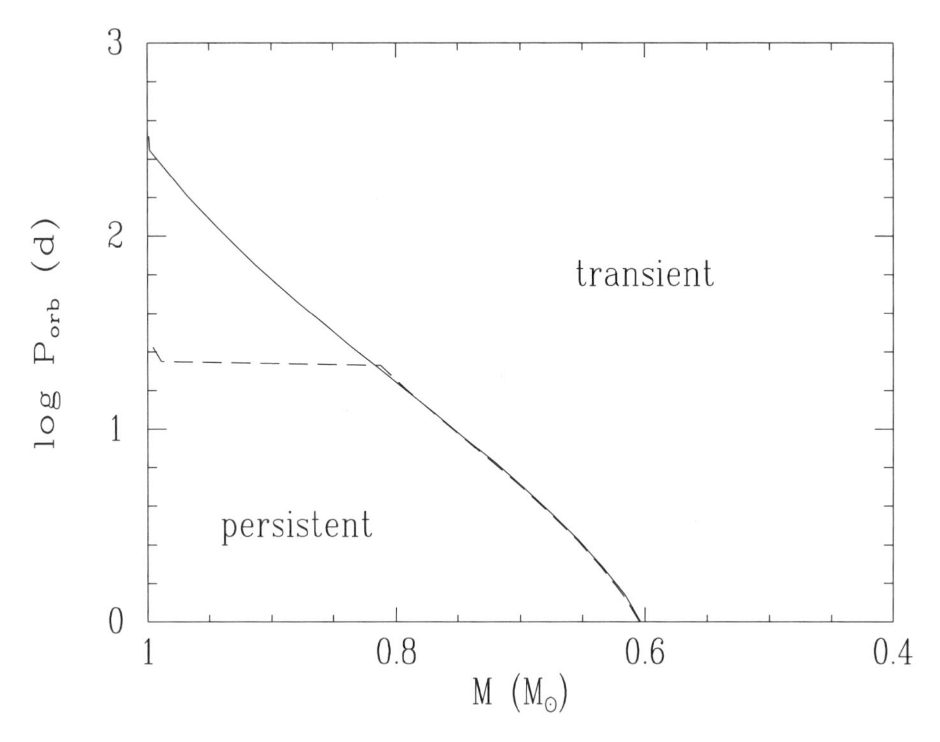

Figure 1. The critical orbital period against the donor mass for an LMXB. The solid and dashed curves give the boundaries between persistent and transient behavior with conservative and non-conservative mass transfer, respectively (Li & Wang 1998).

on the binary orbital period $P_{\rm orb}$, and is strongly influenced by irradiation from the inner region of the accretion disks in LMXBs (van Paradijs 1996). With X-ray heating of the disk included, King, Kolb & Burderi (1996) and King et al. (1997) showed that most of neutron star low-mass X-ray binaries with $P_{\rm orb} \gtrsim 1 - 2$ d will become transient, whereas short-period LMXBs with a main-sequence donor ($P_{\rm orb} < 1$ d) are transient only if the companion star is significantly nuclear-evolved before mass transfer begins.

With typical transient duty cycles $\lesssim 10^{-2}$ observed in soft X-ray transients (Tanaka & Shibazaki 1996), it follows that the mass flow rate through the disk during outbursts must be at least $\sim$ a few $10^{-8} - 10^{-7} M_{\odot}\,{\rm yr}^{-1}$, above the Eddington limit for a $1.4 M_{\odot}$ neutron star. For longer orbital periods or shorter duty cycles, the rates are still higher – a large fraction of the accreted mass is ejected. Thus one may expect that only during the stable mass transfer phase can the neutron star accrete efficiently. Indeed in transient LMXBs the accreted mass by a neutron star is considerably less than in stable accretion case (Li & Wang 1998). However, for relatively narrow ($P_{\rm orb} \sim 1 - 10$ d) systems, the influence of disk instability is not significant, and the neutron stars still accretes a large amount of mass (see Fig. 1).

Besides unstable disk accretion, there may exist other ways to prevent the neutron stars from accreting mass efficiently from their companions. For example, the matter from the donor star flowing towards the neutron star could be ejected from the system, when the neutron star's spin is accelerated to be so rapid that the magnetic dipole radiation pressure becomes greater than the ram pressure of the infalling plasma (Ruderman, Shaham, & Tavani 1989). But this requires very tight orbits, which could not be fulfilled in the evolution of most LMXBs.

It has also been suggested that irradiation of the secondary by the accretion luminosity of the neutron star can considerably alter the secular evolution of LMXBs (Podsiadlowsky 1991; Harpaz & Rappaport 1991). Based on the results in Podsiadlowsky (1991), Frank et al. (1992) propose that X-ray heating of the secondary causes rapid mass-transfer episodes followed by detached phase. More recent calculations by Ritter, Zhang, & Kolb (2000) with anisotropic irradiation taken into account, however, show that LMXBs are very probably stable against this type of irradiation-induced run-away mass transfer. As also noted by King et al. (1997), systems in which accretion is intermittent rather continuous, i.e., transient LMXBs, are even more stable.

3. Evolution of IMXBs

Besides LMBPs there exist a group of intermediate-mass binary pulsars (IMBPs) with relatively heavy white dwarf companions, slow spin periods (up to a few hundred milliseconds), and high period derivatives (or magnetic fields). Their formation mechanism is not well understood. It has been suggested that such systems evolved through an unstable mass transfer and common envelope phase from high-mass X-ray binaries (van den Heuvel 1994). More recent work on the evolution of the LMXB Cyg X-2 (King & Ritter 1999; Podsiadlowsky & Rappaport 2000; Kolb et al. 2000) indicates that the mass of the donor star in this system must have been substantially larger ($\sim 3.5M_{\odot}$) than its current value ($\sim 0.6M_{\odot}$), suggesting that a large fraction of the observed LMXBs may actually have started their lives as intermediate-mass X-ray binaries (IMXBs). Tauris et al. (2000) have shown that the evolution of close X-ray binary systems with intermediate-mass donor stars may survive a spiral-in and a highly super-Eddington mass transfer phase on a sub-thermal timescale. Thus IMXBs may be the major progenitors of low- and intermediate-mass binary pulsars (van den Heuvel 1995; Davies & Hansen 1998; Podsiadlowski et al. 2002).

According ro the recent calculations of the evolution of low- and intermediate-mass X-ray binaries (Tauris & Savonije 1999; Tauris et al. 2000), the intermed--iate-mass donor branch of evolution is more suitable to explain the recycled LMBPs. When the donor is more massive than the accretor by a factor of $\sim 1.5 - 3$, the mass transfer could be stable, proceeding on a thermal timescale. The intermediate-mass systems tend to have shorter periods of accretion, often at much higher rates, and hence smaller accreted mass.

Can an IMXB model solve the two puzzles discussed above? In Fig. 2 we present the evolutionary sequences for a binary system initially consisting of a $3.0M_{\odot}$ main-sequence star and a $1.3M_{\odot}$ neutron star. We choose the orbital periods at onset of mass transfer to be 0.72 day, which reach 1.33 day when the

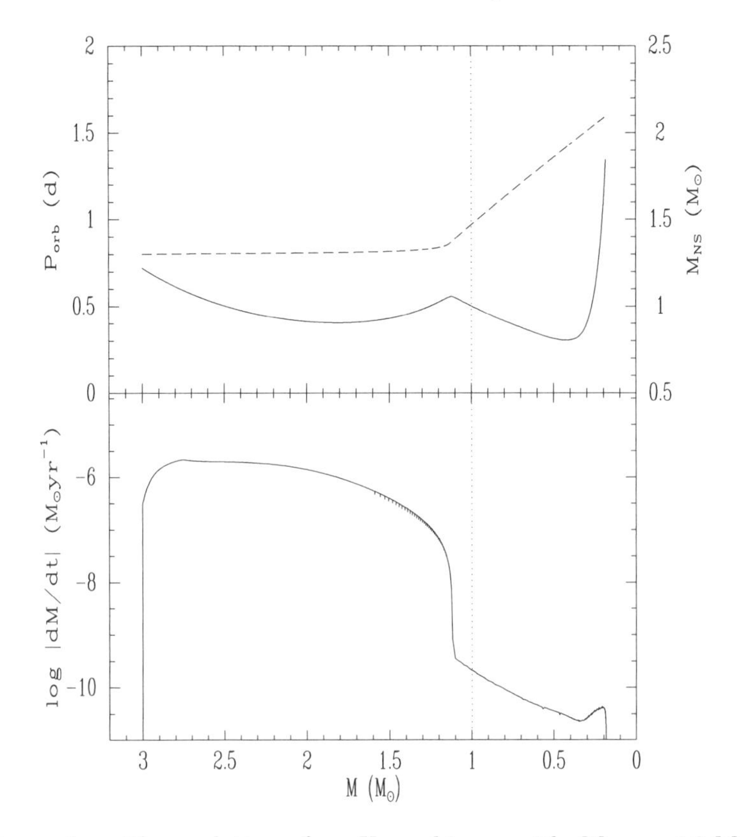

Figure 2. The evolution of an X-ray binary with $M_{\rm NS} = 1.3\,M_\odot$, $M = 3.0\,M_\odot$ and $P_{\rm orb} = 0.72$ day. Shown are the orbital period (solid curve, upper panel), the neutron star mass (dashed curve, upper panel), and the mass transfer rate (solid curve, lower panel), as a function of the donor mass. The dotted line denotes the occurrence of the thermal-viscous instability in the accretion disk (Li 2002).

mass transfer stops. The final white dwarf mass is correspondingly $\sim 0.19\,M_\odot$. As we see in the figure, after a short ($\sim 5\times10^6$ yr), rapid phase of mass transfer, the donor star masses decrease to $\sim 1\,M_\odot$. The following evolution is similar to that of a close LMXB. The mass transfer proceeds on a time scale as long as $\sim 10^9$–10^{10} yr, the rates of which are usually lower than the Eddington accretion rate – a large amount of mass is then transferred onto the neutron stars, leading to the final masses of the neutron stars around $2.1M_\odot$, comparable to those in the case of LMXBs. Sutantyo & Li (2000), and Podsiadlowski et al. (2002) have performed systematic evolution calculations with various choices of the donor mass, and reached a similar conclusion for LMBPs. This suggests that the standard LMXB and IMXB scenarios are *unable* to explain the recycled LMBPs.

There are still some fairly obvious discrepancies between observations and predications from binary-population synthesis (Podsiadlowski et al. 2002). First, there are too many short-period systems to be consistent with the observed period distribution, although transient behavior in LMXBs is not considered. Second, while the predicted distribution of mass-accretion rate (and hence X-ray luminosity) is probably too low by about an order of magnitude than observed in LMXBs. This becomes more severe due to the following fact. The standard formulas used for angular momentum loss by magnetic stellar winds is somewhat of an extrapolation, based on loss rates inferred from stars rotating much more slowly than the secondaries of CVs and LMXBs. Recent work by Sills et al. (2000) indicates that this extrapolation overestimates the angular momentum loss rate in CVs by as much as 2 orders of magnitude. If this is correct, the observed mass transfer rates in CVs and in narrow LMXBs can not be due to magnetic braking.

4. Conclusions

Up to now no theory can consistently reproduce the properties of L/IMXBs and the related LMBPs. We are lacking some important factors in binary evolution considerations, for example, the response of the secondary to X-ray heating. As noted by Podsiadlowski et al. (2001), IMXBs spend most of their X-ray active lifetime as low-mass systems, and consequently the duration of the X-ray active life is generally not much shorter than for true LMXBs. Thus the inclusion of IMXBs as progenitors of millisecond pulsars does not solve the birth rate problem. However, as Fig. 2 shows, mass transfer in close IMXBs may become unstable when they evolve to low-mass systems. The time they spend as X-ray emitters would be significantly reduced, thus alleviating the birthrate problem (Li 2002). Spruit & Taam (2001) and Taam & Spruit (2001) suggest that a circum-binary (CB) disk would be formed in CVs. This idea may also be applied to LMXBs. The CB disk could be a remnant from a late stage of the CE evolution phase in which the LMXB was formed, or form as a by-product of mass loss from the system. A CB disk would be an additional source of angular momentum loss for the binary. The tides raised in the CB disk by the orbiting binary transfers angular momentum to it, and prevent it from spreading inward. The increased mass transfer rates caused by CB disks would be systematic, since a CB disk would dissipate only very slowly by spreading. This might explain the apparent low mass transfer rate predictions. A higher mass transfer rate would also help to explain the short X-ray lifetime and mass loss in LMXBs.

Acknowledgments. This work was supported by NSFC and NKBRSF.

References

Bhattacharya, D. & van den Heuvel, E. 1991, Phys. Rep., 203, 1

Cannizzo, J. K. 1993, in Accretion Disks in Compact Stellar Systems, ed. Wheeler, J. C. (Singapore: World Science), 6

Cannizzo, J. K., Chen, W., & Livio, M. 1995, ApJ, 454, 880

Chakrabarty, D. & Morgan, E. H. 1998, Nature, 294, 346

Davies, M. B. & Hansen, B. M. S. 1998, MNRAS, 3031, 15
Frank, J., King, A. R., & Lasota, J. P. 1992, ApJ, 385, L45
Galloway, D. K., Chakrabarty, D., Morgan, E. H.;, & Remillard, R. A. 2002, ApJ, 576, L137
Grindlay, J. E. & Bailyn, C. 1988, Nature, 336, 48
Harpaz, A. & Rappaport, S. 1991, ApJ, 383, 739
Joss, P. C. & Rappaport, S. A. 1983, Nature, 304, 419
King, A. R., Frank, J., Kolb, U., & Ritter, H. 1997, ApJ, 482, 919
King, A. R., Kolb, U., & Burderi, L. 1996, ApJ, 464, L127
King, A. R. & Ritter, H. 1999, MNRAS, 309, 253
Kolb, U., Davies, M. B., King, A. R., & Ritter, H. 2000, MNRAS, 317, 438
Kulkarni, S. & Narayan, R. 1988, ApJ, 335, 755
Li, X.-D. 2002, ApJ, 564, 930
Li, X.-D. & Wang, Z.-R. 1998, ApJ, 500, 935
Markwardt, C. B., Swank, J. H., Strohmayer, T. E., in't Zand, J. J. M., & Marshall, F. E. 2002, 575, L21
Mineshige, S. & Wheeler, J. C. 1989, ApJ, 343, 241
Paczynski, B. 1983, Nature, 304, 421
Podsiadlowski, Ph. 1991, Nature, 250, 136
Podsiadlowski, Ph. & Rappaport, S. 2000, ApJ, 529, 946
Podsiadlowski, Ph., Rappaport, S., & Pfahl, E. 2002, 565, 1107
Ritter, H., Zhang, Z.-Y., & Kolb, U. 2000, A&A, 360, 969
Ruderman, M. A., Shaham, J., & Tavani, M. 1989, ApJ, 336, 507
Savonije, G. J. 1983, Nature, 304, 422
Sills, A., Pinsonneault, M. H., & Terndrup, D. M. 2000, ApJ, 540, 489
Spruit, H. C. & Taam, R. E. 2001, ApJ, 548, 900
Sutantyo, W. & Li, X.-D. 2000, A&A, 360, 633
Taam, R. E. 1983, ApJ, 270, 694
Taam, R. E. & Spruit, H. C. 2001, ApJ, 561, 329
Tanaka, Y. & Shibazaki, N. 1996, ARA&A, 34, 607
Tauris, T. & Savonije, G. J. 1999, A&A, 350, 928
Tauris, T., van den Heuvel, E. P. J., & Savonije, G. J. 2000, ApJ, 530, L93
Thorsett, S. E. & Chakrabarty, D. 1999, ApJ, 512, 288
van den Heuvel, E. P. J. 1994, A&A, 291, L39
van den Heuvel, E. P. J. 1995, JA&A, 16, 255
van der Klis, M. 2000, ARA&A, 38, 717
van Paradijs, J. 1996, ApJ, 464, L139
Webbink, R. F., Rappaport, S. & Savonije, G. J. 1983, ApJ, 270, 678
Wijnands, R. & van der Klis, M. 1998, Nature, 394, 344

IAU 8th Asia-Pacific Regional Meeting
ASP Conference Series, Vol. 289, 2003
S. Ikeuchi, J. Hearnshaw & T. Hanawa, eds.

Raman Scattering and Accretion Disks in Symbiotic Stars

Hee-Won Lee
Department of Astronomy, Sejong University,
98 Gunja-dong Gwangjin-gu, Seoul 143-747, Korea

Abstract. Symbiotic stars are generally known to be well-detached systems of a mass-losing giant and a hot white dwarf. It has been suspected that some fraction of the slow stellar wind from the giant component may be captured by the white dwarf component forming an accretion disk. Here, it is argued that in symbiotic stars the existence of an accretion disk is strongly implied by the characteristic features exhibited by the Raman-scattered O VI lines around 6830 Å and 7088 Å. Double of triple-peaked profiles in the Raman-scattered lines and single-peak profiles in other emission lines are interpreted as line-of-sight effects, where H I scatterers near the giant view an incident double-peaked profile and an observer with a low inclination sees single-peak profiles. In this accretion disk model, it is proposed that an inhomogeneous mass concentration around the accretion disk formed by a dusty wind may lead to the disparate ratios of the blue peak strength to the red counterpart observed in the 6830 Å and 7088 Å features. We present a (Canada-France-Hawaii Telescope) CFHT spectrum of the symbiotic star V1016 Cyg, which supports the accretion disk model.

1. Introduction

Symbiotic stars are believed to be binary systems consisting of a mass-losing giant and a hot white dwarf with an emission nebula. Some fraction of the slow stellar wind from the giant may be accreted onto the white dwarf resulting in the formation of an accretion disk, which is consistent with the SPH computation by Mastrodemos & Morris (1998). The accretion disk model is of particular importance in that a large fraction of symbiotic stars exhibit bipolar nebular morphology, which is exhibited by only 10% of planetary nebulae (Corradi & Schwarz 1995). The bipolar morphology may be closely related with the binarity of the central star system and accretion disk formation (Soker 1998, Livio et al. 1986).

The symbiotic emission bands around 6830 Å and 7088 Å were identified by Schmid (1989), who proposed that they are Raman scattered features of the resonance O VI 1032, 1038 doublet by atomic hydrogen. The scattering hydrogen atom initially in the ground $1s$ state de-excites into the excited $2s$ state, resulting in emission of an optical photon redward of Hα. The scattering cross section for this Raman process is around $\sigma_{\rm Ram} \sim 10^{-22}$ cm^2, which requires the existence of highly thick H I scattering region with the column density $H_{\rm HI} \sim 10^{22}$ cm^{-2}.

These Raman features exhibit various structures including multiple peak profiles with strong polarization. Schmid (1986, 1996) attributed them to the stellar wind around the giant, treating the emission source as a point-like source. Adopting this model, the blue part of the O VI Raman flux is obtained from the part of the stellar wind that is approaching the white dwarf, whereas the red part is formed in the wind region that is receding from the white dwarf. From this model, no profile difference in 6830 and 7088 features is expected and therefore no considerable observational investigations of the weak 7088 feature have been performed (Harries & Howarth 1996, Lee & Lee 1997).

Motivated by the SPH numerical work of Mastrodemos & Morris (1998), Lee & Park (1999) proposed the accretion disk model, where the O VI emission regions are located around the accretion disk and the giant provides the scattering site. Because the scattering H I component views the emission region in the edge-on direction, the double peak profiles of the Raman features are naturally expected. The red part of the Raman features will be stronger than the blue part, because the accretion flow may tend to get stronger in the red emission region due to the comparable speeds of the binary orbital motion and the accretion gas flow. This also leads to more optically thick red emission region than the blue emission region, which is consistent with the SPH computations. In this paper, we argue that the profile difference of the two O VI Raman features can be strong evidence supporting the accretion disk model and present our CFHT spectroscopic observation of the symbiotic star V1016 Cyg.

2. Accretion Disk Model

We note that the O VI emission regions have different optical depths, where the red emission region is more optically thick than the blue emission region (see the left panel of Fig. 1). Arising from the $S_{1/2} - P_{1/2,3/2}$ transitions, the oscillator strength of the 1032 line ($S_{1/2} - P_{3/2}$) is twice larger than the 1038 line ($S_{1/2} - P_{1/2}$). From our photoionization computation using 'Cloudy' developed by Ferland (2001), the emission region is characterized by the line center optical depth $\tau_0 \sim 10^5$, which implies that a typical O VI photons will be resonantly scattered around 10^6 times before escape.

Since 1032 photons are twice more optically thick than 1038 photons, 1032 photons are expected to travel almost twice longer paths than 1038 photons. The slow stellar wind from the giant component may contain a significant amount of dust, which subsequently joins the accretion disk emission region. Therefore, the emission region may be contaminated by dust, which suppress UV flux in a large amount contributing a significant IR excess. With the IR reprocessing, 1032 photons are subject to destruction by dust two times more efficiently than 1038 photons, since the destruction probability is proportional to the path length before escape. This leads to the doublet line ratio of 1:1 in a very optically thick region.

However, in a much less optically thick region, the ratio will remain 2:1, that is, the ratio of the oscillator strengths, because pure resonance scatterings will not destroy line photons and hence there is no change in the number of emerging line photons. IUE observations of many symbiotic stars show that many resonance doublets including C IV 1548, 1551 show flux ratios between

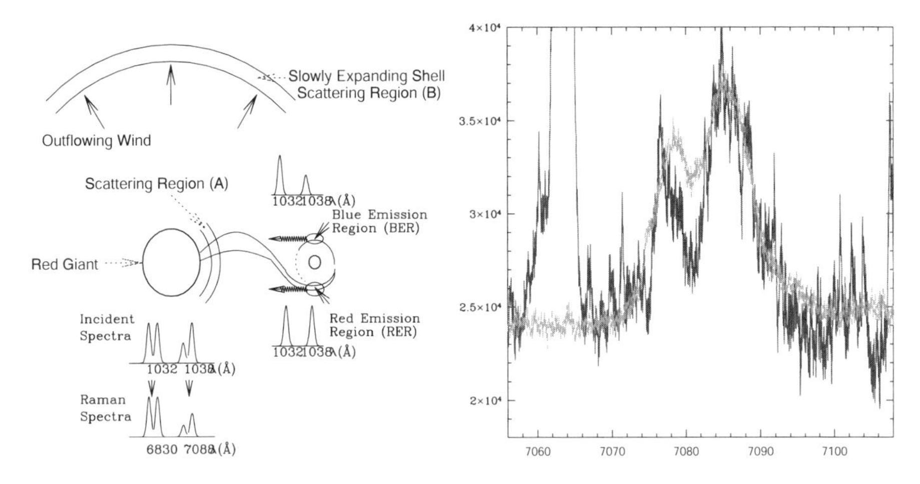

Figure 1. A conceptual diagram of the scattering geometry with an accretion disk for a typical symbiotic star and a CFHT spectrum of the symbiotic star V1016 Cyg. There are two emission regions around the hot star denoted by 'RER (red emission region)' and 'BER (blue emission region)', which provide red-shifted and blue-shifted photons to the direction of the giant, respectively. The scatterers near the giant see an emission source characterized by a double peak profile. As is explained in the text, it is expected to obtain the stronger blue component relative to the red part in the 6830 feature than in the 7088 feature. In the right panel, the 6830 feature represented by the solid line is moved and reduced by a factor of 5 in order to make a direct profile comparison with the 7088 feature shown by the dotted line.

1:1 and 2:1, which is in strong support of our theoretical considerations (e.g. Yoo, Lee & Ahn 2002).

Therefore, the relative strength of the blue part compared to the red part is systematically larger in the 6830 feature than in the 7088 feature, since the red emission region is more optically thick than the blue emission region (see Fig. 1). The observational verification of the systematic profile difference can be regarded as strong evidence toward the existence of an accretion disk in a symbiotic star. Some hint of profile differences has been proposed by Schmid et al. (1999), but low quality spectra prevented from definite conclusions. It is essential to secure the high quality spectra around the much weaker 7088 feature, which requires a use of large telescopes.

3. CFHT Observations

In the right panel of Fig. 1, we show a CFHT spectrum of the symbiotic star V1016 Cyg, which was obtained on the night of 2002 May 26, using the Gecko spectrometer. The 6830 Å feature shown by the solid line is moved and reduced by a factor of 5 in order to make a direct profile comparison with that of the

7088 Å feature. We immediately recognize the the relative weakness of the blue part of the 7088 Å feature compared with that for the 6830 Å feature, in strong support of the accretion disk model (Lee & Park 1999) and the SPH calculations performed by Mastrodemos & Morris (1998).

The O VI Raman features show typical widths of 20 Å, which corresponds to the bulk velocity $v_{\rm OVI} \sim 150$ km s^{-1} of the O VI emission region. This velocity scale is inconsistent with the speed $v_w \sim 10$ km s^{-1} of the slow stellar wind from the giant component. However, if the accretion disk model is adopted, this velocity scale is attributed to the rotation speed of the main O VI emission region. Adopting a typical white dwarf mass 0.7 $M_\odot$, the O VI emission region is formed at a distance $\sim 10^{12}$ cm. This implies that the emission region is fairly extended from the ionizing source.

The largest relative flux ratio is seen near the line center, which may correspond to the blue emission region facing the giant component. In the accretion disk picture, this region has the smallest dust optical depth $\tau_{\rm d} \ll 1$. Therefore, in this particular region the accretion stream is colliding with the slow stellar wind from the giant component.

Acknowledgments. The author is grateful to the organizing committee for support of my travel.

References

Corradi, R. L. M. & Schwarz, H. E. 1995, A&A, 293, 871
Ferland, G. 2001, Hazy, a brief introduction to Cloudy 94.00
Harries, T. J. & Howarth, I. D. 1996, A&AS, 119, 61
Lee, H.-W. & Park, M.-G. 1998, ApJ, 515, L89
Lee, H.-W., Kang, Y.-W. & Byun, Y.-I. 2001, ApJ, 551, L121
Lee. H.-W. & Park, M.-G. 1999, ApJ, 515, L89
Lee, K. W., & Lee, H.-W. 1996, MNRAS, 292, 573
Livio, M. & Soker, N., de Kool, M., Savonije, G. J. 1986, MNRAS, 222, 250
Mastrodemos, N. & Morris, M. 1998, ApJ, 497, 303
Schmid, H. M. 1989, A&A, 211, L31
Schmid, H. M. 1996, MNRAS, 282, 511
Schmid, H. M. et al. 1999, A&A, 348, 950
Soker, N. 1998, ApJ, 496, 833
Yoo, J. J., Lee, H.-W. & Ahn, S.-H. 2002, MNRAS, 334, 974

IAU 8th Asia-Pacific Regional Meeting
ASP Conference Series, Vol. 289, 2003
S. Ikeuchi, J. Hearnshaw & T. Hanawa, eds.

X-ray Observations of the X-ray Binary Pulsar Centaurus X-3 with *RXTE*

Takayoshi Kohmura and Shunji Kitamoto

Rikkyo University, 3-34-1, Nishi-Ikebukuro, Toshima-ku, Tokyo, 171-8501, Japan

Abstract. We present a study of aperiodic variability from the X-ray binary pulsar Centaurus X-3. We have analyzed all archival data of Cen X-3 observed with the *Rossi X-Ray Timing Explorer* (RXTE).

We extracted light curves both in the energy band which included the iron K lines ("iron-band") and in the continuum X-ray bands which did not include the iron K lines. Then, we calculated cross spectrum functions between the iron-band and the remaining continuum bands. We discovered a significant time delay of the temporal variation of iron-band emission as compared with that of the other energy X-rays by (0.98±0.45) ms, (0.39±0.10) ms, (1.24±1.08) ms, and (0.52±0.20) ms (1-σ error). We also discovered the time difference showed a general trend such that the higher energy X-rays advanced in comparison with the lower energy X-rays, with the exception of the iron-band.

1. Introduction

The accreting X-ray pulsars are thought to be magnetized neutron stars with a normal star companion, the majority of which are high mass, early-type stars (Rappaport & Joss 1983; see Nagase 1989 for a review). X-ray observations has revealed the existence of strong iron K line emission from many X-ray binary pulsars, including Cen X-3 (Nagase et al. 1992; Ebisawa et al. 1996).

So far, many authors have investigated the origin of the iron K lines. In the case of Cen X-3, this line emission is not observed during the eclipse, it must originate near to the neutron star (Nagase et al. 1992; Ebisawa et al. 1996). Further, the iron emission region must cover a large solid angle near to the neutron star (although not the full 4π steradian) in order to explain the large equivalent width of the line (Nagase et al. 1992). Nagase et al. (1992) derived an upper limit for the distance, between the iron line emitter and the neutron star, of $\sim 3\times10^{10}$ cm, based on the *Ginga* observation that showed the 6.4 keV line intensity to drop by a factor of 20 in less than 10 min. If the iron K-lines originate from the magnetosphere and the magnetosphere covers a large fraction of the solid angle at the neutron star, the line photons must be delayed in comparison with the X-rays observed directly from the neutron star. If this iron K lines are considered to come from a re-processing matter around the neutron star, e.g. the Alfven shell, we expect that iron-band has a certain delay from the higher energy X-rays.

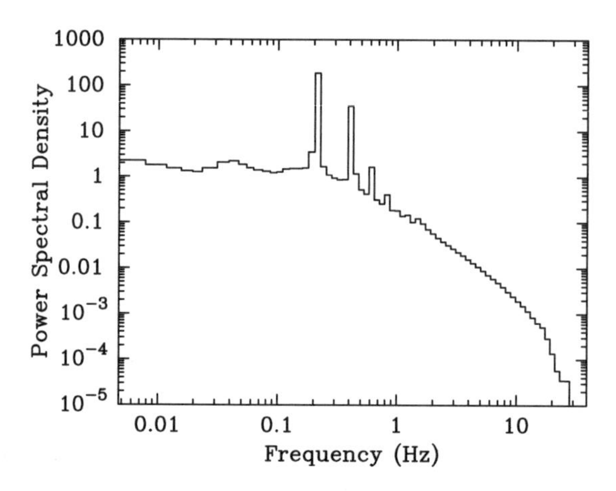

Figure 1. The power spectral density obtained for P20104 data in the energy range from 3.3 keV to 13.1 keV. The time resolution of the data is 15.625 ms. This was calculated by adding the 316 separate power spectral density derived from 256 s exposure. The Poisson statistics component has been subtracted.

2. Analysis and Results

To detect a small delay time, *RXTE* is the best experiment, which provides us high time resolution data with excellent statistics and reasonable energy resolution.

Figure 1 shows the power spectral density of P20104 observation data. The QPO component (Takeshima et al. 1991) was confirmed with the center frequency of 41±3 mHz. The noise due to Poisson statistics component has been subtracted. The first, second, and higher order harmonics of the pulsed component were detected. Aperiodic time variation is significantly detected up to ~25 Hz. We are attempting to analyze the aperiodic time variation rather than the coherent pulsations. The best method to use is the cross spectral analysis (van der Klis et al. 1987), since the cross spectra make it possible to distinguish the pulse component from the interesting aperiodic component.

We first made light curves of some energy bands. One energy band contained the iron lines and the other did not contained (e.g. 3.3–4.7, 4.7–5.8, 5.8–7.2, 7.2–8.7, 8.7–10.9, and 10.9–13.1 keV at P20104 observation). We refer to the light curve which contain the iron lines as the "iron-band". We calculate the cross spectra between the X-ray light curve in the iron-band and the other energy bands, and studied the coherence and the phase lag between them. (the detail of this analysis can be seen in Kohmura et al. 2001, 2002). Figure 2 shows an example of the coherence and phase lag of 10.9–13.1 keV band from the iron-band. Here we pay attention to the phase lag between 3 Hz and 15 Hz, where the coherence is high enough and the effect of the coherent pulse shape is small. If there is a constant time delay which is independent of frequency, the phase lag must be a linear function with no offset. The data between 3 Hz and 15 Hz were fitted with a linear function. The best fit lines is plotted in figure

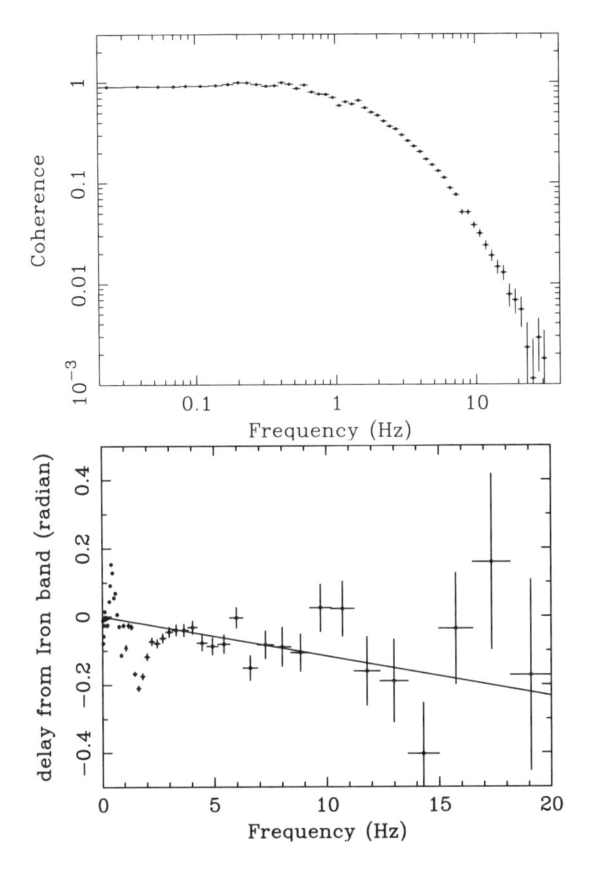

Figure 2. Top: The coherence, calculated between the iron-band and 10.9–13.3 keV band, is plotted as a function of the frequency. Bottom: The phase lags of time variations between iron-band and 10.9–13.3 keV band are plotted as a function of the frequency.

2. Figure 3 shows the result of our cross spectral analysis as a function of the energy. We applied a simple linear function to these data, and the result was $-(2.6\pm0.4)\times10^{-4}\ E$ (keV) + $(1.3\pm0.4)\times10^{-3}$ s. The deviation of the iron-band data at 6.5 keV from this best-fit model is (0.39±0.10) ms, where error is the one sigma level. Therefore, we detected the delayed iron-band intensity variation with almost the four sigma confidence level.

We also applied this cross spectral analysis to the other observation data of Cen X-3, and we summarized of these results in Table 1.

3. Discussion

Using P20104 data, we derived a 0.39±0.10 ms time delay of the iron-band relative to that of the general trend of the other energy bands. However, this value may not necessarily represent an actual delay of the reprocessed X-rays for

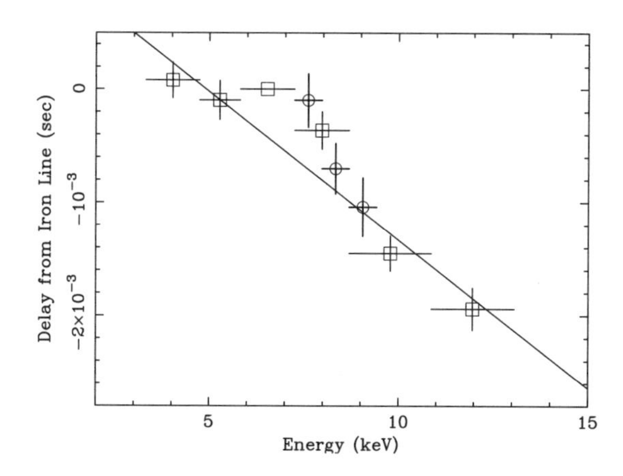

Figure 3. The time delay in the time variation from that of the iron-band as a function of the X-ray energy. The five squares are the data obtained from the original six energy bands and the line around 6.5 keV is marked for reference. The circles are the data derived by using more finely binned energy bands. Errors are one sigma levels. The straight line is the best fit model of the linear function. The positive direction of the vertical axis represents the delay from the variation of the iron-band.

Table 1. Summary of the Observations and Cross Spectral Analysis

Obs ID	Obs mid-time (MJD)	Delay time (ms)	$A \times 10^{-4}$ [s keV^{-1}]*
P10132	50301.7	-(0.98±0.45)	-(5.0±3.0)
P10134	50345.2	-(0.14±0.39)	-(1.0±1.0)
P20104	50508.9	-(0.39±0.10)	-(2.6±0.4)
P20105	50554.8	-(0.13±0.40)	-(2.7±2.5)
P20106	50720.7	-(1.24±1.08)	-(0.5±4.3)
P30083	50934.0	-(0.03±4.95)	-(6.3±10.3)
P30084	51043.0	-(0.52±0.20)	-(3.0±0.5)
P40072	51579.4	+(0.41±0.70)	-(2.9±1.6)

* The time difference is assumed to be a liner function with $A \times (E - 6.4[\mathrm{keV}]) +$ constant.

two possible reasons. The first reason is that the fraction of iron line photons in the iron-band is only ~9 %, and the second is that the value is the deviation of the iron-band from the general trend and may not be the delay from the original temporal variation. For the first reason, we consider the effect of the mixture of the delayed and the non-delayed components. For the second reason, we calculate the mean time difference between non-reprocessed X-rays in the iron-band and the X-rays above the iron K absorption edge, because the photons responsible for iron fluorescence are those above the K-absorption edge. As a result, we derived 6.5±1.6 ms for the time delay of the temporal variation of the iron line from the average temporal variation in the energy band above the iron absorption edge. This result leads to a determination of $(2.0\pm0.5)\times10^{8}$ cm for the distance between the original X-ray source and reprocessor, which emits the iron lines. The radius derived from the above discussion is consistent with the Alfven radius derived from QPO analysis (Kohmura et al. 2001, 2002) within a factor of two.

The advance of the hard X-ray variation was also discovered from all data of Cen X-3. We think this advanced trend of the continuum X-ray is related to the X-ray emission mechanism on the neutron star as follows (Kohmura et al. 2001, 2002). The short term intensity variation can be interpreted as being due to the variation of the amount of material falling onto the polar cap of the neutron star. The intensity must become high when the large amount of matter falls on the polar cap. The kinetic energy of the falling matter is converted into thermal energy almost instantaneously, resulting in an extremely high temperature plasma. This plasma must cool via radiation and conduction, and consequently the hard X-ray advance is expected. Although the quantitative consideration is beyond the aim of this paper, this discovery of the hard advance may be related to the interaction between the accretion matter and the neutron star.

Acknowledgments. T.K. is supported by a JSPS Research Fellowship for Young Scientists. This research has made use of data obtained through the High Energy Astrophysics Science Archive Research Center Online Service, provided by the NASA/Goddard Space Flight Center.

References

Ebisawa, K., Day, C. S. R., Kallman, T.R., Nagase, F., Kotani, T., Kawashima, K. & Kitamoto, S. 1996, PASJ, 48, 425

Kohmura, T., Kitamoto, S. & Torii, K. 2001, ApJ, 562, 943

Kohmura, T. 2002, PhD. Thesis, Osaka University

Nagase, F. 1989, PASJ, 41, 1

Nagase, F., Corbet, R. H. D., Day, C.S.R., Inoue, H., Takeshima, T., Yoshida, K. & Mihara, T. 1992, ApJ, 396, 147

Rappaport, S. & Joss, P. C. 1983, in Accretion-Driven Stellar X-Ray Sources, ed. W.H.G. Lewin & E.P.J. van den Heuvel, Cambridge Univ. Press, p1

Takeshima, T., Dotani, T., Mitsuda, K., & Nagase, F. 1991, PASJ, 43, L43

van der Klis, M., Hasinger, G., Stella L., Langmeier, A., van Paradijs, J. & Lewin, W. H. G. 1987, ApJ, 319, L13

IAU 8th Asia-Pacific Regional Meeting
ASP Conference Series, Vol. 289, 2003
S. Ikeuchi, J. Hearnshaw & T. Hanawa, eds.

Peculiar Characteristics of a Hyper-luminous X-ray Source in M82

Hironori Matsumoto, Takeshi Go Tsuru

Department of Physics, Kyoto University, Kyoto 606-8502, Japan

Ken-ya Watarai

Department of Astronomy, Kyoto University, Kyoto 606-8502, Japan

Shin Mineshige

Yukawa Institute for Theoretical Physics, Kyoto University, Kyoto 606-8502, Japan

Satoki Matsushita

Harvard-Smithsonian Center for Astrophysics, 60 Garden Street, Cambridge, MA 02138, USA

Abstract. M82 X-1, which is the brightest X-ray source in M82, is a strong candidate for an intermediate mass black hole of $10^3 - 10^6\ M_\odot$. Since M82 X-1 is much brighter than ultra-luminous X-ray sources (ULXs), we call it a hyper-luminous X-ray source (HLX). We analyzed the ASCA spectra of M82 X-1 by using the multi-color disk (MCD) black-body model, and found that the MCD model can describe the spectra of M82 X-1 as well as those of ULXs. Not only the bolometric luminosity ($L_{\rm bol}$) but also the innermost disk temperature ($T_{\rm in}$) of M82 X-1 is higher than those of ULXs. The most striking feature is that there is no correlation between $T_{\rm in}$ and $L_{\rm bol}$, while there is a positive correlation between those of ULXs. A lower limit on the mass of M82 X-1 obtained by assuming that $L_{\rm bol}$ is sub-Eddington is much larger than an upper limit obtained by assuming that the innermost disk radius ($R_{\rm in}$) is larger than the gravitational radius of a black hole (BH). These characteristics are not consistent with theoretical models such as a standard accretion disk model with a Kerr BH, a slim disk model, and a beaming model, all of which might be able to explain ULXs. This may suggest either that some important processes we overlook play an important role in such luminous objects, or that the HLX is a different kind of object from ULXs.

1. Introduction

The ASCA spectrum of M82 consists of soft, medium, and hard components (Tsuru et al. 1997). By monitoring M82 with ASCA, we found that the luminosity of the hard component in the 0.5 – 10 keV band changed between

4.5×10^{40} erg s^{-1} and 1.6×10^{41} erg s^{-1} at various time scales from a few hours to a month, while the soft and medium components showed no evidence for time variability (Matsumoto & Tsuru 1999; Ptak & Griffiths 1999). Although the soft and medium components clearly originate in hot gas due to starburst activity, the origin of the hard component was not clear.

We then observed M82 with the Chandra HRC twice, and found that the ASCA hard component comes from M82 X-1, which is the brightest X-ray source in M82 (Kaaret et al. 2001; Matsumoto et al. 2001). Considering its peak luminosity ($\sim 10^{41}$ erg s^{-1}) and its off-center position ($\sim$ 170 pc away from the dynamical center of M82), M82 X-1 is a strong candidate for an intermediate mass black hole (IMBH) of $10^3 - 10^6\ M_\odot$.

Since M82 X-1 is much brighter than enigmatic ultra-luminous X-ray sources (ULXs) (e.g. Makishima et al. 2000), we call it a hyper-luminous X-ray source (HLX). To investigate the emission mechanism of M82 X-1, we analyzed the ASCA spectra of M82 X-1, and compared the HLX M82 X-1 with the ULXs.

Throughout this paper, errors and uncertainties refer to 90% confidence limits. We assumed that the distance to M82 is 3.9 Mpc (Sakai & Madore 1999).

2. Analysis and Results

ASCA observed M82 ten times from 1993 to 1996, and we analyzed the data of all observations. Detailed descriptions about the observations and the data reductions are given in Matsumoto & Tsuru (1999).

We fitted the ASCA spectra with a 2 RS + MCD model described as

$$N_{\rm H}^{\rm whole} \times ({\rm RS}^{\rm soft} + N_{\rm H}^{\rm medium} \times {\rm RS}^{\rm medium} + N_{\rm H}^{\rm hard} \times {\rm MCD}^{\rm hard}), \qquad (1)$$

where $N_{\rm H}$ is the absorbing column density, and RS is a thin thermal plasma model (Raymond & Smith 1977), and MCD is a multicolor disk blackbody (MCD) model (Mitsuda et al. 1984; Makishima et al. 2000). The MCD model is a good approximation to X-ray emission predicted by a standard accretion disk model (Shakura & Sunyaev 1973), and is characterized by the innermost disk temperature ($T_{\rm in}$) and the bolometric luminosity ($L_{\rm bol}$). The innermost disk radius ($R_{\rm in}$) can be calculated from them. We fixed the parameters of the soft and medium components other than their normalizations to the best-fit values of Tsuru et al. (1997). Free parameters are thus the normalizations of the soft, medium, and hard components, $N_{\rm H}^{\rm hard}$, and $T_{\rm in}$.

The 2RS + MCD model can describe the ASCA spectra quite well. The obtained parameters are $kT_{\rm in} = 1.70 - 2.91$ keV, $L_{\rm bol} \cos i = 1.8 - 7.4 \times 10^{40}$ erg s^{-1}, and $R_{\rm in}\sqrt{\cos i} = 64 - 205$ km, where i is the inclination angle of the disk.

3. Discussion

Assuming $L_{\rm bol}$ is sub-Eddington, the mass of M82 X-1 must be larger than $490\,(\cos\ i)^{-1}\ M_\odot$. We also estimated an upper limit on the mass to be $7\,(\cos\ i)^{-0.5}\ M_\odot$ by assuming that $R_{\rm in}$ is larger than the last stable orbit around a Schwarzschild black hole (BH), which is $3R_{\rm s} = 6GM/c^2$. Then, we face a

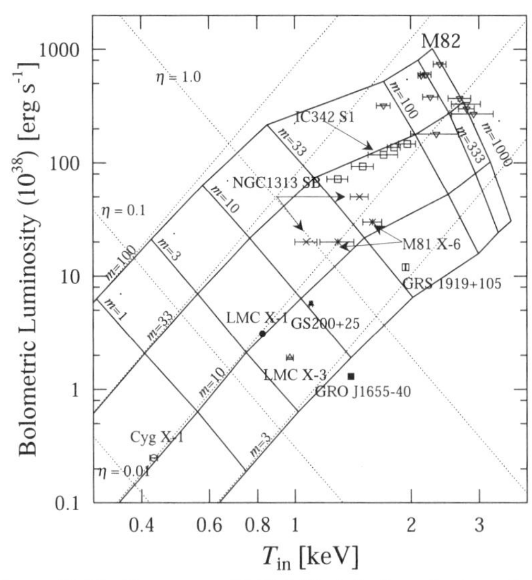

Figure 1. X-ray H-R diagram of M82 X-1 (triangles) and other ULXs (other symbols). Solid lines represent the constant black hole mass ($m = M/M_\odot$) and constant mass accretion rate ($\dot{m} = \dot{M}/\dot{M}_{crit}$) predicted by a slim-disk model. Dotted lines are the same but based on the standard accretion disk theory ($\eta \equiv L_{bol}/L_{Edd}$). We assume face-on geometry ($i = 0°$).

serious problem that the upper limit is much smaller than the lower limit. This problem is essentially the same as the "too hot accretion disk problem" seen in ULXs (Makishima et al. 2000). The problem of ULXs may be solved by invoking a Kerr BH, since the last stable orbit is $0.5R_s$ (Makishima et al. 2000). However, even if we assume a Kerr BH, the upper limit on the mass of M82 X-1 is $42\,(\cos\, i)^{-0.5}\, M_\odot$, which is still much smaller than the lower limit.

Figure 1 is the X-ray H-R diagram of M82 X-1 and ULXs (Makishima et al. 2000; Mizuno, Kubota, & Makishima 2001), which shows the relation between T_{in} and L_{bol}. Not only L_{bol} but also T_{in} of M82 X-1 is higher than those of ULXs. The most striking feature is that there is no correlation between T_{in} and L_{bol}, while there is a positive correlation between those of ULXs.

If a mass accretion rate is close to or larger than the critical rate ($\dot{M}_{crit} \equiv L_{Edd}/c^2$, where L_{Edd} is the Eddington luminosity), theory predicts that an accretion disk becomes a slim disk (e.g. Abramowicz et al. 1988). Although the MCD formalism was originally developed to approximate X-rays from standard disks, Watarai et al. (2000) found that the MCD model is also a good approximation to X-ray spectra from slim disks, if we limit the energy of the X-rays to the ASCA band (0.5 – 10 keV). Watarai, Mizuno, & Mineshige (2001) then found that the H-R diagram of ULXs can be explained by the slim disk model with constant BH mass and a varying mass accretion rate. The solid lines in Figure 1 show the predictions of the slim disk model. Clearly the slim disk model with constant BH mass is inconsistent with the behavior of M82 X-1.

M82 X-1 may be a beaming object (King et al. 2001). However, if M82 X-1 is a jet source and the jet points toward us, we would expect synchrotron radio emission from the jet. Although there are two radio sources at the position of M82 X-1, they are likely supernova remnants rather than a jet source (Matsumoto et al. 2001). Furthermore, the hard X-ray emission of M82 cannot be fitted by a power-law model (Tsuru 1992; Cappi et al. 1999). We therefore think M82 X-1 is unlikely to be a jet source. Simultaneous X-ray and radio observations will make it much clear.

4. Conclusion

The HLX M82 X-1 is a strong candidate for an IMBH, and we analyzed its ASCA spectra using the MCD model. We can estimate the upper and lower limits on the mass of M82 X-1 using the MCD parameters, but the lower limit is much larger than the upper limit. We compared M82 X-1 with ULXs, and found that M82 X-1 shows not only larger $L_{\rm bol}$ but also higher $T_{\rm in}$ than ULXs. Although there is a positive correlation between $L_{\rm bol}$ and $T_{\rm in}$ of ULXs, M82 X-1 does not show such a correlation. These characteristics are not consistent with theoretical models which might be explained ULXs. This may suggest either that some important processes we overlook play an important role in such luminous objects, or that the HLX M82 X-1 is a different kind of object from ULXs.

References

Abramowicz, M. A., Czerny, B., Lasota, J. P. & Szuszkiewicz, E. 1988, ApJ, 332, 646

Cappi, M. et al. 1999, A&A, 350, 777

Kaaret, P., Prestwich., A. H., Zezas, A., Murray, S. S., Kim, D.-W., Kilgard, R. E., Schlegel, E. M. & Ward, M. J. 2001, MNRAS, 321, L29

King, A. R. Davies, M. B., Ward, M. J., Fabbiano, G. & Elvis, M. 2001, MNRAS, 552, L109

Makishima, K. et al. 2000, ApJ, 535, 632

Matsumoto, H. & Tsuru, T. G. 1999, PASJ, 51, 321

Matsumoto, H., Tsuru, T. G., Koyama, K., Awaki, H., Canizares, C. R., Kawai, N., Matsushita, S. & Kawabe R. 2001, ApJ, 547, L25

Mitsuda, K. et al. 1984, PASJ, 36, 741

Mizuno, T., Kubota, A. & Makishima, K. 2001, ApJ, 554, 1282

Ptak, A. & Griffiths, R. 1999, ApJ, 517, L85

Raymond, J. C. & Smith, B. W. 1977, ApJS, 35, 419

Sakai, S. & Madore, B. F. 1999, ApJ, 526, 599

Shakura, N. I. & Sunyaev, R. A. 1973, A&A, 24, 337

Tsuru, T. G. 1992, Ph.D. thesis, the University of Tokyo

Tsuru, T. G., Awaki, H., Koyama, K. & Ptak, A. 1997, PASJ, 49, 619

Watarai, K., Fukue, J., Takeuchi, M., & Mineshige, S. 2000, PASJ, 52, 133

Watarai, K., Mizuno, T., & Mineshige, S. 2001, ApJ, 549, L77

IAU 8th Asia-Pacific Regional Meeting
ASP Conference Series, Vol. 289, 2003
S. Ikeuchi, J. Hearnshaw & T. Hanawa, eds.

X-ray Study of Thermal Composite Supernova Remnants 3C391 and G349.7+0.2

Yang Chen

Dept. of Astronomy, Nanjing University, Nanjing 210093, P.R. China

Patrick Slane

Center for Astrophysics, 60 Garden Street, Cambridge, MA 02138, USA

Abstract. Supernova remnants 3C391 and G349.7+0.2 are both thermal composites, with shell-like radio emission, interior brightened thermal X-rays, and hydroxyl radical masers characterizing a shock interaction with dense clouds. The normal abundances and high X-ray emitting mass of the two SNRs are indicative of ISM-dominance in the hot gas. This inference, the inner X-ray enhancement, and the molecular lines are in agreement with the notion that both the remnants have been evolving in dense cloudy mediums. Both of their morphologies suggest a scenario in which the remnants have broken out of dense regions into an adjacent region of lower density. The variation of the absorption across 3C391 supports this scenario. G349.7+0.2 is found to be one of the most X-ray-luminous shell-type SNRs known in the Galaxy.

1. Introduction

Beyond the well-known classification of shell-like, Crab-like, and composite supernova remnants (SNRs), the so-called "thermal composite" or "mixed morphology" remnants (such as W28, IC 443, G349.7+0.2, and 3C 391 [Green et al. 1997]) are found to have radio shells (which indicate the position of the supernova blastwave), OH (1720 MHz) masers (which are signposts of shock interaction with molecular clouds), and interior enhanced thermal X-ray emission (the nature of which remains unclear). Here we summarize our recent X-ray studies of 3C 391 (Chen & Slane 2001) and G349.7+0.2 (Slane et al. 2002) based on the *ASCA* observations.

2. SNR 3C 391 (G31.9+0.0)

3C 391 shows an elongation from northwest (NW) to southeast (SE) in the radio band, and the centroid of the soft X-ray emission lies in the SE region. Reynolds & Moffett (1993) pointed out that the radio and X-ray morphologies can be explained with a gas breakout from a molecular cloud in the NW into a lower density region in the SE. CO and other molecular line observations confirm the location of the remnant at the southwestern edge of a molecular cloud (Wilner et al. 1998; Reach & Rho 1999). However, the *ROSAT* X-ray study (Rho & Petre 1996) could not actually distinguish between an increase in hydrogen column density or a decrease in temperature across the remnant from SE to NW from the spectra alone.

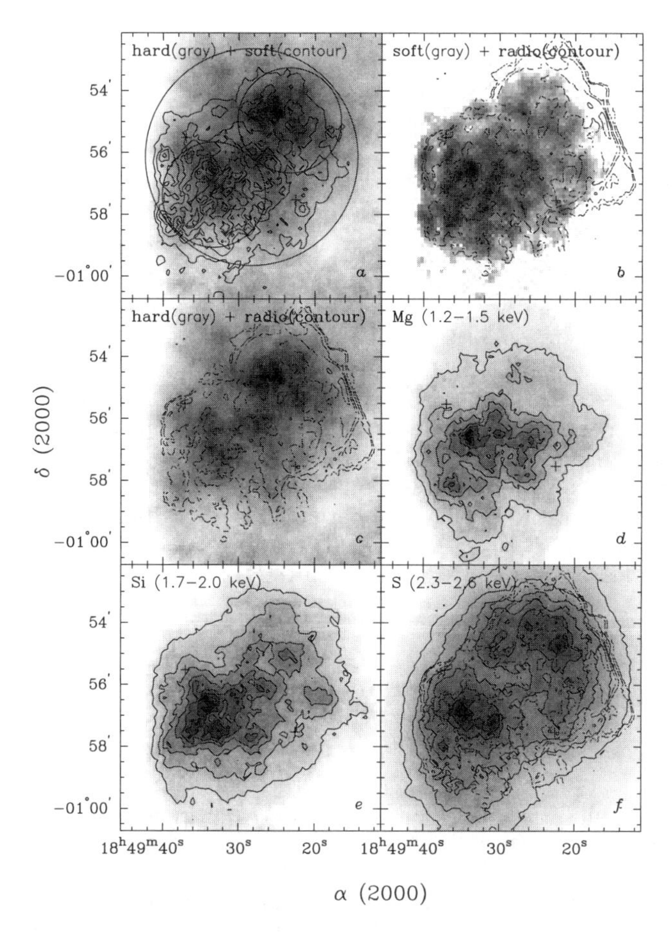

Figure 1. The broad band (0.5–10 keV) *ASCA* SIS0 images of 3C391. Panel *a* is the hard (2.6–10 keV) emission gray-scale image overlaid with the soft (0.5–2.6 keV) emission contours. The three circles in panel *a* designate the areas from which the SIS spectra are extracted. Panels *b* and *c* display the gray-scale images of soft and hard emission, respectively, overlaid with the dashed contours of 1.5 GHz radio emission. Panels *d*, *e*, and figure*f* are the narrow band gray-scale images and solid contours of Mg Heα (1.2–1.5 keV), Si Heα (1.7–2.0 keV), and S Heα (2.3–2.6 keV) emissions. In panel *f* the image is also overlaid with the dashed radio contours. The seven levels of solid contours are linear between the maximum and 20% maximum brightness. The two cross labels in each panel mark the OH maser points (Frail et al. 1996).

Figure 1 displays the X-ray images of 3C 391 using *ASCA* SIS0 data. The hard X-ray image (Fig. 1a,c) looks relatively bright in the NW compared with that in the SE, in contrast to the soft X-ray image which looks faint in the NW. This fact can be explained by increased extinction from the NW cloud and is consistent with a model in which the gas in the SE region has broken out into a lower density environment.

The X-ray spectra of 3C 391 show prominent Mg Heα (1.35 ± 0.01 keV), Si Heα (1.85 ± 0.01 keV), and S Heα (2.46 ± 0.02 keV) lines, and can be fit with a model of an optically thin thermal plasma. The metal abundances yielded are essentially consistent with solar values and thus seem to be consistent with an interstellar composition. The hot plasma in the remnant is basically in ionization equilibrium (the ionization parameter is $n_e t > 8 \times 10^{12}\,\mathrm{cm^{-3}s}$). A spectro-spatial analysis gives similar temperatures ($kT_\mathrm{x} \sim 0.5$ keV) for the NW and SE regions, but higher N_H for the NW than for the SE (with a difference $\sim$ 4–$6 \times 10^{21}\,\mathrm{cm^{-2}}$). The variation of the hydrogen column density across the remnant is in agreement with the presence of a molecular cloud to the NW. This variation implies a mean ambient molecular density of order $\sim$ 10–20 $\mathrm{cm^{-3}}$. Adopting a distance $d = 8d_8$ kpc, the X-ray luminosity is $L_\mathrm{x}(0.5\text{–}10.0\,\mathrm{keV}) \sim 2.7 \times 10^{36} d_8^2$ ergs $\mathrm{s^{-1}}$. The mean hot gas density is $\sim 1.5 f^{-1/2} d_8^{-1/2}\mathrm{cm^{-3}}$ and the X-ray emitting mass is $\sim 92 f^{1/2} d_8^{5/2} M_\odot$, where f is the filling factor of the hot plasma. The lower density within the SNR, as compared with the ambient cloud density, and the centrally brightened X-ray morphology can be explained either by the evaporation of engulfed cloudlets or by a radiative stage of evolution for the remnant. Using a mean radius $7d_8$ pc, the evaporation model yields a remnant age $(4.0\text{–}4.6) \times 10^3 d_8$ yr and an explosion energy $E \sim (1.3\text{–}3.4) \times 10^{50}$ ergs. However, at least the NW part of the remnant may have already approached the radiative phase, this model yielding a remnant age $\sim 1.9 \times 10^4 (E/10^{51}\,\mathrm{ergs})^{-31/42} d_8^{10/3}$ yr.

3. SNR G349.7+0.2

G349.7+0.2 is the Galactic SNR with the third highest radio surface brightness next to Cas A and the Crab Nebula. Its non-thermal radio emission with a roughly circular morphology (as well as a hint of beak-out to the NW), classifies it as a shell-type SNR (Shaver 1985). However, it has an emission peak near the southeastern edge rather than a prominent limb-brightened structure and central cavity typical of shell–type remnants. OH (1720 MHz) masers have been recently found interior to G349.7+0.2 with radial velocities $\sim +16\,\mathrm{km\,s^{-1}}$ indicating a kinematic distance $d \approx 22.4$ kpc (Frail et al. 1996). CO observations toward G349.7+0.2 revealed a molecular cloud associated with the SNR, and with the masers (Reynoso & Mangum 2001).

As shown in Figure 2, the X-ray morphology is consistent with that at radio wavelengths, with a distinct enhancement in the south where the remnant is interacting with a molecular cloud. The X-ray spectrum is dominated by line features from Si Heα ($1.86^{+0.02}_{-0.01}$ keV), S Heα ($2.42^{+0.02}_{-0.01}$ keV), and Fe Kα ($6.53^{+0.09}_{-0.10}$ keV). The X-ray emission from the SNR is well described by a model of a thermal ($kT_\mathrm{x} \sim 1.3$ keV) plasma which has yet to reach ionization equilibrium ($n_e t \sim 1.7^{+0.5}_{-0.3} \times 10^{11}\,\mathrm{cm^{-3}s}$). The hydrogen column of $\sim 6 \times 10^{22}\,\mathrm{cm^{-2}}$ is consistent with

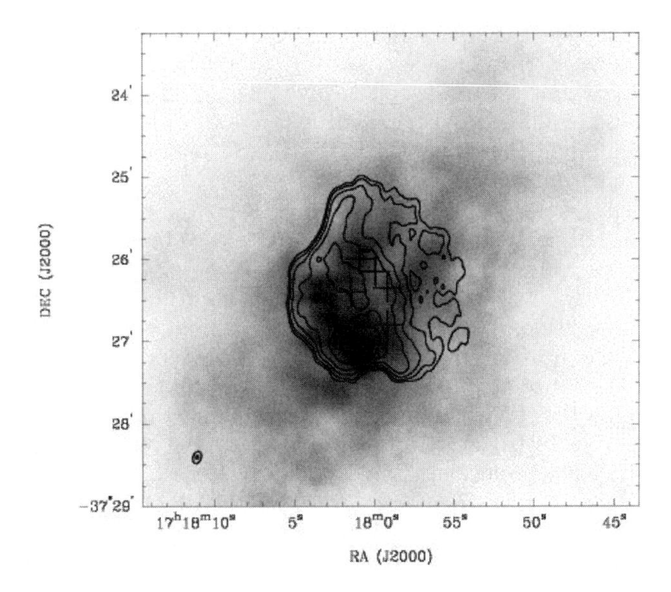

Figure 2. The broad band (1.0–10.0 keV) *ASCA* SIS image of G349.7+0.2. Contours are from the ATCA 18-cm continuum image, with levels at 12, 17, 34, 68, 136, 204 and 272 mJ beam $^{-1}$. Crosses denote the OH maser points.

a substantial absorption below $\sim$ 1.5 keV and the large distance to the remnant of $\sim$ 22 kpc estimated from the maser velocities. The normal solar abundances of the hot plasma indicates that it is dominated by the swept-up interstellar material. We derive an X-ray luminosity of L_x(0.5-10.0keV) $\sim 1.8 \times 10^{37}$ ergs s^{-1}, which makes G349.7+0.2 one of the most X-ray luminous shell-type SNRs known in the Galaxy. The mean hydrogen density of the hot gas is estimated to be $\sim 4.2f^{-1/2}\,\mathrm{cm}^{-3}$, the X-ray emitting mass is $\sim 160f^{1/2}M_{\odot}$, the age of the remnant is estimated to be $\sim$ 2800 yr, and the explosion energy is $E \sim 5 \times 10^{50}f^{-1/2}$ ergs.

The C-shock condition for the OH (1720 MHz) maser emission gives an upper limit of thermal gas pressure of maser portions ($\sim 9 \times 10^{-9}$ ergs cm^{-3}) that is lower than the pressure of the hot plasma ($\sim 1.6 \times 10^{-8}f^{-1/2}$ ergs cm^{-3}). The discrepancy implies that there must be a contribution from the magnetic pressure and/or cosmic ray pressure in the shocked molecular gas.

The small radius and high X-ray luminosity make G349.7+0.2 to some extent be a Galactic counterpart to SNR N132D in the Large Magellanic Cloud (Hughes 1987). The ambient density and pressure conditions appear similar to those inferred for luminous compact SNRs found in starburst regions of other galaxies, and provides support for the notion that these may be the result of SNR evolution in the vicinity of dense molecular clouds.

References

Chen, Y. & Slane, P. 2001, ApJ, 563, 202
Frail, D. A. et al. 1996, AJ, 111, 1651
Green, A. J., Frail, D. A., Goss, W. M., & Otrupcek, R. 1997, AJ, 114, 2058
Hughes, J. P. 1987, ApJ, 314, 103
Reach, W. T. & Rho, J. H. 1999, ApJ, 511, 836
Reynolds, S. P. & Moffett, D. A. 1993, AJ, 105, 2226
Reynoso, E. M. & Mangum, J. G. 2001, AJ, 121, 347
Rho, J. H. & Petre, R. 1996, ApJ, 467, 698
Shaver, P. A. et al. 1985, Nature, 313, 113
Slane, P., Chen, Y., Lazendic, J. S. & Hughes, J.P. 2002, ApJ, accepted
Wilner, D. J., Reynolds, S. P. & Moffett, D. A. 1998, ApJ, 115, 247

IAU 8th Asia-Pacific Regional Meeting
ASP Conference Series, Vol. 289, 2003
S. Ikeuchi, J. Hearnshaw & T. Hanawa, eds.

Time-Dependent Properties of Black-Hole Accretion Flow

S. Mineshige

Yukawa Institute, Kyoto University, Sakyo-ku, Kyoto 606–8502, Japan

Abstract. Time-dependent structure and emission properties of black-hole hot accretion flow are discussed from the theoretical point of view. Special attention is paid to the roles of magnetic fields in the formation of high-energy spectra and variability.

1. Introduction

Accretion onto black holes (BHs) is one of the most important keys to understanding the various activities and formation processes of astrophysical objects. Regarding the dynamics of hot accretion flow, important theoretical and observational developments have been made in the past decade, although there still remain fundamental questions, such as the structure of hot accretion flow, the role of magnetic fields, jet formation mechanisms, and so on. Especially, recent theoretical efforts are rather focused on the role of convection, outflow, and magnetic fields in forming disk structure, spectra, and variability.

Nearly three decades have passed since the so-called standard disk model was proposed by Shakura & Sunyaev (1974). This model is very successful in describing the properties of the flow during the soft state of Galactic BH candidates (GBHCs), while it cannot account for the hard power-law X-ray emission from GBHCs in the hard state nor AGNs. There should be another type of accretion flow in hot, optically thin regimes. Here, 'hot' accretion flows refer to the ones which seem to appear in the adiabatic (cooling-inefficient) regimes of flow. Note that such hot flow contrasts with the standard disk, in which radiative cooling is substantial. New models describing hot accretion flow had been anticipated and this is a reason why the idea of the advection-dominated accretion flow (ADAF) by Narayan & Yi (1995) has attracted so much attention among researchers (see Kato, Fukue, & Mineshige 1998 for a review). However, the original ADAF model now faces serious problems.

2. Various Models for Hot Accretion Flow

The ADAF model is constructed in a vertically one-zone approximation and the resultant differential equations with respect to r (radial distance to the center) are solved either by the self-similar technique or numerically. Recently made 2D/3D, adiabatic hydrodynamical simulations, however, revealed distinct forms of accretion flows. Igumenshchev & Abramowicz (2000), for example, claimed that nearly one-dimensional ADAF appears only when the viscosity parameter

Table 1. ADAF, CDAF, and MHD flow

accretion mode	$\rho(r)$	$T(r)$	$v_r(r)$	$B^2(r)$
ADAF	$\propto r^{-\frac{3}{2}}$	$\propto r^{-1}$	$\propto r^{-\frac{1}{2}}$	...
outflow	$\propto r^{-1}$	$\propto r^{-1}$	$\propto r^{-1}$	...
CDAF	$\propto r^{-\frac{1}{2}}$	$\propto r^{-1}$	$\propto r^{-\frac{3}{2}}$	...
MHD flow	$\propto r^{-0.5}$	$\propto r^{-1.0}$	$\propto r^{-1.5}$	$\propto r^{-1.5}$

α is moderate, $0.01 < \alpha \leq 0.1$. When α is smaller, large-scale circulation or convection occurs and it largely modifies the flow structure. For example, the density profile is significantly flatter than in ADAF (see table 1). Such flow is known as convection-dominated accretion flow (CDAF). If α is large ($\alpha \sim 1$), on the other hand, strong outflows result. Again, the density profile is distinct, slightly flatter than in ADAF (table 1).

The critical test of these flows are to examine their spectral properties. Figure 1 displays the typical ADAF (left) and CDAF (right) spectra calculated in a similar way to that of Ball, Narayan, & Quataert (2001). Distinct spectra of these flow models are due to their different density profiles (see table 1): they are $\rho \propto (r/r_{\rm S})^{-p}$ with $p = 1.5$ (ADAF) and $p = 0.5$ (CDAF), respectively, In this plot, we took the same proportionality constant, ρ_0, determined by the adopted mass-accretion rate (see below), $\dot{M}$, in the case of ADAF also in CDAF, so, figure 1 can be used only for demonstration purpose. The other adopted parameters are: the inner-edge radius is 1.0 $r_{\rm S}$ (with $r_{\rm S}$ being the Schwarzschild radius), the outer-edge radius is $10^{2.5}$ $r_{\rm S}$, the mass-flow rate is $\log \dot{m} \equiv \log(\dot{M}c^2/L_{\rm E}) = -3.29$ (with $L_{\rm E}$ being the Eddington luminosity), the mass of black hole is $M_{\rm BH} = 10^8 M_\odot$, the ion temperature profile is $10^{12}(r/r_{\rm S})^{-1.0}$ K, the electron temperature profile is $10^{10}(r/r_{\rm S})^{-0.6}$ K, and magnetic field strengths are taken to be the equipartition values; i.e., $B^2 \propto (r/r_{\rm S})^{-(1+p)}$. Note that the results are rather insensitive to the black-hole masses.

In the case of ADAF, emission from the innermost ring dominates over the contribution from the outer parts at all wavelengths because of its steeper density profiles. Note that the density profile of $\rho \propto r^{-3/2}$ leads to the bremsstrahlung emissivity of $dE \propto \rho^2 T^{1/2} r^2 dr \propto r^{-1.5} dr$. In the case of CDAF, in contrast, X-ray emission comes mainly from the outer parts due to its somewhat flatter density profile ($\rho \propto r^{-1/2}$ yields $dE \propto r^{0.5} dr$). This is well demonstrated in figure 1. We also understand that bremsstrahlung is a dominant process in CDAF even in X-rays, while Compton up-scattering of synchrotron emission is more important in ADAF.

The CDAF picture does not match the observations of GBHCs during their hard state in two senses: (1) CDAF predicts a rather flat ($f_\nu \propto \nu^0$) spectrum, unless we assume significant electron heating as well as ion heating, but the observations clearly show a power-law decline. (2) Rapid variability seems to arise via time-dependent processes, probably associated with magnetic flares in the innermost region, but in CDAF most of X-ray emission is from the outer

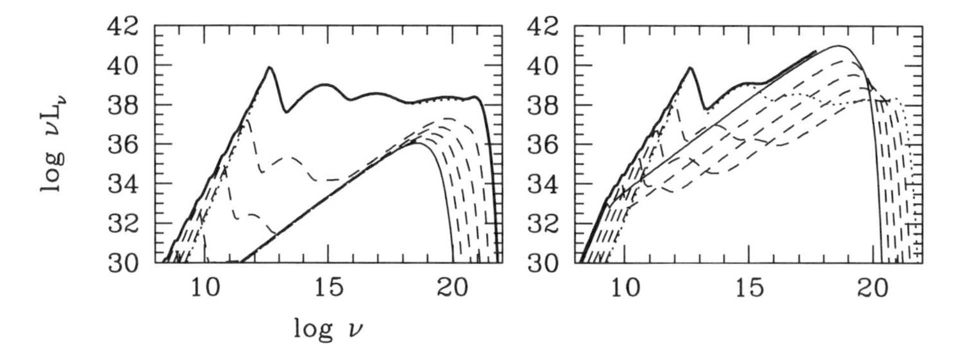

Figure 1. Typical ADAF (upper left) and CDAF (upper right) spectra, together with that of MHD flow (lower). For model parameters, see text. The thick solid line represents the total spectra, while thin solid, dashed, and dotted lines represent the contributions from the outermost, middle, and innermost rings, respectively. The MHD flow spectrum is more like that of CDAF.

part, where rotation is slow, thus being able to produce rapid variability. To fit the observations, the ADAF picture is better.

Hydrodynamical simulations are, however, of limited use here, since the magnitude of the viscosity (or magnetic field strengths) is treated as a free parameter there, although it should be determined in accordance with magnetic field amplifications within the disk. In other words, magneto-hydrodynamical (MHD) simulations are indispensable.

3. MHD Accretion Flow

There are various roles of magnetic fields in accretion flows, which include

- source of viscosity,
- disk corona (and ADAF) heating,
- cause of flares, producing variability,
- source of radiation via synchrotron,
- jet & outflow formation.

These processes have been successively simulated through global MHD simulations of the inner accretion disk (e.g. Machida, Hayashi, & Matsumoto 2000; Hawley & Krolik 2001). They have found that magnetic fields can be amplified sufficiently enough to explain the observations; the estimated viscosity parameters are in a range of $0.01 < \alpha < 1.0$. A large corona can be constructed by MHD processes. Also, it is possible to produce aperiodic variability, which seems to be closely related to spatially inhomogeneous (probably fractal) structure.

There is a good reason to believe that MHD simulations are indispensable to understand the hot accretion flow. Let us compare the standard-type cool disk and hot disk. In the former, gravitational energy is efficiently converted to radiation energy. On the other hand, magnetic field energy is suppressed below gas energy (but not by many orders below), since otherwise magnetic fields will leave the system in forms of bubbles. We thus obtain an equality,

$$E_{\rm mag} \le E_{\rm gas} \ll E_{\rm grav} \sim E_{\rm rad}. \tag{1}$$

This situation is just like the solar photosphere. From the observational point of view, therefore, magnetic fields are not strong enough to affect the dynamical flow structure.

In hot accretion flow and corona, in contrast, gravitational energy turns into fluid energy with little being radiated away. Again, magnetic energy is comparable to (but a bit less than) the gas energy. Thus, we have an equality,

$$E_{\rm mag} \le E_{\rm gas} \sim E_{\rm grav} \gg E_{\rm rad}. \tag{2}$$

The situation is more like X-ray images of the solar corona, in which magnetic fields are known to form filamentary (or loop) structures emerging from below the photosphere. This is the main reason why magnetic fields are likely to produce observable effects in the hard state of BHs.

We are thus motivated to analyze the flow patterns within the simulated 3D MHD flow (Machida, Hayashi, & Matsumoto 2000), finding, however, large-scale convective motions dominating in accretion flows (Machida, Matsumoto, & Mineshige 2001). Further, we also noticed the density and temperature profiles of MHD flow similar to those of CDAF (see table 1). That is, the spectra should be more like that of CDAF (see figure 1) and not like that of ADAF, to our disappointment. Then, such MHD flow models cannot account for the observed spectra nor presence of rapid variability. Then, what is a loophole?

Here we conjecture that the dissipated magnetic energy by reconnection can partly go directly to radiation. So far, all the MHD simulations postulate no radiative cooling (since we are now concerned with the adiabatic accretion regimes), but if dissipated energy could be radiated away by enhanced emissivity as a consequence of magnetic reconnection, entropy production in electrons would be largely reduced, leading to a suppression of convective motions. Then, MHD flow structure would be more like that of ADAF, thus explaining both of the spectral shape and the presence of X-ray variability in BHs. This is still a hypothesis, thus requiring further study.

References

Ball, G. H., Narayan, R. & Quataert, E. 2001, ApJ, 552, 221

Hawley, J. F. & Krolik, J. H. 2001, ApJ, 548, 348

Igumenshchev, V. & Abramowicz, M. A. 2000, ApJS, 130, 463

Kato, S., Fukue, J. & Mineshige, S. 1998 *Black-Hole Accretion Disks* (Kyoto Univ. Press, Kyoto), Chap. 10

Machida, M., Hayashi, M. R. & Matsumoto, R. 2000, ApJ, 532, L67

Machida, M., Matsumoto, R. & Mineshige, S. 2001, PASJ, 53, L1

Narayan, R. & Yi, I. 1995, ApJ, 452, 710

Shakura, N. I. & Sunyaev, R. A. 1973, A&A, 24, 337

IAU 8th Asia-Pacific Regional Meeting
ASP Conference Series, Vol. 289, 2003
S. Ikeuchi, J. Hearnshaw & T. Hanawa, eds.

Results in Gamma-Ray Burst Observations and the Use of GRBs to Explore the Early Universe

Toshio Murakami, Hisayo Izawa, Daisuke Yonetoku

Faculty of Science, University of Kanazawa, Kakuma, Kanazawa, Ishikawa 920-1192, Japan

Abstract. The most recent results and puzzles in Gamma-Ray Burst observations (GRB), except the HETE-II are shortly reviewed in this paper. We focus on the use of GRBs to explore the early universe. We found a very weak correlation between the radiation hardness and the known GRB distances in the BATSE catalogue. Using a relation in the correlation, we give distances to the unknown bright 344 BATSE GRBs and estimate a star formation rate (SFR) up to $z \sim 5$. We compare our result with the most popular SFR of Madau et al. (1996) and the pioneering work of Fenimore et al. (2000).

1. Discovery of an X-ray Afterglow

Since the discovery of a fading X-ray afterglow with BeppoSAX in 1997 from GRB 970228 (Costa et al. 1997, Piro et al. 2001), Optical Transients (OTs) were also discovered simultaneously with the fading X-ray afterglows (Paradijs et al. 1997, 2000). Based on the observed OT redshifts, the distances to GRBs are now firmly established as being cosmological. In-fact, the mean of the 25 known redshifts is $z \sim 1$. A good review in the era of X-ray afterglows can be found in the paper by Paradijs et al. (2000). Most OTs were localized in the disk of a host galaxy, and thus these locations lead us to believe that the progenitors of the GRBs have a disk population and may be a collapse of a massive star in a host galaxy at cosmological distances (Djorgovski et al. 1998). Intensities of the X-ray and OT afterglows faded in a power-law with time, and thus these decay slopes strongly support a fireball scenario which is presented as relativistically expanding internal and/or external shocks to the ambient matter. However the production mechanism of a fireball is completely unknown (Rees & Mészáros 1992, Mészáros & Rees 1997, Woosley 1999, Piran 2001).

2. Still Many Puzzles

The most enigmatic mystery, namely the locations of GRBs, is now solved, following the discovery of the fading X-ray afterglows. The GRBs and the OTs are really at cosmological distances. However there are still a few questions or puzzles to be solved to understand the GRB phenomenon. The most serious one is a progenitor and/or production mechanism of the fireballs. These are still completely unknown. Even if a formation of a black-hole is related to the

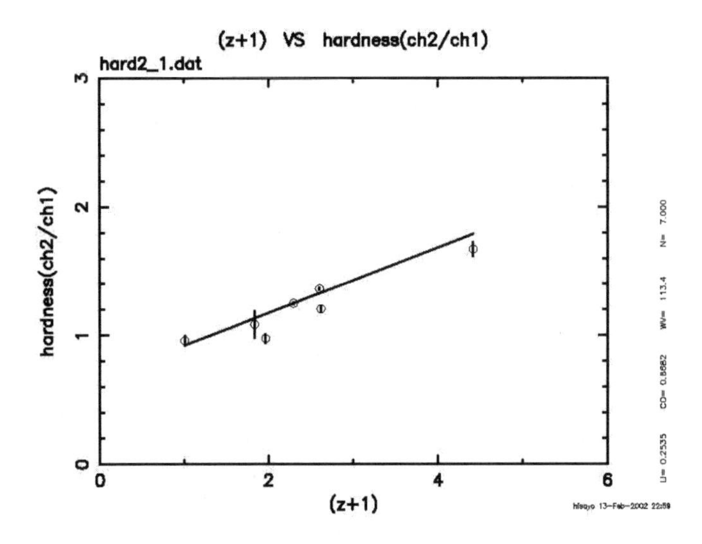

Figure 1. Correlation between the hardnesses (CH2/CH1) in spectra of the BATSE archive and the known redshifts. The number of GRBs with redshift is only seven in the BATSE catalogue. We assume a straight line to our fit.

GRBs, the production mechanism of an extremely relativistic mass ejection from a black-hole is unknown (Woosley 1999). The second puzzle is that two classes of GRBs are known (Kouveliotou et al. 1993). One is a class with longer durations of bursts whose durations are much longer than several seconds and the mean is about 30 s. The short class is less than 1 s in duration and the mean is nearly 0.1 s. The spectra of the shorter class are a little bit harder than the longer class (Kouveliotou et al. 1993). However the shorter class of GRBs have not been localized yet and thus no OT has been identified.

Are the two distinct classes the same in origin? The relation of the shorter to the longer GRBs is not known. The detection and the localization of short class GRBs with HETE-II is a key project.

Finally, the most recent puzzle is that ASCA and BeppoSAX observed evidence of a heavy iron emission line in the fading X-ray afterglows, which is ionized to the hydrogen-like state (Piro et al. 1998, 2000, Yoshida et al. 2001, Yonetoku et al. 2001, Murakami 2001). However XMM (Newton) recently reported only the light elements except iron (Reeves et al. 2002). Moreover, two independent groups opposed the Newton results and reported no detection of the lines from the same data-set (Borozdin & Trudolybov 2002, Rutledge & Sako 2002). Since the heavy elements are the key to accept a massive progenitor and the star forming environment, the conflict should be solved.

3. GRBs to Explore the Early Universe

During the six years of operation of BeppoSAX, 25 X-ray afterglows were observed among the 30 detected and roughly 12 OTs were measured for their redshifts with the mean of $z \sim 1$. The most distant one was at $\sim$4.5 of GRB 000131 (ESO 2000). The BATSE detector on-board Compton was much more sensitive in detecting the weaker GRBs because of the larger collection area than BeppoSAX. So BATSE should have more distant GRBs in the catalogue. In spite of the more sensitive BATSE detectors, the number of known distances of GRBs is much more limited than those from BeppoSAX. In fact, the known redshifts in the BATSE data with parameters such as an intensity and a fluence are only 7. We intend to give distances to all BATSE GRBs.

If there was a relation in the observed BATSE parameters such as the hardness and the known distance, we can estimate and give distance to the rest of the GRBs. We have done searches of many combinations of parameters to find an empirical relation and finally found one correlation whose origin is not understood. The archive used was the 4th BATSE data, which is open to be public from the site (BATSE site 2002). Figure 1 shows the correlation, we found. There are only 7 data points, but there is a very weak correlation between the hardnesses of the fluence in CH2/CH1 and the 7 known redshifts (table: Greiner 2002). Although the number of GRBs and the significance of the fit are very much limited, a probability of the correlation by chance is 0.15%. Assuming a straight line of $\mathrm{HR} = 0.254(z+1)+0.668$, where HR is the hardness ratio in CH2/CH1 and z is the redshift, we give distances to the BATSE GRBs using the formula.

First, we selected 344 GRBs from the BATSE catalogue. These GRBs are rather bright with 0.1 cts/s in the catalogue and the longer duration class. These GRBs have good S/N ratios to minimize the errors in estimating the redshifts. Using the empirical formula shown above, we gave redshifts to the 344 GRBs and the results are shown in Figure 2. It is clear from the figure that there are a few bursts which have redshift of more than $z = 10$. Several GRBs with a redshift of less than $z = 0$ are the result of statistical fluctuations of the hardness ratio, with mainly poor S/N. Much weaker GRBs were not used in this analysis, because although the weaker bursts should be more distant, the rather poor S/N ratios impede estimating the hardness.

Next we estimated the burst rates of GRBs or relative star formation rates (SFR) for the data set with different distance of the five groups in units of comoving volume of a standard universe with the cosmological parameters of $\Omega_\lambda = 0.7$ and $\Omega_m = 0.3$. The method used to estimate the SFR is the same as the one used by the pioneering work of Fenimore et al. (2000). The derived GRB rates are plotted in Figure 3 together with the results of Madau et al. (1996) and Fenimore et al. (2000) normalizing our data at $z \sim 1$ with others. Because we cannot derive the true SFRs in this analysis, the result is therefore a relative value.

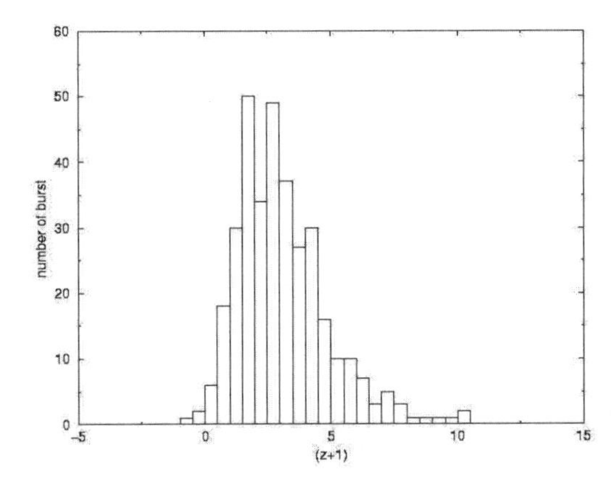

Figure 2. Histogram of number of events vs. derived redshifts from the relation in Figure 1. There are 344 GRBs used in this plot. There are a few GRBs with $z \gtrsim 10$.

4. Discussion

The present methods by us and by Fenimore et al. (2000) to explore the early universe using the GRBs rate look fruitful and advantage compared with other wave bands such as the radio, optical and ultra-violet. In gamma-rays, there is no absorption by a host and also the intergalactic matter, thus we can explore the deeper and the younger universe. Especially the GRBs are the brightest observable events which enable us to observe the deepest universe. It is clear from Figure 3, the relative SFR derived from the GRBs does not show saturation beyond $z \sim 2$ unlike Madau's result (Madau et al. 1996). In this report, we do not estimate and extrapolate the SFR to more than $z \geq 5$, but the GRBs have a capability to explore the SFR to more than $z \geq 10$. In fact, in the analysis of the Fenimore et al. (2000), they derived the SFRs up to z $\sim$ 10 based on the variability-luminosity relation in the observed redshifts.

In the previous section, we mentioned that the origin of the observed correlation between the hardnesses and the redshifts cannot be understood. However a similar relation between the spectral slopes and the redshifts and/or the peak energy, $E_{\rm p}$, and the redshifts was found independently using the BeppoSAX data and reported by Amati et al. (2002). This encourages the validity of our results. The origin of this correlation should be understood.

In 2002 and 2003, more powerful gamma-ray satellites such as Integral and Swift will be launched and will distribute the locations of GRBs within several seconds of detection. This will give us a more complete sample of distant GRBs and enables us to explore the much earlier universe of more than redshift 10. At present and within 10 years, the measurement of distance using the redshifted

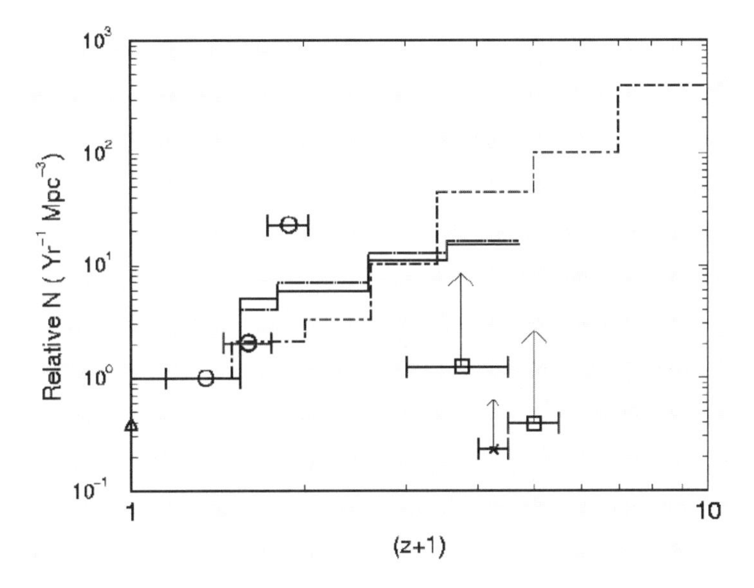

Figure 3. Relative SFR vs redshifts together with the previous works of Madau et al. (1996) and Fenimore et al. (2000). The bars with an open circle and an open rectangle are by Madau and the broken line histogram is by Fenimore. The solid histogram up to $z = 5$ is ours. There are no turn-overs in the present result and Fenimore's result which used the GRBs. This is the advantage in using GRBs compared with the UV observations.

X-ray iron line of GRBs and the detections of a redshifted Lyman-alpha break in the near infrared band of GRBs look like being the only methods to measure redshifts of more than 10. These methods are of key importance to explore the earliest universe before the start of observations with ALMA in the sub-mm wave band and Xeus in X-rays, which are now in preparation.

Acknowledgments. We thank the support of grants by the Japanese Ministry of Education, Culture and Sports (12640302, 14204024). This work was done under the contracts of these grants.

References

Amati, L. 2002, Astro-ph/0205230

Borozdin, K. N. & Trudolybov, S. P. 2002, Astro-ph/0205208

BATSE data, 2002, http://gammaray.nsstc.nasa.gov/batse/grb/catalog/

Costa, E. 1997, Nature, 387, 783

Djorgovski S. G. et al. 1998, ApJ, 508, L17

ESO press release 2000, 20/00

Fenimore, E. E. & Ramirez-Ruiz, E. 2000, Astro-ph/0004176

Greiner's site http://www.aip.de/People/JGreiner/grbgen.html

Kouveliotou, C. et al. 1993, ApJ, 413, L101

Madau, P. et al. 1996, MNRAS, 283, 1388

Mésárous, P. & Rees, M. 1997, ApJ, 476, 232

Murakami, T. et al. 2001, GRBs in the Afterglow Era, edited by E. Cost, F. Frontra and J. Hjorth, ESO-Springer, 115

Piran, T. & Granot, J. 2001, GRBs in the Afterglow Era, edited by E. Cost, F. Frontra and J. Hjorth, ESO-Springer, 300

Piro, R. et al. 1998 A&A, 331, L41

Piro, R. et al. 2000, Science, 290, 955

Piro, L. et al. 2001, GRBs in the Afterglow Era, edited by E. Cost, F. Frontra and J. Hjorth, ESO-Springer, 97

Rees, M. & Mézárous, P. 1992, MNRAS, 258, L41

Reeves, J. N. 2002, Nature, 416, 512

Rutledge, R. E. & Sako, M. 2002, Astro-ph/0206073

van Paradijs, J. et al. 1997, Nature, 386, 686

van Paradijs J. et al. 2000, ARA&A, 38, 379

Woosely et al. 1999, ApJ, 516, 788

Yoshida, A. et al. 2001, ApJ, 557, L27

Yonetoku, D. et al. 2001, ApJ, 557, L23

IAU 8th Asia-Pacific Regional Meeting
ASP Conference Series, Vol. 289, 2003
S. Ikeuchi, J. Hearnshaw & T. Hanawa, eds.

A Deep Optical Observation of an Enigmatic Unidentified Gamma-Ray Source 3EG J1835+5918

Tomonori Totani

Princeton University Observatory, Peyton Hall, Princeton, NJ08544-1001, USA and Theory Division, National Astronomical Observatory, Mitaka, Tokyo 181-8588, Japan

Wataru Kawasaki

Department of Astronomy, The University of Tokyo, 7-3-1, Hongo, Bunkyo-ku, Tokyo, 113-0033, Japan

Nobuyuki Kawai

Department of Physics, Tokyo Institute of Technology, 2-12-1, Ookayama, Meguro, Tokyo 152-0033, Japan

Abstract. We report on a deep optical imaging observation using the Subaru telescope for a very soft X-ray source, RX J1836.2+5925, which has been suspected to be an isolated neutron star associated with the brightest as-yet unidentified EGRET source outside the Galactic plane, 3EG J1835+5918. An extended source having a complex, bipolar shape is found at $B \sim 26$, which might be an extended pulsar nebular whose flux is about 5–6 orders of magnitude lower than gamma-ray flux, although finding a galaxy of this magnitude by chance in the error circle is of order unity. We have found two even fainter, possible point sources at $B \sim 28$, although their detections are not firm because of a low signal-to-noise ratio. If the extended object of $B \sim 26$ is a galaxy and not related to 3EG J1835+5918, a lower limit on the X-ray/optical flux ratio can be set as $f_X/f_B \gtrsim 2700$, giving further strong support for the neutron-star identification of 3EG J1835+5918. Interestingly, if either of the two sources at $B \sim 28$ is the real counterpart of RX J1836.2+5925 and thermal emission from the surface of an isolated neutron star, the temperature and distance to the source become $\sim 4 \times 10^5$ K and ~ 300 pc, respectively, showing a striking similarity of its spectral energy distribution to the proto-type radio-quiet gamma-ray pulsar Geminga. No detection of nonthermal hard X-ray emission is consistent with the ASCA upper limit, if the nonthermal flux of 3EG J1835+5918/RX J1836.2+5925 is at a similar level to that of Geminga.

1. Introduction

Over half of the GeV gamma-ray sources detected by the EGRET experiment have not yet been identified as known astronomical objects (Hartman et al.

1999), and understanding their origin is one of the most important issues in high-energy astrophysics. Many of the unidentified sources located in the Galactic plane are believed to be associated with either pulsars, supernova remnants, or massive stars, as suggested by statistically significant correlation between unidentified EGRET sources and tracers of these objects (see, e.g., Romero 2001 for a review). Another Galactic population of unidentified EGRET sources at intermediate Galactic latitude ($|b| \lesssim 40°$) has been reported by Gehrels et al. (2000), which are apparently associated with the Gould belt; they might be off-beam gamma-ray pulsars (Harding, Zhang 2001). There are also unidentified sources at even higher latitude ($|b| \gtrsim 40°$), suggesting the extragalactic origin. Variable sources among them are likely to be undetected blazars, while some stable sources might be dynamically forming or merging clusters of galaxies (Totani, Kitayama 2000; Waxman, Loeb 2000; Kawasaki, Totani 2002).

One of the unidentified EGRET sources, 3EG J1835+5918, has been paid particular attention in recent years, since it is the brightest unidentified source outside the Galactic plane ($l = 88.7°$, $b = 25.1°$). Its gamma-ray properties, i.e., steady flux and hard spectrum are well consistent with other gamma-ray pulsars observed by EGRET (Reimer et al. 2001), rather than blazars. No strong radio counterpart also argues against a blazar origin. Intensive multi-wavelength studies of this source (Mirabal et al. 2000; Mirabal, Halpern 2001; Reimer et al. 2001) have revealed about a dozen X-ray sources in or around the error circle of 3EG J1835+5918, and most of them are coronal emission from stars or quasars without any special or blazar-like characteristics, which are not likely the counterpart of 3EG J1835+5918. Then, the only one unidentified X-ray source, RX J1836.2+5925, remains as the most likely counterpart. It has a very soft X-ray spectrum ($T < 5 \times 10^5$ K) and has no optical counterpart down to $V \sim 25$ (Mirabal, Halpern 2001), giving a high X-ray/optical ratio. This and the lack of strong radio emission are reminiscent of the characteristics of isolated neutron stars or radio-quiet pulsars (e.g., Brazier, Johnston 1999; Neuhäuser, Trümper 1999), which is why both Mirabal & Halpern (2001) and Reimer et al. (2001) concluded that the most likely identification of 3EG J1835+5918 / RX J1836.2+5925 is an isolated radio-quiet pulsar, which is similar to the proto-type object Geminga.

However, with the X-ray flux of $(2\text{–}6)\times 10^{-13}$ erg cm^{-2}s^{-1} (Mirabal, Halpern 2001), the optical upper limit ($V > 25.2$ at 3σ) corresponds to a X-ray/optical flux ratio of $f_X/f_V \gtrsim 300$, and is not large enough compared with the flux ratio expected for isolated pulsars or neutron stars, although all X-ray sources other than isolated neutron stars and low-mass X-ray binaries have $f_X/f_V \lesssim 80$ (Stocke et al. 1991). Here, we report on a deep optical imaging observation of 3EG J1835+5918 / RX J1836.2+5925 by the 8.2-m Subaru telescope to examine the proposed identification and the origin of this enigmatic source.

2. The Subaru Observation

The observation was made on 2001 July 17, by the Subaru/FOCAS, whose field of view is a circle with $6'$ diameter. We took images in the B and U bands, with total exposure times of 2400 and 6000 s, respectively. The 5σ peak level corresponds to $B = 27.5$ and $U = 25.9$. Figure 1 shows the B band

Table 1. Result of Observation around RX J1836.2+5925

object	profile	α(J2000)	δ(J2000)	B mag
A	point	18h36m13.49s	59d25m30.1s	$29.1^{+1.9}_{-0.7}$
B	point	18h36m13.90s	59d25m29.1s	$28.3^{+0.6}_{-0.4}$
C	extended	18h36m13.87s	59d25m31.6s	
D	extended	18h36m13.93s	59d25m32.0s	
C+D				26.1 ± 0.1

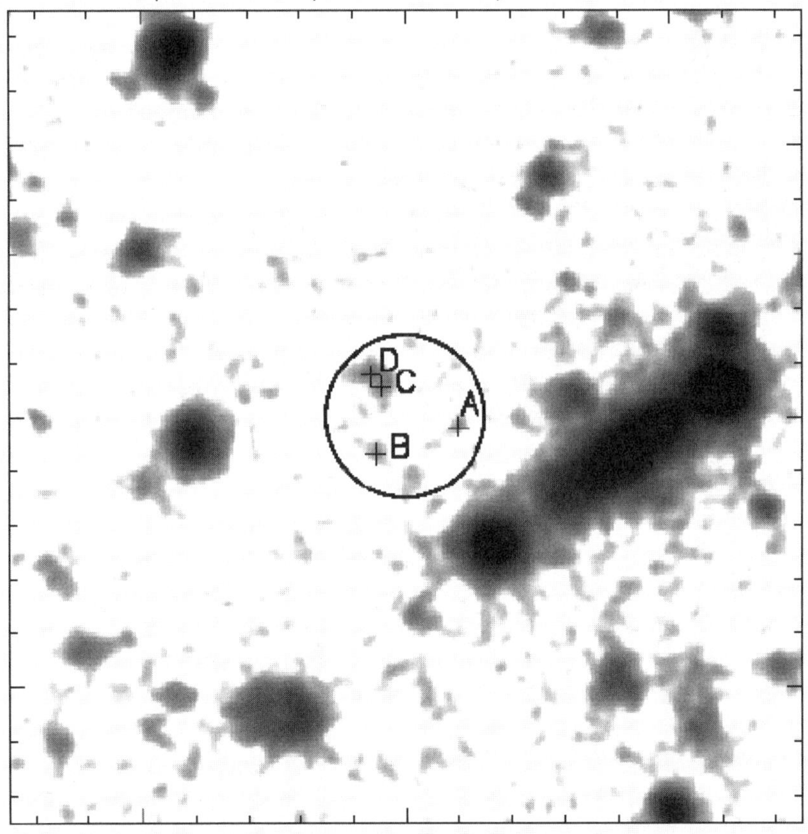

Figure 1. The $30'' \times 30''$ image in the B band around the region of RX J1836.2+5925 whose error circle is shown at the center of the figure. The four objects found in the circle are indicated as A–D. North is up.

images centered on the position of RX J1836.2+5925 determined by Mirabal and Halpern (2001) with the error radius of $3''$ (J2000, $18^h36^m13^s.77$, $+59°25'30''.4$). The Galactic extinction to this direction is $A_B = 0.22$ and $A_U = 0.28$ (Schlegel et al. 1998) giving upper bounds on extinction in any Galactic objects. The seeing size in the B band is $0.9''$ FWHM. On the other hand, the quality of U band image is not as good as the B band and no object is found in the error circle; we thus set a rough upper limit of $U < 24.8$, including systematic uncertainty of photometry.

In the B band image we detected two extended sources connected with each other (labelled as C and D in the images). Their peak are 5.1 and 5.8σ of the sky fluctuation level, respectively. We found two point-like sources with peak levels higher than 3σ levels, which are labelled as A and B in the images. The results of photometry as well as the source locations for the above four objects are summarized in Table 1. In the following of this paper, we will refer to the sources C and D as one source (C+D). We checked that the noise fluctuation is symmetric above and below the zero point of the background, and we did the same detection procedure for the inverted image to estimate the number of spurious objects. This indicates that sources A and B are not firm detections; the probabilities that they are spurious detection are estimated to be ~ 60 and 30%, respectively, from the counts of spurious objects in the inverted frame. An even fainter, extended source can be seen in the bottom of the error circle, whose peak level is less than 3σ; we do not consider this source in this letter.

If 3EG J1835+5918 is a pulsar with high proper motion, it might have moved to outside of the error circle determined by ROSAT data. Therefore examination only within the ROSAT error circle is not sufficient. Our observation is 3.5 yr later than the ROSAT HRI observation (Mirabal, Halpern 2001), and assuming a distance of 200 pc and transverse velocity of 500 km/s, the proper motion becomes $\sim 2''$. Therefore, we examined the outer region $2''$ beyond the ROSAT error circle, but we found no significant sources like A–D.

3. Implications and Discussion

For detailed discussions of these results, see Totani, Kawasaki & Kawai (2002), where a complete list of references is given.

References

Totani, T., Kawasaki, W. & Kawai, N. 2002, PASJ, 54, L45

IAU 8th Asia-Pacific Regional Meeting
ASP Conference Series, Vol. 289, 2003
S. Ikeuchi, J. Hearnshaw & T. Hanawa, eds.

Magnetic Reconnection in the Accretion Disk Corona of Compact Stars

I. V. Oreshina, A. V. Oreshina and B. V. Somov

Sternberg Astronomical Institute, Moscow State University, Universitetskii prospekt, 13, Moscow, Russia 119992

Abstract. We present a new model of the magnetosphere for a compact star surrounded by an accretion disk. It is assumed that the magnetic axis of the star does not coincide with the rotation axis. The shape of the accretion disk, the magnetic field strength, and its gradient inside the disk corona are estimated. From these results, we describe an analytical model of a reconnecting current sheet and estimate its characteristics as well as the total power released by current sheets in the disk corona. It is shown that magnetic reconnection quantitatively explains the observed X-ray emission from compact stars.

1. Introduction

There is some similarity in physical conditions between the system "disk–corona" of compact stars (such as white dwarfs or neutron stars) and the system "photo--sphere–corona" of the Sun (e.g. Somov 2000; Liu 2002). This is a basic idea of our present study.

In the accretion disk, plasma motions result in the formation of magnetic loops (Heyvaerts 1991) that emerge from the disk surface and interact with the magnetic field of compact stars. As in solar flares, we presume that magnetic-field reconnection generates thin current sheets, where the magnetic-field energy is converted into thermal and kinetic energy of plasma and accelerated particles.

The aim of this work is to examine whether reconnection of magnetic- field lines may explain the hard X-ray emission of compact stars. For this, we address two main problems. Number one is the shape and the size of the magnetosphere, the magnetic-field strength and its gradient across the accretion-disk corona. Number two is the role of magnetic reconnection as a possible mechanism of X-ray emission from compact stars.

2. The Magnetosphere of a Compact Star with an Accretion Disk

Real conditions are approximated by using the following mathematical model. The compact star is characterized by its magnetic dipole moment:

$$\mathbf{m} = m\, e^{i\psi},$$

Inside the magnetosphere, the magnetic field is defined by (Somov, 2000)

$$\operatorname{div} \mathbf{B} = 0, \quad \operatorname{rot} \mathbf{B} = 0.$$

Outside the magnetosphere, $\mathbf{B} = 0$. On the magnetopause S, the magnetic pressure is equal to the pressure p of the gas dynamic:

$$\left.\frac{B^2}{8\,\pi}\right|_S = p.$$

It is assumed that the magnetic field $\mathbf{B}$ does not penetrate through the magnetopause S and the accretion disk Γ, so that

$$\mathbf{B} \cdot \mathbf{n}|_{S,\Gamma} = 0.$$

Note that the shape of the magnetosphere and that of the disk are computed self consistently.

The results of computations are given in Figure 1.

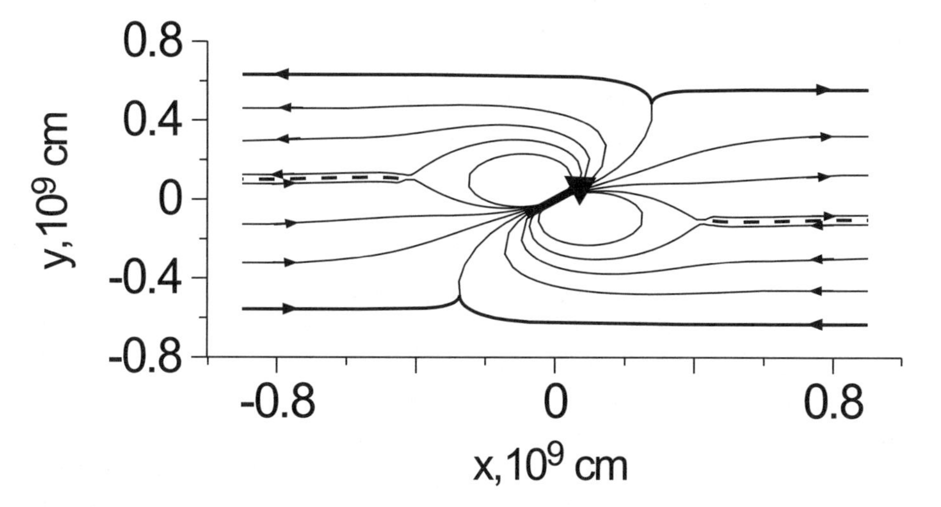

Figure 1. Cross section of the magnetic-field structure

For parameters $m = 10^{30}$ G · cm^3 , $\psi = \pi/4$, $p_0 = 1.38 \cdot 10^6$ dynes/cm^2, we obtain the inner radius of the accretion disk is about $4 \cdot 10^8$ cm. The half size of the magnetosphere is about $6 \cdot 10^8$ cm. These values are in good agreement with those inferred for the 4U 1907+09 neutron star and similar objects (Mukerjee 2001). At a distance of $5 \cdot 10^8$ cm from the star, the magnetic-field strength is $(1-2) \cdot 10^4$ G while the magnetic-field gradient is $10^{-6} - 10^{-2}$ G · cm^{-1}.

3. Contribution of Magnetic Reconnection to the Observed Radiation of Compact Stars

Let us consider a reconnecting current sheet (see Figure 2). Together with the plasma, magnetic field lines flow in at a low velocity v along the y direction, then they reconnect, and subsequently flow out at high velocity V along the x direction.

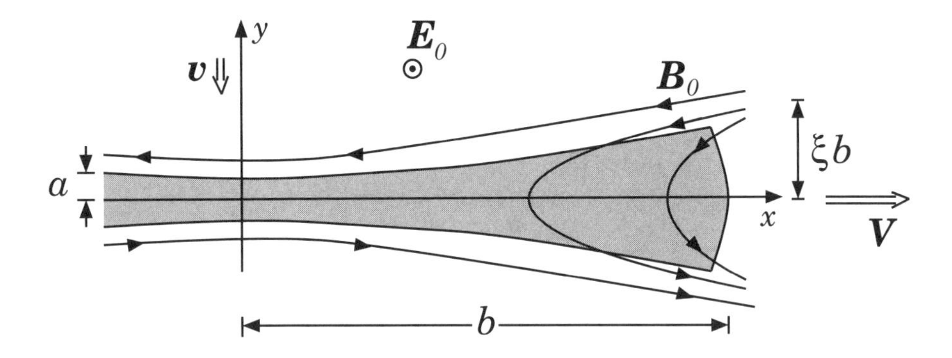

Figure 2. Reconnecting current sheet

Here, the sheet is described by order-of-magnitude relations containing only the main terms (Oreshina & Somov 2000):

$n_0\, v\, b = n\, V\, \xi\, b$ – the mass conservation law;

$\frac{B_0^2}{8\pi} = n\, k_{\rm B}\, T$ – the momentum conservation law along the y-axis;

$n\, k_{\rm B}\, T = \frac{1}{2}\, M\, n\, V^2$ – the momentum conservation law along the x-axis;

$\frac{c\, B_0}{4\pi\, a} = \sigma\, E_0$ – the Ohm law;

$\frac{B_0^2}{4\pi}\, v\, b = \frac{1}{2}\,(MnV^2 + 5nk_{\rm B}T)\, V\, \xi b + C_{\|}$ – the energy conservation law.

Here $v = cE_0/B_0$ is the plasma drift velocity into the sheet, $B_0 = h_0\, b$ is the magnetic field outside the sheet, $\sigma = \sigma_1\, T^{1/2}\, n/E_0$ is the anomalous plasma conductivity, $C_{\|} = [n\,(k_{\rm B}\, T)^{3/2}/M^{1/2}]\, \xi\, b$ is the anomalous heat flux along magnetic field lines.

The main input parameters are n_0 – the plasma number density outside the sheet, h_0 – the magnetic field gradient in the vicinity of the zero line, E_0 – the electric field strength, and $\xi \equiv B_y/B_0$ – the relative transverse component of the magnetic field inside the sheet.

We have computed the values of
a – the half thickness of the sheet,
b – the half width of the sheet,
T – the plasma temperature inside the sheet,
n – the plasma number density inside the sheet, and
V – the velocity of plasma that flows out from the sheet.
From these results we can find the power released per current sheet. For example, for input parameters $n_0 = 10^{13}$ cm^{-3}, $h_0 = 10^{-2}$ G/cm, $E_0 = 10^3$ CGSE units, and $\xi = 0.1$, we obtain $b = 5 \cdot 10^6$ cm and the power released per sheet length $P_{\rm s}/l = 3 \cdot 10^{24}$ erg/(sec cm).

Now, a lower limit of the energy release from the whole accretion disk can be estimated. Let us assume that the sheet length l has the same order of magnitude as the width $2b$. The power released by a single current sheet is P_s. The sheets are continually forming in the disk corona as the result of permanently emerging magnetic-loops. Let us consider the inner part of the ring-shaped accretion disk. The inner radius is $R_1 \sim 4 \cdot 10^8$ cm while the outer radius is $R_2 \sim 8 \cdot 10^8$ cm (see Figure 1). Its area is thus $S_r = \pi(R_2^2 - R_1^2)$, while the area of a single current sheet is $S_s = l \cdot 2b$. Thus, in the inner part of the accretion disk, a number $N \sim 2S_r/S_s$ of current sheets exist simultaneously. The total energy release per second is

$$P \sim N \cdot P_s = \frac{2\, S_r}{S_s} \cdot P_s = 7 \cdot 10^{35} \text{ erg/sec.}$$

This estimate is consistent with the total power released by some neutron stars such as Aql X-1, SLX1732-304, 4U0614+09, 4U1915-05, SAX J1808.4-3658 (Barret et al. 2000).

4. Conclusion

The main conclusion of this work is:
1) The lower limit of the power radiated by the accretion-disk corona of the compact stars is $P \sim 7 \cdot 10^{35}$ erg/sec.
2) Magnetic reconnection in the accretion-disk corona is a powerful mechanism which may explain the observed X-ray emission from compact stars.

Acknowledgments. Anna V. Oreshina would like to express her sincere thanks to the Organization Committee for the financial support and the possibility to take part in the Meeting.

References

Somov, B.V. 2000, Cosmic Plasma Physics (Dordrecht: Kluwer Academic Publishers)

Liu, B. F., Mineshige, S., & Shibata, K. 2002, ApJ, 572, L173

Heyvaerts, J. 1991, in Advances in Solar System Magnetohydrodynamics, ed. E. R. Priest & A. W. Hood (Cambridge Univ. Press.)

Mukerjee, K. et al. 2001, ApJ, 548, 368

Oreshina, A. V. & Somov, B. V. 2000, Astronomy Letters, 26(11), 750

Barret, D., Olive, J. F. & Boirin, L. 2000, ApJ, 533, 329

IAU 8th Asia-Pacific Regional Meeting
ASP Conference Series, Vol. 289, 2003
S. Ikeuchi, J. Hearnshaw & T. Hanawa, eds.

General Relativistic Simulation of Magnetohydrodynamic Energy Extraction of a Rotating Black Hole

Shinji Koide

Department of Engineering, Toyama University, 3190 Gofuku, Toyama 930-8555, Japan

Abstract. We have developed the numerical method for general relativistic magnetohydrodynamic simulations in Kerr space-time. The method is applied to the basic astrophysical problem of the Kerr black hole activity in a large-scale strong magnetic field. The numerical result shows that the magnetic field extracts the rotational energy of the black hole with negative energy-at-infinity.

1. Introduction

Relativistic jets are sometimes found around active regions in the universe. For example, they have been observed by superluminal motion not only from the quasars and active galactic nuclei (AGNs) (Biretta, Sparks, & Macchetto 1999), but also from the binary systems in our Galaxy such as GRS1915+105 (Mirabel & Rodriguez 1994). Recently the biggest explosions in the universe (of course, except for the 'Big Bang'), gamma-ray bursts, have been suggested to contain extremely high Lorentz factor jets (Kulkarni 1999). It is believed that such highly relativistic jets are formed around extremely rapidly rotating black holes (Kerr black holes). Especially, the interaction between the Kerr black hole and the strong magnetized plasma is one of the most promising models for the relativistic jet central engines. Among the violent phenomena due to the interaction, the magnetic extraction of the rotational energy of the black hole is one of the most powerful processes. This is also regarded as a fundamental physical problem of the activity of an astrophysical black hole.

Blandford and Znajek presented the force-free, static solution of the electromagnetic field around the Kerr black hole (Blandford & Znajek 1977). The solution shows the electromagnetic energy is radiated from the black hole horizon directly. The power is so large if the magnetic field is strong enough that it is applicable to the astrophysical jet engine. When we consider the dynamic process of the magnetic extraction of the rotational energy of the Kerr black hole, the direct energy radiation from the black hole appears strange from the view point of the original meaning of the horizon: any material, energy, and information can not pass through the horizon outwards.

We have developed the general relativistic MHD (GRMHD) simulation code (Koide, Shibata, & Kudoh 1998, 1999; Koide et al. 2000, 2001). To investigate the dynamic process of the electromagnetic extraction of the rotational energy of the Kerr black hole within the causality, we applied the GRMHD code to a simulation of a rather simple system of the strong magnetic field, thin plasma, and Kerr black hole (Koide et al. 2002).

2. Numerical Result

To understand the basic physics of rotational energy extraction from a black hole with finite magnetic field, we have investigated a somewhat simpler system using the GRMHD numerical calculations (Koide et al. 2002). Initially the system consists of a Kerr black hole with a uniform magnetic field, uniform plasma, and no accretion disk. We set the rotational parameter of the Kerr black hole, $a = 0.99995$, which corresponds to a nearly maximally rotating black hole. Around the hole, we initialize the plasma to a uniform mass density, ρ_0 and low pressure, $p_0 = 0.06\rho_0 c^2$. The initial momentum of the plasma is zero everywhere, and the initial magnetic field is uniform (Wald 1974) with the magnetic field strength, $B_0 = 33.3\sqrt{\rho_0 c^2}$. This is the magnetic-field-dominated case, with the Alfven velocity, $v_{\rm A} = 0.985c$ close to the speed of light. We assume axisymmetry with respect to the z-axis and reflection symmetry with respect to the equatorial plane. We perform simulations in the region $0.51 r_{\rm S} \leq r \leq 20 r_{\rm S}$ and $0.01 \leq \theta \leq \pi/2$, where $r_{\rm S}$ is the Schwarzschild radius.

Figure 1 shows the time evolution of the system where $\tau_{\rm S} = r_{\rm S}/c$ is the unit of time. The azimuthal component of the magnetic field has begun to increase due to the azimuthal twisting of the magnetic field lines. In the ergosphere, the plasma rotates the same direction of the black hole rotation due to the frame dragging effect in any case. The magnetic field lines then are twisted azimuthally in the direction of the black hole rotation by the rotation of the plasma in the ergosphere. The twist of the magnetic field lines propagates outward along the magnetic field lines against the infalling plasma flow as a torsional Alfven wave (Fig. 1).

We show the energy transport of the system of the strong magnetic field, thin plasma, and Kerr black hole (Fig. 2). The total energy flux density, $\mathbf{S}_{\rm tot}$ shows that the net energy flows out along the magnetic field from the ergosphere. We found the net power from the ergosphere is $L_{\rm tot} = 0.186 B_0^2 r_{\rm S}^2 c/\mu_0$. This energy flux is so large, that the total energy-at-infinity density, e^∞ reduces quickly and eventually becomes negative in ergosphere at $t = 6.53\tau_{\rm S}$. When the net negative energy-at-infinity is swallowed by the black hole, the net energy (total mass) of the black hole decreases. So the ultimate result of the generation of an outward Alfven wave is the magnetic extraction of rotational energy of the Kerr black hole. The negative energy-at-infinity also appears in the Penrose process to extract the rotational energy of the black hole within the causality (Penrose 1969). Then we call the extraction mechanism of rotation energy of the black hole the 'magnetohydrodynamic (MHD) Penrose process'. The energy flux from the ergosphere is dominated by the electromagnetic component (Fig. 2, right panel). The electromagnetic energy flux is transported by the torsional Alfven wave and the negative energy-at-infinity is responsible to the plasma. The power of the Alfven wave from the ergosphere is $L_{\rm EM} = 0.259 B_0^2 r_{\rm S}^2 c/\rho_0$. The electromagnetic power of the Blandford-Znajek mechanism is estimated by $L_{\rm BZ} = (\pi/16)(a^2 c/v_{\rm A}) B_0^2 r_{\rm S}^2 c/\mu_0 \sim 0.2 B_0^2 r_{\rm S}^2 c/\mu 0$ (Blandford & Znajek 1977), which is almost the same as the value we obtained from our numerical simulation.

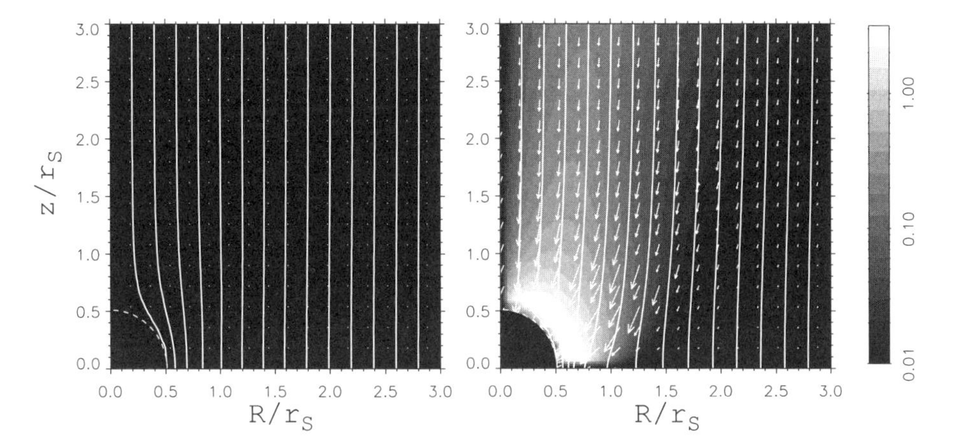

Figure 1. Time evolution of a simple system of a large-scale strong magnetic field, thin plasma, and a Kerr black hole at $t = 0$ (left panel) and $t = 6.53\tau_S$ (right panel). The grey-scale shows the value of $-B_\phi/B_0$. The arrows show the poloidal velocity of the plasma. Solid lines are magnetic field lines (surfaces). The black quarter-circle at the origin indicates the event horizon of the black hole. The dotted line shows the inner boundary of the calculation region at $r = 0.505r_S$.

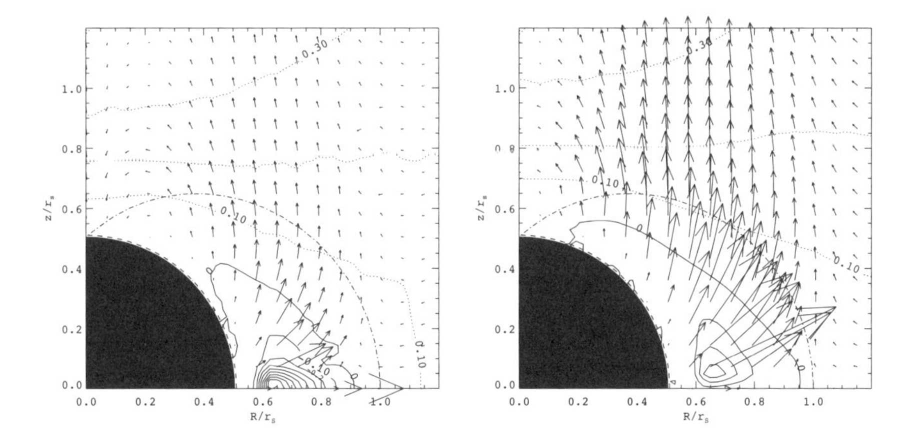

Figure 2. The total (left panel) and electromagnetic (right panel) energy transport at $t = 6.53\tau_S$. The dotted lines show the positive value of energy-at-infinity, $e^\infty_{\rm tot}$, $e^\infty_{\rm EM}$ and the solid lines show the non-positive (zero or negative) value. The arrows show the energy flux density, $\mathbf{S}_{\rm tot}$, $\mathbf{S}_{\rm EM}$. The black region indicates the inside of the black hole horizon. The dashed line is the boundary of the calculation region near the horizon. The chain line shows the boundary of the ergosphere.

3. Summary

We have applied the GRMHD code to investigate the basic mechanism of the energy extraction of the Kerr black hole by the magnetic field. The numerical result shows that the rotational energy of the Kerr black hole can be extracted by the strong magnetic field with negative energy-at-infinity within the causality at the horizon. The extracted energy is transported outward from the ergosphere as the torsional Alfven wave. The electromagnetic power of the Alfven wave is almost the same as that of Blandford-Znajek mechanism: $L_{\rm EM} = (\pi/16)(a^2 c/v_{\rm A}) B_0^2 r_{\rm S}^2 c/\mu_0$. We call this mechanism MHD Penrose process because the negative energy-at-infinity plays an essential role just as in the Penrose process.

Acknowledgments. We thank Kazunari Shibata, Takahiro Kudoh, David L. Meier, and Mika Koide for their crucial help for this study.

References

Biretta, J. A., Sparks, W. B., & Macchetto, F. 1999, ApJ, 520, 621

Blandford, R. D. & Znajek, R. 1977, MNRAS, 179 433

Koide, S., Shibata, K. & Kudoh, T. 1998, ApJ, 495, L63

Koide, S., Shibata, K. & Kudoh, T. 1999 ApJ, 522, 727

Koide, S., Meier, D. L., Shibata, K. & Kudoh, T. 2000 ApJ, 536, 668

Koide, S., Shibata, K., Kudoh, T. & Meier, D. L. 2001, Journal of the Korean Astronomical Society, 34, S215

Koide, S., Shibata, K., Kudoh, T. & Meier, D. L. 2002, Science 295, 1688

Kulkarni, S. R. 1999, Nature 398, 389

Mirabel, I. F. & Rodriguez, L. F. 1994, Nature 374, 141

Penrose, R. 1969, Nuovo Cimento, 1, 252

Wald, R. M. 1974, Phys. Rev. D, 10, 1680

IAU 8th Asia-Pacific Regional Meeting
ASP Conference Series, Vol. 289, 2003
S. Ikeuchi, J. Hearnshaw & T. Hanawa, eds.

High Energy Astrophysical Neutrinos

H. Athar
Physics Division, National Center for Theoretical Sciences, 101 Section 2, Kuang Fu Road, Hsinchu 300, Taiwan

Abstract. High energy neutrinos with energy typically greater than tens of thousands of GeV may originate from several astrophysical sources. The sources may include, for instance, our Galaxy, the active centers of nearby galaxies, as well as possibly the distant sites of gamma ray bursts. I briefly review some aspects of production and propagation as well as prospects for observations of these high energy astrophysical neutrinos.

1. Introduction

During nearly the past 50 years, the empirical search for neutrinos has spanned roughly six orders of magnitude in energy, from approximately 10^{-3} GeV up to approximately 10^3 GeV. The lower energy limit corresponds to solar neutrinos, whereas the upper energy limit corresponds to atmospheric neutrinos. The intermediate energy range include the terrestrial and supernova neutrinos. This search has already given us remarkable insight into neutrino interaction properties as well as their intrinsic properties, such as mixing and mass. Here, I briefly review the possibility of having neutrinos with energy greater than 10^3 GeV. The upper energy limit for high energy astrophysical neutrinos is limited only by the experiments concerned. A main motivation to search for such high energy astrophysical neutrinos is to get more accurate information about the origin of observed high energy photons (and ultra high energy cosmic rays) that is presently not possible through conventional gamma ray astronomy. For instance, the observation of high energy gamma ray flux alone from active centers of nearby galaxies (AGNs) such as M87 and distant sites of gamma ray bursts (GRBs) does not allow us to identify its origin in purely electromagnetic or purely hadronic interactions unambiguously. Sizable high energy astrophysical neutrino flux is expected if latter interactions are to play a dominant role. Search for high energy astrophysical neutrinos will thus provide us a complementary and yet unexplored view about some of the highest energy phenomena occurring in the known universe. For a general introduction of the subject of high energy astrophysical neutrinos, see Bahcall & Halzen (1996), Protheroe (1999), Bahcall (2001). See, also Battiston (2002).

2. High Energy Astrophysical Neutrinos

2.1. Production

The purely hadronic (such as pp or pn) and photo hadronic (such as γp or γn) interactions taking place in the cosmos currently represent the main source interactions for the production of high energy astrophysical neutrinos. Examples of the astrophysical sites where these interactions (may) take place include our Galaxy, the AGNs and the GRBs. In some model calculations for high energy astrophysical neutrino flux, the proton acceleration mechanism is considered to be the same as for electron acceleration at the astrophysical sites.

The accelerated protons in the above interactions in these sites produce unstable hadrons such as $\pi^{\pm}$ and $D_s^{\pm}$ that decay into neutrinos of all three flavors. The same interactions also produce π^0 that can contribute dominantly towards the observed high energy photons, whereas the escaping accelerated protons may (or may not) dominantly constitute the observed ultra high energy cosmic rays depending upon the finer details of the relevant astrophysical site. The absolute normalization of the high energy astrophysical neutrino flux is obtained by assuming that a certain fraction of the observed high energy photon flux has (purely) hadronic origin and (or) that the observed ultra high energy cosmic ray flux can dominantly originate from that class of astrophysical sites. Typically, the muon neutrino flux is twice the electron neutrino flux with essentially negligible tau neutrino flux at the production site. For a recent review article, see Halzen & Hooper (2002). High energy astrophysical neutrino production is in principle also conceivable in purely electromagnetic (such as $\gamma\gamma$) interactions taking place in cosmos, see Athar & Lin (2002).

The dedicated high energy neutrino detectors also provide us a clue as well as a check for the absolute normalization of the high energy neutrino flux. For instance, the present upper bound from Antarctic Muon and Neutrino Detector Array (AMANDA) give value of $9.8 \cdot 10^{-6}$ cm^{-2} s^{-1} sr^{-1} GeV for absolute flux of diffuse high energy neutrinos for the energy range between $5 \cdot 10^3$ GeV to $3 \cdot 10^5$ GeV, see Ahrens et al. (2002). The AMANDA (B10) is at the south pole and its current upper bound is based on non-observation. Its present cylindrical configuration searches for upwards going high energy (muon) neutrinos covering the northern hemisphere with an effective area of ~ 0.01 km^2 for (muon) neutrino energy $\sim 10^4$ GeV.

2.2. Propagation

With three light stable neutrinos, as suggested by standard model of particle physics, neutrino flavor mixing is a dominant effect during high energy astrophysical neutrino propagation, once they are produced, see Athar, Jeżabek, & Yasuda (2000). Since the average interstellar matter density is rather low, therefore the neutrino nucleon deep inelastic scattering (DIS) effects are usually negligible. The high energy astrophysical neutrinos thus restore the arrival direction and energy information at the production. On the other hand, because of rather large unobstructed distances traversed by these neutrinos, typically greater than 1 pc, where 1 pc $\simeq 3 \cdot 10^{18}$ cm, the neutrino flavor mixing equally distribute the three neutrino flavors in the mixed high energy astrophysical neutrino flux. The present neutrino oscillation data implies that the deviations

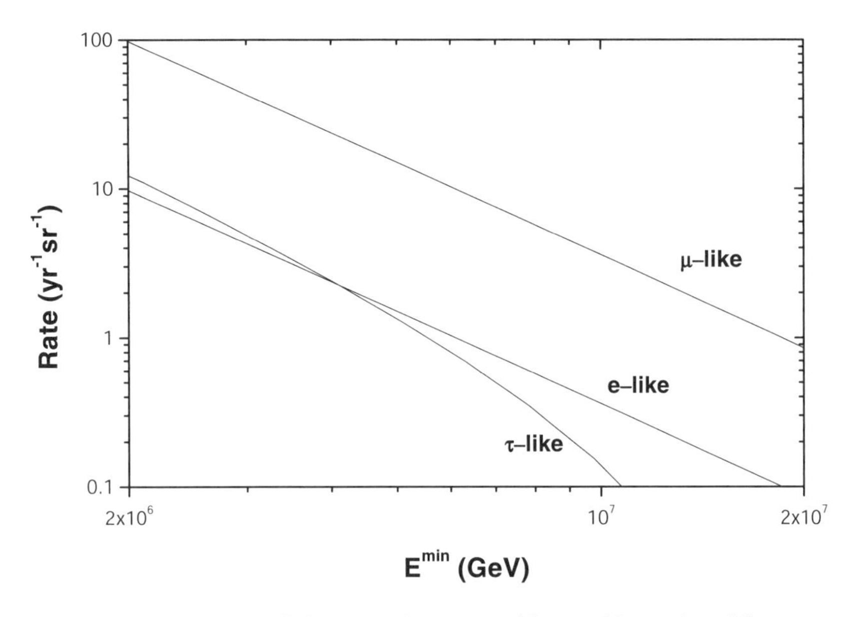

Figure 1. Expected downward going $e-$like, $\mu-$like and $\tau-$like event rate produced by AGN neutrinos as a function of minimum energy of the corresponding charged lepton in a large km^3 volume ice or water neutrino detector. Three flavor neutrino mixing is assumed.

from these symmetric distributions are not more than a few percent. The absolute level of (downwards going) high energy astrophysical electron neutrino flux, arriving at the detector, is essentially independent of the relative ratio at the production, as it is least effected by neutrino oscillations, see Athar & Lin (2001).

2.3. Prospects for Observations

Typical high energy astrophysical neutrino observation can be achieved by attempting to observe the Cherenkov radiation from the associated charged leptons and/or showers produced in DIS occurring near or inside the detector. For simplicity, here I ignore a possible observational difference between neutrinos and anti neutrinos.

The mixed high energy astrophysical neutrino flux arrives at an Earth based detector in three general directions. The downward going neutrinos do not cross any significant Earth cord while reaching the detector. In Fig. 1, the downward going event rate for the three neutrino flavors is displayed under the assumption of neutrino flavor mixing in units of yr^{-1} sr^{-1} as a function of minimum energy of the corresponding charged lepton produced in charged current DIS occurring near or inside the detector. The different levels and energy dependencies of the rates reflect the differences in the associated charged lepton ranges, in a large (under construction) km^3 volume ice or water detector such as the proposed extension of AMANDA (B10), although as mentioned in the last subsection, the three neutrino flavors arrive at the detector in almost equal proportion. The contained $e-$like event rate is obtained by rescaling the $\mu-$like event rate for illustration. A diffuse AGN neutrino flux model is used here as an example where *pp* interactions are considered to play an important role, (see Szabo & Protheroe 1994). The event topology for each flavor can possibly be identified separately in km^3 volume water or ice detector within the energy range shown in Fig. 1. For details, see Athar, Parente, & Zas (2000).

The near horizontal neutrino flux crosses a small Earth cord before reaching the detector. Several proposals are under study to construct a specific detector for such type of neutrinos, see Hou & Huang (2002). The upward going neutrino flux crosses a significant Earth cord before reaching the detector, and is therefore absorbed by the Earth to a large extent for energy typically greater than $5 \cdot 10^4$ GeV, see, for instance, Hettlage & Mannheim (2001). Above this energy, the Earth diameter exceeds the charged current DIS length. In addition to attempting to measure the Cherenkov radiation, several other alternatives are also presently being explored. See, for instance, Chiba et al. (2001).

3. Conclusion

Several astrophysical sources such as center(s) of our as well as other nearby galaxies may produce high energy astrophysical neutrino flux with energy around or above 10^4 GeV, compatible with the observed high energy gamma ray and ultra high energy cosmic ray flux above the atmospheric neutrino flux. This implies need for a more meaningful search.

The author thanks Physics Division of National Center for Theoretical Sciences for financial support.

References

Ahrens et al. [The AMANDA Collaboration] 2002, arXiv:astro-ph/0206487

Athar, H., Jeżabek, M. & Yasuda, O. 2000, Phys. Rev. D, 62, 103007

Athar, H. & Lin, G.-L. 2001, arXiv:hep-ph/0108204

Athar, H. & Lin, G.-L. 2002, arXiv:hep-ph/0203265

Athar, H., Parente, G. & Zas, E. 2000, Phys. Rev. D, 62, 093010

Bahcall, J. N. 2001, Int. J. Mod. Phys. A, 16, 4955

Bahcall, J. N. & Halzen, F. 1996, Phys. World, 9, 41

Battiston, R. 2002, arXiv:astro-ph/0208108

Chiba, M. et al. 2001 in AIP Conf. Proc. 579, Radio Detection of High Energy Particles, ed. D. Saltzberg & P. Gorham (New York: AIP), 204

Halzen, F. & Hooper, D. 2002, Rept. Prog. Phys., 65, 1025

Hettlage, C. & Mannheim, K. 2001, Nucl. Phys. Proc. Suppl., 95, 165

Hou, G. W. & Huang, M. A. 2002, arXiv:astro-ph/0204145

Protheroe, R. J. 1999, arXiv:astro-ph/9907374

Szabo, A. P. & Protheroe, R. J. 1994, Astropart. Phys., 2, 375

QSOs, AGNs and IGM

IAU 8th Asia-Pacific Regional Meeting
ASP Conference Series, Vol. 289, 2003
S. Ikeuchi, J. Hearnshaw & T. Hanawa, eds.

Associated Absorption in Radio-loud Quasars and the Growth of Radio Sources

Richard W. Hunstead

School of Physics, University of Sydney, NSW 2006, Australia

Joanne C. Baker

Department of Astrophysics, University of Oxford, Denys Wilkinson Building, Keble Road, Oxford OX1 3RH, UK

Abstract. We have carried out a detailed study of associated C IV $\lambda\lambda1548, 1550$ absorption lines in the spectra of radio-loud quasars drawn from the low-frequency selected Molonglo Quasar Sample in two redshift ranges, $0.7 < z < 1.0$ and $1.5 < z < 3.0$. We find that associated C IV absorption occurs preferentially in steep-spectrum quasars, and the absorption strength is anti-correlated with the projected linear size, suggesting an evolutionary sequence. We also find that heavily absorbed quasars are systematically redder, implying that radio sources are triggered in gas- and dust-rich environments, as expected in a post-merger starburst.

1. Introduction

Absorption lines seen against background quasars have long been recognized as valuable probes of interstellar and intergalactic gas throughout the universe. Associated absorbers, those occurring very close to the quasar redshift, sample the local quasar environment, potentially intercepting gas from a range of locations along the line of sight. For instance, absorption can arise in gas flows near the quasar central engine, the interstellar medium (ISM) of the quasar host galaxy, or the ISM of neighbouring galaxies.

Narrow-line associated absorption systems – usually defined by an absorption redshift, z_a, lying within $5000\,\mathrm{km\,s^{-1}}$ of the emission redshift, z_e, and with FWHMs of tens to hundreds $\mathrm{km\,s^{-1}}$ – are more common than can be accounted for by randomly distributed galaxies along the quasar sightline (Foltz et al. 1986, 1988; Richards et al. 1999, 2001). The absorption systems occur most frequently in radio-loud quasars, especially those with steep radio spectra (Anderson et al. 1987; Foltz et al. 1988; Richards et al. 1999, 2001), favouring a direct link with active galactic nuclei (AGN) activity. This trend can be interpreted in part as an orientation effect, consistent with the basic framework of unified schemes. However, orientation cannot explain the 20%–30% of compact radio sources with steep rather than flat spectra, since a steep spectrum implies that the radio jets are not beamed strongly towards us. Instead, the small projected sizes of compact steep-spectrum (CSS) sources are believed to be intrinsic, and arise because the sources are young and will eventually grow into large-scale sources.

2. Observations

To build up a picture of the rate of occurrence of associated absorption, and its implications for radio source evolution, we need to sample sightlines that span a wide range of viewing angles to the radio axis. Complete and unbiased samples are therefore crucial. We focus here on associated C IV absorption in a complete sample of low-frequency-selected quasars from the well-defined Molonglo Quasar Sample (MQS; Kapahi et al. 1998), which in turn is drawn from the 408 MHz Molonglo Reference Catalogue (MRC; Large et al. 1981). The MQS is defined by $S_{408} > 0.95$ Jy, $-30° < \delta < -20°$, and $|b| > 20°$, and has no optical magnitude limit. It is relatively unbiased with respect to radio-jet orientation, since isotropically emitting components dominate the radio emission at 408 MHz.

Details of the sample selection, VLA radio images and low-resolution spectra are given in earlier papers (Kapahi et al. 1998, Baker et al. 1999). To study associated absorption in the MQS we sought additional spectra for two redshift-limited subsamples, a low-redshift sample with $0.7 < z < 1.0$ and a high-redshift sample with $1.5 < z < 3.0$. In order to measure linear sizes and morphologies for the CSS quasars, we obtained radio images at resolutions $\sim$0.1 arcsec with MERLIN (Multi-Element Radio-Linked Interferometer Network; Thomasson et al. 1994). A detailed account of the spectroscopic observations and analysis is given by Baker et al. (2002), and the MERLIN observations are discussed by de Silva et al. (2002).

For consistency with previous work on the MQS we adopt the following cosmological parameters: $H_0 = 50$ km s^{-1} Mpc^{-1}, $q_0 = 0.5$ and $\Lambda = 0$.

Ground-based spectroscopy of $1.5 < z < 3.0$ *quasars* Most of the 20 spectra were obtained in four observing runs with the Anglo-Australian Telescope (AAT) at Siding Spring Observatory, Australia, between 1995 and 1999. We used the RGO spectrograph and Tek 1024 × 1024 CCD, and gratings to give resolutions of either 1.2 Å or 2.4 Å FWHM, with a wavelength range spanning the Lyα to C IV region. Because of the faintness of the targets ($19 < b_J < 22$) exposure times were long, typically several hours at 1.2 Å and an hour or more at 2.4 Å.

Further observations were made in 2000 with the ESO 3.6-m telescope at La Silla, Chile, using EFOSC-2 and grating 7 to give a resolution of $\sim$6 Å FWHM. The faintest targets were observed in 1999 at a resolution of 2.4 Å FWHM with FORS1 on the ESO VLT (UT1) at Paranal, Chile.

HST spectroscopy of $0.7 < z < 1.0$ *quasars* UV spectroscopy was carried out for 19 MQS quasars using the STIS instrument on *HST* over the period 1999 May to 2001 February. The NUV-MAMA detector was used with the G230L grism, giving a resolution of 3.0 Å over the wavelength range 1570–3180 Å. The data were reduced using the standard STIS data pipeline.

MERLIN imaging of CSS quasars Because of their declinations, the MQS targets were visible to MERLIN only at low elevations. This led to some problems with ground-spill, atmospheric absorption and phase calibration. Nevertheless, the sensitivity and *uv* coverage were adequate to obtain useful images for the majority of targets. Observations at 1.7 GHz were carried out in 1997 at an an-

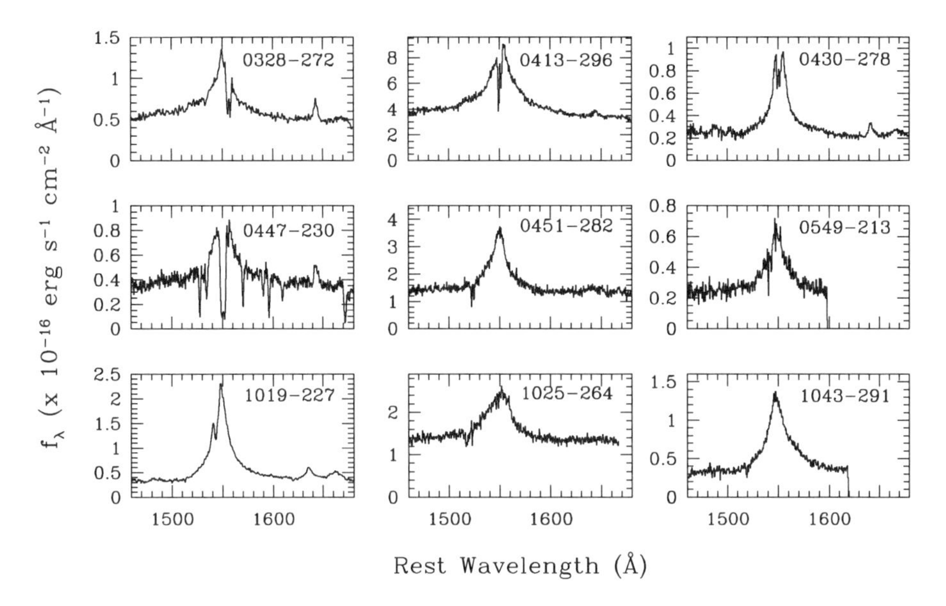

Figure 1. Spectra of the C IV region (1460–1680 Å) for a subset of the $z > 1.5$ quasars observed from the ground. Of these, three (0430−278, 0447−230, 1019−270) are CSS quasars, four (0328−272, 0413−296, 0549−213, 1025−264) are large steep-spectrum (LSS) quasars, and two (0451−282, 1043−291) are compact flat spectrum (CFS) quasars.

gular resolution of $\approx$0.1″, with further observations at 5 GHz at 0.04″ resolution being made the following year (de Silva et al. 2002).

3. Results

A montage of nine spectra from the ground-based sample in the region of C IV is shown in Figure 1. The full spectrum of one of these, the CSS source MRC B0447−230, is shown in Figure 2, together with the MERLIN 5 GHz contour image.

The strongest C IV absorption system lying within 5000 km s^{-1} of the C IV emission line was identified in each spectrum. The C IV emission-line redshift was measured from the peak position of a fitted Lorentzian profile, and the relative velocity Δv between the absorption and emission redshifts was defined as $\Delta v = c(z_{\rm e} - z_{\rm a})/(1 + z_{\rm e})$. Equivalent widths in the absorber rest frame, $W_{\rm abs}$, were measured for both lines of the C IV $\lambda\lambda$1548, 1550 doublet together, so that blended and unblended lines could be compared.

The equivalent widths of C IV are plotted in Figure 3 (*left*) as a function of the projected linear sizes of the radio sources for both low- and high-redshift datasets. Core-dominated quasars are excluded because they are severely foreshortened. It is clear that the strongest absorption occurs preferentially in the smallest sources. Notably, all but one of the CSS sources show associated ab-

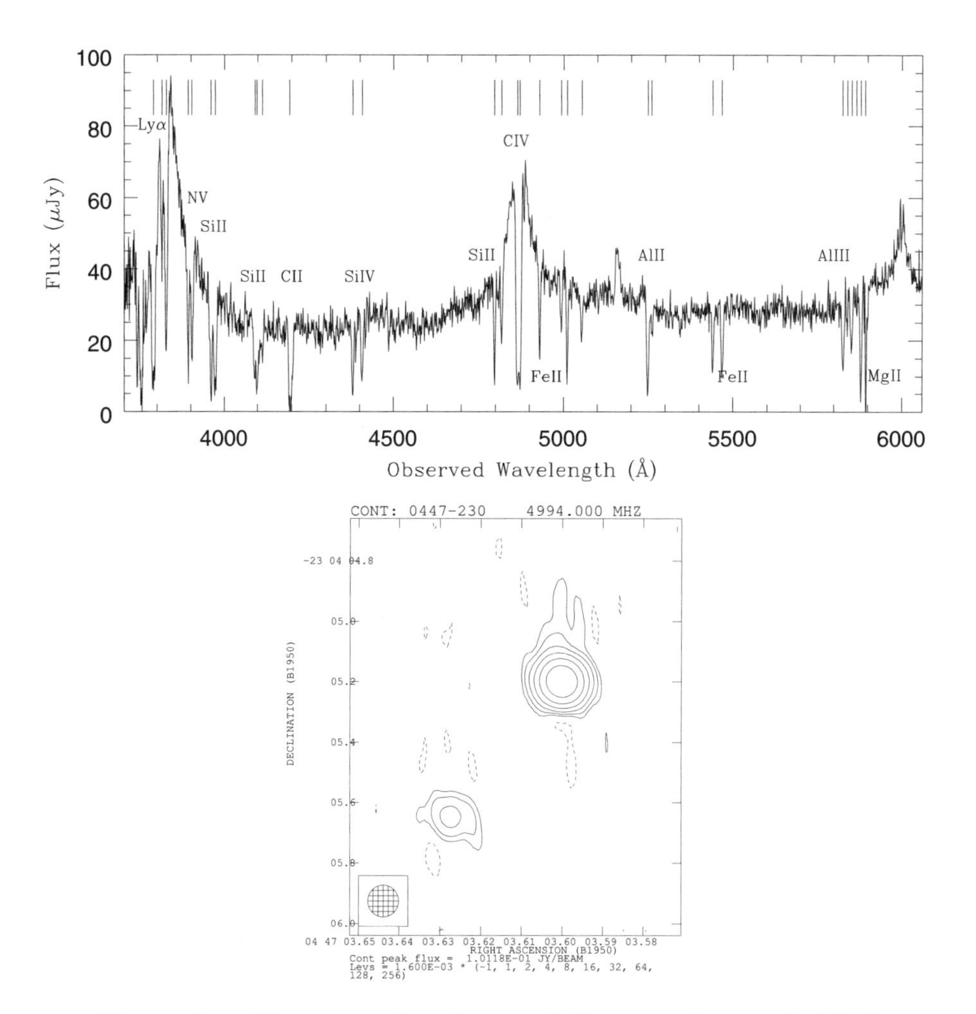

Figure 2. *Upper panel*: Low-resolution spectrum of MRC B0447−230, a CSS quasar at $z = 2.145$, obtained with FORS1 on the VLT (UT1). Lines in the strong associated-absorption system are labelled above the spectrum and lines in an intervening Mg II and Fe II absorption system at $z \sim 1$ are labelled below. The weaker lines immediately redwards of the Si II lines have been identified as Si II* fine structure lines. *Lower panel*: MERLIN contour image of MRC B0447−230 at 5 GHz showing a clear double structure with a projected linear separation of 5 kpc. Both components have a steep spectral index; there is no detectable core.

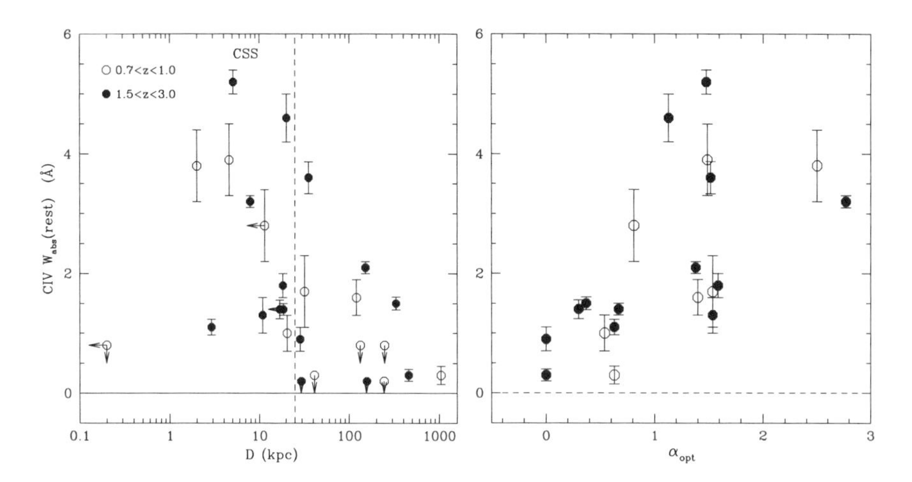

Figure 3. *Left*: Equivalent width of the C IV absorption as a function of the radio source size D (kpc). The two redshift intervals are plotted with different symbols. The dotted line at 25 kpc marks the size limit used to define CSSs. *Right*: Equivalent width as a function of optical spectral index, $\alpha_{\rm opt}$ (where $S_\nu \propto \nu^{-\alpha}$), for all quasars with detected absorption.

sorption stronger than $W_{\rm abs} = 1$ Å. The plot shows a continuous decline in absorption strength with increasing linear size, demonstrating that CSS quasars do not form a separate class, as has been claimed previously (e.g. Antonucci 1993). It is also worth noting that the low- and high-redshift quasars appear to follow the same relationship, at least for the cosmological model assumed here.

For quasars where C IV absorption was detected, $W_{\rm abs}$ is plotted in Figure 3 (*right*) against the power-law slope of the optical continuum, $\alpha_{\rm opt}$, measured over the range 4000–10 000 Å. There is a strong correlation between C IV absorption-line strength and continuum slope in the sense that heavily absorbed quasars are systematically redder. In an earlier study, Baker (1997) showed that at low redshift, $z < 0.5$, there was a tight correlation between $\alpha_{\rm opt}$ and the Hα/Hβ Balmer decrement, suggesting the presence of a dust screen lying outside the broad-line region. The simplest interpretation of this new correlation is that the absorbing gas clouds contain dust, and they lie well outside the nuclear continuum source.

4. Discussion

4.1. Orientation and Evolution

The evidence presented in Figure 3 points to the strongest C IV absorbers being found more often in steep-spectrum quasars, particularly those of small linear size, D. The results are summarized in Table 1. We now discuss whether these results can be explained by one or more of the following hypotheses:

1. The distribution of absorbing clouds is orientation dependent,
2. The absorption strength correlates with the density of the gaseous (or cluster) environment surrounding the quasar, and
3. The absorption column density changes with time.

Table 1. Fractions of MQS quasars with C IV absorption equivalent widths $W_{abs} > 1$ Å in three subsets: compact steep spectrum (CSS), large steep spectrum (LSS) and compact flat spectrum (CFS). Spectral index α is defined here by $S_\nu \propto \nu^{-\alpha}$.

CSS $\alpha > 0.5$ $D < 25$ kpc	LSS $\alpha > 0.5$ $D > 25$ kpc	CFS $\alpha < 0.5$ $D < 25$ kpc
11/12	8/19	0/6

In the orientation-based unified schemes for radio sources, flat-spectrum and core-dominated sources have their nuclear radio emission strongly boosted on account of being viewed at a small angle to the axis of the approaching relativistic jet. The cores of more highly inclined steep-spectrum quasars are less strongly boosted, so their overall radio emission is dominated by the isotropic radiation from their diffuse lobes. In this picture, the prevalence of absorption in steep- rather than flat-spectrum quasars (and its decrease with increasing core dominance; Baker et al. 2002) implies that the absorbing material lies away from the radio jet axis, increasing in column density the greater the inclination angle. Earlier studies (Anderson et al. 1987; Foltz et al. 1988; Barthel et al. 1997; Richards et al. 1999, 2001) arrived at similar conclusions. However, orientation alone cannot explain the range of sizes of steep-spectrum quasars, including CSS quasars, so another process is required to explain the anticorrelation of absorbing column density with increasing radio size.

If the absorbing material arises in dense quasar environments, its signature should be evident in studies of the host galaxies or the cluster environment. There is currently no evidence that small (CSS) sources inhabit denser, more gas-rich regions (O'Dea 1998; Wold et al. 2000, Saikia et al. 2001). Indeed, it seems that quasars of all types are found in environments of widely varying richness. Theoretically, the pressure confinement of radio lobes argues that environmental density must affect source size to some extent, and asymmetric sources provide a clear illustration. However, to constrict sources to scales of tens rather than hundreds of kpc would require overdensities of several orders of magnitude, which are not observed at other wavelengths.

If neither orientation nor environmental density can account for the full range of source sizes, then age/source evolution must be important. Recent radio VLBI observations (e.g. Taylor et al. 2000) do indeed infer ages as young as 10^2–10^4 yr for some CSS sources, compared with the more usual 10^7–10^8 yr for larger sources. To explain the results shown in Figure 3, the absorption column density must decrease over the lifetime of an expanding radio source.

4.2. Dust in the Absorbing Clouds

The strong correlation between absorption strength and continuum slope in Figure 3 suggests, perhaps surprisingly, that dust and highly ionized gas exist in close proximity, possibly in the same clouds. The mere presence of reddening indicates that small dust grains are present in the quasar emission-line regions. De Young (1998) has shown that dust grains in the range 100–2500 Å can survive for no longer than 10^4–10^6 yr, otherwise dust replenishment is necessary. Dust may be created by new and extensive star formation, and entrained by turbulent flows within the radio cocoon, but the net outcome of the dust recycling will be gradual clearing, at least along the radio axis. We can set an approximate timescale of 10^5–10^6 yr for the emergence of quasars from their dusty cocoons and the opening up of their ionization cones, comparable with the lifetimes of CSS sources.

4.3. Redshift

An important outcome from this study is that the results are all independent of redshift, i.e., in a flux-limited sample we found no differences between the global properties of low- and high-redshift quasars. Redshift independence also indicates that the absorbing material originates in structures that do not evolve significantly with cosmic epoch, thereby ruling out many galactic origins for the absorbing clouds, such as surrounding clusters and star-forming dwarf galaxies. This tends to point to the absorbing material being linked directly to processes intrinsic to the AGN and its triggering.

4.4. Origin of the Absorbing Clouds

If we interpret the anticorrelation of absorption strength with source size (Figure 3) in terms of age, this implies that radio sources are triggered in an environment rich in dust and gas. Mergers have long been believed to be likely causes of AGN activity, and the accompanying massive starbursts then provide a ready explanation for the origin of the dusty material. The location of the absorbing clouds is still open to question, however, with strong constraints being imposed by the presence of excited-state fine-structure lines (Hamann et al. 2001).

5. Summary

The independence with redshift, and the clearing of absorbing material with time, suggest that the presence of dusty absorbing gas must be closely related to the onset of radio activity. Such a build-up of dusty material would occur naturally following a major burst of star formation, either during or following a merger or close interaction with a gas-rich galaxy. Over the lifetime of a source, the enshrouding material would be gradually cleared, primarily along the radio jet axis. The absence of any absorption close to the radio jet axis (in CFS quasars) suggests that the direct interaction of the radio beam with the ISM clouds is effective in punching a hole in the enshrouding gas and dust, while turbulent entrainment at the edges of the cocoon may be effective in replenishing the dust and gas for off-axis sightlines.

The basic test for this picture is that CSS sources should show more evidence of recent star formation than larger sources. In at least one case, the CSS quasar 3C 48 (Canalizo & Stockton 2000), the estimated ages of starburst knots and the radio source are consistent. More accurate dating of starburst and radio source ages will help to confirm these links and allow us to estimate better the time delays between a starburst and the onset of radio emission.

References

Anderson, S. F., Weymann, R. J., Foltz, C. B. & Chaffee, F. H., Jr. 1987, AJ, 94, 278

Antonucci, R. 1993, ARA&A, 31, 473

Baker, J. C. 1997, MNRAS, 286,23

Baker, J. C., Hunstead, R. W., Kapahi, V. K. & Subrahmanya, C. R. 1999, ApJS, 122, 29

Baker, J. C., Hunstead, R. W., Athreya, R. M., Barthel, P. D., de Silva, E., Lehnert, M .D. & Saunders, R. D. E. 2002, ApJ, 568, 592

Canalizo, G. & Stockton, A. 2000, ApJ, 528, 201

de Silva, E., Baker, J.C., Saunders, R.D.E. & Hunstead, R.W. 2002, in preparation

De Young, D.S. 1998, ApJ, 507, 161

Foltz, C. B., Chaffee, F. H., Weymann, R. J. & Anderson, S. F. 1988, in QSO Absorption Lines, ed. J. C. Blades, D. A. Turnshek & C. A. Norman (Cambridge: CUP), 53

Foltz, C. B., Weyman, R. J., Peterson, B. M., Sun, L., Malkan, M. A., & Chaffee, F. H. 1986, ApJ, 307, 504

Hamann, F. W., Barlow, T. A., Chaffee, F. C., Foltz, C. B., & Weymann, R. J. 2001, ApJ, 550, 142

Kapahi, V. K., Athreya, R. M., Subrahmanya, C. R., Baker, J. C., Hunstead, R. W., McCarthy, P. J. & van Breugel, W. 1998, ApJS, 118, 327

Large, M. I., Mills, B. Y., Little, A. G., Crawford, D. F. & Sutton, J.M. 1981, MNRAS, 194, 693

O'Dea, C. P. 1998, PASP, 110, 493

Richards, G. T., Laurent-Muehleisen, S. A., Becker, R. H. & York, D. G. 2001, ApJ, 547, 635

Richards, G. T., York, D. G., Yanny, B., Kollgaard, R. I., Laurent-Muehleisen, S. A. & vanden Berk, D. E. 1999, ApJ, 513, 576

Saikia, D. J., Jeyakumar, S., Salter, C. J., Thomasson, P., Spencer, R. E., & Mantovani, F. 2001, MNRAS, 321, 37

Taylor, G. B., Marr, J. M., Pearson, T. J. & Readhead, A. C. S. 2000, ApJ, 541, 112

Thomasson, P., Garrington, S. T., Muxlow, T. W. B. & Leahy, J. P. 1994, MERLIN User Guide, Version 1.1 (www.merlin.ac.uk/user_guide/)

Wold, M., Lacy, M., Lilje, P. B. & Serjeant, S. 2000, MNRAS, 316, 267

IAU 8th Asia-Pacific Regional Meeting
ASP Conference Series, Vol. 289, 2003
S. Ikeuchi, J. Hearnshaw & T. Hanawa, eds.

Very Extended Emission-Line Region around the Seyfert 2 Galaxy NGC 4388

M. Yoshida[1], M. Yagi[1], S. Okamura[2], K. Aoki[1], Y. Ohyama[1], Y. Komiyama[1], N. Yasuda[1], M. Iye[1], N. Kashikawa[1], M. Doi[2], H. Furusawa[2], M. Hamabe[3], M. Kimura[2], M. Miyazaki[2], S. Miyazaki[1], F. Nakata[2], M. Ouchi[2], M. Sekiguchi[2], K. Shimasaku[2], and H. Ohtani[4]

[1]*National Astronomical Observatory of Japan, Mitaka, Japan*
[2]*Tokyo University, Tokyo, Japan*
[3]*Japan Women's University, Tokyo, Japan*
[4]*Kyoto University, Kyoto, Japan*

Abstract. We found a very large, ~ 35 kpc, emission-line region (VEELR) around the Seyfert type 2 galaxy NGC 4388, using deep narrow-band imaging with the Suprime-Cam of the Subaru telescope. This region consists of many faint gas clouds or filaments, and extends northeastwards from the galaxy. The total ionized gas mass calculated from the $L_{\mathrm{H}\alpha}$ is $\sim 4 \times 10^6 f_{\mathrm{V}}^{-1/2} M_{\odot}$. Deep spectroscopy of the VEELR suggests that most of the clouds in the VEELR is ionized by the nuclear ionizing radiation. Shock heating plays an important role in ionizing some VEELR clouds outside the nuclear radiation cone. The velocity field of the VEELR is complicated and almost all the clouds are blue-shifted ralative to the galaxy. Most plausible origin of the VEELR gas is the interstellar medium of NGC 4388, stripped by the ram pressure of the hot intracluster medium of the Virgo cluster. The VEELR may provide us an important clue to investigate the evolutional link between galaxy activity and ram pressure stripping.

1. Introduction

The evolution of galaxies in clusters of galaxies has been under long debate. It is well known that HI gases in spiral galaxies in rich clusters are systematically deficient relative to field spirals (e.g. Bothun 1982). Coupled with morphology segregation of the distribution of cluster galaxies (e.g. Dressler 1980), many authors have suggested that some gas stripping mechanisms, such as ram pressure stripping (e.g. Schulz and Struck 2001) or starburst wind triggered by galaxy harassment (e.g. Moore et al. 1996), are very efficient in clusters core.

An HI deficient galaxy, NGC 4388, is a spiral galaxy in the vicinity of the core of the Virgo cluster (Cayatte et al. 1990). Its nucleus exhibits type 2 Seyfert activity. We made deep narrow-band imaging observations of NGC 4388 and found a very large emission-line region (hereafter, referred to as the "VEELR") which is extended well outside of the previously found ionized gas (Yoshida et al. 2002). Subsequently, we performed a deep spectroscopy of the VEELR.

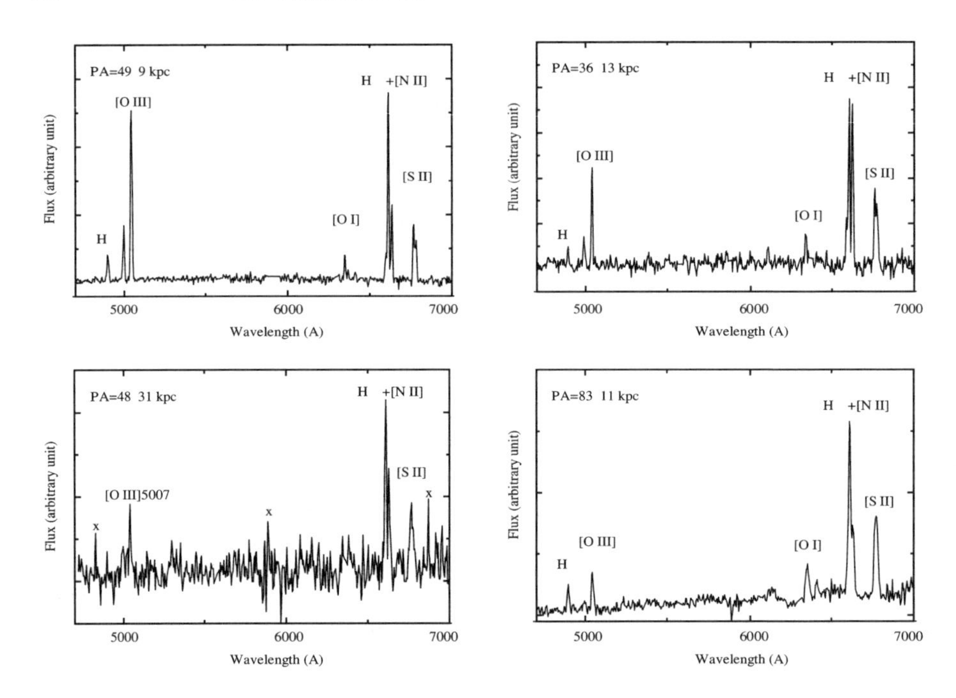

Figure 1. Samples of the spectra of VEELR clouds.

We present here the results of these observations and discuss the origin of the VEELR. Detailed investigation on the VEELR of NGC 4388 provides us an important clue to understand gas stripping mechanism of cluster spirals.

2. Observations and Results

We made broad band (V and R_c) and narrow band (Hα+[N II] and [O III]) observations of NGC 4388 with Suprime-Cam of the Subaru Telescope in 2001 March and April (Yoshida et al. 2002).

Complex structure of an ensemble of many clouds or filaments whose size reaches up to 35 kpc from the nucleus of NGC 4388 is clearly seen in our Hα+[N II] image (Fig. 1 of Yoshida et al. 2002). We call this structure the VEELR. The total Hα luminosity $L_{\mathrm{H}\alpha}$ of the VEELR is $\sim 2 \times 10^{38}$ erg s^{-1} and the ionized gas mass calculated from the $L_{\mathrm{H}\alpha}$ is $\sim 4 \times 10^6 f_{\mathrm{v}}^{-1/2} M_{\odot}$, where f_{v} is the volume filling factor of the gas.

The emission-line intensity ratio map of [O III]/Hα of the VEELR shows a clear evidence of the existence of the ionization cone which is formed by nuclear anisotropic ionizing radiation (Yoshida et al. 2002).

We performed a deep optical spectroscopy of the VEELR of NGC 4388 with FOCAS (Kashikawa et al. 2000) attached to the Subaru Telescope in 2002 March. In order to obtain spectra of complicatedly distributed filaments/clouds of the VEELR, we used multi-slit spectroscopy mode of FOCAS, and obtained the spectra of 40 filaments/clouds.

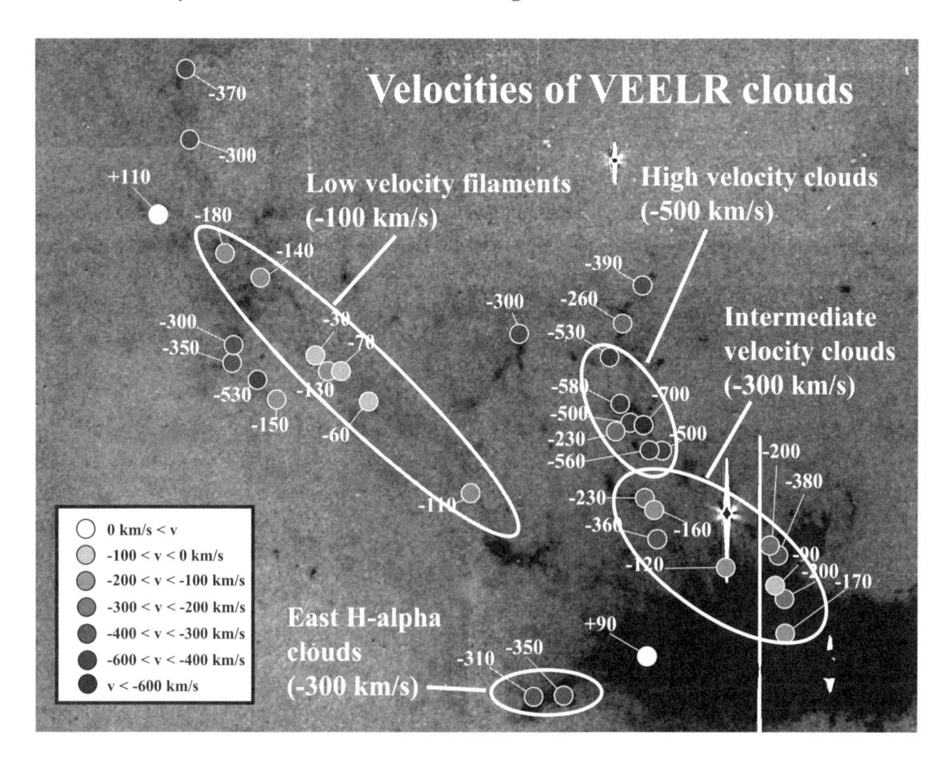

Figure 2. Relative velocities of the VEELR clouds against the systemic velocity of NGC 4388 overplotted on the Hα+[N II] image (grey scale). This figure shows a area of $3.3' \times 2.3'$ of the north east region of the galaxy.

Some samples of spectra of the VEELR clouds are shown in Figure 1. The emission-line spectra of the inner VEELR clouds show high excitation properties which is very similar to the spectra of the nuclear narrow-line region of NGC 4388 indicating that photoionization by the nuclear power-law UV continuum is the dominant ionization mechanism of most region of the VEELR. The most distant filament we made spectroscopy also shows power-law photoionization like spectrum (bottom left panel of Fig. 1). On the other hand, low ionization forbidden lines such as [O I] λ6300 dominates in the emission-line spectra of some bright Hα clouds outside the nuclear radiation cone, which indicates shock heating plays an important role in their ionization (bottom right panel of Fig. 1).

The velocity field of the VEELR is quite complicated (Fig. 2). Almost all the filaments measured is blue-shifted relative to the galaxy itself. The relative velocities ranges from -100 km s^{-1} to over -700 km s^{-1} and the mean velocity is -350 km s^{-1}. There seems several groups of clouds/filaments which are divided by both of spatial distribution and velocities (see Fig. 3). The most remarkable group is the high velocity clouds extended along PA$\approx 35°$.

3. Discussion

We examined some hypotheses (AGN wind, starburst superwind, tidal debris of galaxy-galaxy interaction, or ram pressure stripping) concerning the origin of the VEELR gas. As a consequence, its large size, highly asymmetric morphology and blue-shifted high velocity gas flow of the VEELR lead us to a conclusion that most of the gas is the interstellar medium stripped by interaction between NGC 4388 and the hot intracluster medium (ICM) of the Virgo cluster.

According to the numerical simulations made by Schulz and Struck (2001), ram pressure stripped gas has a velocity of $\sim 100 - 200$ km s^{-1} in the case of their tilted model which is appropriate for NGC 4388 case. This value is consistent with the velocities of the low velocity filaments extended along PA $\approx 50° - 60°$ (Fig. 2). Thus these filaments may be the disk gas stripped purely by ram pressure of the hot ICM.

On the other hand, the high velocity clouds of the VEELR have much higher velocities than those predicted by the simulations, and it is suggested that additional acceleration mechanisms exist. For the north-east inner ($r < 4$ kpc) gas clouds of NGC 4388, Veilleux et al. (1999) concluded that these gas are the ejecta of nuclear outflow accelerated by the radio jet. The high velocity clouds are located on the extension of the Veilleux's NE clouds, thus one possible acceleration mechanism of the clouds is the nuclear outflow. Another possible mechanism is outflow from the disk star-forming regions. Schulz and Struck (2001) suggested that compression of the disk gas at early stage of ram pressure stripping triggers gravitational instability and global starbursts. In fact, active star-formation occurs in the disk of NGC 4388. Supernova explosions or stellar winds in these star-forming regions must blow out the disk gas into intergalactic space. The high velocity clouds and/or the east Hα clouds may be accelerated by this activity.

Anyway, the stripped gas is accelerated to a speed of nearly 1000 km s^{-1} in the VEELR. It is well beyond the escape speed of the galaxy and will not return to the disk, unless the galaxy has unusually massive dark halo. If galactic outflows induced by AGN or starburst and ram pressure stripping occurred at the same time, interstellar gas of massive spiral galaxies is efficiently removed and the remnants may evolve into gas poor S0 galaxies very rapidly. The VEELR of NGC 4388 is a good example for investigating above process and evolutional link between galactic activity and ram pressure stripping.

References

Bothun, G. D. 1982, ApJS, 50, 39
Cayatte, V. et al. 1990, AJ, 100, 604
Dressler, A. 1980, ApJ, 235, 351
Kashikawa, N. et al. 2000, Proc. SPIE., 4008, 104
Moor, B. et al. 1996, Nature, 379, 613
Schulz, S. and Struck, C. 2001, MNRAS, 328, 185
Veilleux, S. et al. 1999, ApJ, 520, 111
Yoshida, M. et al. 2002, ApJ, 567, 118

IAU 8th Asia-Pacific Regional Meeting
ASP Conference Series, Vol. 289, 2003
S. Ikeuchi, J. Hearnshaw & T. Hanawa, eds.

Evidence for the Evolutionary Sequence of Blazars: Different Types of Accretion Flows in BL Lac Objects

Xinwu Cao

Shanghai Astronomical Observatory, Chinese Academy of Sciences, Shanghai, 200030, China

Abstract. The distribution of line luminosity $L_{\mathrm{H}\beta}$ of 23 BL Lac objects suggests a bimodal nature. We found that standard thin disks are probably in the sources with $L_{\mathrm{H}\beta} > 10^{41}$ erg s^{-1}. For the sources with $L_{\mathrm{H}\beta} < 10^{41}$ erg s^{-1}, the accretion flows have transited from the standard thin disk type to the ADAF type. The results support the evolutionary sequence of blazars as being: FSRQ→LBL→HBL.

1. Introduction

Ghisellini et al. (1998) used a large sample of blazar broadband spectra to study the blazar sequence. They suggested a sequence: HBL→LBL→FSRQ. This sequence represents an increasing energy density of the external radiation field that leads to an increasing amount of Compton cooling. The decrease of the maximum energy in the electron distribution causes the synchrotron and Compton peaks to shift to lower frequencies. Georganopoulos, Kirk, & Mastichiadis (2001) argued that the radiating jet plasma is outside the broad line scattering region in weak sources and within it in powerful sources, and the model fits to the spectra of several blazars proposed a sequence: FSRQ→LBL→HBL. The evolutionary sequence: FSRQ→LBL→HBL, has recently been suggested by D'Elia & Cavaliere (2000). In this evolutionary sequence, less gas is left to fuel the central engine for BL Lac objects, and the advection dominated accretion flows (ADAFs) may be in most BL Lac objects.

There are several tens of BL Lac objects in which one (or more) broad emission line has been detected. It is therefore possible to infer the central ionizing luminosity through their broad line emission for these BL Lac objects. The limits on the central black hole mass can be obtained from its line emission.

2. Estimate on the Ionizing Luminosity

The observed continuum emission from the jets in BL Lac objects is strongly beamed to us. We have to estimate the ionizing continuum luminosity $L_{\lambda,\mathrm{ion}}$ at the given wavelength λ_0 as

$$L_{\lambda,\mathrm{ion}}(\lambda_0) = \frac{L_{\mathrm{line}}}{EW_{\mathrm{ion}}}, \tag{1}$$

where $EW_{\rm ion}$ is the equivalent width of the broad emission line corresponding to the ionizing continuum emission (different from the observed continuum emission).

The average value of $EW_{\rm H\beta}$ is 100Å for 70 radio-quiet quasars in the PG sample. We will take $EW_{\rm ion}$ = 100Å and use the broad emission line Hβ to estimate the ionizing continuum luminosity of BL Lac objects. If the $EW_{\rm ion}$ of BL Lac objects deviates from that of quasars systematically, then the estimated black hole mass could be modified with $EW_{\rm ion}$ (see further discussion in Cao 2002).

3. Estimate of the Black Hole Mass

For a standard thin disk, the ionizing continuum luminosity is mainly determined by the central black hole mass and accretion rate. For a low accretion rate, the accretion flow will transit from the standard thin disk type to the ADAF type.

3.1. Standard Thin Accretion Disks

The standard thin accretion disks are thought to be in most quasars, and the luminosity at optical wavelength can be related with the disk luminosity. The central black hole mass $M_{\rm bh}$ is then estimated by $M_{\rm bh} \simeq 10^{-38}(L_{\rm d}/\dot{m})M_{\odot}$, where $\dot{m} = \dot{M}/\dot{M}_{\rm Edd}$. The ionizing luminosity $L_{\lambda,{\rm ion}}$ can be inferred from $L_{\rm H\beta}$, and we can finally estimate the black hole mass from the broad line luminosity $L_{\rm H\beta}$, if $\dot{m}$ is known.

3.2. ADAFs

The transition of the accretion flow from the thin disk type to the ADAF type occurs while $\dot{m}$ decreases to a value below $\dot{m}_{\rm crit}$ (Narayan & Yi 1995; Yi 1996). The spectrum of an ADAF: $L_{\lambda}(M_{\rm bh}, \dot{m}, \alpha, \beta)$ can be calculated if the parameters $M_{\rm bh}$, $\dot{m}$, α, and the fraction of the magnetic pressure β are specified. We can calculate the spectra of ADAFs using the approach proposed by Mahadevan (1997). Our numerical results show that the maximal optical luminosity $L_{\lambda}^{\rm max}$ always requires $\beta = 0.5$, i.e., equipartition case, if all other parameters are fixed. We plot the maximal value of $\lambda L_{\lambda}^{\rm max}(\lambda_0)$ varying with black hole mass $M_{\rm bh}$ in Fig. 1, for $\alpha = 0.3$ and $\beta = 0.5$. We can obtain a lower limit on the mass of the black hole from $L_{\rm H\beta}$ using the relation $\lambda L_{\lambda}^{\rm max} - M_{\rm bh}$ plotted in Fig. 1.

For BL Lac objects, most accretion power may be carried by strong jets. In this case, the flow is fainter than the pure ADAF considered here. The maximal $\lambda L_{\lambda}^{\rm max}(\lambda_0)$ derived here is therefore still valid.

4. Masses of Black Holes in BL Lac Objects

We search the literature for all BL Lac objects with broad emission line fluxes, and this leads to a sample of 23 sources. One can find the list of the sources in Cao (2002). Most of them are LBLs, except three HBLs: 0651+428, Mkn 421, and Mkn 501. The distribution of the line luminosity $L_{\rm H\beta}$ is plotted in Fig. 2.

We use Eq. (1) and the line luminosity $L_{\rm H\beta}$ to estimate the ionizing continuum luminosity at 4861 Å. The central black hole masses $M_{\rm bh,1}$ and $M_{\rm bh,2}$

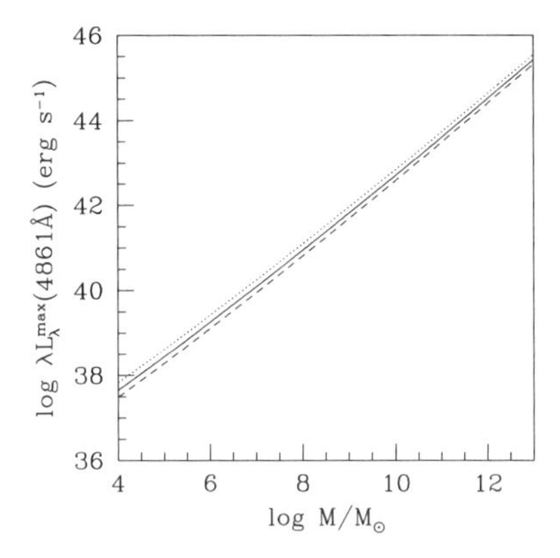

Figure 1. The maximal optical luminosity of ADAFs at 4861 Å as functions of black hole mass for different values α: $\alpha = 0.1$ (dashed line), 0.3 (solid), and 1 (dotted).

can be estimated in the cases of a thin disk and an ADAF, respectively. If the transition of the accretion flow from the thin disk type to the ADAF type occurs at $\dot{m} \sim \dot{m}_{\rm crit}$, we can have an upper limit on $M_{\rm bh,1}$ setting $\dot{m} = 0.025$ for $\alpha = 0.3$. For ADAFs, $M_{\rm bh,2}$ is the lower limit, since the maximal optical continuum luminosity is calculated for the given black hole mass. The values of the derived black hole masses can be found in Cao (2002).

5. Discussion

The distribution of line luminosity $L_{\rm H\beta}$ of all these BL Lac objects suggests a bimodal nature (see Fig. 2). We define the sources with $L_{\rm H\beta} < 10^{41}$ ergs s^{-1} as population A, and all others are in population B.

For thin disk cases, the lower limits on the hole mass would be in 10^{4-8} $M_\odot$, while the upper limits on the hole mass would be in 10^{6-10} $M_\odot$ if $\dot{m} \sim 0.025$. For accretion rate $\dot{m} < 0.025$, the accretion flow would be in ADAF state. In this case, the lower limits on the black hole mass are in: 10^{8-12} $M_\odot$. Noting that this is the lower limit, the black hole mass could be much higher if the accretion rate is low or/and the viscosity α is small.

If the accretion flows in these sources of population A are in thin disk state, their central black holes would have masses 10^{4-6} $M_\odot$ for any value of $\dot{m}$. The lower limits on the mass of the black hole in these sources are in the range of $1.66 - 24.5 \times 10^8 M_\odot$, if the accretion flows are in ADAF state. It may probably that the sources in population A have already been in ADAF state.

The sources in population B may have standard thin accretion disks surrounding the black holes, otherwise some black holes should be at least as huge as $10^{12} M_\odot$. If the accretion rate $\dot{m}$ of these sources is as small as ~ 0.025, slightly higher than the critical value: $\dot{m}_{\rm crit}$, the black hole mass would be in

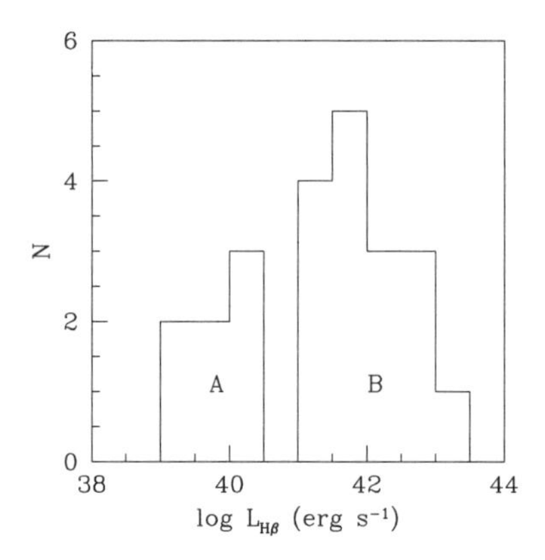

Figure 2. The distribution of the broad emission line luminosity $L_{\mathrm{H}\beta}$.

the range $10^{8-10} M_{\odot}$. This is compatible with the unified models of radio-loud quasars.

There are three HBLs in our sample in population A, and all sources in population B are LBLs. It may imply that the accretion flows in all HBLs are in ADAF state. We speculate that these LBLs in population B will finally exhaust the gas near the hole and the disks will transit to ADAFs. These LBLs are most probably in the intermediate state of the evolutionary sequence from FSRQ to BL Lac object. Most other LBLs and HBLs without any broad emission line detected may be in population A, and the accretion flows have already been in ADAF state. The fact that no HBL is in population B may imply that the evolutionary sequence of BL Lac objects should be LBL→HBL. The results present in this work support the evolutionary sequence FSRQ→LBL→HBL suggested by D'Elia & Cavaliere (2000).

Acknowledgments. This work is supported by NSFC (No. 10173016) and the NKBRSF (No. G1999075403).

References

Cao, X. 2002, ApJ, 570, L13

D'Elia, V., & Cavaliere, A. 2000, PASP, 227, 252

Dibai, E. A. 1981, Soviet Astron., 24, 389

Georganopoulos, M., Kirk, J. G. & Mastichiadis, A. 2001, in ASP Conf. Ser. 227, Blazar Demographics and Physics, ed. P. Padovani & C. M. Urry (San Francisco: ASP)

Ghisellini, G., Celotti, A., Fossati, G., Maraschi, L. & Comastri, A. 1998, MNRAS, 301, 451

Mahadevan, R. 1997, ApJ, 477, 585

Narayan, R. & Yi, I. 1995, ApJ, 452, 710

Yi, I. 1996, ApJ, 473, 645

IAU 8th Asia-Pacific Regional Meeting
ASP Conference Series, Vol. 289, 2003
S. Ikeuchi, J. Hearnshaw & T. Hanawa, eds.

3–4 μm Spectroscopy of 17 Seyfert 2 Nuclei: Quantification of the Compact Starburst Contribution

Masatoshi Imanishi

National Astronomical Observatory, Mitaka, Tokyo 181-8588, Japan

Abstract. We report on 3–4 μm slit spectroscopy of 17 Seyfert 2 nuclei. The 3.3 μm polycyclic aromatic hydrocarbon (PAH) emission was used to estimate the magnitude of compact (< a few hundred pc) nuclear starbursts. For three of the selected Seyfert 2 nuclei, the magnitudes of compact nuclear starbursts estimated from the 3.3 μm PAH emission feature (with no extinction correction) were in satisfactory quantitative agreement with those based on UV observations after extinction correction, indicating that the *observed* 3.3 μm PAH emission luminosity can be used to estimate the magnitude of compact nuclear starbursts in the nuclei of Seyfert 2s. We found that (1) except in one case, the observed nuclear 3–4 μm fluxes were dominated by AGNs and not by starbursts, (2) compact nuclear starbursts were detected in 6 out of 17 Seyfert 2 nuclei, but cannot dominate the infrared dust emission luminosities in the majority of the observed Seyfert 2 galaxies, and (3) more powerful AGNs tended to be related to more powerful compact nuclear starbursts. In addition to those showing 3.3 μm PAH emission, 3.4 μm carbonaceous dust absorption was detected in two Seyfert 2 nuclei, and 3.05 μm H_2O ice absorption was detected in one nucleus, and possibly detected in one more.

1. Introduction

According to unified models of active galactic nuclei (AGNs), Seyfert 2s are believed to contain obscured AGNs behind dusty tori. At the nuclei of Seyfert 2s, in addition to the obscured AGNs, compact nuclear starbursts may also be energetically important. These nuclear starbursts are thought to occur at the outer edges of the dusty tori (Heckman et al. 1997). Their sizes are less than a few hundred pc, or less than 2 arcsec at $z > 0.005$, so ground-based *slit* spectroscopy, with slits less than a few arcsec in width, is best suited to investigate their properties, as this minimizes the contamination from extended (> kpc) emission from the host galaxies.

Our understanding of the compact nuclear starbursts in Seyfert 2s is, however, still incomplete. It has been argued that they are detected in optical spectra (Gonzalez Delgado et al. 2001), but they are undetected in near-infrared K-band spectra (Ivanov et al. 2000). Unfortunately, these spectra can only be used to make *qualitative* statements about their presence or absence, and cannot be used to quantify their absolute magnitudes (luminosities). Their luminosities can be quantified if the UV data are corrected for dust extinction correction using an empirical method, but this method has so far been applied only to four UV-bright Seyfert 2s (Gonzalez Delgado et al. 1998). New methods are required in order to quantify the magnitudes of compact nuclear starbursts in a larger sample of Seyfert 2s.

Ground-based 3–4 μm slit spectroscopy can provide an excellent tool for making quantitative determinations of the luminosities of compact nuclear starbursts, for several reasons. First, the 3.3 μm polycyclic aromatic hydrocarbon (PAH) emission feature is detected only in starbursts and not in AGNs, making its luminosity a good measure of starburst activity. Second, dust extinction at these wavelengths is much lower than in the UV and optical, making the uncertainty in the dust extinction correction much smaller. Third, since the 3.3 μm PAH emission from starbursts is intrinsically very strong (the equivalent width is $\sim$0.1 μm), signatures of even weak compact nuclear starbursts in the nuclei of Seyfert 2s are detectable in normal (S/N $\sim$ 20) spectra. We have therefore carried out 3–4 μm slit-spectroscopic observations of 17 Seyfert 2s. Results on 13 of these objects were reported in Imanishi (2002). For Mrk 686, we present a new 2.8–3.1 μm spectrum to demonstrate the possible presence of 3.05 μm H_2O ice absorption. Four more Seyfert 2s have been recently observed with IRTF and Subaru, and their spectra are shown here for the first time. Although the current sample is not statistically complete, the observations of these Seyfert 2s can provide useful information on compact nuclear starbursts.

2. Results

Figure 1 shows the observed 3–4 μm spectra. The 3.3 μm PAH emission is detected in six Seyfert 2s (Mrk 78, Mrk 266SW, Mrk 273, Mrk 477, NGC 3227, and NGC 5135), suggesting that some fraction of the observed nuclear 3–4 μm fluxes originate in starbursts. Mrk 463 and NGC 1068 show detectable 3.4 μm carbonaceous dust absorption features, but no detectable 3.3 μm PAH emission. The 3.05 μm H_2O ice absorption is detected in NGC 4388 (and possibly also in Mrk 686). The others show nearly featureless 3–4 μm spectra.

3. Discussion

3.1. The Luminosities of Compact Nuclear Starbursts; PAH vs UV

For three selected Seyfert 2s, whose nuclear starburst magnitudes had previously been estimated based on UV data (IC 3639, Mrk 477, NGC 5135; Gonzalez Delgado et al. 1998), we found that our estimates based on the *observed* 3.3 μm PAH emission luminosities (with no extinction correction) agreed quantitatively very well with the previous estimates from the UV after dust extinction correction had been applied (Imanishi 2002). This agreement suggests that dust extinction is negligible at 3–4 μm in the case of compact nuclear starbursts in Seyfert 2s, and thus that we have a basis to argue that the *observed* 3.3 μm PAH emission luminosities can be used to measure the magnitudes of compact nuclear starbursts.

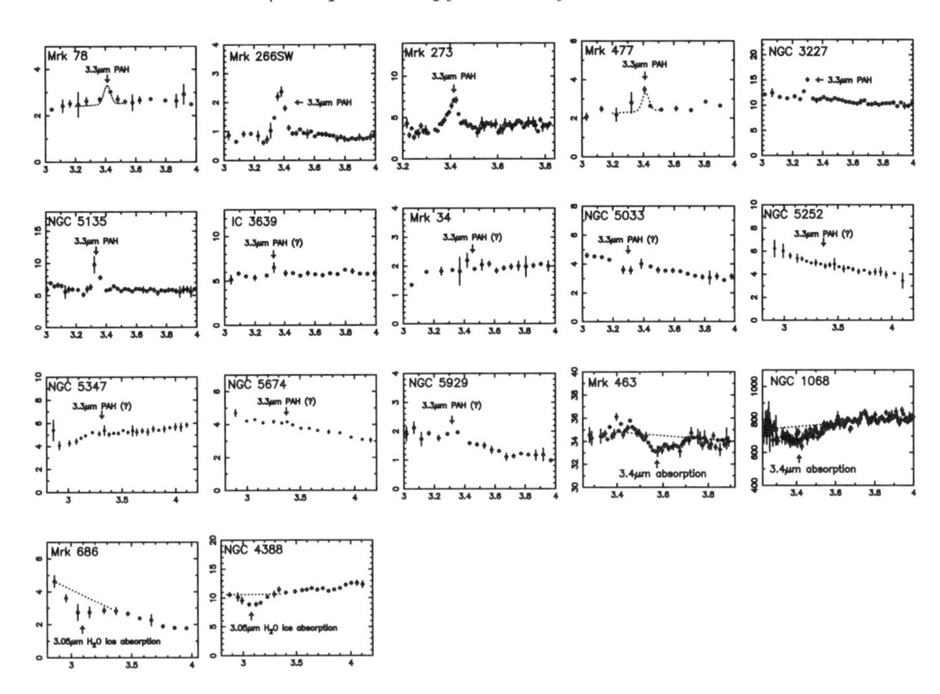

Figure 1. 3–4 μm spectra of 17 Seyfert 2s. The abscissa and ordinate are the observed wavelength in μm and F_λ in 10^{-15} W m^{-2} μm^{-1}, respectively.

3.2. The Origin of the Observed 3–4 μm Fluxes from Seyfert 2 Nuclei

If the observed 3–4 μm fluxes from Seyfert 2 nuclei were dominated by compact nuclear starbursts, then the rest-frame equivalent widths of the 3.3 μm PAH emission should be as large as $\sim$ 0.1 μm, the value for starburst-dominated galaxies (Imanishi & Dudley 2000). However, with the exception of one object (Mrk 266SW), they are all much lower than $\sim$0.1 μm. We can thus conclude that the observed 3–4 μm fluxes from the majority of Seyfert 2 nuclei are dominated by AGNs and not by starbursts.

3.3. The Energetic Importance of Compact Nuclear Starbursts

We compare the 3.3 μm PAH emission luminosities measured with our slit spectra with infrared dust emission luminosities of Seyfert 2s measured with *IRAS*, to estimate the energetic contribution of the compact nuclear starbursts. If the compact nuclear starbursts dominated the infrared luminosities of Seyfert 2s, then the PAH to infrared luminosity ratios should be as large as 10^{-3}, the value for starburst-dominated galaxies (Imanishi 2002). However, the actual measured ratios are much smaller than this value. We conclude that compact nuclear starbursts do not contribute significantly to the infrared dust emission luminosities of Seyfert 2s, which must instead be dominated by AGNs and/or extended (> kpc) starbursts in the host galaxies.

3.4. The Correlation between Compact Nuclear Starbursts and AGNs

Figure 2 compares the 3.3 μm PAH emission luminosities measured with our slit spectra with *IRAS* 12 μm luminosities. We applied the generalized Kendall's tau rank correlation statistic (Isobe, Feigelson, & Nelson 1986) to the data points in Figure 2 and estimated the probability that a correlation is not present to be 0.07. Since the PAH luminosities measured with slit spectra reflect the magnitudes of the compact starbursts, if the *IRAS* 12 μm luminosities are a good measure of AGN power (Gonzalez Delgado et al. 2001), then the magnitudes of compact nuclear starbursts and AGNs are correlated in the present sample.

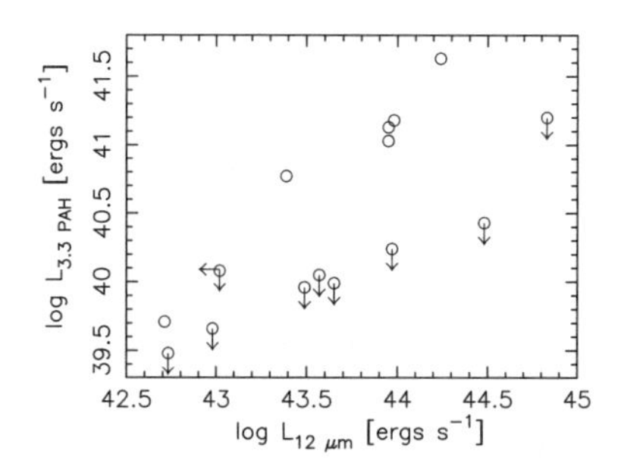

Figure 2. The relation between the logarithm of the *IRAS* 12 μm luminosity (νF_ν) in ergs s^{-1} (abscissa) and the 3.3 μm PAH emission luminosity in ergs s^{-1} measured with our slit spectra (ordinate). NGC 5033 is not included (see Imanishi 2002). NGC 5674 is also not included because no *IRAS* data are available.

4. Summary

Our three main conclusions are stated in the abstract of the paper. A detailed discussion can be found in Imanishi (2002). We are planning to apply this successful method to a statistically complete sample of Seyfert 2s.

References

Gonzalez Delgado, R. M., Heckman, T., Leitherer, C., Meurer, G., Krolik, J., Wilson, A. S., Kinney, A. & Koratkar, A. 1998, ApJ, 505, 174

Gonzalez Delgado, R. M., Heckman, T. & Leitherer, C. 2001, ApJ, 546, 845

Heckman, T. M., Gonzalez Delgado, R., Leitherer, C., Meurer, G. R., Krolik, J., Wilson, A. S., Koratkar, A. & Kinney, A. 1997, ApJ, 482, 114

Imanishi, M. 2002, ApJ, 569, 44

Imanishi, M.,& Dudley, C. C. 2000, ApJ, 545, 701

Isobe, T., Feigelson, E. D. & Nelson, P. I. 1986, ApJ, 306, 490

Ivanov, V. D., Rieke, G. H., Groppi, C. E., Alonso-Herrero, A., Rieke, M. J. & Engelbracht, C. W. 2000, ApJ, 545, 190

IAU 8th Asia-Pacific Regional Meeting
ASP Conference Series, Vol. nnn, 2003
S. Ikeuchi, J. Hearnshaw & T. Hanawa, eds.

A High Resolution Imaging Survey of CO, HCN and HCO^+ Lines towards Nearby Seyfert Galaxies

Kotaro Kohno

Institute of Astronomy, University of Tokyo, Mitaka, Tokyo, 181-0015, Japan

Abstract. We have conducted a high resolution imaging survey of mm-wave molecular lines, i.e., CO(1–0), HCN(1–0), and HCO^+(1–0) towards nearby Seyfert galaxies using the Nobeyama Millimeter Array and the RAINBOW interferometer. Some of Seyfert galaxies show extremely high HCN/CO and HCN/HCO^+ line ratios, which are not observed in nuclear starburst galaxies. These molecular line ratios can be a new diagnostic tool to investigate the "AGN – starburst connection" in active galaxies.

1. Introduction

The dense molecular medium plays various roles in the vicinity of active galactic nuclei (AGNs). The presence of a dense and dusty interstellar matter (ISM), which obscures the broad line regions in AGNs, is inevitable on scales of < 1 pc to a few tens of parsecs, according to the proposed unified model of Seyfert galaxies (e.g., Antonucchi 1993). A circum-nuclear dense ISM could be a reservoir of fuel for the active nuclei, and also be a site of massive star formation.

We (the radio astronomy group in the Institute of Astronomy, University of Tokyo and Nobeyama Radio Observatory/National Astronomical Observatory in Japan) have conducted a high resolution imaging survey of CO(1–0), HCN(1–0), and HCO^+(1–0) lines towards nearby Seyfert galaxies using the Nobeyama Millimeter Array and the RAINBOW interferometer. In this paper, we describe the sample and survey status, along with preliminary results. Some of the results and images are presented in Kohno et al. (2001), Kohno et al. (2003), Sofue et al. (2003) and so on.

2. Sample and Survey Status

The Seyfert sample is mainly taken from the Palomar Northern Seyfert galaxies (52 objects), which are selected based on a spectroscopic survey of 486 northern bright galaxies (Ho et al. 1997; Ho & Ulvestad 2001). Some famous Seyfert galaxies in the southern sky, which is out of the survey area of the Palomar Seyfert sample, are also included in our sample. We tabulate the sample and survey status in Table 1. Our sample consists of 20 nearby Seyfert galaxies in total. It contains 16 Seyfert galaxies which are listed in the Palomar sample (i.e., $\sim 30\%$ of the Palomar sample). Typical spatial resolutions are about $2''$

to 5″, and typical rms noise levels in channel maps are about 20 to 40 mJy beam^{-1} for CO (velocity resolutions are about 5 to 21 km s^{-1}), and about 5 to 10 mJy beam^{-1} for HCN and HCO$^+$ (velocity resolutions are about 27 or 54 km s^{-1}). Some of the data will also be taken from the literature (e.g., NGC 1275 from Inoue et al. 1995; NGC 3031 from Sakamoto et al. 2000). In addition to the Seyfert survey, we are also conducting a simultaneous imaging survey of HCN and HCO$^+$ emissions in nuclear starburst galaxies (e.g., Shibatsuka et al. in preparation). Comparison of the line ratios in Seyfert galaxies with those in nuclear starburst galaxies will give a clew to find a similarity and/or difference in the nature of dense molecular material in the centers of galaxies with and without nuclear activity.

Table 1. The Seyfert sample

Name	Class	Morphology	*D*	CO	HCN	HCO$^+$
Seyfert galaxies listed in Palomar northern sample						
NGC 1068	S1.9	(R)SA(rs)b	14.4		○	○
NGC 1275	S1.5	Pec	70.1	○		
NGC 1667	S2	SAB(r)c	61.2	○	○	○
NGC 3031	S1.5	SA(s)ab	1.4	○		
NGC 3079	S2	SB(s)c	20.4	○	○	○
NGC 3227	S1.5	SAB(s)a pec	20.6	△	△	△
NGC 3982	S1.9	SAB(r)b:	17.0	○	△	△
NGC 4051	S1.2	SAB(rs)bc	17.0	○	△	△
NGC 4258	S1.9	SAB(s)bc	6.8	△	△	△
NGC 4151	S1.5	(R')SAB(rs)ab:	20.3	○		
NGC 4501	S2	SA(rs)b	16.8	○		
NGC 4579	S1.9/L1.9	SAB(rs)b	16.8	○		
NGC 5033	S1.5	SA(s)c	18.7	○	○	○
NGC 5194	S2	SA(s)bc pec	7.7	○	○	○
NGC 6951	S2	SAB(rs)bc	24.1	○	○	
NGC 7479	S1.9	SB(s)c		○	○	○
Southern Seyfert galaxies						
NGC 1097	S1	(R')SB(r'l)b	14.0	○	○	○
NGC 5135	S2	SB(l)ab	54.8	○		
NGC 6764	S2	SB(s)bc	32.2	○	○	○
NGC 6814	S1.5	SAB(rs)bc	20.8	○		
NGC 7465	S2	(R')SB(s)0	26.2	○		
NGC 7469	S1.2	(R')SAB(rs)a	65.2	○	○	○

(1) Galaxy name; (2) Seyfert class from Ho et al. 1997 for Palomar sample, and from NED for the rest; (3) Distance in Mpc. From Tully 1988 for Palomar sample, and determined by using radial velocities in NED and $H_0 = 75$ km s^{-1} Mpc^{-1} for the southern Seyferts; (4), (5), (6) The survey status. ○ = observed; △ = observations in progress; blank = not observed.

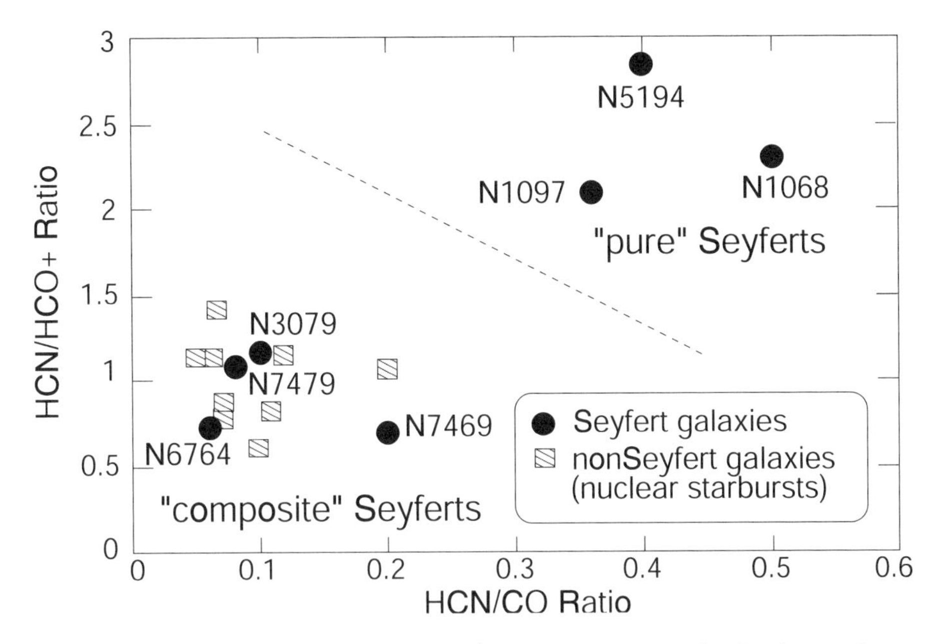

Figure 1. HCN/CO and HCO+/HCN ratios in nearby Seyfert and nuclear starburst galaxies.

3. Results and Discussion

Preliminary results of our survey on the CO, HCN, and HCO^+ molecular line ratios in Seyfert and non-Seyfert galaxies are displayed in Figure 1. We can see that there is a segregation; i.e., some of the Seyfert galaxies show extremely strong nuclear HCN emission at a few 100 pc scale. What is the nature of these "HCN-enhanced Seyferts"?

Here we compare the observed line ratios in Seyferts with those in nuclear starburst galaxies, which were also measured with similar angular resolutions. In Figure 1, it is immediately evident that Seyferts without abnormal HCN enhancements, i.e., NGC 3079, NGC 6764, NGC 7479, and NGC 7469, show $R_{\rm HCN/CO}$ and $R_{\rm HCN/HCO^+}$ values just comparable to those in nuclear starbursts; they have $R_{\rm HCN/CO}$ less than 0.3, and $R_{\rm HCN/HCO^+}$ ranging from 0.5 to 1.5. On the other hand, HCN-enhanced Seyferts, i.e., NGC 1068, NGC 1097, and NGC 5194 also have very high $R_{\rm HCN/HCO^+}$ values (> 2). Note that Nguyen-Q-Rieu et al. (1992) reported a very high $R_{\rm HCN/HCO^+}$ in NGC 3079 and Maffei 2 (> 3), yet our new high resolution measurements gave moderate (~ 1) ratios. We can measure $R_{\rm HCN/HCO^+}$ accurately because these two lines can be observed simultaneously thanks to the Ultra Wide-Band Correlator (Okumura et al. 2000).

We propose that these two groups in our "HCN diagram" (Figure 1) can be understood in terms of the "AGN - nuclear starburst connection" (note that this should not be confused with "AGN - starburst cohabitation", which often refers to the association of AGN with star formation on galactic scales in AGN hosts). In the Seyferts with line ratios comparable to those in nuclear starburst galaxies, it seems likely that a nuclear starburst (presumably in the dense molecular torus) is associated with the Seyfert nucleus (i.e., "composite"). In the nuclear regions of composite Seyferts, the HCO^+ fractional abundance is expected to increase as a result of frequent supernova (SN) explosions. In fact, in evolved starbursts such as M82, where large scale outflows have occurred as a result of numerous SN explosions, HCO^+ emission is often stronger than that of HCN (e.g., Nguyen-Q-Rieu et al. 1992). On the other hand, the HCN-enhanced Seyferts, which shows $R_{\rm HCN/CO} > 0.3$ and $R_{\rm HCN/HCO^+} > 2$, would host "pure" AGNs, where there is no associated nuclear starburst activity. In such a condition, the HCN line can be very strong because it has been predicted that the fractional abundance of HCN is enhanced by strong X-ray radiation from AGN (Leep & Dalgarno 1996), resulting in abnormally high $R_{\rm HCN/CO}$ and $R_{\rm HCN/HCO^+}$ values. Our interpretation is supported by other wavelength data; for instance, NGC 1068 has been claimed as a pure Seyfert (Cid Fernandes et al. 2001 and references therein), whereas NGC 6764 (Schinnerer et al. 2000) and NGC 7469 (Genzel et al. 1995) have a composite nature. We need further analysis to validate the proposed interpretation, but if it is the case, this will serve as a new way to investigate the nature of AGNs; although this technique requires high angular resolution observations in order to avoid contamination from extended circum-nuclear star-forming regions, it has some advantages (e.g., not being affected by dust extinction).

References

Antonucchi, R. 1993, ARAA, 31, 473

Cid Fernandes, R., Heckman, T. M., Schmitt, H., González Delgado, R. M. & Storchi-Bergmann, T. 2001, ApJ, 558, 81

Genzel, R., Weitzel, L., Tacconi-Garman, Blietz, M., Krabbe, A., Lutz, D. & Sternberg, A. 1995, ApJ, 444, 129

Ho, L. C., Filippenko, A. V. & Sargent, W. L. W. 1997, ApJS, 112, 315

Ho, L. C. & Ulvestad, J. S. 2001, ApJS, 133, 77

Kohno, K., Matsushita, S., Vila-Vilaró, B., Okumura, S. K., Shibatsuka, T., Okiura, M., Ishizuki, S. & Kawabe, R. 2001, in The Central Kiloparsec of Starbursts and AGN: The La Palma Connection, ed. J. H. Knapen, J. E. Beckman, I. Shlosman, & T. J. Mahoney (ASP, San Francisco), 672

Kohno, K., Ishizuki, S., Vila-Vilaró, B., Matsushita, S. & Kawabe, R. 2003, PASJ, 55, in press (astro-ph/0210579)

Nguyen-Q-Rieu, Jackson, J. M., Henkel, C., Truong-Bach & Mauersberger, R. 1992, ApJ, 399, 521

Okumura, S. K., Momose, M., Kawaguchi, N., Kanzawa, T., Tsutsumi, T., Tanaka, T., Ichikawa, T., Suzuki, K. et al. 2000, PASJ, 52, 393

Sakamoto, K., Fukuda, H., Wada, K. & Habe, A. 2001, AJ, 122, 1319

Schinnerer, E., Eckart, A. & Boller, T. 2000, ApJ, 545, 205

Sofue, Y., Koda, J., Nakanishi, H., Onodera, S., Kohno, K., Tomita, A. & Okumura, S. K. 2003, PASJ, submitted

Tully, R. 1988, Nearby Galaxy Catalog (Cambridge Univ. Press, Cambridge)

IAU 8th Asia-Pacific Regional Meeting
ASP Conference Series, Vol. 289, 2003
S. Ikeuchi, J. Hearnshaw & T. Hanawa, eds.

Starburst-AGN Connections from High Redshift to the Present Day

Yoshiaki Taniguchi

Astronomical Institute, Graduate School of Science, Tohoku University, Japan

Abstract. We give a review of possible starburst-AGN (active galactic nuclei) connections from high redshift to the present day. First, we give an historical review on some basic ideas related to the starburst-AGN connection published in the literature. Secondly, we focus our attention on the so-called Magorrian relation, which is the close relationship between the nuclear black hole mass and the bulge mass, established in nearby galaxies. If the Magorrian relation is universal, we obtain an important implication that any supermassive black holes were made through successive merging processes of starburst remnants (i.e., neutron stars and stellar-sized black holes), providing a channel for the starburst-AGN connection. Thirdly, we briefly discuss a possible new scenario for the formation of quasar nuclei at very high redshift, based on an idea that successive mergers of starburst remnants formed in sub-galactic gaseous clouds.

1. General Introduction

1.1. Introduction

Nuclear (or circum-nuclear) gas is often ionized at some level in most nearby galaxies (in particular, disk galaxies). It is known that there are two fundamental types of nuclear activity: (1) the nuclear starburst activity and (2) the nonthermal nuclear activity. Nuclear gas in the former class of galaxies is photoionized by massive OB stars while that in the latter class of galaxies is photoionized by nonthermal ionizing continuum emission from the central engine of active galactic nuclei (AGN) (e.g., Rees 1984). According to the recent extensive spectroscopic study of nuclear regions of 486 nearby galaxies promoted by Ho, Filippenko, & Sargent (1997), both types of galactic nuclei share approximately 40%, respectively, if we include objects with low emission-line luminosity or low activity. However, typical luminous starburst nuclei share approximately several per cent of nearby galaxies (e.g., Balzano 1983). Also, Seyfert nuclei, typical AGNs in the nearby universe, are found in approximately 10 per cent of nearby galaxies (Ho et al. 1997). Therefore, roughly speaking, $\approx$ 10% of galactic nuclei experience nuclear starburst activity, $\approx$ 10% of galactic nuclei experience nonthermal activity, and the remaining $\approx$ 80% of galactic nuclei show little evidence for significantly high level of such activities (i.e., nearly normal galactic nuclei, hereinafter NGNs).

Here several questions arise, namely: Why do some galactic nuclei experience a nuclear starburst? Why do some galactic nuclei have an AGN? Why do the majority of nearby galaxies show little evidence for such activities in their nuclear regions? These fundamental questions can be replaced by a fascinating question: Are there any evolutionary connections among the three types of galactic nuclei? If this is the case, we have further important questions: How are they connected? What are the important physical processes in such connections? Indeed, many astronomers, including the author, have been enslaved by the so-called starburst-AGN connection. We will introduce the main ideas proposed up to now in the next section.

1.2. Proposed Starburst-AGN Connections in the Literature

Although many ideas on the starburst-AGN connections have been proposed up to now, they may be broadly classified as follows.

[1] *From starburst to AGN through the formation of a supermassive black hole*: This idea suggests that a supermassive black hole (SMBH), which is believed to be the key ingredient of the central engine of AGNs, is made through successive mergers among starburst remnants (e.g., Weedman 1983; Norman & Scoville 1988; Taniguchi, Ikeuchi, & Shioya 1999; Ebisuzaki et al. 2001; see also Taniguchi et al. 2002b; Mouri & Taniguchi 2002b).

[2] *From starburst to AGN resulting from starburst-driven gas fuelling a supermassive black hole*: This idea suggests that the gas fuelling can be supplied either from the gaseous envelopes of supergiant stars near the supermassive black hole (Scoville & Norman 1988), from supernova ejecta (Taniguchi 1992), or from the gas associated with the nucleus of a merging partner (Taniguchi 1999; see also Taniguchi & Wada 1996).

[3] *From starburst to AGN-like phenomena*: This idea is completely different from the above two ideas because the photoionization of nuclear gas is attributed to some descendants of massive stars; e.g., hot Wolf-Rayet stars (Warmers: Terlevich & Melnick 1985), supernovae in dense gas media (Terlevich et al. 1992), shock heating by superwinds (Heckman 1980; Taniguchi 1987; Taniguchi et al. 1999), or hot planetary nebula nuclei (Taniguchi, Shioya, & Murayama 2000a; see also Shioya et al. 2002). Note that the shock heating is thought to work in some LINERs (Low Ionization Nuclear Emission-line Regions) and ULIRGs (Ultraluminous Infrared Galaxies, or ULIGs). See also Rieke et al. (1988) for a possible evolutionary path of a nuclear starburst.

[4] *From ULIRGs to quasars*: This idea suggests an evolutionary link between ULIRGs and quasars; i.e., ULIRGs are precursors of quasars in the local universe (Sanders et al. 1988). In this model, mergers between two or more gas-rich galaxies are crucially important for initiating very luminous nuclear starbursts in the central region of the merger remnant (see also Taniguchi & Shioya 1998; Taniguchi 1999).

[5] *From ULIRGs and/or LIRGs through S2s to S1s*: In this idea, type 2 Seyferts (S2s) are considered as a possible missing link between ULIRGs and/or LIRGs (luminous infrared galaxies) and type 1 Seyferts (S1s) (Heckman et al. 1989; Mouri & Taniguchi 1992, 2002a). The reason for this is that S2s tend to have circum-nuclear starburst regions more often than S1s, and their starburst ages appears older than those of typical nuclear starbursts (e.g., Cid Fernandes et al. 2001; Storchi-Bergman et al. 2001; Mouri & Taniguchi 2002a and references therein).

Although all the above ideas may not always work in actual galaxies, it seems better to keep in mind the following points: (a) Massive stars formed in a nuclear starburst evolve through hot phases (i.e., Wolf-Rayet stars, planetary nebula nuclei, and so on) to supernova explosions inevitably. Therefore, we have to take account of all the evolutionary phases when we discuss the evolution of starburst nuclei. (b) Compact remnants (i.e., stellar-sized black holes and neutron stars) also inevitably remain in the nuclear starburst region. Therefore, we have to think about the dynamical evolution of such remnants under a realistic gravitational potential, together with dynamical interactions with existing stars in the region concerned. Careful consideration of these two points makes it possible to discuss the starburst-AGN connection.

1.3. Toward a Simple Unified Model for Triggering AGNs

As outlined briefly, there are some possible evolutionary connections among AGNs, starburst galactic nuclei (SGNs), and NGNs. Prior to a discussion on the starburst-AGN connection, let us consider why some galactic nuclei are AGNs in which the central SMBH plays an important role. If only galaxies with an AGN could have an SMBH in their nucleus, it would be easily understood why some galaxies have AGNs. However, recent high-resolution optical spectroscopy of a sample of nearby, normal galaxies has shown that most nucleated galaxies have an SMBH in their center (e.g., Richstone et al. 1998; Magorrian et al. 1998). Furthermore, the relationship between the SMBH mass and the bulge (or spheroid) mass is basically similar for AGNs and NGNs (Gebhardt et al. 2000; Ferrarese et al. 2001; Wandel 2002). Therefore, the presence of an SMBH in the nucleus is not a crucial discriminator between AGNs and NGNs.

The frequency of occurrence of luminous AGNs (i.e., Seyfert nuclei) in nearby galaxies (i.e., $\approx 10\%$) implies that a typical lifetime of such nuclear activity is $\sim 10^9$ yr. Therefore, it is suggested strongly that some NGNs could be triggered to evolve to AGNs and then die after a duration of $\sim 10^9$ yr. The dead nuclei should be regarded again as NGNs. From this point of view, it seems reasonable to imagine that SGNs may provide a missing link between AGNs and NGNs. We think that this is indeed the starburst-AGN connection which we want to understand. In order to explore the whole evolutionary links among AGNs, SGNs, and NGNs, we have to take account of both the dynamical structure of host galaxies of all types of nuclei and the environmental effect (see Figure 1).

Recent systematic studies for large samples of Seyfert nuclei have shown that; 1) Seyfert nuclei do not prefer barred galaxies as their hosts (e.g., Mulchaey & Regan 1997; Hunt et al. 1999), and 2) only $\simeq 10\%$ of Seyfert galaxies have companion galaxies (e.g., De Robertis, Hayhoe, & Yee 1998a; De Robertis et

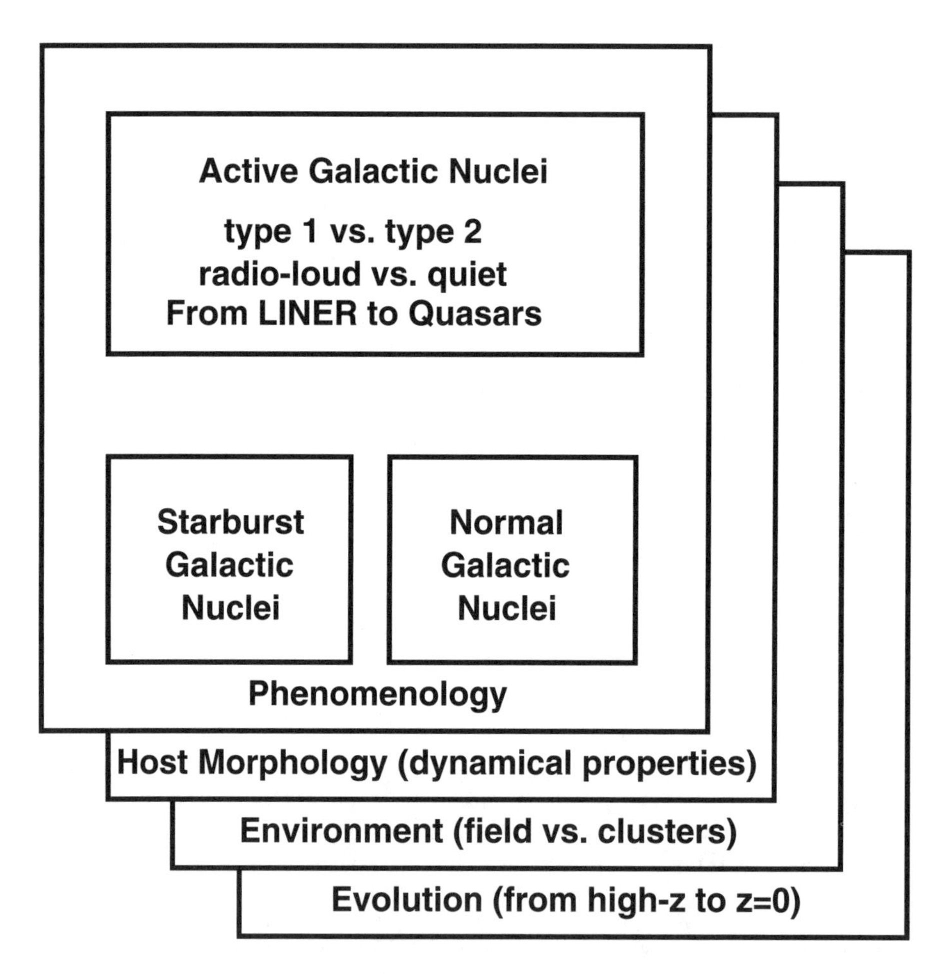

Figure 1. A grand unified model toward the understanding of starburst-AGN connections from high redshift to the present day.

al. 1998b). Therefore, it is suggested that both the dynamical effect by non-axisymmetric structures and the interaction with a companion galaxy give no simple triggering mechanism for AGNs. On the other hand, it is quite likely that any galaxies have been experiencing minor mergers during their lives (e.g., Ostriker & Tremaine 1975; Tremaine 1981; see also Zaritsky et al. 1997). Accordingly, Taniguchi (1999) suggested that the minor-merger-driven fueling appears consistent with almost all important observational properties of Seyfert galaxies (see also Taniguchi & Wada 1996). Nucleated (i.e., either an SMBH or a dense nuclear star cluster) galaxies and satellites seem necessary to ensure that the gas in the host disk is surely fueled into the very inner region (e.g., $\ll 1$ pc). Taking account that local quasars may be formed by major mergers between/among galaxies (e.g., Sanders et al. 1988; Taniguchi & Shioya 1998; Taniguchi et al. 1999a), we may have a simple unified formation mechanism of AGN in the local universe; i.e., all AGNs in the local universe are triggered by minor or major mergers between/among (nucleated) galaxies, including satellite galaxies.

2. Lesson from the Magorrian Relation

As briefly introduced in Section 1.3, the recent high-resolution optical spectroscopy of a sample of nearby, normal galaxies have shown that most nucleated galaxies have an SMBH in their center (e.g., Richstone et al. 1998; Magorrian et al. 1998); i.e., the Magorrian relation, hereinafter MR. The most important point of MR in the context of the starburst-AGN connection is that the ratio between the SMBH mass ($M_\bullet$) and the bulge (or spheroid) mass ($M_\circ$), $\approx 0.001 - 0.002$, is basically similar for AGNs and NGNs[1] (Gebhardt et al. 2000; Ferrarese et al. 2001; Wandel 2002). Therefore, we note again that the presence of an SMBH in the nucleus is not a discriminator between AGNs and NGNs. Namely, this implies that AGN phenomena are not associated with the formation process of an SMBH itself.

Recently, Merrifield et al. (2000) investigated a relationship between the $M_\bullet/M_\circ$ ratio and the age of the spheroidal system for a sample of nearby galaxies and found that galactic bulges with younger stellar populations tend to have smaller $M_\bullet/M_\circ$ ratios. This suggests that MR is slightly affected by the recent past starburst. Since the age spread in their sample galaxies is over several Gyr, one may estimate the growth timescale of an SMBH, $\tau_\bullet \sim 1$ Gyr. However, the above tendency appears weak and thus the $M_\bullet/M_\circ$ ratio can be regarded as constant for galaxies with various bulge ages.

In summary, what we have learned from MR can be summarized as follows.

[1] Since MR means that the $M_\bullet/M_\circ$ ratio is almost constant for many nearby galaxies, it is suggested that SMBH formation is linked physically to the spheroidal formation. A natural implication seems that an SMBH comes from the coalescence of nuclear star clusters which formed at the major epoch of spheroidal formation; i.e., an SMBH comes from mergers of compact remnants of massive stars born in the spheroidal formation.

[1]The Magorrian relation is now considered as the relationship between $M_\bullet$ and the central velocity dispersion of the spheroidal system (e.g., Tremaine et al. 2002 and references therein). However, for simplicity, we use the original MR between $M_\bullet$ and $M_\circ$.

[2] The same MR is found for both AGNs and NGNs. This suggests that AGN phenomena are not directly linked to the SMBH formation. Another implication is that the triggering process is much more important to turn on the nuclear activity.

[3] The $M_\bullet/M_\circ$ ratio appears to be constant for galaxies with various bulge ages, suggesting that the MR is approximately universal for many galaxies with the spheroidal component.

[4] The longer growth timescale of an SMBH ($\tau_\bullet \sim 1$ Gyr) suggests that SMBHs in AGNs may not grow up through the gas accretion process which helps the mass growth to some extent. This also results in the same implication as that of Item [1].

As mentioned before, MR provides us some important implications for the understanding of starburst-AGN connection. Here, adopting a working hypothesis that MR is universal for galaxies from high redshift to the present day, we further discuss what MR means.

Suppose that new gas fuelling the spheroidal component of a galaxy occurred at epoch t_1 with the fuelled gas mass, $\Delta M_\circ$. At this epoch, an SMBH with mass of $M_\bullet(t_1)$ was present in the galaxy center. We observe this system at epoch t_2. The spheroidal mass and the SMBH mass are $M_\circ(t_2)$ and $M_\bullet(t_2)$, respectively. At this epoch, we assume that the SMBH increases in mass, $\Delta M_\bullet$. Therefore, we have the following two relations.

$$M_\bullet(t_2) = M_\bullet(t_1) + \Delta M_\bullet, \qquad (1)$$

and,

$$M_\circ(t_2) = M_\circ(t_1) + \Delta M_\circ. \qquad (2)$$

If MR is universal, we find

$$\frac{M_\bullet(t_2)}{M_\circ(t_2)} \approx \frac{M_\bullet(t_1)}{M_\circ(t_1)}. \qquad (3)$$

In order to achieve the condition given in equation (3), there are two alternative cases.

[1] $\Delta M_\bullet \ll M_\bullet(t_1)$, and $\Delta M_\circ \ll M_\circ(t_1)$.

[2] $\Delta M_\bullet \gg M_\bullet(t_1)$, and $\Delta M_\circ \gg M_\circ(t_1)$.

In the first case, the mass increases in both the SMBH and the spheroidal component are negligibly small compared to their original masses. This case may be applicable to new gas supply onto the nucleus of a typical disk galaxy in which an SMBH already exists; i.e., the starburst-AGN connection for SGNs and AGNs like Seyfert nuclei.

On the other hand, in the second case, the mass increases in both the SMBH and the spheroidal component are significantly larger than their original masses. The conditions [2] require the following relation:

$$\frac{M_\bullet(t_2)}{M_\circ(t_2)} \approx \frac{\Delta M_\bullet}{\Delta M_\circ} \equiv f_{\rm BH}. \quad (4)$$

This case may be applicable both to the ULIRG-quasar connection at low and intermediate redshift and to the forming galaxy-quasar connection at high redshift. It is noted that the universal MR means $f_{\rm BH} \simeq 0.001 - 0.002$ from high redshift to the present day. Nice examples for this case are ULIRGs which are believed to be made from mergers between or among gas-rich galaxies. The nearest ULIRG, Arp 220, has a number of nuclear super star clusters (SSCs). Such SSCs will fall into the nuclear region via the dynamical friction within $\sim 10^{7-9}$ yr (Shaya et al. 1994; Taniguchi et al. 1999a), probably making an SMBH with $M_\bullet \sim 10^9 M_\odot$ (see also Norman & Scoville 1988; Ebisuzaki et al. 2001; Mouri & Taniguchi 2002b). Indeed, based on their elaborate numerical simulations, Bekki & Couch (2001) found $f_{\rm BH} \simeq 0.003$ for the aftermath of an ultraluminous starburst occurring in a ULIRG.

The formation of SMBHs has been a long-standing problem in modern astrophysics (Rees 1978, 1984). It is still uncertain how such SMBHs could be born in the nucleus of high-redshift quasars up to $z \sim 6$. However, recent high-resolution X-ray imaging studies have discovered possible candidates of intermediate-mass black holes (IMBHs) with masses of $M_\bullet \sim 10^{2-4} M_\odot$ in circum-nuclear regions of many (disk) galaxies (e.g., Colbert & Mushotzky 1999; Matsumoto & Tsuru 1999; Makishima et al. 2000; Strickland et al. 2001; Zezas & Fabbiano 2002). It is known that a large number of massive stars are formed in a circum-nuclear giant H II region. Therefore, Taniguchi et al. (2000b) proposed that a continual merger of compact remnants left from these massive stars is responsible for the formation of such an IMBH within a timescale of $\sim 10^9$ yr. A necessary condition is that several hundreds of massive stars are formed in a compact region with a radius of a few parsecs. Then Ebisuzaki et al. (2001) proposed that the runaway merging is an important dynamical process in such a star cluster; its timescale may be as short as $\sim 10^7$ yr (see also Mouri & Taniguchi 2002b). They also proposed that circum-nuclear star clusters themselves could merge into one during the course of dynamical evolution in the galaxy potential within a timescale of $\sim 10^9$ yr.

This idea can be applied to the formation of SMBHs in the hearts of quasars at high redshift. The number density of quasars peaks at $z \simeq 2$. Since the growth timescale of an SMBH with $M_\bullet \sim 10^9 M_\odot$ may be $\sim 10^9$ yr, it is required that episodic massive star formation could occur $\sim 10^9$ yr before $z \simeq 2$; i.e., $z_{\rm SF} \sim 15$. Even in this dark age, sub-galactic gas clumps with mass of $\sim 10^{7-8} M_\odot$ could form in the context of cold dark matter scenarios (e.g., Gnedin & Ostriker 1997). If numerous sub-galactic gas clumps were formed in a localized region with a dimension of ~ 10 kpc, piling up compact remnants could make it possible to form an SMBH with mass of $\sim 10^9 M_\odot$. According to the universal MR, the spheroidal mass of a quasar host is $\sim 10^{12} M_\odot$, being comparable to massive galaxies in the present day. Although recent optical deep imaging surveys have shown that bulge formers at high redshift (e.g., Lyman break galaxies at $z \sim 2-4$) are very small systems and thus it is unlikely that they are massive galaxies. On the other hand, recent submillimeter deep surveys have revealed that massive galaxies are really present at high redshift although they are basically hidden in the optical because of heavy extinction by a lot of dust grains (e.g., Frayer et al. 1998, 1999; Genzel et al. 2002). Indeed, a high redshift quasar BR 1202−0725 at $z = 4.7$ is associated with a massive gaseous system with mass of $\sim 10^{11-12} M_\odot$ (Ohta et al. 1996; Omont et al. 1996). The first ULIRG beyond $z = 2$, IRAS F10214+4724, is also a very massive system (Downes et al. 1992). Therefore, since it is likely that precursors of high-redshift quasars are dust enshrouded massive objects, we can apply the universal MR to such high-redshift quasars.

3. Formation of Quasar Nuclei at High Redshift

Now let us consider the formation of quasars at high redshift. In the nearby universe, quasars are associated with the nuclei of (giant) galaxies (e.g., Bahcall et al. 1997). Therefore, another interesting issue is to investigate a causal relationship between quasars and galaxies in the early universe (e.g., Ikeuchi 1981; Ostriker & Cowie 1981; Turner 1991; Silk & Rees 1998; Madau & Rees 2001). Here we give an outline of a new scenario for the formation of quasar nuclei (i.e., supermassive black holes with mass of $\sim 10^8 M_\odot$) at high redshift ($z \approx 2-5$) proposed by Ikeuchi & Taniguchi (2002).

Step I: The formation of sub-galactic gas clumps occurs at $z \approx 15$ as predicted by cold dark matter models. We assume that the total mass of the clump is $\sim 10^8 M_\odot$ and the gas mass is $\sim 2 \times 10^7 M_\odot$. Approximately one thousand clumps located within a radius of 10 kpc will be used to build up a galaxy with a mass of $\sim 10^{11} M_\odot$.

Step II: The gravitational instability in clumps could lead to the formation of massive stars in them. Given the star formation efficiency of 50%, $\sim 10^6$ stars with a mass of 10 $M_\odot$ are formed in each clump. These stars ionize about one third of the gas in the clump.

Step III: All these massive stars evolve within a timescale of $\sim 10^7$ yr and then explode as supernovae (SNe). These SNe overlap each other and then blow out as a superbubble. Superbubbles arising from one thousand clumps also overlap and then evolve into one huge superbubble. This superbubble can expand at a radius of ~ 500 kpc within a duration of $\sim 5 \times 10^8$ yr. Since this radius is larger than the mean separation among galaxies, the IGM is completely ionized by these superbubbles; i.e., the reionization of the universe. They also contribute to the metal enrichment up to a level of $Z \sim 0.01 Z_\odot$.

Step IV: Compact remnants left from massive stars after the supernova explosions can merge into one within a duration of $\sim 10^9$ yr. This leads to the formation of a seed SMBH with a mass of $\sim 2 \times 10^6 M_\odot$.

Step V: Approximately fifty seed SMBHs located within a radius of 500 pc of the galaxy merge into one within a duration of $\sim 10^9$ yr. Thus an SMBH with a mass of $\sim 10^8 M_\odot$ is made a few 10^9 yr after the initial starbursts in the sub-galactic clumps. This means that quasar nuclei (i.e, SMBHs with $M_{\rm BH} \sim 10^8 M_\odot$) can be made at $z \approx 2-5$.

Acknowledgments. I would like to thank my colleagues, in particular, Satoru Ikeuchi, Hideaki Mouri, Yasuhiro Shioya, Takashi Murayama, Tohru Nagao, Youichi Ohyama, and Neil Trentham for useful discussions.

References

Bahcall, J. N., Kirhakos, S., Saxe, D. H. & Schneider, D. P. 1997, ApJ, 479, 642

Balzano, V. A. 1983, ApJ, 268, 602

Bekki, K., & Couch, W. J. 2001, ApJ, 557, L19

Cid Fernandes, R., Heckman, T., Schmitt, H., Gonzáres Delgado, R.M. & Storchi-Bergman, T. 2001, ApJ, 558, 81

Colbert E. J. M., Mushotzky R. F. 1999, ApJ 519, 89

De Robertis, M. M., Hayhoe, K. & Yee, H. K. C. 1998a, ApJS, 115, 163

De Robertis, M. M., Yee, H. K. C. & Hayhoe, K. 1998b, ApJ, 496, 93

Downes, D., Radford, J. E., Greve, A., Thum, C., Solomon, P. M. & Wink, J. E. 1992, ApJ, 398, L25

Ebisuzaki, T. et al. 2001, ApJ, 562, L19

Ferrarese, L., Pogge, R. W., Peterson, B. M., Merritt, D., Wandel, A. & Joseph, C. L. 2001, ApJ, 555, L79

Frayer, D. T. et al. 1998, ApJ, 506, L7

Frayer, D. T. et al. 1999, ApJ, 514, L13

Gebhardt, K. et al. 2000, ApJ, 543, L5; Erratum, ApJ, 555, L75

Genzel, R. et al. 2002, ApJ, in press (astro-ph/0210449)

Gnedin, N. Y. & Ostriker, J. P. 1997, ApJ, 486, 581

Heckman, T. M. 1980, A&A, 87, 142

Heckman, T. M., Blitz, L., Wilson, A. S., Armus, L. & Miley, G. K. 1989, ApJ, 342, 735

Ho, L. C., Filippenko, A. V. & Sargent, W. L. W. 1997, ApJ, 487, 591

Hunt, L. K., Malkan, M. A., Moriondo, G. & Salvati, M. 1999, ApJ, 510, 637

Ikeuchi, S. 1981, PASJ, 33, 211

Ikeuchi, S. & Taniguchi, Y. 2002, in preparation

Madau, P. & Rees, M. J. 2001, ApJ, 551, L27

Magorrian, J. et al. 1998, AJ, 115, 2285

Makishima, K. et al. 2000, ApJ, 535, 632

Matsumoto, H. & Tsuru, G. T. 1999, PASJ 51, 321

Merrifield, M. R., Duncan, F. A. & Terlevich, A. I. 2000, MNRAS, 313, L29

Moles, M., Márquez, I. & Pérez, E. 1995, ApJ, 438, 604

Mouri, H., & Taniguchi, Y. 1992, ApJ, 386, 68

Mouri, H., & Taniguchi, Y. 2002a, ApJ, 565, 786

Mouri, H., & Taniguchi, Y. 2002b, ApJ, 566, L17

Mulchaey, J. S. & Regan, M. 1997, ApJ, 482, L135

Norman, C. & Scoville, N. 1988, ApJ, 332, 124

Ohta, K. et al. 1996, Nature, 382, 426
Omont, A. ct al. 1996, Nature, 382, 428
Ostriker, J. P. & Cowie, L. L. 1981, ApJ, 243, L127
Ostriker, J. P., & Tremaine, S. 1975, ApJ, 202, L113
Rees, M. J. 1978, Observatory, 98, 210
Rees, M. J. 1984, ARA & A, 22, 471
Richstone, D. et al. 1998, Nature, 395A, 14
Rieke, G. H., Lebofsky, M. J. & Walker, C. E. 1988, ApJ, 325, 679
Sanders, D. B. et al. 1988, ApJ, 325, 74
Scoville, N. & Norman, C. 1988, ApJ, 332, 163
Shaya, E. J., Dowling, D. M., Currie, D. G., Faber, S. M. & Groth, E. J. 1994, AJ, 107, 1675
Shioya, Y., Taniguchi, Y., Murayama, T., Nishiura, S., Nagao, T. & Kakazu, Y. 2002, ApJ, 576, 36
Silk, J., & Rees, M. J. 1998, A&A, 331, L1
Storchi-Bergman, T., Gonzáles Delgado, R. M., Schmitt, H. R., Cid Fernandes, R. & Heckman, T. 2001, ApJ, 559, 147
Strickland, D. K. et al. 2001, ApJ, 560, 707
Taniguchi, Y. 1987, ApJ, 317, L57
Taniguchi, Y. 1992, Astron. Soc. Pacific Conf. Ser., 31, Taipei Astrophysics Workshop on Relationships between Active Galactic Nuclei and Starburst Galaxies, Ed. A.V. Filippenko, p. 357
Taniguchi, Y. 1999, ApJ, 524, 65
Taniguchi, Y., Ikeuchi, S. & Shioya, Y. 1999a, ApJ, 514, L9
Taniguchi, Y. & Shioya, Y. 1998, ApJ, 501, L167
Taniguchi, Y., Shioya, Y., & Murayama, T. 2000a, AJ, 120, 1265
Taniguchi, Y., Shioya, Y. Tsuru, G. T., & Ikeuchi, S. 2000b, PASJ, 52, 533
Taniguchi, Y. & Wada, K. 1996, ApJ, 469, 581
Taniguchi, Y., Yoshino, A., Ohyama, Y. & Nishiura, S. 1999b, ApJ, 514, 660
Terlevich, R. & Melnick, J. 1985, MNRAS, 213, 841
Terlevich, R., Tenorio-Tagle, G., Franco, J. & Melnick, J. 1992, MNRAS, 255, 713
Tremaine, S. 1981, in The Structure and Evolution of Normal Galaxies, ed. S. M. Fall & D. Lynden-Bell (Cambridge: Cambridge University Press), 67
Tremaine, S. et al. 2002, ApJ, 574, 740
Turner, E. L. 1991, AJ, 101, 5
Wandel, A. 2002, ApJ, 565, 762
Weedman, D. W. 1983, ApJ, 266, 479
Zaritsky, D., Smith, R., Frenk, C. & White, S. D. M. 1997, ApJ, 478, 39
Zezas, A. & Fabbiano, G. 2002, ApJ, 577, 726

IAU 8th Asia-Pacific Regional Meeting
ASP Conference Series, Vol. 289, 2003
S. Ikeuchi, J. Hearnshaw & T. Hanawa, eds.

A New Picture of QSO Formation

Masayuki Umemura

Center for Computational Physics, University of Tsukuba, Tsukuba, Ibaraki 305-8577, Japan

Nozomu Kawakatu

Institute of Physics, University of Tsukuba, Tsukuba, Ibaraki 305-8577, Japan

Abstract. Based on a novel mechanism to build up a supermassive black hole (SMBH), we propose a new scenario for QSO formation. In the present scenario, the SMBH-to-bulge mass ratio is basically determined by the nuclear energy conversion efficiency from hydrogen to helium, $\varepsilon = 0.007$. It is predicted that a host luminosity-dominant "proto-QSO phase" exists before the QSO phase. The proto-QSO phase is preceded by an optically-thick ultraluminous infrared galaxy (ULIRG) phase.

1. Introduction

The recent discovery of high redshift quasars with $z > 6$ (Fan et al. 2001) implies that the formation of QSO black holes proceeded in less than 10^9 yr. Also, the demography of massive dark objects (MDOs) have revealed intriguing correlations as follows. The SMBH mass exhibits a linear relation to the bulge mass with the ratio of $f_{\rm BH} \equiv M_{\rm BH}/M_{\rm bulge} = 0.001 - 0.006$ as a median value (Richstone et al. 1998; Magorrian et al. 1998; Gebhardt et al. 2000; Merritt & Ferrarese 2001). The $f_{\rm BH}$ tends to grow with the age of youngest stars in a bulge until 10^9 yr (Merrifield et al. 2000). As for quasars, $f_{\rm BH}$ is of a similar level to that for elliptical galaxies (Laor 1998). On the other hand, the observations of the X-ray emission (Brandt et al. 1997) or Paα lines (Veilleux, Sanders, & Kim 1999) intrinsic for active nuclei have been detected in roughly one forth of ultraluminous infrared galaxies (ULIRGs). Furthermore, it has been revealed that QSO host galaxies are mostly luminous and well evolved early-type galaxies (Hooper, Impey, & Foltz 1997; McLure, Dunlop, & Kukula 2000). These findings on QSO hosts and SMBHs suggest that the formation of QSO, bulge, and SMBH is mutually related.

In this paper, based on a recently proposed radiation-hydrodynamical model for black hole formation (Umemura 2001; Kawakatu & Umemura 2002), we attempt to construct a new picture of QSO formation, which predicts the physical relation among QSO, bulge, and SMBH, and also "proto-QSO phase".

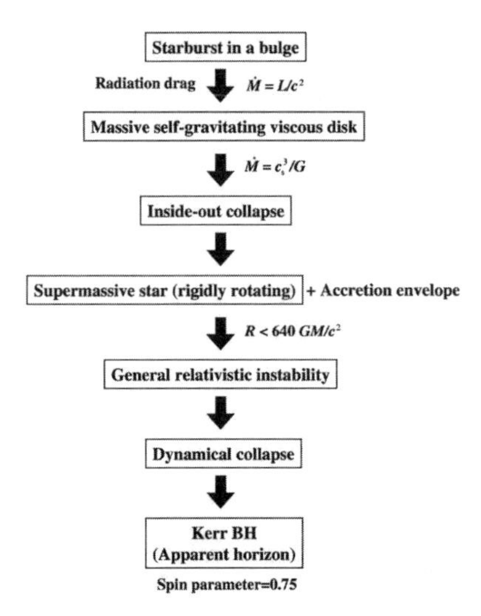

Figure 1. SMBH formation from a galactic scale to a horizon.

2. SMBH Formation

Recently, a novel mechanism by radiation drag is proposed for the SMBH formation (Umemura 2001). The drag force by the radiation from bulge stars can extract the angular momentum from interstellar gas and allow the gas to accrete onto the central BH. For the total luminosity L_* of a uniform bulge, the angular momentum loss rate by the radiation drag is given by $d\ln J/dt \simeq -(L_*/c^2M_{\rm g})(1-{\rm e}^{-\tau})$, where τ is the total optical depth of the system and $M_{\rm g}$ is the total mass of gas. Then, the mass accretion rate is estimated to be $\dot{M} = -M_{\rm g}d\ln J/dt \simeq (L_*/c^2)(1-{\rm e}^{-\tau})$. The timescale of radiation drag-induced mass accretion is given by

$$t_{\rm drag} \simeq \frac{c^2R^2}{\chi L_*} = 8.6\times 10^7 \left(\frac{R}{1\,{\rm kpc}}\right)^2 \left(\frac{L_*}{10^{12}\,L_\odot}\right)^{-1} \left(\frac{Z}{Z_\odot}\right)^{-1} {\rm yr}, \tag{1}$$

where $R_{\rm kpc} = R/{\rm kpc}$ is the bulge radius and Z is the metallicity of gas. In the optically-thick regime, the mass accretion rate is simply $\dot{M} = L_*/c^2$, and then the mass of a central MDO is estimated by

$$M_{\rm MDO} = \int_0^t \dot{M}dt \simeq \int_0^t L_*/c^2 dt. \tag{2}$$

If we invoke the instantaneous recycling approximation for the star formation, the MDO mass to bulge mass ratio is given by

$$\frac{M_{\rm MDO}}{M_{\rm bulge}} = 0.14\varepsilon\alpha^{-1} = 0.002\alpha_{0.5}^{-1}, \tag{3}$$

where ε is the energy conversion efficiency of from hydrogen to helium, which is 0.007, and $\alpha_{0.5} = \alpha/0.5$ is the net efficiency of the conversion into stars after subtracting the mass loss. Hence, the final mass is basically determined by ε.

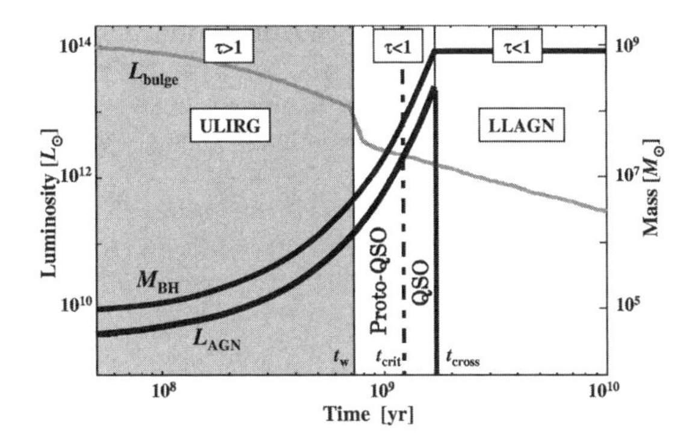

Figure 2. A scenario for QSO formation. The ordinary QSO phase is preceded by a host luminosity-dominant "proto-QSO phase".

In the MDO, the viscosity is expected to work effectively. Tsuribe (1999) provided a self-similar solution for the inside-out mass accretion rate in a self-gravitating viscous disk, $\dot{M}_\alpha = 3\alpha_\mathrm{v} c_\mathrm{s}^3/QG$, where α_v is the viscous parameter, c_s is the sound velocity, and Q is the Toomre's Q. Through this inside-out collapse, a rigidly rotating supermassive star is expected to form. Then, it evolves into a Kerr black hole (Shibata & Shapiro 2002). The present picture for the formation of a SMBH is summarized in Figure 1.

3. QSO Formation

If the mass accretion driven by the viscosity on to the BH horizon is determined by an order of Eddington rate, the BH mass grows according to

$$M_\mathrm{BH} - M_0 e^{\nu t/t_\mathrm{Edd}}, \tag{4}$$

where ν is the ratio of BH accretion rate to the Eddington rate and t_Edd is the Eddington timescale, $t_\mathrm{Edd} = 1.9 \times 10^8$ yr. M_0 is the mass of a seed BH with $\sim 10^5 M_\odot$ (Shibata & Shapiro 2002). In order to construct a model for the chemical evolution of host galaxy, we use an evolutionary spectral synthesis code 'PEGASE' (Fioc & Rocca-Volmerange 1997), and also employ a galactic wind model with the wind epoch of $t_\mathrm{w} = 7 \times 10^8$ yr to match the present-day color-magnitude relation (Arimoto & Yoshii 1987). The system is assumed to change from optically-thin to optically-thick phase at t_w. Based on the present coevolution model, the evolution of bulge luminosity (L_bulge), the AGN luminosity (L_AGN), and the mass of SMBH (M_BH) are shown in Figure 2, assuming the constant Eddington ratio ($\nu = 1$). The M_BH reaches M_MDO at a time t_cross, so that the BH fraction becomes $f_\mathrm{BH} \simeq 0.001$, which is just comparable to the observed ratio.

In the optically-thin phase after the galactic wind ($t > t_\mathrm{w}$), M_BH still continues to grow until t_cross and therefore the AGN brightens with time. After L_AGN exhibits a peak at t_cross, it fades out abruptly due to exhausting the fuel of the MDO. This fading nucleus could be a low luminosity AGN (LLAGN). It is found that the era of $t_{\rm w} < t < t_{\rm cross}$ can be divided into two phases with a transition time $t_{\rm crit}$ when $L_{\rm bulge} = L_{\rm AGN}$; the earlier phase is the host luminosity-dominant phase and the later phase is the AGN luminosity-dominant phase. The lifetimes of both phases are comparable to each other, which is about 10^8 yr. The AGN-dominant phase is likely to correspond to ordinary QSOs, while the host-dominant phase is obviously different from observed QSOs so far. We define this phase as "a proto-QSO". The observable properties of proto-QSOs are predicted as follows: (1) The width of broad emission line is narrower, which is less than 1500 km/s. (2) $f_{\rm BH}$ rapidly increases from 10^{-5} to $10^{-3.5}$ in $\approx 10^8$ years. (3) The colors of $(B-V)$ at rest bands and $(V-K)$ at observed bands are about 0.5 magnitude bluer than those of QSOs. (4) In both proto-QSO and QSO phases, the metallicity of gas in galactic nuclei is $Z_{\rm BLR} \simeq 8Z_{\odot}$, and that of stars weighted by the host luminosity is $Z_* \simeq 3Z_{\odot}$, which are consistent with the observations for QSOs and the elliptical galaxies. (5) A massive dusty disk ($> 10^8 M_{\odot}$) surrounds a massive BH, and it may obscure the nucleus in the edge-on view to form a type 2 nucleus. The predicted properties of proto-QSOs are quite similar to those of radio galaxies. Thus, radio galaxies are a key candidate for proto-QSOs.

The proto-QSO phase is preceded by a bright and optically thick phase, which may correspond to a ultraluminous infrared galaxy (ULIRG) phase. Also, the precursor of ULIRGs is an optically-thin and very luminous phase with the lifetime of $\sim 10^7$ years. This may correspond to the assembly phase of LBGs or Lyα emitters. In this phase, the metallicity is subsolar ($Z_* < 0.1Z_{\odot}$), and the hard X-ray luminosity is $L_{\rm x} \sim 5 \times 10^8 L_{\odot}$ if $L_{\rm x} = 0.1L_{\rm AGN}$.

References

Arimoto, N. & Yoshii, Y. 1986, A&A, 164, 260
Brandt, W. N. et al. 1997, MNRAS, 290, 617
Fan, X. et al. 2001, AJ, 122, 2833
Fioc, M. & Rocca-Volmerrange, B. 1997, A&A, 326, 950
Gebhardt, K. et al. 2000, ApJ, 539, L13
Hooper, E. J., Impey, C. D. & Foltz, C. B. 1997, ApJ, 480, L95
Kawakatu, N. & Umemura, M. 2002, MNRAS, 329, 572
Laor, A. 1998, ApJ, 505, L83
Magorrian, J. et al. 1998, AJ, 115, 2285
McLure, R. J., Dunlop, J. S. & Kukula, M. J. 2000, MNRAS, 318, 693
Merrifield, M. R. et al. 2000, MNRAS, 313, L29
Richstone, D. et al. 1998, Nature, 395A, 14
Shibata, M. & Shapiro, S. 2002, ApJ, 572, L39
Tsuribe, T. 1999, ApJ, 527, 102
Umemura, M. 2001, ApJ, 560, L29
Veilleux, S., Sanders, D. B. & Kim, D.-C. 1999, ApJ, 522, 139

IAU 8th Asia-Pacific Regional Meeting
ASP Conference Series, Vol. 289, *2003*
S. Ikeuchi, J. Hearnshaw & T. Hanawa, eds.

Distribution of the Faraday Rotation Measure in AGN Jets as a Clue for Deciding a Valid Theoretical Model

Y. Uchida, M. Nakamura, H. Kigure, S. Hirose

Department of Physics, Science University of Tokyo, 1-3 Kagurazaka, Shinjuku-ku, Tokyo 162-8601, Japan

Abstract. We note that extremely valuable observational clues to the formation mechanism of AGN jets are provided from the distribution of the Faraday Rotation Measure (FRM). The results of recent observations suggest that there is quite a systematic magnetic structure in the jet, indicating that some magnetodynamic process is essential in the jet formation.

We have developed our MHD model (Uchida & Shibata 1985) for the formation of AGN jets, based on the magnetic action in the gravitational contraction forming the central gravitator with an accretion disk. The weak primordial intergalactic magnetic field is amplified in the contraction of frozen-in gas into the central object, and the braking of the disk rotation by amplified magnetic field allows further accretion of the disk. The reaction of this process produces magnetic twists which dynamically pinch the large scale field into a thin jet shape in propagating out along the large scale magnetic field. The gas is driven out in the direction of the axis as spinning jets in the propagation of the twist with large Alfvén velocity.

We show the model counterpart of the observed FRM distribution from our 3D MHD model seen from various directions, and such comparisons can determine which of the proposed models is correct.

1. Introduction

Polarization observations of radio jets from Active Galactic Nuclei (AGN) from the 1970s revealed that the directions of the projected magnetic field derived from the directions of the radio polarization of the synchrotron emission have some remarkably systematic distribution in the jets, typically systematically tilted slightly from the axis of the jet (Miley & Hartsuijker 1978 etc.).

On the other hand, the FRM is given by the integral of $n_e B_{\|}$ along the lines-of-sight between the emitter and the observer (where $B_{\|}$ is the line-of-sight component of the magnetic field, and n_e is the electron density there), and it is, in principle, *not* possible to specify which part on the line-of-sight the contribution comes from. It was thus generally *assumed* that the contribution may come from some unrelated magnetized clouds in the foreground distant from the source, and the positive significance of the FRM was not pursued.

Examining the distribution of the FRM in some jets, however, we noted that the general assumption that the Faraday rotator is an unrelated magnetized cloud far from the source is *not* necessarily true. This is because there are some cases in which the distribution of the FRM shows a clear correlation with the structure of the jet itself.

2. Examples of FRM Observations

One example is a recent result for the 3C273 jet obtained by Asada et al. (2002) by using the VLBA Archive Data. In this example, the distribution of the FRM (Fig. 1a) has a systematic gradient in the direction perpendicular to the jet axis, and the distribution of the projected magnetic field (Fig. 1b) is systematically tilted somewhat from the axis of the jet. The FRM distribution can be represented as the sum of the base-value (a constant over the source) and an antisymmetric distribution.

It is not likely that the foreground magnetized cloud at a large distance from the jet has a very sharp variation of either magnetic field (in the magnitude or in sign) or of the density just along the projected very thin jet. One likely interpretation may be that the antisymmetric contribution is due to the toroidal component of the helical magnetic field, whereas the base-value may come either from the foreground large scale magnetized clouds at large distances, or the longitudinal component in the helical magnetic field in the jet.

Another example is a result for the 3C449 jet obtained by Feretti et al. (1999) (Fig. 1e). Subtracting the base-value due to the foreground contribution, the distribution of the FRM has a reasonably clear correlation with the shape of the jets. It indicates that there exists an ordered longitudinal magnetic field in (or at least in the outer shell of) the jets and that the shape of the jets, seen as wiggles in projection, is actually a helix in 3D.

3. How to Calculate the Model Counterpart of the FRM Distribution

The comparison with observations is meaningful only if it is done with a realistic model, and that, in turn, is made possible by the computer simulation approach. We computed the distribution of the FRM by integrating $B_{\parallel} n_{\rm e}$ along the lines-of-sight by using the numerical data of the simulation. We calculated the FRM for specific orientation, by rotating the model, to compare with the observation. When the quantity to be compared is a vector quantity like the magnetic field, this is the only way to *really compare* the model with the observations, since we are looking at the astronomical objects from a given particular direction (none of them are in the same orientation), but we can not easily determine the tilt angle of the object from the plane of the sky.

In computing the FRM distribution from the model, we use in the present first trial a simplified two-layer model. We assume that the synchrotron radiation is emitted from the skin of the columnar core part of the jet, and the Faraday rotation screen is the part surrounding it.

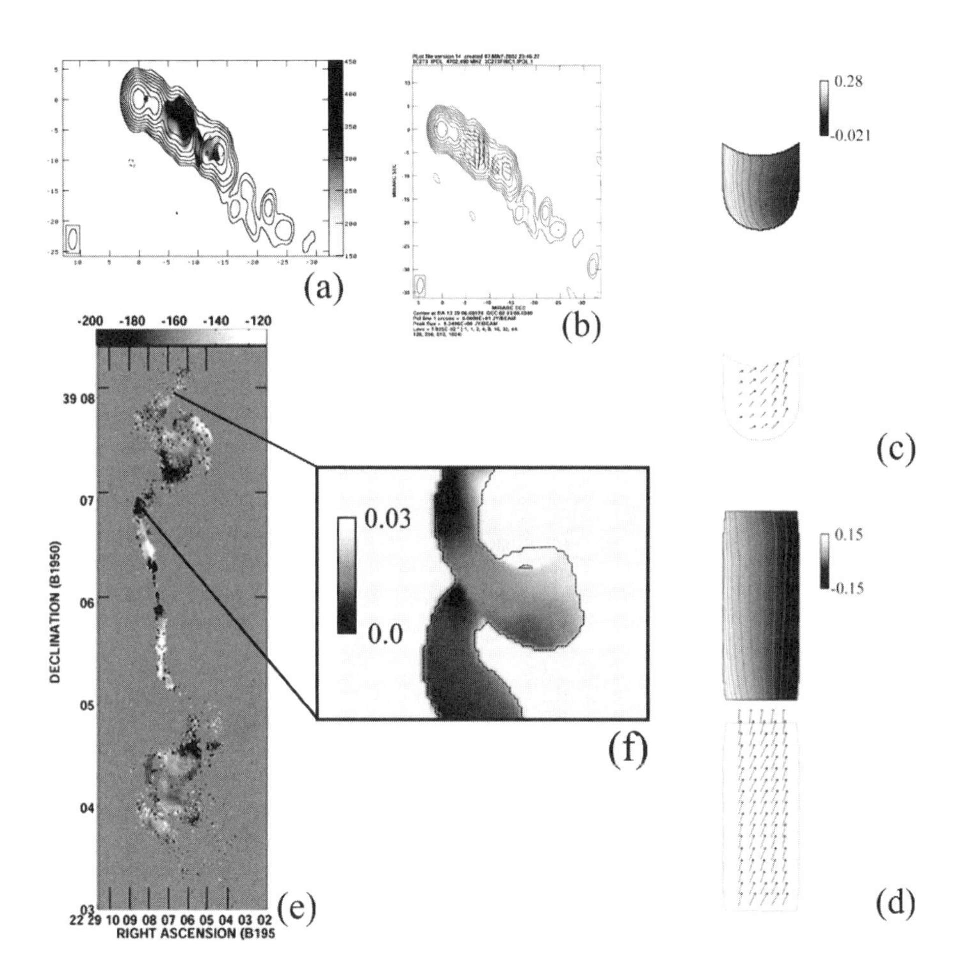

Figure 1. The distribution of (a) the FRM, and (b) the direction of intrinsic polarization, in the 3C273 jet (Asada et al. 2002). The model counterpart of the FRM distribution and the projected magnetic field, when seen at (c) 30 degrees, (d) 90 degrees, from the jet axis. (e) The distribution of the FRM in the 3C449 jets, and (f) its model counterpart

4. Results

4.1. Comparison with the Observation for the Straight Part of the Jet

We performed the 2.5D axisymmetric simulation in which the interaction between the large scale magnetic field and the accretion disk causes a jet formation. The rotation of the accretion disk continuously generates the toroidal component of the magnetic field. It propagates along the large scale poloidal field as the torsional Alfvén waves (TAWs). The propagation of TAWs pinches the large scale magnetic field and collimates it into a slender 'jet' shape (Sweeping Magnetic Twist Model). We compare the distribution of the FRM and the projected magnetic field calculated from the result of MHD simulation with the observation of the 3C273 jet. As for the projected magnetic field, its model counterpart is calculated from the magnetic field on the emitting surface.

When seen from the direction perpendicular to the axis, the result of integration is nearly antisymmetric with respect to the axis since in this case we see the toroidal component of the helical field (Fig. 1d). It is known that the jet axis is not on the plane of the sky, so we calculated the distribution when seen from another direction. The distribution of the FRM is distorted from antisymmetry. But the systematic gradient still remains. The change in the base-value reflects the increase of the contribution from the longitudinal field component (Fig. 1c). The projected field derived from the model is somewhat tilted as observed. Therefore our model can explain the observation.

4.2. Comparison with the Observation for the Wiggled Part of the Jet

We performed the full 3D simulation in which a wiggled structure of the jet is made owing to the helical instability (Nakamura et al. 2001). A circular motion is imposed near the lower boundary to represent the continuous injection of TAWs. In this simulation, the instability grows after TAWs encountering the edge of the "cavity" from which the mass fell into the central core. We calculated the model counterpart of the FRM distribution for the wiggly part of the jet. The result of comparison shows that the model-counterpart of the FRM calculated from the result of the simulation of a wiggly part of the jet in our "Sweeping Magnetic Twist Model" can explain nicely the observed distribution of the FRM in 3C449 (Fig. 1f). This suggests that the wiggly shape is actually a helix in three dimensions.

References

Asada, K., Inoue, M., Uchida, Y., Kameno, S., Fujisawa, K., Iguchi, S. & Mutoh, M. 2002, PASJ, 54, L39

Feretti, L., Perley, R., Giovannini, G. & Andernach, H. 1999, A & A, 341, 29.

Miley, G. K. & Hartsuijker, A. P. 1978, A&AS, 34, 129

Nakamura, M., Uchida, Y. & Hirose, S. 2001, New Astronomy, 6, 61

Uchida, Y. & Shibata, K. 1985, PASJ, 37, 515

IAU 8th Asia-Pacific Regional Meeting
ASP Conference Series, Vol. 289, 2003
S. Ikeuchi, J. Hearnshaw & T. Hanawa, eds.

Cosmic Rays in the Large Scale Structure of the Universe

Dongsu Ryu

Chungnam National University, Korea

Abstract. Clusters of galaxies and large scale structures contain a significant amount of cosmic rays (as well as magnetic fields) embedded in the hot gas detectable in X-rays; the energy output from growing black holes at the centers of active galactic nuclei, supernovae and gamma ray bursts leads naturally to a scenario where the intracluster medium in analogy to the interstellar medium is filled with energetic particles and permeated by magnetic fields. Those cosmic rays may be energetically as important as the hot gas in clusters; the two components could be in pressure equipartition. Here, we present an exploration of large scale structure formation through numerical simulations, where we have included dynamical effects of cosmic rays, in order to study how much structure formation is influenced by them. Cosmic rays are assumed to be injected from individual sources which form in the course of large scale structure formation. Their advection along with compression has been followed. Preliminary results are presented, and the cosmological implications are discussed.

1. Introduction

Clusters of galaxies are important probes for cosmology, since they are the largest bound systems in the universe. While galaxies are the most obvious constituents of clusters in visible light, most of the matter in clusters is in the form of dark matter. Even the baryonic matter is primarily contained within the diffuse intracluster medium, rather than in galaxies. In addition, the intracluster medium contains a significant amount of energy in the forms of cosmic rays (see e.g., Sarazin 1999 and references therein) and magnetic field (see e.g., Clarke et al. 2001).

Extended regions populated by cosmic ray *electrons* have been observed in some clusters for more than thirty years through diffuse, nonthermal radio emissions (see e.g., Kim et al. 1989). Although cosmic ray *protons* can be detected by γ-rays via π^0 decay following inelastic collisions with gas nuclei, such γ-rays have not yet been detected from clusters (Sreekumar et al. 1996). The observations of nonthermal radiations from cosmic ray electrons in clusters, however, suggest that cosmic ray protons could be in energy equipartition with the gas (See e.g., Blasi 1999).

In this contribution, we study the dynamical importance and influence of cosmic ray protons, which were injected to the intracluster medium from sources formed in the course of large scale structure formation in the universe.

2. Numerical Simulations

For simulations, an Eulerian hydro+N-body cosmological code (Ryu et al. 1993) has been used. The cold dark matter cosmology with a cosmological constant (ΛCDM) has been employed with the following parameters: rms density fluctuations on a scale of $8h^{-1}$ Mpc, $\sigma_8 = 0.8$, spectral index for the initial power spectrum of perturbations, $n = 1$, normalized Hubble constant, $h \equiv H_0/(100 \text{ km s}^{-1} \text{ Mpc}^{-1}) = 0.7$, total mass density, $\Omega_M = 0.27$, and gas mass density, $\Omega_{\text{gas}} = 0.043$. A cubic comoving region of size $75h^{-1}$ Mpc has been set for computational domain, and 512^3 cells for gas and 256^3 dark matter particles have been used. Two simulations have been performed, one without cosmic rays and the other with cosmic rays.

The following equation for the cosmic ray pressure has been solved in addition to the equations for dark matter and gas (Ryu et al. 1993),

$$\frac{\partial p_{\text{CR}}}{\partial t} + \frac{1}{a}u_k\frac{\partial P_{\text{CR}}}{\partial x_k} + \frac{\gamma_{\text{CR}}}{a}P_{\text{CR}}\frac{\partial u_k}{\partial x_k} = -3(\gamma_{\text{CR}} - 1)\frac{\dot{a}}{a}p_{\text{CR}}. \quad (1)$$

Here, a is the expansion parameter, x_k is the comoving length, p_{CR} is the comoving pressure of cosmic rays, u_k is the proper peculiar velocity, and γ_{CR} is the adiabatic index of cosmic rays which has been assumed to be 4/3. Diffusion of cosmic rays has been ignored, since the diffusion length is much shorter than the computational cell size. The dynamical feedback of cosmic rays to the gas has been incorporated by including the $(1/a)(\partial P_{\text{CR}}/\partial x_k)$ term in the momentum equation and the $(1/a)u_k(\partial P_{\text{CR}}/\partial x_k)$ term in the energy equation.

Cosmic ray sources were assumed to form at 40 different epochs after the redshift $z \leq 10$, if the following criteria are satisfied in each computational cell

$$M_{\text{gas}} \geq \frac{3 \times 10^{10}}{1+z}h^{-1}M_{\odot}, \quad \frac{\partial u_k}{\partial x_k} < 0, \quad (2)$$

where M_{gas} is the total gas mass inside the computational cell. We further assume that each source ejects the the following amount of cosmic ray energy into the intracluster medium

$$E_{\text{CR}} = 3 \times 10^4 h^{-1} M_{\odot} \times c^2. \quad (3)$$

Note that at $z = 0$ this translates into the cosmic ray efficiency, $E_{\text{CR}}/M_{\text{gas}}c^2 \approx 10^{-6}$.

3. Results

The right panel of Fig.1 shows that the power of density perturbation has been decreased by $\sim 25\%$ at $z = 2$, by $\sim 35\%$ at $z = 1$, and by $\sim 45\%$ at $z = 0$ on the cluster scale of $\sim 1h^{-1}$ Mpc due to cosmic ray pressure. However, the structures of scales larger than the cluster scale have not been disturbed, since sources form mostly at the highest density peaks within the clusters.

Detailed results of simulations will be reported elsewhere.

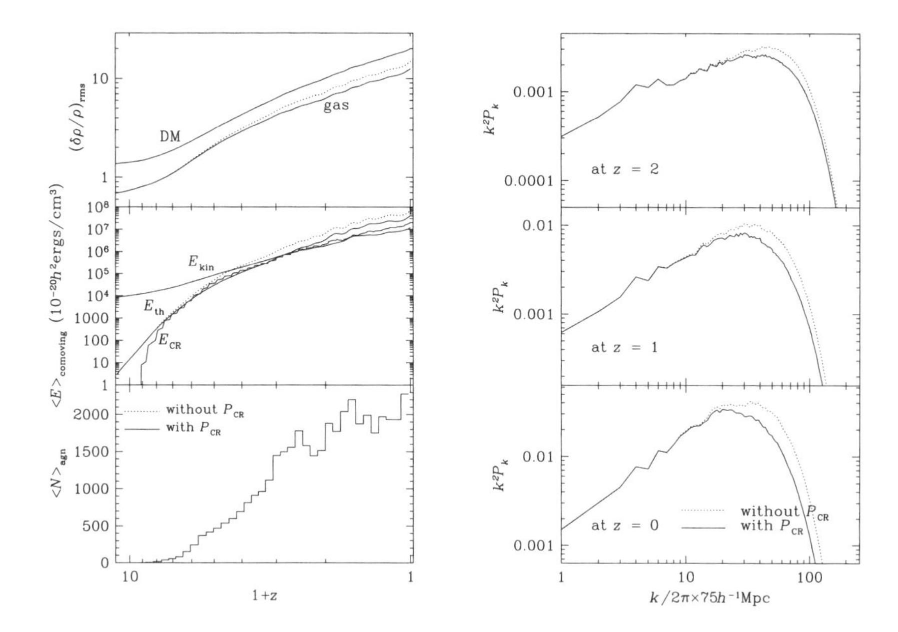

Figure 1. *Left*: Top - Temporal evolution of the root-mean-square density perturbation of dark matter and gas. Middle - Temporal evolution of the mass-averaged energy per unit comoving volume. $E_{\rm kin}$ is the gas kinetic energy, $E_{\rm th}$ is the gas thermal energy, and $E_{\rm CR}$ is the cosmic ray energy. Bottom - Formation history of cosmic ray sources. Sources are assumed to form at 40 logarithmically spaced epochs after $z = 10$. *Right*: Power spectrum of gas density perturbations at three different epochs. Solid lines are for the case with cosmic rays included and dotted lines are for the case without cosmic rays.

The left-bottom panel of Fig.1 shows that the source formation rate, which was realized by the criteria (2) in our simulation, increases first, and then stays more or less constant after $z \sim 2$. Total of $\sim 4 \times 10^4$ sources have formed in the simulation box of $(75h^{-1}{\rm Mpc})^3$. The left-middle panel shows the *mass averaged densities* of gas kinetic energy, gas thermal energy and cosmic ray energy, which can be interpreted as the averaged energy densities in clusters with high mass density. The cosmic ray energy in clusters dominated over the gas thermal energy before $z \sim 2$ during which the gas thermal energy was still small. Then, the ratio has decreased, and the cosmic ray energy density has become and stays $\sim 50\%$ of the gas thermal energy density after $z = 1$. With such amount of cosmic ray energy, the gas density perturbation has reduced by $\sim 20\%$ at the present epoch, while the density perturbation of dark matter has been hardly affected, as shown in left-top panel.

Acknowledgments. The work was supported in part by grant No. R01-1999-00023 from the Korea Science & Engineering Foundation. Simulations were performed through the support by "The 3nd Supercomputing Application Support Program of KISTI".

References

Blasi, P. 1999, ApJ, 525, 603

Clarke, T. E., Kronberg, P. P. & Böhringer, H. 2001, ApJ, 547, L111

Kim, K.-T., Kronberg P. P., Giovannini, G. & Venturi, T. 1989, Nature, 341, 720

Ryu, D., Ostriker, J. P., Kang, H. & Cen, R. 1993, ApJ, 414, 1

Sarazin, C. L. 1999, ApJ, 520, 529

Sreekumar, P. et al. 1996, ApJ, 464, 628

IAU 8th Asia-Pacific Regional Meeting
ASP Conference Series, Vol. 289, 2003
S. Ikeuchi, J. Hearnshaw & T. Hanawa, eds.

Highlights from the First Five Years of the VSOP Mission

H. Hirabayashi, P. G. Edwards, Y. Murata, Y. Asaki, D. W. Murphy

Institute of Space and Astronautical Science, 3-1-1 Yoshinodai, Sagamihara, Kanagawa 229-8510, Japan

H. Kobayashi, M. Inoue, S. Kameno, T. Umemoto

National Astronomical Observatory, Ohsawa 2–21–1, Mitaka, Tokyo 181-8588, Japan

Abstract. The HALCA satellite was launched by ISAS in 1997 as the main element of the VLBI Space Observatory Programme (VSOP). By the fifth anniversary of HALCA's launch, over 700 VSOP observations have been carried out, predominantly, but not exclusively, of the relativistic jets and accretion disks surrounding super-massive black holes in active galactic nuclei (AGN). Objects studied include nearby AGN such as M87, where the high angular resolution corresponds to high linear resolution, out to high redshift ($z > 3$) quasars. Observations are made at 1.6 GHz (18 cm) and 5 GHz (6 cm). Achievements by VSOP in these five years and future works will be reviewed. The lessons learned by VSOP are also both scientifically and technically important for the planning of the next generation of space-VLBI missions.

1. Introduction

Very Long Baseline Interferometry (VLBI), which offers milli-arcsecond (mas) scale resolution at radio wavelengths, was pioneered in the late 1960s. The extension of VLBI into space, to obtain even higher angular resolution, was first demonstrated in a series of experiments with a Tracking and Data Relay Satellite System (TDRSS) communications satellite in the late 1980s. These experiments and other studies paved the way for the development of the MUSES-B satellite (Hirabayashi et al. 2000a) by the Institute of Space and Astronautical Science.

The primary goals of this satellite were to test the technologies required for VLBI in space: a deployable radio telescope antenna, stable two-way communications between the satellite and ground tracking stations, and accurate orbit determination, etc. The satellite, renamed HALCA after launch, is the orbital element of the VLBI Space Observatory Programme (VSOP), which is a large international collaboration of space agencies and ground observatories which have combined resources to create the first dedicated space VLBI mission. An overview of the mission and early results are presented in Hirabayashi et al. (1998, 2000a).

2. Observations

Before launch, the mission lifetime was expected to be five years, with cosmic-ray degradation of the solar panels providing the main limitation on the lifetime. After three years in orbit had become clear that the solar panel degradation was not as severe as had been anticipated, and that sufficient power will be available for several more years of operation (Murata et al. 2000). The mission lifetime may, instead, be determined by other factors: only a small amount of hydrazine thruster fuel (used for safe-holds and recoveries) remains, and one of the four reaction wheels has been taken out of operation.

Until 2002 February, excluding periods where observing has been stopped, approximately 55% of the time was been used for maneuvers, calibration, maintenance and testing of the satellite systems. About 30% of HALCA's in-orbit time was used for observing projects selected by international peer-review from open proposals submitted by the astronomical community in response to Announcements of Opportunity. This part of the mission's scientific programme constitutes the General Observing Time (GOT), with GOT observations lasting typically 10 hours and using an array of ~10 ground telescopes. The final 15% of the in-orbit time was devoted to a mission-led systematic survey of active galactic nuclei: the VSOP Survey Program (Hirabayashi et al. 2000b; Edwards et al. 2002). Survey observations are in general of shorter duration and made with fewer telescopes than the GOT observations. Nevertheless, the Survey will provide a large complete sample of homogeneous data on sub–milli-arcsecond radio structures, which is essential for studying cosmology and statistics of AGN, and for designing future Space VLBI missions. From 2002 February, VSOP observations have been devoted almost exclusively to Survey program observations, with all 294 sources expected to be observed by mid-2003.

3. Highlights

The longer baselines lengths, and gain in resolution, enabled by VSOP observations allow a range of unique scientific studies to be conducted, examples of which are given in this section.

The measurement of brightness temperature depends explicitly on the physical baseline length, and so the limits placed by space VLBI observations cannot be achieved by ground-based VLBI at other frequencies. The $\sim 10^{12}$ K inverse-Compton limit to the brightness temperature (see Kellermann 2002 for a discussion of various limits on brightness temperature) provides an important method for constraining the Doppler factor of the jet, which in turn provides information on the angle to the line of sight and the intrinsic jet speed. Both VSOP Survey Program observations and General Observing Time programs have found that many extragalactic radio sources have brightness temperatures in excess of the inverse-Compton limit (e.g., Hirabayashi et al. 2000b; Tingay et al. 2001).

HALCA only detects left circular polarized radiation. However, it is possible to record LCP with HALCA and both LCP and RCP at some ground radio telescopes, enabling polarization observations to be conducted (Kemball et al. 2000; Gabuzda 2002). The structure of the BL Lac object OJ287 seen in a VSOP 5 GHz image closely resembles that of a contemporaneous 22 GHz VLBA image. However, comparison of the two (similar resolution) images revealed a roughly 90° rotation in the polarization position angle for the core between the two frequencies. This is the first time the theoretically predicted transition in polarization angle between the optically thick (5 GHz) to the optically thin (22 GHz) regimes has been directly observed, and the detection would not have been possible without the ability to obtain high resolution observations at 5 GHz with VSOP (Gabuzda & Gomez 2001).

VSOP observations allow "matched-resolution" spectral index maps of radio sources to be made. With ground-only arrays the resolution of spectral index maps always degrades to that of the lower frequency, however with space VLBI these losses are compensated for by the gains in baseline length. Spectral index maps allow wide-ranging studies of the parsec-scale radio emission to be made, including the presence or absence of synchrotron self-absorption in the core, and the evolution of the spectral index of jet components, with implications for the aging of the components and the shocks believed to generate them (e.g., Piner et al. 2000; Edwards et al. 2000).

The effects of "free–free" absorption are largest at lower frequencies and only the longest possible baselines provide adequate angular resolution. The physical conditions in the inner parsec of accretion disks in galactic nuclei can be directly probed by imaging the absorption of background radio emission by ionized gas in the disk (e.g., Jones et al. 2000; Kameno et al. 2000).

Observations of high redshift source have important implications for our understanding of the evolution of structures in the early Universe. A VSOP observation of the $z = 3.57$ quasar PKS 2215+020 was made at 1.6 GHz, corresponding to 7.6 GHz in the source frame. An optically faint, radio-loud quasar, PKS 2215+020 has, remarkably, also been detected in X-rays by ROSAT. The VSOP image of PKS 2215+020 revealed a rich core-jet morphology and unusually large jet, which can be traced beyond 300 parsecs from the core — by far the longest jet observed at redshifts above 3. The direction of the jet coincides with the elongations visible in the VLA and ROSAT images (Lobanov et al. 2001). More recently, two-epoch VSOP observations of the $z = 3.707$ quasar PKS 1351−018 have resulted in the tentative detection of apparent superluminal motion in this high-redshift source (Frey et al. 2002).

VSOP provides the only way to gain improved resolution for spectral line maser sources. VSOP observations have demonstrated that OH maser spots in the star-forming region OH 34.26+0.15 are only partially resolved, in contrast to expectations that interstellar scattering would result in a large degree of angular broadening at 1.6 GHz. The strongest peak has a brightness temperature in excess of 6×10^{12} K – exceeding the minimum value predicted by some models by an order of magnitude (Slysh et al. 2000). Results such as this have implications for our understanding of both maser physics and the interstellar medium.

4. The Future

Following the success of the VLBI Space Observatory Programme (VSOP), a next generation space VLBI mission, VSOP-2, is currently being planned in Japan. Higher observing frequencies (up to 43 GHz), cooled receivers, increased bandwidths and downlink bit-rate (1 Gbps or more) and a larger telescope diameter will result in gains in resolution and sensitivity by factors of ~10 over the VSOP mission (Hirabayashi et al. 2001; Murata et al. 2002). Development studies for the deployable antenna, cooled low noise receivers, high speed digitization etc., are in progress.

Acknowledgments. The VSOP Project is led by the Institute of Space and Astronautical Science, with significant contributions from the National Astronomical Observatory, the Jet Propulsion Laboratory, the U.S. National Radio Astronomy Observatory, the Canadian Dominion Radio Astrophysical Observatory, the Australia Telescope National Facility, the European VLBI Network, the Joint Institute for VLBI in Europe, and the directors and staff of many of the world's radio observatories.

References

Edwards, P. G., Giovannini, G., Cotton, W. D., Feretti, L., Fujisawa, K., Hirabayashi, H., Lara, L., Venturi, T. 2000, PASJ, 52, 1015

Edwards, P. G. et al. 2002, in Proc. 8th IAU Asian-Pacific Regional Meeting, vol. II, ed. S. Ikeuchi et al. (Tokyo: Astron. Soc. Japan), 375

Frey, S. 2002, in Proceedings of the 6th European VLBI Network Symposium ed. E. Ros, R .W. Porcas, A. P. Lobanov, J. A. Zensus (Bonn: MPIfR), 89

Gabuzda, D. C. & Gomez, J. L. 2001, MNRAS, 320, L49

Gabuzda, D. C. 2002, in Proceedings of the 6th European VLBI Network Symposium ed. E. Ros, R. W. Porcas, A. P. Lobanov, J. A. Zensus (Bonn: MPIfR), 83

Hirabayashi, H. et al. 1998, Science, 281, 1825 and erratum 282, 1995

Hirabayashi, H. et al. 2001, in ASP Conf. Ser. 251, New Century of X-ray Astronomy, ed. H. Inoue & H. Kunieda (San Francisco: ASP), 540

Hirabayashi, H. et al. 2000a, PASJ, 52, 955

Hirabayashi, H. et al. 2000b, PASJ, 52, 997

Jones, D. L., Wehrle, A. E., Piner, B. G., Meier, D. L. 2000, in Astrophysical Phenomena Revealed by Space VLBI, ed. H. Hirabayashi, P. G. Edwards, D. W. Murphy (Sagamihara: ISAS), 71

Kameno, S., Horiuchi, S., Shen, Z.-Q., Inoue, M., Kobayashi, H., Hirabayashi, H., Murata, Y. 2000, PASJ, 52, 209

Kellermann, K. I. 2002, PASA, 19, 77

Lobanov, A. P. et al. 2001, ApJ, 547, 714

Murata, Y. et al. 2000, in Astrophysical Phenomena Revealed by Space VLBI, ed. H. Hirabayashi, P. G. Edwards, & D. W. Murphy (Sagamihara: ISAS), 9

Piner, B. G., Edwards, P. G., Wehrle, A. E., Hirabayashi, H., Lovell, J. E.J., Unwin, S. C. 2000, ApJ, 537, 91

Slysh, V. I. et al. 2000, MNRAS, 320, 217

Tingay, S. J. et al. 2001, ApJ, 549, L55

Solar and Stellar Activities, Binaries

IAU 8th Asia-Pacific Regional Meeting
ASP Conference Series, Vol. nnn, 2003
S. Ikeuchi, J. Hearnshaw & T. Hanawa, eds.

Progress on Numerical Simulations of Solar Flares and Coronal Mass Ejections

Kazunari Shibata

Kwasan Observatory, Kyoto University, Yamashina, Kyoto 607-8471, Japan

Abstract. Recent progress of numerical simulations of solar flares and coronal mass ejections is discussed with emphasis on MHD simulations of magnetic reconnection and their application to recent space observations such as those by Yohkoh.

1. Introduction

Now is the golden age of solar observations. Recent space observations of the Sun with Yohkoh, SOHO, and TRACE have made a revolution in solar physics, and revealed that the magnetic field, especially its non-steady dynamics associated with reconnection, plays a central role in the production of flares and coronal mass ejections (CMEs). Since these non-steady dynamics are well described by magnetohydrodynamic (MHD) equations, we have to solve non-steady (resistive) MHD equations to understand the basic physics of flares and CMEs. This is practically impossible with an analytical approach, because of the intrinsic non-linearity of MHD equations. Thus the only way to attack this problem is by the numerical approach. Recent rapid progress in computers has enabled us for the first time to simulate these complicated non linear, non-steady MHD processes in flares and CMEs in a realistic situation. From this point of view, we are now also in the golden age of solar MHD simulations: we can now compare excellent observational movies (e.g., Yohkoh X-ray movies) with MHD simulation movies, so that we can discuss the physics of flares and CMEs quantitatively and in detail, and we can even estimate unobservable quantities from such a comparison.

In this article, we first briefly review recent developments in the observation of solar flares and CMEs by spacecraft, in particular by Yohkoh, and then discuss recent progress of numerical simulations of solar flares and CMEs, with emphasis on MHD reconnection and applications to these observations.

2. Observations of Flares and Coronal Mass Ejections

Recent space observations have revealed a variety of evidence for reconnection in flares and the solar corona, such as cusp-shaped loops, loop top hard X-ray sources, giant arcades, plasmoid ejections, X-ray jets, and so on (e.g., see Tsuneta 1996; Shibata 1999; Aschwanden et al. 2001 and references therein).

It has also been revealed that plasmoid (or flux rope) ejections are much more common in flares than had been thought, and that not only LDE (long duration event) flares but also impulsive flares are very similar to CME-related flares. On the other hand, it has been found that many non-flare CMEs are associated with giant arcades, which are physically quite similar to cusp-shaped flare loops/arcades. Even tiny microflares or nanoflares often produce X-ray jets, Hα surges, or EUV jets.

Table 1. Comparison of various "flares"

"flare"	micro-flares	implusive flares	LDE flares	giant arcades
size (L) (10^4 km)	$0.5 - 4$	$1 - 10$	$10 - 40$	$30 - 100$
time scale (t) (s)	$60 - 600$	$60 - 3 \times 10^3$	$3 \times 10^3 - 10^5$	$10^4 - 2 \times 10^5$
energy (erg)	$10^{26} - 10^{29}$	$10^{29} - 10^{32}$	$10^{30} - 10^{32}$	$10^{29} - 10^{32}$
mass ejection	X-ray jet/ H_α surge	X-ray/H_α filament eruption	X-ray/H_α filament eruption	X-ray/H_α filament eruption
B (G)	100	100	30	10
n_e (cm^{-3})	10^{10}	10^{10}	2×10^9	3×10^8
V_A (km s^{-1})	3000	3000	2000	1500
$t_A = L/V_A$	5	10	90	400
t/t_A	$12 - 120$	$6 - 300$	$30 - 10^3$	$25 - 500$

Hence, a unified view has emerged from these new observations (Table I), and unified model has been proposed (Shibata 1999). Though the total energy of these "flares" ranges widely, from 10^{26} ergs for microflares to 10^{32} ergs for LDE flares and giant arcades, it is easy to explain such an observed variation of "flare" energies by the stored magnetic energy in a corresponding flare volume $E_{\text{flare}} \simeq \frac{B^2}{8\pi} L^3 \simeq 4 \times 10^{32} \left(\frac{B}{100\text{ G}}\right)^2 \left(\frac{L}{10^{10}\text{ cm}}\right)^3$ erg, where B is the average magnetic field strength in a flare volume with a characteristic length L. The time scale of flares shows an even wider dynamic range; i.e., from a few tens of seconds for microflares, to a few days for giant arcades associated with CMEs. However, if we normalize the time scale by the Alfven time $t_A = L/V_A$, all time scales of various "flares" become comparable to 10–1000. (In other words, the non-dimensional reconnection rate is 0.001–0.1.)

Furthermore, the observations of ubiquity of plasmoid (flux rope) ejections suggest the concept of *plasmoid-induced-reconnection* (Shibata and Tanuma 2001), in which magnetic reconnection is strongly coupled with plasmoid ejections, because of the following two roles of plasmoids in reconnection: i.e., (1) to store energy by inhibiting reconnection, and (2) to induce strong inflow into the reconnection region when plasmoids are ejected out of the current sheet. This, as well as the highly time dependent behavior of flare emissions in hard X-rays and microwaves, further suggests the *fractal reconnection* (Tajima and Shibata 1997, Shibata and Tanuma 2001), in which many plasmoids (flux ropes or filaments) with different size are created in a current sheet, and are coalesced into each other and ejected out of the sheet to induce reconnection with various size and time scales. This is favorable to connect on a macro-scale ($\sim 10^9$ cm) and on a micro-scale (~ 100 cm) where anomalous resistivity or collisionless conductivity can eventually work.

3. Simulations of Magnetic Reconnection

At first, we should remember that it is still not possible to take fully realistic physical parameters for modelling solar flares and CMEs with present numerical simulations. The biggest difficulty is in the magnetic Reynolds number $R_{\rm m} = LV_{\rm A}/\eta \sim 10^{13}$ for classical Spitzer resistivity and typical coronal conditions, $L = 10^9$ cm, $V_{\rm A} = 10^8$ cm/s, and $\eta \sim 10^4$ cm^2/s (for $T = 10^6$ K). With the present supercomputers, the maximum magnetic Reynolds number in simulations is of the order of $10^3 - 10^4$. Thus there is a huge gap between realistic and numerical magnetic Reynolds numbers.

One of the recent achievements in numerical simulations of magnetic reconnection is the following: *If uniform resistivity is assumed, the magnetic reconnection becomes the Sweet-Parker type, i.e., slow reconnection with a time scale* $t_{\rm Sweet-Parker} \simeq R_{\rm m}^{1/2} t_{\rm A}$. *In order to get fast reconnection such as Petschek type,* $t_{\rm Petschek} \simeq 10 - 100 R_{\rm m}^0 t_{\rm A}$, *we need to assume spatially localized resistivity.* (Biskamp 1986, Scholer 1989, Ugai 1992, Yokoyama and Shibata 1994, Schumacher and Kliem 1996, Magara and Shibata 1999). The so called anomalous resistivity model, $\eta = \eta_0 |v_{\rm d} - v_{\rm c}|$ (for $v_{\rm d} = j/\rho > v_{\rm c}$) and otherwise $\eta = 0$, satisfies this condition.

4. Simulations of Solar Flares and CMEs

4.1. Reconnection Driven by Emerging Flux

Shibata, Nozawa, and Matsumoto (1992) carried out time-dependent MHD numerical simulations of reconnection occurring in a current sheet between an emerging flux and a pre-existing horizontal coronal magnetic field, by assuming the anomalous resistivity model and self-consistent emerging flux model (as a result of the Parker instability). They found that the reconnection proceeds in a very time-dependent manner: The tearing instability occurs in the current sheet, creating magnetic islands (plasmoids). These islands coalesce with each other, making bigger and bigger islands (plasmoids), and are eventually ejected out of the sheet. After the plasmoid ejection, the Petschek-type fast reconnection occurs. They predicted coexistence of both cool and hot plasma ejections.

Yokoyama and Shibata (1995, 1996) extended this simulation significantly. They studied both horizontal and oblique field cases for coronal magnetic field geometry, and confirmed that the basic physics is common in both cases. They found that not only hot jets but also cool jets are accelerated (Fig. 1), and predicted the coexistence of both X-ray jets and Hα surges. Indeed, such coexistence of both jets has been found by simultaneous observations by Yohkoh and ground based Hα observations (Canfield et al. 1996).

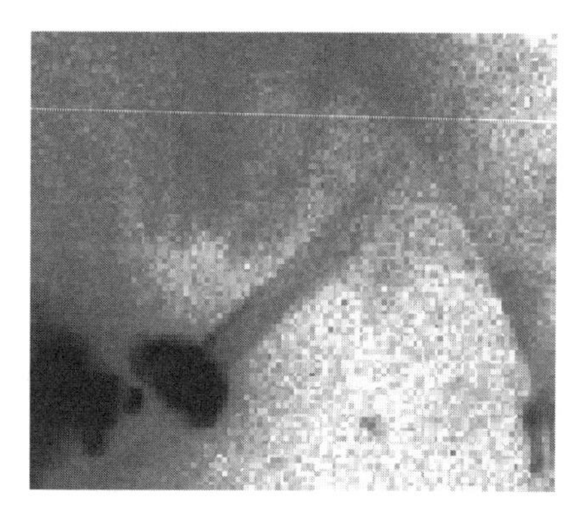

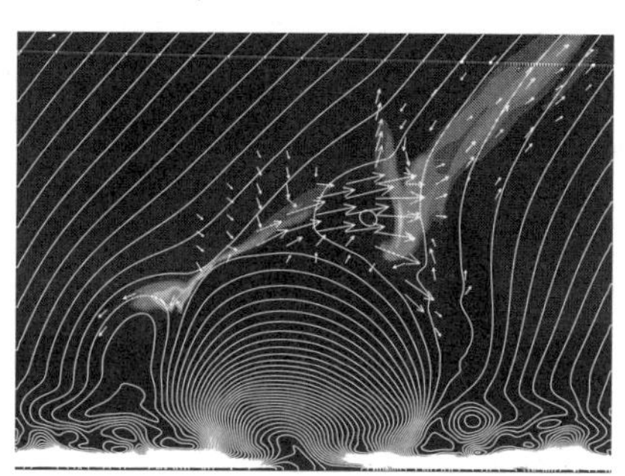

Figure 1. left: X-ray jet observed with the Yohkoh soft X-ray telescope (Shibata et al. 1992). Right: MHD simulation of X-ray jet (Yokoyama and Shibata 1995, 1996). The color map shows the temperature distribution.

4.2. Effect of Heat Conduction

The typical flare temperature is of order of 10^7 K. In such a high temperature, the conduction cooling time becomes very short:

$$t_{\rm cond} \simeq \frac{3nkTL^2}{\kappa_0 T^{5/2}} \simeq 1.3 \text{ s} \left(\frac{T}{10^7 \text{ K}}\right)^{-5/2} \left(\frac{n}{10^9 \text{ cm}^{-3}}\right) \left(\frac{L}{10^9 \text{ cm}}\right)^2, \qquad (1)$$

whereas the radiative cooling time is much longer ($\sim 10^5$ s for flare plasma in the early phase with $T \sim 10^7$ K, $n \sim 10^9$ cm^{-3}). Here, $\kappa_0 \simeq 10^{-6}$ is Spitzer's thermal conductivity along magnetic field lines. The Alfven time, $t_{\rm A} \simeq L/V_{\rm A} \sim$ 10 s, is longer than the conduction cooling time. Hence, the heat conduction is very important in the dynamics and thermodynamics of flare plasma. In particular, the heat conduction causes *chromospheric evaporation* (ablation of dense chromospheric plasma), providing flare loops with hot dense plasma, which becomes the source of strong soft X-ray emission. If we want to model soft X-ray emission from flare loops, we have to take into account heat conduction and chromospheric evaporation, as already studied by one-dimensional flare loop modelling (e.g., Hori et al. (1997) for a flare loop model and Shimojo et al. (2000) for a jet model). Nevertheless, the MHD simulations including both reconnection and conduction have never been performed for more than the 20 years since it was first recognized that such simulations are really necessary for realistic flare modelling. This was the result of various numerical difficulties, such as the short conduction time and anisotropic heat conductivity due to the magnetic field, which requires implicit treatment and large computational power.

Yokoyama and Shibata (1997) succeeded, for the first time, to perform MHD numerical simulations including both reconnection and conduction. They confirmed the semi-analytical prediction by Forbes, Malherbe, and Priest (1989) that an *adiabatic slow shock is dissociated into a conduction front and an isothermal slow shock* when $t_{\rm cond} < t_{\rm A}$. Later, Chen et al. (1999) also succeeded in carrying out similar simulations. It should be stressed here that at present the groups that succeeded to perform simulations of reconnection including heat conduction *under solar flare conditions* are only two, i.e., in Japan (Yokoyama and Shibata) and in China (Chen et al.).

Yokoyama and Shibata (1998, 2001) further extended simulations to include chromospheric evaporation. Figure 2 shows a typical example of these simulations. We can find that the adiabatic slow shock is dissociated into a conduction front and an isothermal slow shock. The outer edge of the hot cusp region, which is nearly along the magnetic field lines, is the conduction front. After the conduction front reaches the top of the chromosphere, the evaporation starts and the dense hot loop (flare loop) is created. Figure 2 shows also the theoretical soft X-ray and microwave images, which are made using the temperature and density distribution in MHD simulations and the soft X-ray filter response function, as well as the optically thin thermal free-free emissivity at 17 GHz. The cusp-shaped structure similar to the observed cusp shape of LDE flares is nicely reproduced in these theoretical soft X-ray images ($\log I_{\rm Be}$ and $\log I_{\rm thinAl}$). On the other hand, the theoretical radio image at 17 GHz ($\log I_{\rm freefree}$) does not show a bright cusp shape, which is also consistent with radio observations of LDE flares (e.g., Hanaoka 1994).

One of the most important findings of Yokoyama and Shibata's (1998, 2001) simulation is that the maximum temperature of reconnection-heated plasma (i.e., flare plasma) is given by the following formula:

$$T_{\rm max} \simeq \left(\frac{B^2 V_{\rm A} L}{\kappa_0 2\pi}\right)^{2/7} \simeq 3 \times 10^7 \left(\frac{B}{50\ {\rm G}}\right)^{6/7} \left(\frac{n_0}{10^9\ {\rm cm}^{-3}}\right)^{-1/7} \left(\frac{L}{10^9\ {\rm cm}}\right)^{2/7} \quad {\rm K}, \tag{2}$$

where n_0 is the pre-flare proton number density (= electron density), L is the characteristic length of the (reconnected) magnetic loop. This relation is derived from the balance between conduction cooling $\kappa_0 T^{7/2}/(2L^2)$ (e.g., Hori et al. 1997) and reconnection heating $(B^2/4\pi)(v_{\rm A}/L)$. It was found that this formula can be applied also to stellar flares (Shibata and Yokoyama 1999).

4.3. Plasmoid (Flux Rope) Ejections

What is the triggering mechanism of solar flares and CMEs? Feynman and Martin (1995) reported interesting observations that filament eruptions (i.e., CMEs) tend to occur if emerging flux appears near the quiescent filament and if the polarity distribution of the emerging flux is favorable for magnetic reconnection with the ambient field.

Chen and Shibata (2000) succeeded in reproducing this observational tendency using an implicit MHD code developed by Hu (1989). Figure 3 shows a typical example of their simulations. Initially, they assumed a flux rope in a stable equilibrium in a 2D situation. Then, emerging flux is input from the lower boundary, which makes local reconnection just below the flux rope (filament). This small change of magnetic field configuration leads to loss of equilibrium or instability in the global system, eventually leading to eruption of the whole flux rope system. It is also possible to trigger global eruption even when emerging flux appears in a distant place from the neutral filament, if the polarity distribution is favorable for local reconnection (see Fig. 3b). Here it should be stressed that reconnection is strongly coupled to the eruption of the flux rope (filament

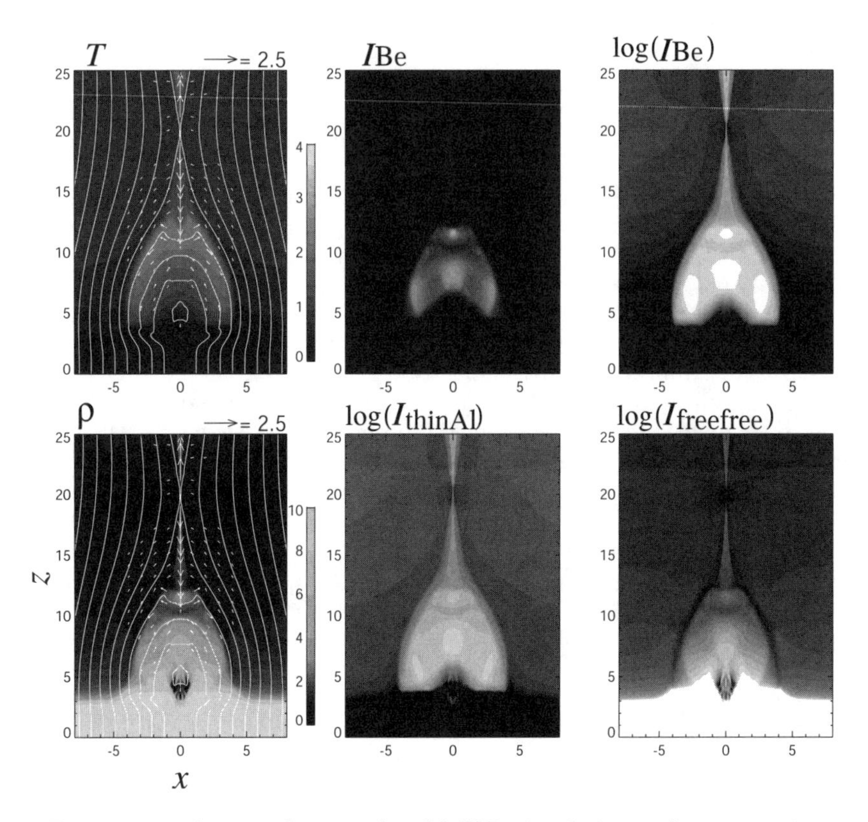

Figure 2. A typical example of MHD simulations of reconnection with heat conduction and chromospheric evaporation (Yokoyama and Shibata 1998, 2001). T and ρ show the temperature and density distributions at $t = 2.5$ (in non-dimensional units) $\simeq 450$ s. In this simulation, initially, an anti-parallel vertical magnetic field is assumed, and a dense plasma (the chromosphere) is located in the lower portion of the computational region. The Petschek-type reconnection is realized by the anomalous resistivity model. Radiative cooling and gravity are neglected, since the time scale treated in the simulations is short, corresponding to the early phase of flares. Because of the high heat conductivity, an adiabatic slow shock is dissociated into a conduction front and an isothermal slow shock, where the density jump is large. A dense thin layer with a reconnection jet emanating from the X-point corresponds to the isothermal slow shock region. Soft X-ray ($I_{\rm Be}$, $\log I_{\rm Be}$, $\log I_{\rm thinAl}$) and radio ($\log I_{\rm free-free}$) maps derived from the simulation results are shown in the middle and right panels. The radio emission is derived as thermal free-free emission. In the X-ray maps, the response of the filter attached to the soft X-ray telescope on board Yohkoh is taken into account. The unit of length, velocity, density, and temperature are 3000 km, 170 km/s, 10^9 cm^{-3} and 2×10^6 K, respectively. Note that the theoretical soft X-ray images show cusp-shaped structure similar to the actual observations of LDE flares (Tsuneta 1996).

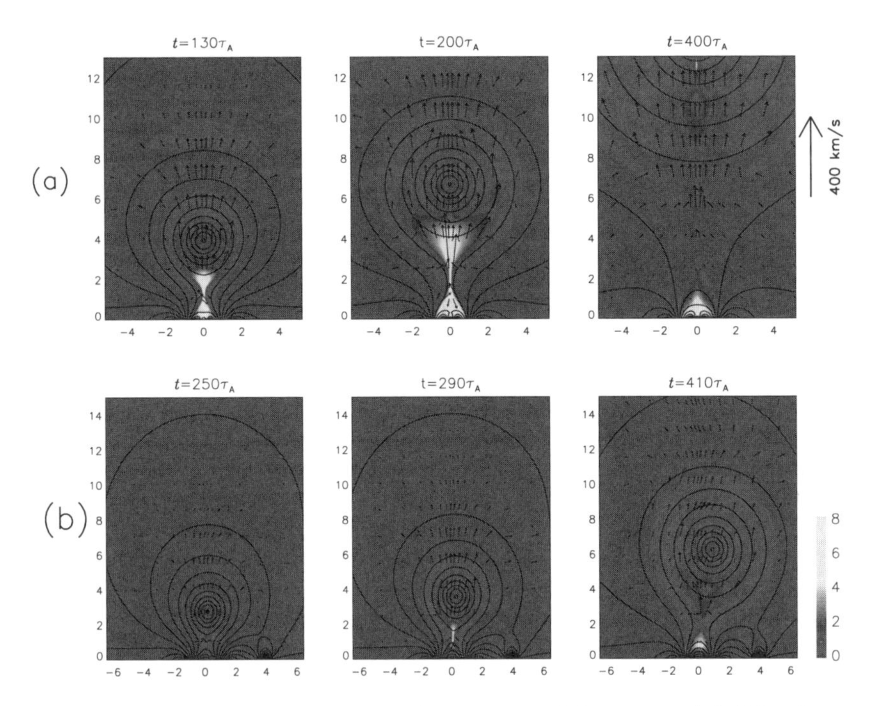

Figure 3. MHD simulations of eruption of flux rope (CME) triggered by emerging flux (Chen and Shibata 2000): two cases are shown, in which emerging flux appeared (a) just below a filament, and (b) in the distant place from the neutral line. The color map shows the temperature distribution.

or plasmoid) as discussed above. If we inhibit reconnection, the fast ejection of the flux rope cannot be possible. Since the flux rope becomes a CME itself, reconnection plays an essential role in CMEs. (See also related simulation studies by Mikic and Linker 1994, Hu 2000, Choe and Chen 2001, Kusano 2002).

5. Remaining Questions

From recent development of space observations and numerical simulations, we can now say that the magnetic reconnection mechanism is established, at least, phenomenologically. However, from physical points of view, there are a number of fundamental questions remaining:
(1) What is the condition of fast reconnection?
(2) Where are slow and fast shocks?
We still do not have firm observational evidence of these shocks. We need better observational data, as well as more realistic simulation models to compare with the observations. Namely, we have not yet developed realistic 3D models of flares/CMEs including reconnection, heat conduction and evaporation.

Furthermore, the following fundamental questions remain in relation to flares/CMEs:

(3) What is the particle acceleration mechanism in flares/CMEs?
(4) What is the energy storage and triggering mechanism of flares/CMEs?

We hope some of above questions will be solved in the near future under collaboration between advanced numerical simulations and new space solar observatories such as Solar B.

References

Aschwanden, M. et al. 2001, ARAA, 39, 175

Biskamp, D. 1986, Phys. Fluids, 29, 1520

Canfield, R. C. et al. 1996, ApJ, 464, 1016

Chen, P. F., et al. 1999, ApJ, 513, 516

Chen, P. F., and Shibata, K. 2000, ApJ, 545, 524

Choe, G. S., and Chen, C. Z. 2000, ApJ, 541, 449

Feynman, J., and Martin, S. F. 1995, JGR, 100, 3355

Forbes, T. G., Malherbe, J. M., Priest, E. R. 1989, Sol. Phys., 120, 285

Hanaoka, Y. 1994, in Proc. Kofu meeting, ed. S. Enome, and T. Hirayama, Nobeyama Radio Observatory, p. 181

Hori, K. et al. 1997, ApJ, 489, 426

Hu, Y. Q., 1989, J. Comp. Phys., 84, 441

Hu, Y. Q. 2000, Sol. Phys., 200, 115

Kusano, K. 2002, ApJ, 571, 532

Magara, T. and Shibata, K. 1999, ApJ, 514, 456

Mikic, Z. and Linker, J. 1994, ApJ, 430, 898

Scholer, M. 1989, JGR, 94, 8805

Schumacher, J. and Kliem, B. 1996, Phys. Plasmas, 3, 4703

Shibata, K., Nozawa, S., and Matsumoto, R. 1992, PASJ 44, 265

Shibata, K. et al. 1992, PASJ, 44, L173

Shibata, K. 1999, Astrophys. Sp. Sci., 264, 129

Shibata, K. and Yokoyama, T. 1999, ApJ, 526, L49

Shibata, K. and Tanuma, S. 2001, Earth, Planets, and Space, 53, 473

Shimojo, M., et al. 2001, ApJ, 550, 1051

Tajima, T., and Shibata, K. 1997, Plasma Astrophysics, Addison-Wesley

Tsuneta, S. 1996, ApJ, 456, 840

Ugai, M. 1992, Phys. Fluids B, 4, 2953

Yokoyama, T. and Shibata, K. 1994, ApJ, 436, L197

Yokoyama, T. and Shibata, K. 1995, Nature 375, 42

Yokoyama, T. and Shibata, K. 1996, PASJ 48, 353

Yokoyama, T. and Shibata, K. 1997, ApJ, 474, L61

Yokoyama, T. and Shibata, K. 1998, ApJ, 494, L113

Yokoyama, T. and Shibata, K. 2001, ApJ, 549, 1160

IAU 8th Asia-Pacific Regional Meeting
ASP Conference Series, Vol. 289, 2003
S. Ikeuchi, J. Hearnshaw & T. Hanawa, eds.

An Investigation of Loop-Type CMEs with a 3D MHD Simulation

J. Kuwabara, Y. Uchida, R. Cameron

Department of Physics, Faculty of Science, Science University of Tokyo, Shinjuku-ku, Tokyo 162-8601, Japan

Abstract. CMEs have erupted filaments (core) and flux loops that lie over the filament (leading edge) and a cavity between the core and leading edge. In one model "Flux loops are pushed up by erupted filaments and expand upwards" which roughly explains much about CMEs. But there are some CMEs with characteristics that are inexplicable with this model, which have a twisted structure along the loop and the structure of a convex lens at the loop top (Illing and Hundhausen 2000). We consider Torsional Alfven Wave (TAW) propagation as the cause of these characteristics for some CMEs, and have studied this with a 3D MHD simulation. As a result, we found that TAW propagation most likely explains these characteristics for some CMEs and that a TAW can carry out plasma from the chromosphere to coronal space along the loop. We propose a new scenario about the occurrence of CMEs based on results of our simulation and observations.

1. Introduction

A CME (Coronal Mass Ejection) is a large-scale phenomenon and releases a huge mass to interplanetary space.[1] Recently, we have come to understand the dynamics of CMEs and magnetic structures in the lower part of the corona with high sensitivity and high resolution satellites, "Yohkoh" and "SoHO". CMEs are related closely to the flare and filament eruption. CMEs which have large flux loops have several characteristics as follows: (i) CMEs have large flux loops containing core, leading edge, and cavity; (ii) During time evolution, each of the two footpoints of CMEs are fixed at specific regions on the chromosphere; (iii) CMEs release a huge mass to interplanetary space; (iv) Before CMEs occur, there are magnetic flux loops connecting the footpoints of CMEs and the future site of an arcade flare may come up. Moreover, some CMEs have extra characteristics as follows: (v) Some CMEs have a twisted structure along the loop; (vi) Some CMEs have the structure of a convex lens (jutting out knots).

[1] Uchida et al. (2001), Cameron and Uchida (2001) proposed there are at least two different types of CMEs. One has a spherical wave frontand is considered to be blast-wave propagation in the corona. The other type has large flux loops. We consider that flux loops are erupted with filaments into interplanetary space when a flare occurs.

One model "Flux loops are pushed up by erupted filaments and expand upwards" explains some characteristics of CMEs, namely items (i), (ii), (iii), (iv). But items (v) and (vi) are inexplicable with this model. We consider that (v) and (vi) are suggestive of the structure of Torsional Alfven Wave (TAW) propagation; therefore we studied how TAW propagation affects deformation of flux loops with a 3D MHD simulation.

2. The Effects of TAW for Deformation of Flux Loops

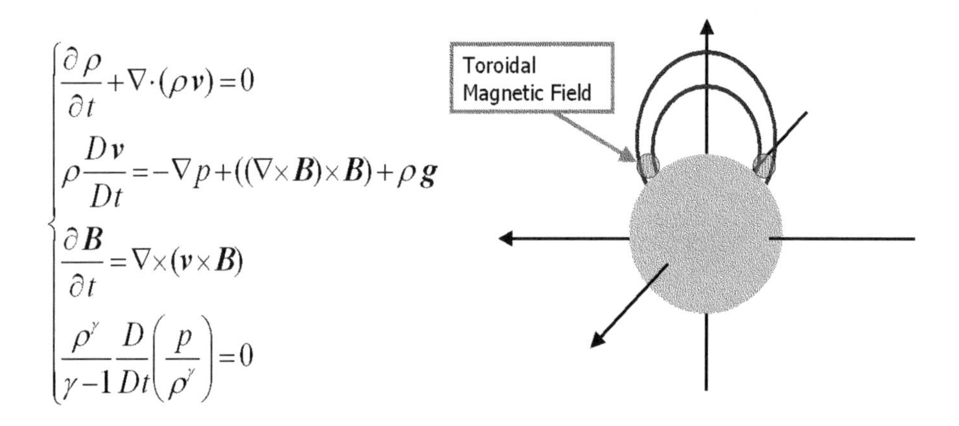

Figure 1. Basic Equations and Initial Conditions

On this simulation we assume ideal MHD, since the magnetic Reynolds number is very high in the corona. We solve the MHD equations numerically with a 2-step Lax-Wendroff scheme. For initial conditions, we assume the corona to be in hydrostatic equilibrium and a current-free magnetic field by some ring current. We inject twists at each of the footpoints of the loop, and then a TAW propagates along the loop. These TAWs at each of the footpoints are injected so as to annihilate each other when they collide at the loop top (Figure 3).

3. Discussion: The Source of TAW

Thus we came to know that a TAW certainly causes the deformation of the loop's twisted structure with knots. But it is not clear how TAW packets are injected at each of the footpoints of the loop. There are some observational results that can explain this question.

3.1. The Magnetic Field Structure Nearby the Site of an Arcade Flare

Before CMEs occur, there were magnetic flux loops connecting the footpoints of CMEs and the future site of an arcade flare may come up. The magnetic field structure near the site of an arcade flare (including the magnetic flux loops connecting the footpoints of CMEs) change dynamically when the arcade flare occurs.

3.2. New Scenario Scenario about Occurrence of CMEs

We consider TAW packets escape from the site of the arcade flare and propose a new scenario about the occurrence of CMEs.

Before CMEs occur, the large-scale magnetic field of the relevant region is composed of a separated pair of sources located at the footpoints of the CMEs which come up and another pair of poles located in the region somewhere between them where the associated arcade flare is to occur. TAW packets escape from the site of the arcade flare and propagate to the footpoints of the CMEs. The TAW packets are then reflected at the footpoints and propagate upwards along the large loop newly formed in the arcade flare. These eventually collide at the loop top, and then the loop is deformed and pushed up by the erupted filaments and expands upwards (Figure 2).

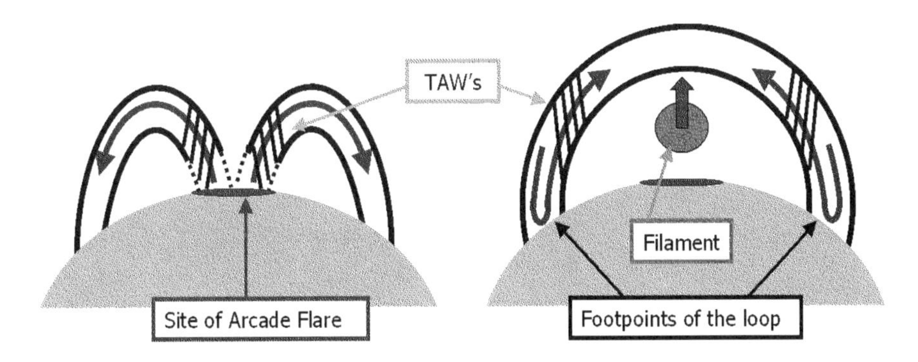

Figure 2. New scenario scenario about occurrence of CMEs

If the TAW packets escape freely from the site of the arcade flare, this is the cause of characteristics (v) and (vi) that some CMEs have. On the other hand, if only a few TAW packets escape from the site of the arcade flare, then non-twisted flux loops expand upwards. This scenario explains each of the cases of CMEs, namely those with properties (v), (vi) and those without. However, we don't know how TAW packets escape from the site of the arcade flare via the dynamics of the magnetic structure. We expect it be clearer in the future.

References

Uchida, Y., Cameron, R., Kuwabara, J., Tanaka, T., Hata, M., Suzuki, I. 2001, Multi-Wavelength Observations of Coronal Structure and Dynamics – Yohkoh 10th Anniversary Meeting. Proceedings of a conference held September 17–20, 2001, at King Kamehameha's Kona Beach Hotel in Kailua-Kona, Hawaii, USA. Ed. by P.C.H. Martens and D. Cauffman. To be published by Elsevier Science on behalf of COSPAR in the COSPAR Colloquia Series

Cameron, R., Uchida, Y. 2001, American Geophysical Union, Spring Meeting 2001, abstract SH42A-09

Uchida, Y., Tanaka, T., Hata, M., Cameron, R. 2001, Publ. Astron. Soc. Australia, 18, 345

Illing, R. M. E. and Hundhausen, A. J. 2000, J. Geophys. Res., 105, 18169

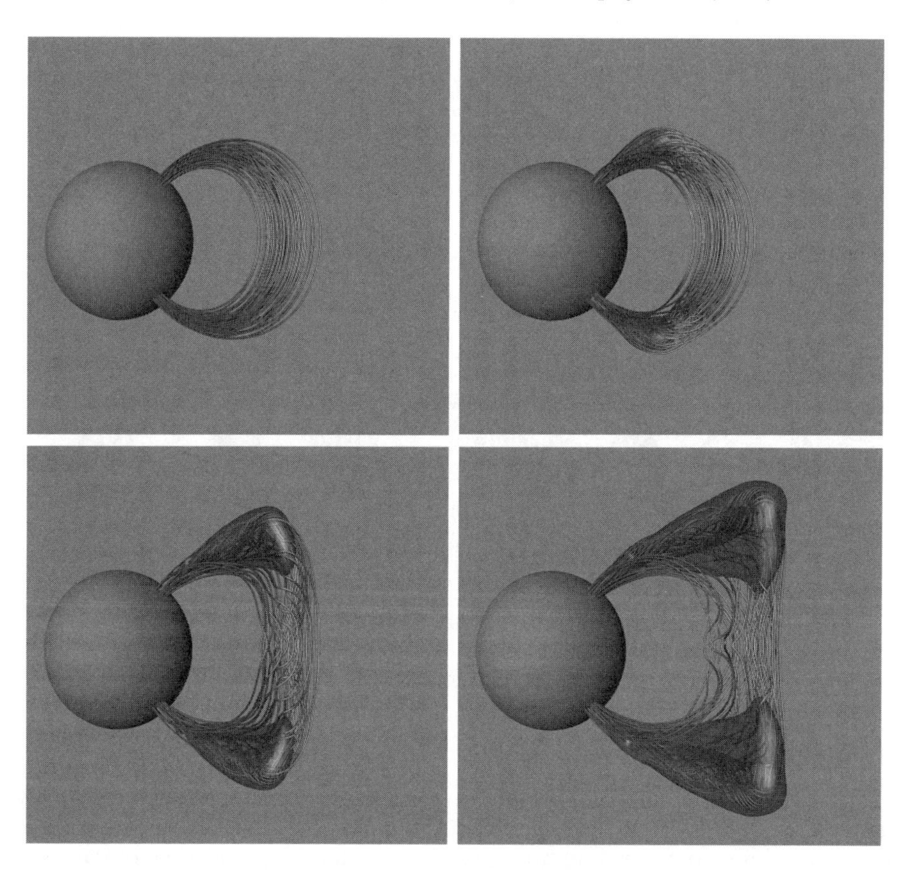

Figure 3. During the flux loops' expansion upwards, TAW propagation causes the loop to deform into a twisted structure. Flux loops are trailed by plasma carried out by the TAW from the chromosphere and come to jut out with knots. The twisted flux loops become relaxed when they collide at the loop top. We consider that this shows a structure suggestive of the observational results.

IAU 8th Asia-Pacific Regional Meeting
ASP Conference Series, Vol. 289, 2003
S. Ikeuchi, J. Hearnshaw & T. Hanawa, eds.

Basic Principles and Examples of Solar-type Flare Modelling

Boris V. Somov

Astronomical Institute, Moscow State University, Universitetskii Prospekt 13, Moscow 119992, Russia

Takeo Kosugi, Taro Sakao

Institute of Space and Astronautical Science, 3-1-1 Yoshinodai, Sagamihara, Kanagawa 229-8510, Japan

Abstract. We review the fundamental ideas which are under current use to model flares and other non-stationary phenomena in the solar atmosphere. Recent multi-wavelength observations of solar flares allow us to improve a theory of solar-type flares, which can be applied to many astrophysical phenomena accompanied by fast plasma ejection, powerful fluxes of radiation, and the acceleration of electrons and ions to high energies.

1. Introduction

A solar flare is a complex phenomenon at the Sun with many different facets. In this paper the aim is to summarize briefly only the large-scale physical processes responsible for accumulation of a flare energy before a flare and for fast release of this energy during a flare. We begin with a flare energy source, the magnetic fields in the solar atmosphere. They dominate the morphology and energetics of solar active regions, where flares occur, because the magnetic energy density is higher there. Then we focus on the cause of fast energy release in flares.

We apply the so-called 'rainbow reconnection' model to the solar observations with the HXT on board Yohkoh (Kosugi et al. 1991), the MDI instrument on the SOHO, the TRACE satellite, and the Solar Magnetic Field Telescope of the Beijing Astronomical Observatory (Liu & Zhang 2001). This allows us to improve a theory of solar flares. In particular, the well-observed flare on 2000 July 14, the Bastille-day flare, is interpreted (Somov et al. 2002). The main large-scale structure and dynamics of this flare is explained in terms of the collisionless reconnection model by Somov, Kosugi, and Sakao (1998).

2. Potential and Non-potential Fields

In order to clarify the role of a magnetic field in flares, let us recall the most important physical properties of the field in a flare-active region (Figure 1). Naturally an actual field consists of two components: (a) the potential or current-free one and (b) the non-potential part related to the electric currents flowing in an active region. Starting from some small height in the chromosphere up to some significant height in the corona, $h \sim 0.5 - 0.7\ R_{\odot}$, the magnetic energy density greatly exceeds that of the thermal, kinetic and gravitational energy of the plasma. So, the magnetic field can be considered in the strong field approximation (see Somov 2000). This means, in fact, that the coronal field is mainly potential. At least, it is potential in a large scale, in which the field determines the *global* structure of an active region.

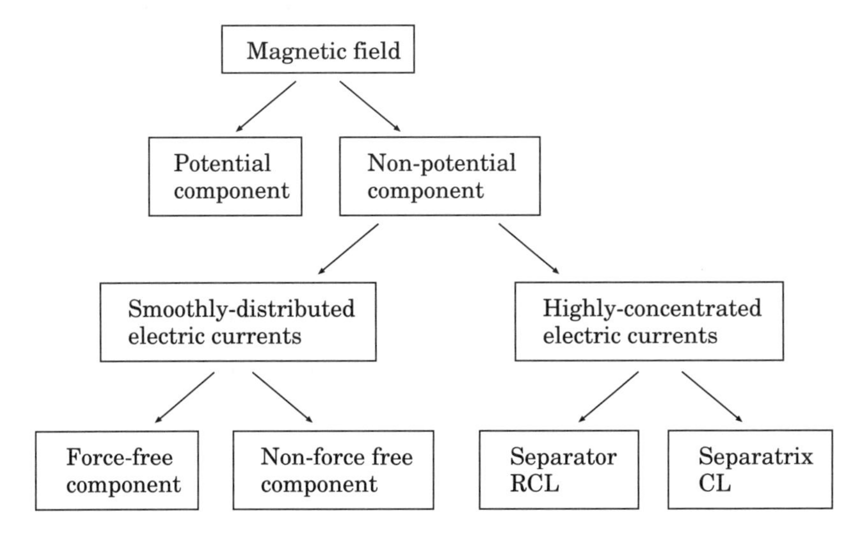

Figure 1. Main types of the magnetic field in an active region.

However the potential field, which satisfies the given boundary conditions in the photoshere and in the solar wind, has the absolute minimum of energy because the potential field is current-free by definition. Two consequences for the physics of flares follow from this fact.

First, being disrupted somewhere, for example by an eruptive prominence, the field lines of the potential field are connected back again via reconnection. In the strong field approximation, the magnetic field, changing in time, sets the solar plasma in motion. Such a motion can be described by equations which are much simpler than the equations of the ordinary MHD (Somov 2000). The potential field is a solution of the simplified equation set. We should not forget about this fact if we do not want to lose a natural simplicity of the actual conditions in the solar atmosphere in favour of the standard MHD computer codes.

Second, since no energy can be taken from the current-free field, the current-carrying components have to be unavoidably introduced in the modelling to explain accumulation of energy before a flare and its release in the flare process. The non-potential parts of the field accumulate the *free* magnetic energy, which is just an excess over the potential field energy. Fast transformation of the free energy into thermal and kinetic energy of superhot plasmas and accelerated particles constitutes the flare phenomenon. It is of principal importance to distinguish the currents of different origin (Figure 1) because they have different physical properties and, as a consequence, different behaviours in the pre-flare and flare processes. The actual currents in the solar atmosphere are conventionally comprised of two different types: (a) smoothly-distributed currents that are necessarily parallel or nearly parallel to the field lines, so the field is locally force-free or nearly force-free; (b) strongly-concentrated currents like reconnecting current layers (RCL) at separators or current layers (CL) at separatrices.

3. Rainbow Reconnection Model

The appearance of separators in the solar atmosphere was initially attributed to the emergence of a new magnetic flux from the photosphere into the region where another magnetic flux already exists. In fact, the presence of separators must be viewed as a much more general phenomenon. Figure 2a exhibits the simplest model of the uniform distribution of the vertical component B_z of the field in the photosphere (Somov 1985). The interface between fields with opposite polarity – the neutral line NL – divides the region of the magnetic field source along the y axis. In accordance with the fact that it is often visible in solar magnetograms, this region is deformed by photospheric flows with the velocity field $\mathbf{v}$ in such a way that the neutral line gradually acquires the shape of the letter S as shown in Figure 2b.

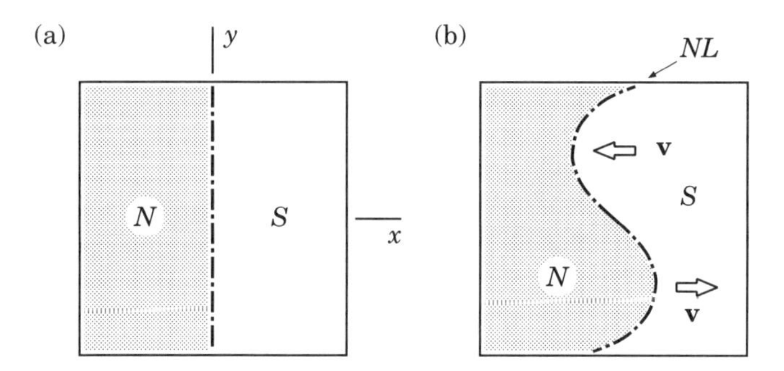

Figure 2. (a) Model distribution of the uniform vertical component of magnetic field in the photospheric plane (x, y). (a) A vortex flow distorts NL so that it takes the shape of the letter S.

Beginning with some critical bending of the neutral line, the field calculated in the potential approximation begins to contain a separator as shown in Figure 3 (Somov 1985, 1986). In this figure, the separator X is located above the photospheric NL like a rainbow above a river which makes a bend. The vortex-type flow, as shown in Figure 2b, generates two components of the velocity field: parallel to NL and directed to NL. The first component provides a shear of magnetic field lines above the photospheric NL. The second one tends to compress the photospheric plasma near the NL and in such a way it can drive reconnection in the corona and in the photosphere.

Reconnection at two levels plays different roles in flares. Photospheric reconnection seems to be mainly responsible for supply of a cold dense plasma upward, into pre-flare filament prominences. Coronal reconnection, being slow

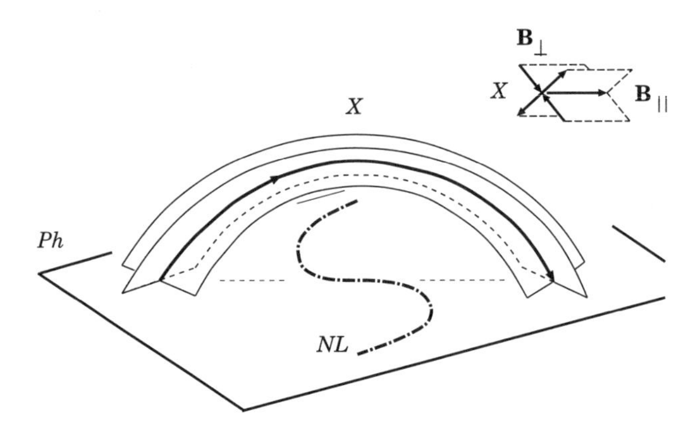

Figure 3. The separator X above the S-shaped bend of the photospheric neutral line NL. The inset in the upper righthand corner shows the structure of the magnetic field near the top of the separator.

before a flare, allows to accumulate a sufficient amount of magnetic energy. During a flare, the fast collisionless reconnection in the corona, converts this excess of energy into kinetic and thermal energies of fast particles and super-hot plasma. As for the physical mechanism of the Bastille-day flare, we assume that it is the collisionless reconnection at the separator in the corona (Somov et al. 1998).

More specifically, we assume that in the large-scale two-ribbon flares with an observed significant decrease of the footpoint separation, like the Bastille-day flare, two conditions are satisfied (Somov et al. 2002). First, before occurrence of a flare, the magnetic field separatrices are involved in the large-scale shear photospheric flow. The second condition is the presence of an RCL generated by large-scale converging flow in the photosphere also before a flare. These two conditions are sufficient ones for an active region to produce a huge flare similar to the Bastille-day flare. Other realizations of large flares are possible, of course, but this one seems to be the most plausible situation. At least, in addition to the flare HXR ribbons and kernels, it explains formation of the twisted filament prominences along the photospheric neutral line before and after the flare.

References

Kosugi, T., Makishima, K., Murakami, T., Sakao, T., Dotani, T., Inda, M., Kai, K., Masuda, S. et al. 1991, Solar Phys., 136, 17

Liu, Y. & Zhang, H. 1991, A&A, 372, 1019

Somov, B. V. 1985, Soviet Physics Uspekhi, 28, 271

Somov, B. V. 1986, A&A, 163, 210

Somov, B. V. 2000, Cosmic Plasma Physics (Dordrecht: Kluwer Academic Publ.)

Somov, B. V., Kosugi, T. & Sakao, T. 1998, ApJ, 497, 943

Somov, B. V., Kosugi, T., Hudson, H. S., Sakao, T. & Masuda, S. 2002, ApJ, in press

IAU 8th Asia-Pacific Regional Meeting
ASP Conference Series, Vol. 289, 2003
S. Ikeuchi, J. Hearnshaw & T. Hanawa, eds.

Energy Accumulation for the Bastille Flare

A. I. Podgorny

Lebedev Physical Institute RAN, Moscow, 119991, Russia

I. A. Bilenko

Sternberg Astronomical Institute of MSU, Moscow, 119899, Russia

S. Minami, M. Morimoto

Osaka City University, Osaka, 588, Japan

I. M. Podgorny

Institute for Astronomy RAN, Moscow, 109017, Russia

Abstract. Magnetic field evolution above the NOAA 9077 active region is considered before the flare on 2000 July 14. The system of 3D MHD resistive equations is solved for a compressible plasma. The anisotropy of thermal conductivity in the magnetic field is taken into account. It is shown that a vertical current sheet (CS) appears in the corona due to disturbances that are produced at the photospheric magnetic field changing before the flare. The $\mathbf{j} \times \mathbf{B}/c$ force in the CS can accelerate plasma flux upwards and produce a coronal mass ejection (CME).

1. Introduction

In our previous papers (Podgorny, 1995; Podgorny and Podgorny, 1992, 1998) the energy storage in a current sheet (CS) above a simple active region has been numerically investigated in a 3D MHD approximation. It is shown that CS creation occurs in the vicinity of a neutral line as a result of a magnetic disturbance accumulation arriving from the photosphere. Results of numerical experiments permit to build the electro-dynamical solar flare model (Podgorny and Podgorny, 1992, 2001) that explains the main flare phenomena. After evolution the CS becomes unstable, and fast energy release occurs as a result of reconnection. The numerical experiments demonstrate that the flare and CME appear in the same powerful event. The aim of this paper is to consider possibilities of a CS creation above the NOOA 9077 active region due to different types of observed photospheric disturbances. Some preliminary results have been published (Bilenko et al. 2000).

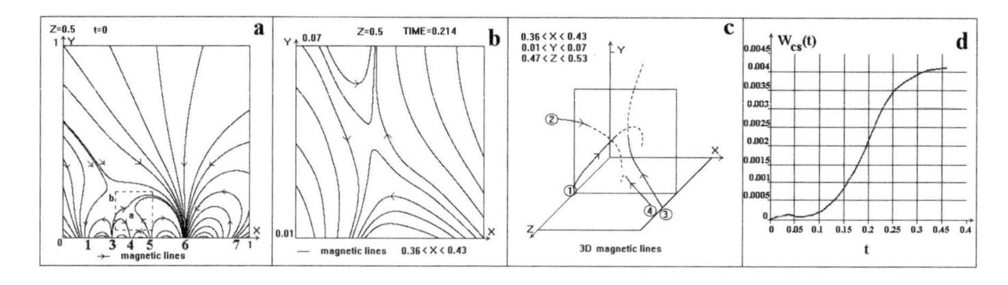

Figure 1. Magnetic field lines above the active region in the initial stage (a); magnetic field in the plane $z = 0.5$ in the extended scale (b); field lines in 3D space, 4 is the singular line (c); the energy accumulation in CS (d).

2. Observation Data and Numerical Method

Magnetic spot positions of the active region NOAA 9077 before the flare were extended along a straight line. This line is referred as the line $y = 0$, $z = 0.5$. The X and Z axes are placed on the photosphere. The most important peculiarity of the magnetic field configuration region can be seen in the plane $z = 0.5$.

The magnetic field of NOAA 9077 active region is approximated by fields of 7 vertical magnetic dipoles placed below the photosphere. The dimension of the active region $L_0 = 260000$ km is taken as the length unit. The average magnetic field above the active region $B_0 = 300$ gauss is taken as the magnetic field unit. Figure 1a shows magnetic field lines in the plane $z = 0.5$ for the initial time ($t = 0$).

For setting the initial conditions the data obtained from the 150-Foot Solar Tower at Mount Wilson Observatory (ftp://helios.astro.ucla.edu) in electronic form are used. Four days prior to the flare the dipole 3 has been increased on two occasions. Dipole 6 has been increased between June 10 and 12. Such evolution of the active region is assumed in the numerical experiment.

The magnetic Reynolds number $\mathrm{Re}_m = 4\pi V_A L\sigma/c^2$ in the corona is of the order of 10^{15}, and frozen-in condition is fulfilled for a very long time. All this time $t_D = \mathrm{Re}_m t_A$ the diffusion of magnetic field does not prevent magnetic energy accumulation in the CS. Here $t_A = L/V_A$. The effective (numerical) Re_m in numerical experiment is order of 50. So, for the investigation of CS creation and its evolution in the frozen-in condition, the time of the photospheric disturbance must not considerably exceed t_A.

The 3D resistive MHD equations for compressive plasma are solved numerically in the calculation region $0 \le x \le 1$, $0 \le y \le 1$, $0 \le z \le 1$. The Y-axis is directed perpendicular to the solar surface. The units of the plasma density ρ_0 and temperature T_0 are taken as its initial values. The units of the plasma velocity, time, and current density are taken respectively as the Alfvenic velocity $V_0 = V_A = B_0/\sqrt{4\pi\rho_0}$, $t_0 = L_0/V_0$, $j_0 = cB_0/4\pi L_0$.

For solving MHD equations the PERESVET code (Podgorny, 1995; Podgorny and Podgorny, 1992) is used with the $41 \times 41 \times 41$ net. The parameters

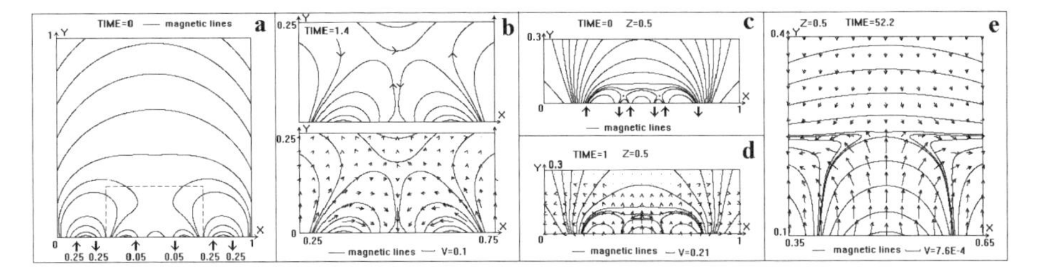

Figure 2. The initial magnetic field before the magnetic flux immersion (a); lines of the vertical CS on an extended scale (b); Initial magnetic field in calculation with strong immersion (c); the beginning of a horizontal CS creation (d); horizontal CS on the extended scale (e).

are chosen according to principle of limited simulation: $\mathrm{Re}_m = 10^5$, $\mathrm{Re} = 100$, $\beta = 2 \times 10^{-5}$.

3. Current Sheet Creation in the Vicinity of the Neutral Line *a*

In the calculations of magnetic energy storage in a CS, that appears in vicinity of the neutral point a, a linear increasing of the magnetic field is set during the time $t = 0.2$. The magnetic field configuration corresponding to a CS is seen in Figure 1b in the expanded scale. This CS, as all CS in the laboratory and in space, possesses a normal magnetic field component. During magnetic lines reconnection the stretched magnetic lines push plasma along the CS. Plasma accelerated upwards can be ejected in the interplanetary space, producing a coronal mass ejection. Fig. 1c presents the CS in 3D space in the extended scale.

The energy accumulation in the CS is presented in Fig. 1d. in dimnsionless units (the unit of energy is $(B_0^2/4\pi)L_0^3 = 1.26 \times 10^{35}$ erg). At $t = 0.3$ the saturation is achieved at 5×10^{32} erg.

The calculations showed, that in the vicinity of another X-point (point b, see Fig. 1a) CS is not created.

4. Current Sheet Creation Due to Magnetic Flux Immersion

The important feature of Zhang et al. (2001) observation of a preflare state of NOAA active region is the compact magnetic flux immersion. Ten hours before the flare the position of the positive magnetic flux moved about 7×10^3 km relative to the negative flux. The positive flux was absorbed by the negative one. We explain this phenomenon as magnetic flux immersion into the photosphere. At flux immersion the legs of the arch approach each other, and the arch disappears under the photosphere.

We investigated such an effect numerically in a symmetrical magnetic configuration. The vertical magnetic dipoles are situated symmetrically under the photosphere in the plane $z = 0.5$. The moment of dipoles 3 and 4 are cancelled during $\Delta t = 1$. Fig. 2a shows lines of the initial magnetic field. The positions of dipole along the X axis are shown under the figure. We calculated the situation that corresponds to magnetic flux decreasing above the photosphere near the point $x = 0.5$. Decreasing of dipoles 3 and 4 produce magnetic field decreasing above the point $x = 0.5$, and the plasma is accelerated downwards. Simultaneously the vertical CS appears. Magnetic lines from dipoles 2 and 5 are stretched along the CS (Fig. 2b). The velocity vectors are presented below. Such CS decay produces a flare, but it cannot produce a very strong CME.

Fig. 2c shows the dipole positions for the second calculation. During this CS evolution and its decay, when dipoles 3 and 4 decrease, the field lines that connect dipoles 2 and 5 are piled above the point $x = 0.5$. The flowing up magnetic field initiates the upwards directed MHD disturbances. The field possesses two neutral lines. The points of their intersection with the figure plane are seen. These neutral points disappear as dipoles 3 and 4 are decreasing. Simultaneously the lines appear that connect dipoles 2 and 5. This magnetic field configuration is show in Fig. 2d. The increasing magnetic flux produces MHD disturbances moving upwards. Velocity vectors produced by these disturbances are shown also. As a result, the horizontal CS is built between the emerging flux and the opposite flux situated above. The latter consists of the lines that connect dipoles 1 and 6. The magnetic field configuration with the horizontal CS is sown in Fig. 2e. The decay of this CS can also produce a flare above NOAA 9077, but apparently plasma ejection from a horizontal CS cannot produce a powerful CME that has been observed for the Bastille flare.

5. Conclusion

MHD numerical simulations demonstrate that there are several possibilities for energy accumulation above the NOAA 9077 active region. In the vicinity of one of the neutral lines the CS is created. The accumulated energy is about 10^{32} erg. Decay of this CS can produce a powerful solar flare that is associated with CME. Zhang at al. (2001) have reported about the cancelling magnetic flux of two spots. Such magnetic disturbance is also simulated. Numerical analysis shows that a horizontal CS in the corona in this case can appear.

Acknowledgments. This work was supported by the RBRF (grants 01-02-16186 and 00-01-00091) grant Astronomiya of the Russian Federation. We are thankful for data from the synoptic program at the 150-Foot Solar Tower of the Mt. Wilson Observatory.

References

Bilenko, I. A., Podgorny, A. I. & Podgorny I. M. 2002, Solar Phys. in press

Podgorny, A. I. 1995, Solar Phys., 156, 41

Podgorny, A. I. & Podgorny, I. M. 1992, Solar Phys., 139, 125

Podgorny, A. I. & Podgorny, I. M. 1998, Astron. Reports, 42, 116

Podgorny, A. I. & Podgorny, I. M. 2001, Astron. Reports, 45, 60

Zhang, J., Wang, J., Deng, Y. & Wu, D. 2001, ApJ, 549, L99

IAU 8th Asia-Pacific Regional Meeting
ASP Conference Series, Vol. 289, 2003
S. Ikeuchi, J. Hearnshaw & T. Hanawa, eds.

Fine Structures Observed by the Chinese Solar Radio Broadband Fast Dynamic Spectrometers

Yihua Yan, Qijun Fu, Yuying Liu, Zhijun Chen

National Astronomical Observatories, Chinese Academy of Sciences, Beijing 100012, China

Hongao Wu, Fuying Xu

Purple Mountain Observatory, Chinese Academy of Sciences, Nanjing 210008, China

Zhihai Qin

Astronomy Department, Nanjing University, Nanjing 210093, China

Min Wang, and Zhiguo Xia

National Astronomical Observatories, Chinese Academy of Sciences, Kunming 620011, China

Abstract. Observations by the newly established Chinese Solar Radio Broadband Fast Dynamic Spectrometers (Fu et al. 1995) in the 0.7–7.6 GHz frequency range are introduced. The instrument has been working properly in the 23rd solar maximum with very high temporal (5, 8, 100 ms) and spectral (1.4, 10, 20 MHz) resolutions simultaneously. The observational results are helpful for a better understanding of particle acceleration and energy conversion in the solar corona.

1. Introduction

The Solar Radio Broadband Dynamic Spectrometer (SRBS) of China is the first instrument in microwave to acquire dynamic spectrums of solar bursts with the combination of wide frequency coverage (0.7–7.6 GHz), high temporal resolution, high spectral resolution, and high sensitivity (Fu et al. 1995). Observations at centimetre and decimetre wavelengths are important for addressing fundamental problems of energy release, particle acceleration and particle transport (Bastian et al. 1998). Tarnstrom and Philip (1972) noted that the duration of spikes is comparable to the electron–ion collision time, and the duration of individual fine structure decreases as frequency increases. Metric and kilometer type III bursts satisfy an empirical relation for the decay time with frequency (Alvarez and Haddock 1973). Therefore, the spectrometer with time resolution of better than 10ms, high spectral resolution and high sensitivity for observing microwave fine structures should be essential. This motivated the development of the Chinese solar radio broadband spectrometers and we will describe the instrument in

§2. Then some new results observed with this instrument are presented and discussed. Finally we draw some conclusions.

2. Instrument Description

Since 1994, a broadband solar radio spectrometer had been developed, with a frequency coverage of 0.7–7.6 GHz, a frequency resolution of 1–10 MHz, and a temporal resolution of 1–10 ms (Fu et al. 1995). This instrument is composed of 5 spectrometers: 0.7–1.4 GHz, 1.0–2.0 GHz, 2.6–3.8 GHz, 4.5–7.5 GHz, and 5.2–7.6 GHz. The three spectrometers at 1.0–2.0 GHz, 2.6–3.8 GHz, and 5.2–7.6 GHz are presently located at Huairou Solar Observing Station of National Astronomical Observatories, Chinese Academy of Sciences (NAOC). The radio environment has been measured and calibration techniques are developed to ensure reliable observations (Yan et al. 2001c,d,e; Sych and Yan 2002). The other two are located in the cities of Kunming and Nanjing, about 2000 km and 1000 km away from Beijing, respectively. The parts of 1.0–2.0 GHz, 2.6–3.8 GHz, and 5.2–7.6 GHz for both polarization and intensity observations have been put into operation since 1994 January, 1996 September and 1999 August, respectively. The 4.5–7.5 GHz spectrometer has been put into operation since August 1999 at Purple Mountain Observatory (PMO), and the 0.7–1.5 GHz part has been put into operation since 2000 June at Yunnan Observatory/NAOC. At present, we routinely provide solar radio observations at 2840 MHz for inclusions in 'Chinese Solar-Geophysical Data' published by NAOC/Beijing, and in part II of 'Solar-Geophysical Data' published by National Geophysical Data Center at Boulder, Colorado. The performance of the spectrometers is very powerful in detecting radio fine structures. Table 1 shows the description of the spectrometers at Huairou (Fu et al. 1995; Ji et al 2000). Table 1 shows the description of the spectrometers at Huairou (Fu et al. 1995; Ji et al 2000), and the other two are listed in Table 2.

3. Observational Results

Observations at centimeter and decimetric wavelengths are very important for addressing fundamental problems of energy release, particle acceleration and particle transport (Bastian et al. 1998). With the newly developed Chinese Solar Radio Broadband Fast Dynamic Spectrometers, many results have been obtained, e.g., zebra structures around 3 GHz (Ning et al. 2000b,c; Ledenev et al. 2001a,b; Chernov et al. 2001a), bi-directional electron beams in the corona indicating energy release site (Fu et al. 1997; Huang et al. 1998; Ning et al. 2000a; Xie et al. 2000), microwave type-U and type-M bursts and other fine structures (Ning et al. 2000d; Wang et al. 2001a,b), microwave spikes (Wang et al. 1999; Wang et al. 2001b; Chernov et al. 2001b), the slow drifting and pulsation structures associated with plasmoid ejections for the famous Bastille-day event (Karlicky et al. 2001; Wang et al. 2001a,c; Yan et al. 2001a,b), and discovery of an unusually large group delay in microwave millisecond oscillating events (Fleishman et al. 2002).

Table 1. The Solar Radio Spectrometers at Huairou

Frequency range:	1.0–2.0 GHz	2.6–3.8 GHz	5.2–7.6 GHz
Temporal resolution:	5 ms (after 2002 June) 20ms (before Dec. 2001)	8 ms	5 ms
Frequency resolution (MHz/chan.):	4/240 (after 2002 June) 20/50 (before 2001 Dec.)	10/120	20/120
Sensitivity:	3%,	2%,	2% $S_{\text{quiet Sun}}$
Dynamic range:	10 dB above 3(2)%$S_{\text{quiet Sun}}$		
Polarizations:	LHCP, RHCP		
Observing time:	22–10UT (Summer), 0–8UT (Winter)		

Table 2. Performance of the 2 Spectrometers in Kunming and Nanjing

Frequency range (GHz):	0.7–1.5	4.5–7.5
Temporal resolution (ms):	5	5
Frequency resolution (MHz):	<3	10
Sensitivity (% $S_{\text{quiet Sun}}$):	3	2
Dynamic Range (dB):	10	10
LHCP & RHCP measurement:	Yes	No
Location	Kunming	Nanjing

3.1. Microwave Type III Bursts

Type III bursts are the most intensively studied form of all radio emission. Frequency drift rates are mostly positive (from low to high frequencies) with 40 MHz/s–22 GHz/s and the duration at one frequency channel ranges from less than 30ms to 200ms. Solar microwave type III bursts have attracted wide attentions as an important signature of energetic electron beam in low corona. The Type III Burst Pair in 1–2 GHz may indicate the reconnection region. Fu, et al. (1997) estimated the separatrix frequency at 1660–1760 MHz height above the photosphere of the effective layer of separatrix frequency to be 3×10^4 km. Huang et al. (1998) analyzed the energetic spectrum of NT electrons in the acceleration region with a spectrum index of 4.5, an electric field of 10^{-4} V/m intensity and 10 Mm in length. Ning et al. (2000a) estimated $\beta \sim 0.01$ in the reconnection region. If the ambient magnetic field is 100G, the beam velocity is deduced as 1.07×10 3 km/s. Xie et al. (2000) estimated the height of the reconnection region.

3.2. Microwave Type U Bursts

Type U bursts are interpreted as the signature of electron beams along magnetic field lines. The total durations of type-U bursts decrease with increasing frequency: in metric bands: 5–40 s; in decimeter bands: 1 s; and in centimeter bands: <1 s. Table 3 summarizes the properties of the Type-U bursts by Fu et al. (1999, Paper I), Wang et al. (2001b, Paper II) and Wang, Yan and Fu (2001a Paper III).

Table 3. The properties of Microwave Type-U bursts.

Parameters from:	Paper I	Paper II	Paper III
Top Frequency:	3.2–3.4 GHz	2.63–2.70 GHz	1.16 GHz
Δf for each mini-U burst:	60–200 MHz	340–360 MHz	~700 MHz
df/dt in rising branch	–7~–28 GHz/s	–1.7~–3.0 GHz/s	–100 MHz/s
Duration of rising branch	8–24 ms	224–264 ms	~10 s
Polarization degree	>80%	80%	

Type U, M and N are sub-classes of Type III burst. They are much rarer than the normal type III bursts (Ning et al. 2000d; Wang et al. 2001c).

3.3. Microwave Millisecond Spike Emission

Solar radio spike emission have been observed for almost forty years, but most of them were in meter and decimeter frequency range. Wang et al. (1999) summarized some of the characteristics: short duration, narrow frequency bandwidth, high polarization degree, irregular sequence in frequency and time delays, etc., as shown in Table 4.

Table 4. The properties of Microwave spike bursts (Wang et al. 1999).

Event Date	Magn. Pol.	Pol. (%)	Time Delay	Band-width (MHz)	Dur. (ms)	Total Int. (s.f.u.)	Freq. Drift (GHz/s)
1997 Sep 29	S	0	r	200	50	200	6.0
1997 Sep 22	N	100R	-	130	70	60	0
1997 Sep 24	N	100R	-	100	30	80	0
1997 Sep 26	N	10R	0	70	40	180	0
1997 Oct 28	N	30L	0	50	60	200	0
1997 Nov 02	N	0	0	100	50	240	0
1997 Nov 03	N	100L	-	200	20	180	0
1997 Nov 17	N	20L	r,l	40	30	250	1-2
1997 Dec 26	N	15L	r	30	30	280	0
1998 Nov 02	N	20L	0	300	60	240	0
1998 Apr 15	N	10L	r	70	30	200	2.0
Average:				117/3.7%	42.7	192	

A microwave millisecond spike cluster was observed at the highest frequency range so far in 5.2–7.6 GHz on 2001 April 10 around 05:03:58UT. The properties are listed in Table 5.

Table 5. The properties of a spike cluster.

Number of spikes:	99
Duration of each spike:	< 10 ms
Center frequency:	5.87 GHz
Average bandwidth:	24.5 MHz, 0.4%

A new model for microwave millisecond spikes has been proposed (Chernov et al. 2001b). The mechanism based on the interaction of Langmuir wave with ion-sound waves: $l + s \rightarrow t$, can operate in shock front, propagating from a magnetic reconnection region. Two events with microwave millisecond spikes on 1997 November 4 and 28 were discussed using this new model. It is considered that microwave millisecond spikes are probably an unique manifestation of flare fast shocks in the radio emission (Chernov et al. 2001b).

3.4. Microwave Zebra Pattern Structure (ZPS) and Fiber Bursts (FB)

ZPS and FB are well-known fine structures in the metric and decimetric continuum emission of type IV solar radio bursts. In the microwave range we could find only some indications of a probable Zebra -pattern in Isliker & Benz (1994). But now with new microwave spectrometer one can observe detailed ZPS and FB at high frequency (Chernov et al. 2001a). Table 6 shows the comparison of microwave ZPS and FB with those in 160–320 MHz metric band (Chernov et al. 2001a, Kuijpers 1980).

3.5. Drifting Pulsation Structure (DPS)

Recently, in the paper by Kliem et al. (2000) and Karlicky et al. (2001) the slowly negatively drifting pulsation structure (DPS) was presented and interpreted as the signature of the dynamic magnetic reconnection .The slow negative drift of the whole DPS is in all these cases caused by the motion of the whole reconnection space and the plasmoid upwards to lower plasma densities (–3 MHz/s $\sim$ –60 MHz/s).

3.6. Microwave Millisecond Quasi-period Pulsation

Microwave millisecond quasi-period pulsation with long time delay between L- and R- polarization components have been observed (Fleishman et al. 2002), and the parameters are estimated as shown in Table 7.

3.7. Microwave "Patches" and Other Fine Structures

According to the previous observations, the patches have a duration between one and some tens of seconds, their circular polarization is usually weak, and

Table 6. The comparison of ZPS and FB properties.

Parameters	Chernov et al. (2001a)	Kuijpers (1980)
Duration:	ZPS: 1–2 s (8 s) FB: 1–2 s (42 s)	FB: 5–10 s
Frequency range:	basically $<$ 4 GHz	
$\Delta f_{emission-absorption}$	20–70 MHz	
Frequency separation	20–70 MHz	
Emission frequency band	$\sim$20–30 MHz	ZPS: in 160–200 MHz: 2–3 MHz in 800–900 MHz: 20 MHz
Number of stripes	2$\sim$>6	FB: 10–30 (up to 300)
Frequency extent	200$\sim$>500 MHz	ZPS: $>$ 40 MHz FB: 30 MHz (up to 120 MHz)

the flux density is not very high. The patches observed on 1998 June 12 around 23:14 UT have the following properties: Very short duration ($\sim$300 ms); very high flux density ($\sim$1000 s.f.u); very high polarization degree ($\sim$100% RCP); extremely narrow bandwidth ($\Delta f/f$ $\sim$5%); very high spectral indexes. The gyrosynchrotron process cannot explain such outstanding characteristics. Plasma and maser emission seem to be the most plausible mechanisms (Wang et al. 2001c). There are some other radio fine structures that have not classified. For example, the quasi-period drifting structures. Possibly, it may reflect the fluctuation (inhomogeneity) of electron density in magnetic loops. Electron beams may be reflected at some places such as magnetic mirror, steep density gradient, shock wave front, and so on.

3.8. Association of Microwave Fine Structures (FS) with Solar Activity

The Bastille-day flare of 2000 July 14 included three successive processes in radio observations (Wang et al. 2001c): firstly, the X-ray rose and reached to maximum at 10:10–10:23 UT, accompanying with the FS only in the range 2.6–7.6 GHz; Secondly, the microwave radio emission reached to maximum accompanying with many FS over the range 1.0–7.6 GHz at 10:23–10:34; Then, a decimetric Type IV burst and its associated FSs, (fibers) in the range 1.0–2.0 GHz appeared after 10:40 UT. These processes are interpreted as due to a flux rope eruption during the flare process, as inferred from extrapolation from observed magnetograms (Yan et al. 2001a,b).

4. Discussion and Conclusions

Are there the elementary unity of fine structures (FSs)? Some FS events consist of spike emission. There are the following two possibilities. 1. There may exist

Table 7. Microwave millisecond quasi-periodic pulsation.

-background plasma density:	$n \sim 1.7 \times 10\ 10\mathrm{cm}^{-3}$
-magnetic field :	$B = 290$ G
-kinetic electron temperature :	$T = 3 \times 10^6$ K
-number density of fast elections :	$n_b \sim 10^5$ cm^{-3}
-the level of plasma turbulence :	$w = W/nT = 10^{-6}$
-brightness temperature:	$T_\mathrm{b} = 5 \times 10^{12}$ K
-linear scale of the source :	$L = 100$ km

two kinds of FS: one consists of spike emission, but other is not. 2. All FSs consist of spike emission, but most of them could not be found because the resolution and sensitivity of present observing instruments are still not high enough.

The Chinese solar radio broadband spectrometers have been working properly in the 23rd solar maximum with very high temporal and spectral resolutions, and high sensitivity in microwave for the first time. These results are helpful for a better understanding of particle acceleration and energy conversion in the solar corona. The observations show that the occurrence of FSs in microwaves are less than that in lower frequency bands, but they are much more complicated, and sometimes it is difficult to distinguish between them. It may reflect that the physical circumstance, magnetic field configuration and dynamic process during flaring time in low corona are intricate. For finding new information, developing and improving the data process, such as wavelet, are significant. The co-operation with other space or ground-based observations is very important and will be further strengthened. Making a comprehensive analysis of the important types of FSs, such as microwave type III bursts, microwave spikes, microwave type U burst with other wavelength observations deserves in-depth study.

Acknowledgments. The work was supported by CAS, NSFC and MOST grants.

References

Alvarez, H., and Haddock, F. T. 1973, Solar Phys., 30, 175

Bastian, T. S., Benz, A. O., and Gary, D. E. 1998, Ann. Rev. Astron. Astrophys. 36, 131

Chernov, G. P., Yasnov, L. V., Yan, Y., and Fu, Q. 2001a, Chin. J. Astron. Astrophys., 1(6), 525

Chernov, G. P., Fu, Q. J., Lao, D. B., and Hanaoka, Y. 2001b, Solar Phys., 201, 153

Fleishman, G. D., Fu, Q. J., Huang, G. L., Melnikov, V. F., and Wang, M. 2002, Astron. Astrophys., 385, 671

Fu, Q., Qin, Z., Ji, H., and Pei, L. 1995, Solar Phys., 160, 97

Fu, Q. et al. 1997, Acta Astrophysica Sinica, 17, 441

Fu, Q. et al. 1999, Publ. Purple Mount. Observ., 18, 159

Huang, G.-L. et al. 1998, ApSS, 259, 317

Ji, H., Fu, Q., Liu, Y., Cheng, C., Chen, Z., Lao, D., Ni, C., Pei, L., Xu, Z., Chen, S., Yao, Q., Qin, Z. and Yang, G. 2000, Acta Astrophys. Sinica, 20, 209

Karlicky, M., Yan, Y. H., Fu, Q. J., Wang, S. J., Jiricka, K., Meszarosova, H., Liu, Y. Y. 2001, Astron. Astrophys., 369, 1104

Kliem, B., Karlicky M. andBenz, A.O. 2000, A&A, 360, 715

Ledenev, V. G., Karlicky, M., Yan, Y. H. and Fu, Q. J. 2001a, Solar Phys., 202, 71

Ledenev, V. G., Yan, Y. and Fu, Q. J. 2001b, Chin. J. Astron. Astrophys., 1, 475

Ning, Z. J., Fu, Q. J. and Lu, Q. K. 2000a, Solar Phys., 194, 137

Ning, Z. J., Fu, Q. J. and Lu, Q. K. 2000b, Publ. Astron. Soc. Japan, 52, 919

Ning, Z. J., Fu, Q. J. and Lu, Q. K. 2000c, A&A, 364, 853

Ning, Z. J., Yan, Y. H., Fu, Q. J. and Lu, Q. K. 2000d, A&A, 364, 793

Ning, Z. J., Fu, Q. J., Yan, Y. H., Liu, Y. Y. and Lu, Q. K. 2001, Ap&SS, 277, 615

Sych, R. A. and Yan, Y. H. 2002, Chin. J. Astron. Astrophys., 2 (2), 183

Wang, M., Fu, Q. J., Xie, R. X. and Huang, G.L. 1999 Solar Phys., 189, 331

Wang, M., Fu, Q. J., Xie, R. X. and Duan, C. C. 2001a, Solar Phys., 203, 145

Wang, M., Fu, Q. J., Xie, R. X., Huang, G. L. and Duan, C. C. 2001b, Solar Phys., 199, 157

Wang, M., Fu, Q. J., Xie, R. X., Huang, G. L. and Duan, C. C. 2001c, A&A, 380, 318

Wang, S. J., Yan, Y. H. and Fu, Q. J. 2001a, Astron. Astrophys. Letters, 370, L13

Wang, S. J., Yan, Y. H. and Fu, Q. J. 2001b, A&A, 373, 1083

Wang, S. J., Yan, Y. H., Zhao, R. Z., Fu, Q. J., Tan, C. M., Xu, L., Wang, S. J. and Lin, H.A. 2001c, Solar Phys., 204, 153

Tarnstrom, G. L. and Philip, K.W. 1972, A&A, 17, 267

Xie, R. X., Fu, Q. J., Wang, M., and Liu, Y. Y. 2000, Solar Phys., 197, 375

Yan, Y. H., Deng, Y. Y., Karlicky, M., Fu, Q. J., Wang, S. J. and Liu, Y. Y. 2001a, ApJL, 551, L115

Yan, Y. H., Fu, Q. J., Aurass, H., Lesovoi, S., Altyntsev, A., Liu, Y. Y., Chen, Z. J. and Ji, H. R. 2001b Recent Insights into the Physics of the Sun and Heliosphere, IAU Symp., 203, 338

Yan, Y. H., Fu, Q. J., Liu, Y. Y. and Chen, Z. J. 2001c, Preserving the Astronomical Sky, IAU Symp., 196, 311

Yan, Y. H., Ji, H. R., Fu, Q. J., Liu, Y. Y. and Chen, Z. J. 2001d, Preserving the Astronomical Sky, IAU Symp., 196, 315

Yan, Y. H., Tan, C. M., Xu, L., Ji, H. R., Fu, Q. J., and Song G. 2001e, Since in China (Series A), 31, 73

IAU 8th Asia-Pacific Regional Meeting
ASP Conference Series, Vol. 289, *2003*
S. Ikeuchi, J. Hearnshaw & T. Hanawa, eds.

Flux Tube Oscillations and Coronal Heating

Y. Sobouti[1,2], K. Karami[1] and S. Nasiri[1,3]

1) Institute for Advanced Studies in Basic Sciences, P.O. Box 45195-159, Zanjan, Iran

2) Center for Theoretical Physics and Mathematics, AEOI, P.O. Box 11345-8486, Tehran, Iran

3) Department of Physics, Zanjan University, Zanjan, Iran

Abstract. Wave transmission in low β magnetic flux tubes has, mathematically, the same structure as the propagation of electromagnetic waves in optical fibers. In both cases the problem is reducible to a single wave equation for the longitudinal component of the perturbed field along the fiber/tube axis. We derive this equation, solve the dispersion relation associated with it, and assign three wave numbers to each mode. In cylindrical coordinates (r, φ, z), for a given φ-wave number, the plane of the r- and z- wave numbers is divided into one "mode zone" in which each grid point is a possible mode of the system and one "forbidden zone" in which no mode may dwell. The cutoff line, the boundary of the two zones, is given both analytically and numerically. Next we introduce weak resistive and viscous dissipation to the system, solve for the decay time of each mode and for the densities of heat generation rates by each dissipative process. The two densities have identical spatial dependencies, but different magnitudes. The resistive heat rate is inversely proportional to the Lundquist number, S, and the viscous one to the Reynolds number, R. The time decay exponent is proportional to the sum $(S^{-1} + R^{-1})$.

1. Introduction

The astronomical literature on waves in magnetic flux tubes has a back log of a quarter of a century. Ionson (1978), Wentzel (1979 a,b), Wilson (1979), Roberts (1981 a,b), Edwin and Roberts (1983), Hollweg (1984), Steinolfson et al. (1986), Davila (1987), Steinolfson and Davila (1993), and Ofman et al. (1994, 1995) have all addressed various aspects of the problem. The analysis of TRACE data by Nakariakov et al. (1999), however, has given a new impetus to such studies. Convincing evidence has emerged that the coronal loops can and do oscillate in matters of few hundreds of seconds and do heat up in the course of damping of the wave motions. Here, we are primarily interested in the mathematical and analytical properties of modes in magnetic flux tubes and of their decay by resistive and viscous processes.

2. Exposition of the Problem

Let the flux tube be a cylinder of radius $r = 1$, of length πL, and lie along the z-axis of a cylindrical coordinate system (r, φ, z). Under coronal conditions assume i) the plasma pressure is negligible in comparison with the magnetic one. ii) The scale height is much larger than the height of the flux tube, so that the density stratification can be neglected. iii) The space is pervaded by a constant magnetic field along the z-axis. iv) The plasma density, ρ, has the constant values ρ_i and ρ_e inside and outside of the cylinder, but varies discontinuously at $r = 1$. Let the system undergo a small perturbation about its equilibrium state. The perturbation induced velocity and magnetic fields, $\delta\mathbf{v}(r, \varphi, z)$ and $\delta\mathbf{B}(r, \varphi, z)$, respectively, are governed by the following equations.

$$\frac{\partial \delta\mathbf{v}}{\partial t} = \frac{1}{4\pi\rho}(\nabla \times \delta\mathbf{B}) \times \mathbf{B}, \qquad \frac{\partial \delta\mathbf{B}}{\partial t} = \nabla \times (\delta\mathbf{v} \times \mathbf{B}). \tag{1}$$

The first of Eq. (1) has no z-component. We write out the z- and transverse components of the second one and eliminate the transverse components of $\delta\mathbf{v}$ and $\delta\mathbf{B}$ in favor of δB_z. For an exponential z-, φ- and time dependence, $e^{i(lz/L+m\varphi-\omega t)}$, ℓ and m integers, one obtains the following Bessel's equation for δB_z:

$$\left(\frac{d^2}{dr^2} + \frac{1}{r}\frac{d}{dr} + k^2 - \frac{m^2}{r^2}\right)\delta B_z(r) = 0, \qquad k^2 = \left(\frac{\omega}{v_A}\right)^2 - \frac{\ell^2}{L^2}, \tag{2}$$

where $v_A^2 = [B^2/4\pi\rho]^{\frac{1}{2}}$ is the Alfven speed. Both v_A and k are different inside and outside of the flux tube; for ρ is. For $k_i^2 = (\omega/v_{Ai})^2 - (\ell/L)^2 > 0$ in $r < 1$ and $k_e^2 = (\ell/L)^2 - (\omega/v_{Ae})^2 > 0$ in $r > 1$, solutions of Eq. (3) are

$$\begin{aligned} \delta B_z(r) &= J_m(k_i r), \quad r < 1 \\ &= AK_m(k_e r), \quad r > 1, \end{aligned} \tag{3}$$

where K_m is the modified Bessel function and A is a constant to be determined. In terms of δB_z the transverse components, $\delta\mathbf{B}_\perp$ and $\delta\mathbf{v}_\perp$ are

$$\delta\mathbf{B}_\perp = -\frac{\ell}{L\omega}B\delta\mathbf{v}_\perp = -i\frac{\ell}{Lk}\nabla_\perp(\delta B_z), \tag{4}$$

where $\nabla_\perp$ is the gradient operator in the (r, φ) plane.

3. Boundary Conditions and Dispersion Relation

To avoid shock waves at $r = 1$, the Lagrangian change, or equivalently in this case of constant B, the Eulerian change in pressure, $\delta p \propto B\delta B_z$, must be continuous. ii) From $\nabla.\delta\mathbf{B} = 0$, δB_r must be continuous at $r = 1$. Applying these conditions to solutions of Eqs. (3) and (4), with a change in notation, $k_i = x$ and $k_e = y$, gives

$$\frac{1}{x}\frac{J'_m(x)}{J_m(x)} = \frac{1}{y}\frac{k'_m(y)}{k_m(y)}, \quad y^2 = c_l^2 - \frac{\rho_e}{\rho_i}x^2, \quad c_l^2 = \left(1 - \frac{\rho_e}{\rho_i}\right)\left(\frac{\ell}{L}\right)^2. \tag{5}$$

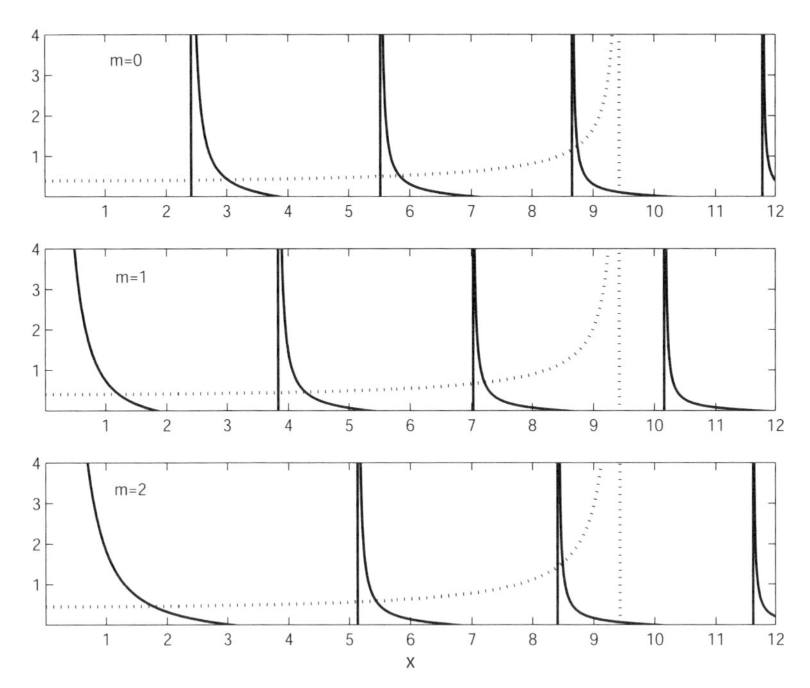

Figure 1. The plots of left and right sides of Eq. (5), solid and dotted curves, respectively, as functions of x for $m = 0, 1$ and 2. Auxiliary parameters are $l = 100$, $C_{100}^2 = 8.9$, radius $= 10^3$ km, height $= 10^5$ km, and $\rho_e/\rho_i = 0.1$. Intersections of solid and dotted curves, x_{nml}, are the eigenvalues.

This is the dispersion relation. Its solution for $x = k_i = [(\omega/v_{Ai})^2 - (\ell/L)^2]^{\frac{1}{2}}$ will give the eigen-frequencies, ω. In Fig. 1, the solid multi-branched curves are the left hand side of Eq. (5). They intersect the x-axis at zeros of $J'_m(x)$ and go to infinity at zeros of $J_m(x)$. The dotted curve is the right hand side of Eq. (5). It tends to infinity at $y = 0$. Intersections of the dotted curve with the multi - branched solid curves are the desired eigenvalues, x_{nml}. They sit between the nth zeros of $J'_m(x)$ and $J_m(x)$, γ'_{nm} and γ_{nm}, respectively. Thus

$$x_{nml} = \frac{1}{2}(\gamma_{mn-1} + \gamma'_{mn}) + \alpha_\ell \pi/2 \approx (2n + m - 2 + \alpha_\ell)\pi/2 < x_{\max}, \tag{6}$$

where $-\frac{1}{2} < \alpha_\ell < \frac{1}{2}$ is to be evaluated numerically, and $x_{\max} = \frac{\ell}{L}(\frac{\rho_i}{\rho_e} - 1)$ is the abscissa of the dotted asymptote in Fig. 1, where $y(x_{\max}) = 0$. The inequality in Eq. (6) utilizes the asymptotic values of the roots of J_m and J'_m.

Cutoffs: From the inequality of Eq. (6) one obtains

$$0 < n < 1 + \frac{2\ell}{\pi L}\sqrt{\frac{\rho_i}{\rho_e} - 1} + \frac{\alpha_\ell}{2} - \frac{m}{2}. \tag{7}$$

For a given m and ℓ there is an upper cutoff to n. Vice versa, for a given n and m there is a lower cutoff for ℓ. Thus, for a given m, the wave number plane (n, ℓ) divides into two regions, a "mode zone" in which the possible modes of the flux tube reside, and a "forbidden zone" in which no mode can dwell. We recapitulate the findings of this sections. The eigenvalue problem for wave propagation in zero-β flux tube reduces to solving a Bessel's equation for the z-components of the perturbed magnetic field. To each mode of oscillation there corresponds a trio of wave numbers (n, m, ℓ) associated with the three directions (r, φ, z). For a given m, there is a lower cutoff for ℓ and an upper cutoff for n.

4. Dissipation

In the presence of viscous and resistive dissipations, the terms $(\eta/\rho)\nabla^2\delta\mathbf{v}$ and $(c^2/4\pi\sigma)\nabla^2\delta\mathbf{B}$ should be added to the right hand sides of Eqs. (1), where η and σ are the bulk viscosity and conductivity of the plasma, and c is the speed of light. The field components, undergo an exponential time decay. For weak dissipations the decay time $\tau_{nm\ell}$ of a mode (nml) is given by

$$\frac{1}{\tau_{nm\ell}} = \frac{\omega_A}{4\pi}\left(\frac{1}{S} + \frac{1}{R}\right)\left(x^2_{nm\ell} + \frac{\ell^2}{L^2}\right) \tag{8}$$

where $\omega_A = v_{Ai}$ is the Alfven frequency, $S = (\frac{4\pi\sigma}{c^2})/(\frac{2\pi}{v_{Ai}})$ and $R = (\frac{\rho_i}{\eta})/(\frac{2\pi}{v_{Ai}})$, Lundquist's and Reynolds' numbers, respectively, are the ratios of resistive and viscous time scales to the Alfven time required to cross the circumference of the tube of radius one. The density of heat generation rates by either process have identical r-dependence. They, however, differ magnitude wise. For the resistive process the heat rate is proportional to S and for the viscous one is proportional to R. More details are given in Karami, Nasiri and Sobouti (2002).

References

Davila, J. M. 1987, ApJ, 317, 514
Edvin, P. M., & Roberts, B. 1983, Sol. Phys., 88, 179
Hollweg, J. V. 1984, ApJ, 277, 392
Ionson, J. A. 1978, ApJ, 226, 650
Karami, K., Nasiri, S., & Sobouti Y. 2002, A&Ato Appear
Nakariakov, V. M. et al. 1999, Science, 285, 862
Ofman, L., Davila, J. M., & Steinolfson, R. S. 1994, ApJ, 421, 360
Ofman L., Davila, J. M., & Steinolfson, R. S. 1995, ApJ, 444, 471
Roberts, B. 1981a, Sol. Phys., 69, 27
Roberts, B. 1981b, Sol. Phys., 69, 39
Steinolfson, R. S. et al. 1986, ApJ, 526
Steinolfson, R. S., & Davila, J. 1993, ApJ, 415, 354
Wentzel, D. G. 1979a, ApJ, 227, 319
Wentzel, D. G. 1979b, ApJ, 227, 319
Wilson, P. R. 1979, A&A, 71, 9

IAU 8th Asia-Pacific Regional Meeting
ASP Conference Series, Vol. 289, 2003
S. Ikeuchi, J. Hearnshaw & T. Hanawa, eds.

Binary Evolution – Problems and Applications

Zhanwen Han
Yunnan Observatory, National Astronomical Observatories of China, Chinese Academy of Sciences, Kunming, 650011, P.R. China

Abstract. Observations of binary stars are posing strong constraints on binary evolution theory. The main problems, as seen from the studies of binary population synthesis, are the criterion for dynamically unstable mass transfer, angular momentum loss during stable Roche lobe overflow, and common envelope evolution. Solving the problems helps us to understand the formation of various binary related objects, and improve the model of binary population synthesis greatly. A good model of binary population synthesis can contribute very much to (or even rewrite the story of) the understanding of the UV upturn in giant elliptical galaxies, while the UV upturn is used to derive ages which are important in cosmological models. The binary population synthesis model also helps us to understand models of type Ia supernovae, which are the best distance indicator for cosmology at the present time.

1. Introduction

About 50 per cent of all stars are in binaries (Petrie 1960) and therefore binary evolution provides natural explanations to the formation of many interesting objects, such as Algols, FK Comae stars, cataclysmic varibles, planetary nebulae, barium stars, CH stars, type Ia supernovae, AM Canum Venaticorum binaries, low-mass X-ray binaries, high-mass X-ray binaries, symbiotic stars, blue stragglers, pulsars, subdwarf B stars, double degenerates, etc. Consequently, binary evolution may help to improve greatly the so-called "evolutionary population synthesis", which is a powerful tool in the study of galaxies.

2. Binary Evolution Problems

Figure 1 shows a typical case of binary evolution. The binary system initially composes two main-sequence stars. As the system evolves, the more massive component, i.e. the primary, expands and may fill its Roche lobe as a red giant, and Roche-lobe overflow (RLOF) begins. If the RLOF is dynamically stable, the remnant is a white dwarf (WD) binary with a long orbital period. If the RLOF is dynamically unstable, the mass transfer rate is so high that the secondary is not able to accrete the transferred mass and a common envelope (CE) is resulted (Paczyński 1976). The CE engulfs the core of the primary and the secondary. Due to the friction between the embedded binary and the envelope, a large amount of orbital energy is released and deposited into the envelope. If the

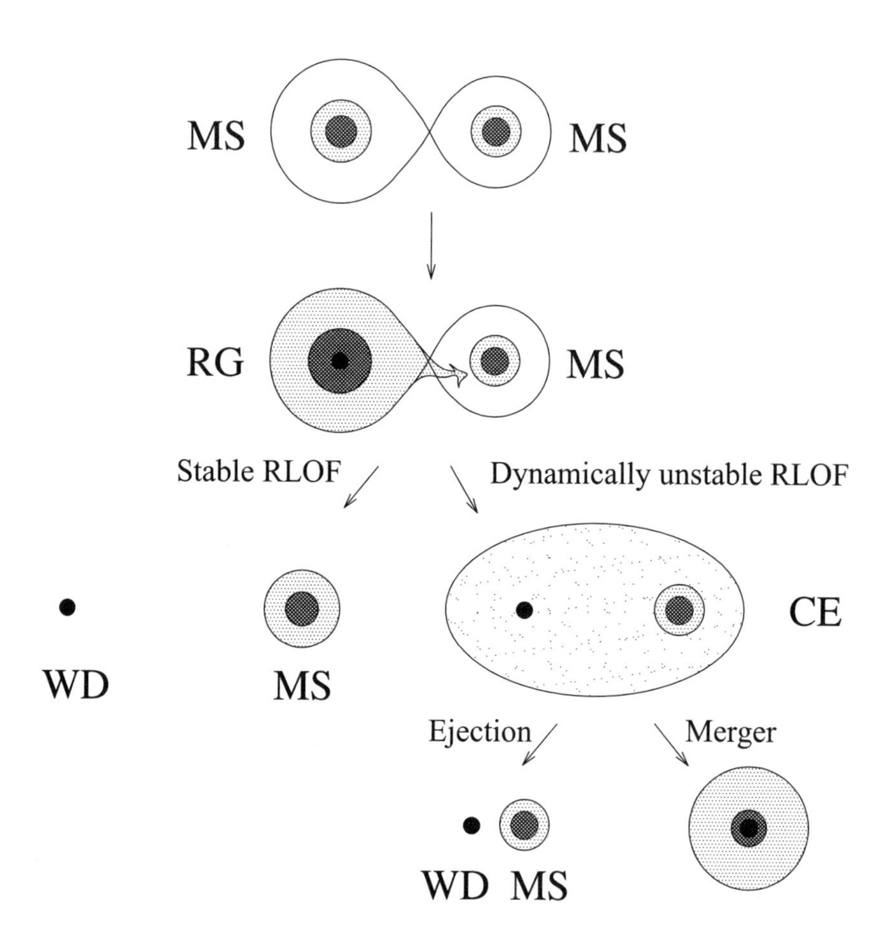

Figure 1. A typical flow-chart of binary evolution. MS is for main-sequence, RG for red giant, WD for white dwarf, CE for common envelope, RLOF for Roche-lobe overflow.

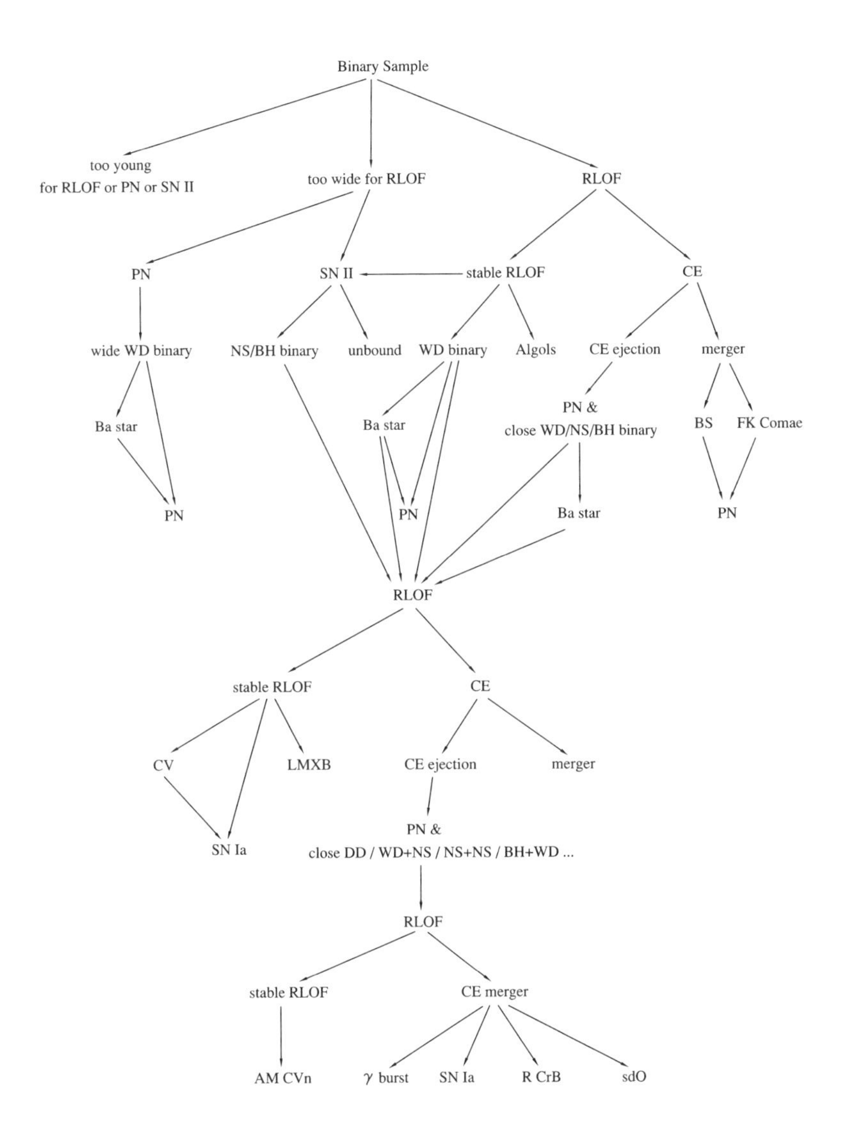

Figure 2. Flow chart of binary evolution. RLOF is for Roche-lobe overflow, CE for common-envelope, PN for planetary nebula, WD for white dwarf, Ba for barium, SN for supernova, NS for neutron star, BH for black hole, BS for blue straggler, CV for cataclysmic variable, LMXB for low-mass X-ray binary, DD for double degenerate, AM CVn for AM Canum Venaticorum star, R CrB for R Coronae Borealis star, sdO for subdwarf O star.

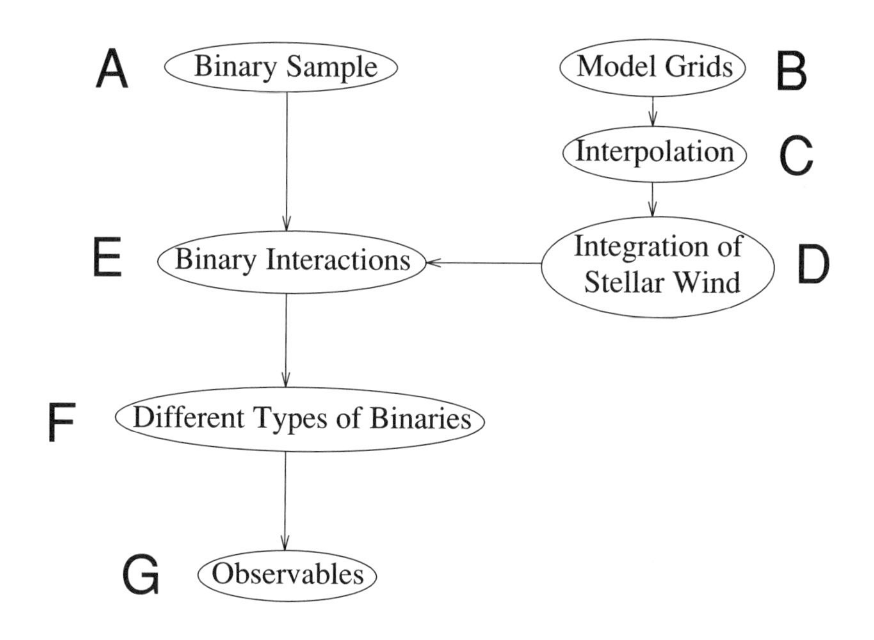

Figure 3. Flow chart of binary population synthesis.

deposited energy can overcome the binding energy of the envelope, the CE is ejected and results in a WD binary with a very short orbital period. Otherwise the system coalesces. Figure 2 gives a detailed but by no means complete flow chart of binary evolution.

Binary population synthesis (BPS) was developed recently and with the approach we evolve millions of binaries simultaneously and the result is compared to observations directly (e.g. Han 1994; Han et al. 1995, Han 1998, Han et al. 2002). In this way we can infer the physics in binary evolution and predict the distributions of the properties of binary-related objects. The flow-chart of BPS is shown is Figure 3. In bock A, we give a big binary sample, e.g. a million binaries. We evolve the sample according to stellar model grids (blocks B, C and D) and consider binary interactions (block E) and obtain various types of binaries (block F). The observables of the binaries are compared to observations (block G).

The criterion for the stability of RLOF is crucial to binary evolution. For a red giant binary, polytropic model gives a critical mass ratio of primary to secondary $q_{\rm crit} = \frac{2}{3}$, above which mass transfer is assumed to be dynamically unstable. The criterion was revised by Hjellming & Webbink (1987), Webbink (1988) and Soberman et al. (1997) with composite polytropes. In Webbink's model, the critical mass ratio depends on the fractional core mass of the giant. However the revised criterion is for conservative RLOF only. Han et al. (2001) revised the criterion by considering non-conservative RLOF and the new criterion depends heavily on the angular momentum lost from binary system.

The above criteria are from polytropic models. Employing the latest version of Eggleton's stellar evolution code (Eggleton 1971, 1972, 1973; Han et al. 1994; Pols et al. 1995), however, Han et al. (2002a) carried out detailed binary evolution calculations for systems composing red giants and white dwarfs. They assume that all the mass transferred is lost and carries away the same specific angular momentum as to the WD. They found that the critical mass ratio is between 1.1 and 1.3. For RLOF with its onset in the Hertzsprung gap, Han et al. (2000) and Chen & Han (2002) found the critical mass ratio is about 3. For a double white dwarf system, the less massive component fills its Roche lobe. Han & Webbink found that the critical mass ratio is $q_{\rm crit} \approx 0.7 - 0.1(M_2/M_\odot)$, where M_2 is the mass of the accretor.

A stable RLOF may result in a WD binary system. The properties of the WD binary depend on how much mass and how much angular momentum is lost during the stable RLOF. Recently, Maxted et al. (2001) found observationally dozens of subdwarf B binaries with short orbital periods (hours to days), most of which have WD companions. The observations provide good constraints to binary evolution models. Han et al. (2002a, 2002b) studied the formation of subdwarf B stars with BPS and concluded that the stable RLOF in red giant binary system needs to be non-conservative, and about half of the transferred mass is lost and carries away specific angular momentum similar to that of the system. Han & Webbink (1999) derived the mass-losing fraction from energy conservation point of view for stable RLOF in double WD systems.

If a RLOF is dynamically unstable, a CE may be formed (Paczyński 1976). The CE can be ejected to give rise to a WD binary with a very short orbital period, or the system may coalesce to result in a merger. We can say that CE evolution is the most important process but also the least known one in binary evolution. The following formulism is usually adopted to determine whether the CE can be ejected or not.

$$\alpha_{\rm CE}\Delta E_{\rm orb} \geq |E_{\rm bind}| \tag{1}$$

where $\Delta E_{\rm orb}$ is the orbital energy released, $E_{\rm bind}$ the binding energy of the envelope, and $\alpha_{\rm CE}$ the common-envelope ejection efficiency, i.e. the fraction of the released orbital energy used to overcome the binding energy. If the left term of equation (1) is greater than the right term, the CE is assumed to be ejected. The formula to calculate $E_{\rm bind}$ adopted in many studies is as follows.

$$E_{\rm bind} = -\frac{GM_1M_{\rm 1e}}{\lambda R_1} \tag{2}$$

where M_1, $M_{\rm 1e}$ and R_1 are the mass, the envelope mass and the radius of the primary at the onset of the RLOF, and the factor λ is taken to be 0.5 or 0.1 in most cases. Obviously, such a calculation or the choice of the value of λ is too simple. Podsiadlowski, Han & Rappaport (2002) carried out full stellar evolution calculation and showed how λ changes as a star evolves. They found that λ is not a constant and changes dramatically from 0.01 to far over 1 (even to infinity). In their calculation, the binding energy includes both gravitational and thermal (including ionization) energy (see Han et al. 1994). The inclusion of thermal energy helps us to avoid unphysical values of $\alpha_{\rm CE}$ in order to explain observations (Dewi & Tauris 2000).

3. Binary Evolution Applications

Binary evolution plays an important role in many aspects of astrophysics. At stellar level, binary evolution provides a natural way in the explanation of many interesting objects. Binary population synthesis (BPS) was developed to evolve millions of binaries simultaneously. We list here some of the previous works on BPS: the distribution of visual binaries (Eggleton, Fitchett & Tout 1989); the formation of cataclysmic variables (de Kool 1990, 1992; de Kool & Ritter 1993; Kolb 1993; Kolb & de Kool 1993; Han, Podsiadlowski, Eggleton 1995; Han, Eggleton, Podsiadlowski, et al. 1995; Kolb, King & Ritter 1998; Han, 1998); the formation of neutron star binaries, low mass x-ray binaries, high mass x-ray binaries (Iben, Tutukov & Yungelson 1995; Kalogera & Webbink 1996, 1998; Kalogera 1997, 1998; Kalogera, Kolb & King 1998; Portegies, Simon & Yungelson 1998); neutron star binary mergers (Bagot, Portegies, Simon, et al. 1998; Belczynski & Bulik 1999); the formation of double degenerates (Tutukov & Yungelson 1996; Iben, Tutukov & Yungelson 1997; Han 1998); the rate of supernovae (Yungelson, Livio & Tutukov 1997; Han, Podsiadlowski, Eggleton 1995; Han, Eggleton, Podsiadlowski, et al. 1995; Han, 1998); the formation of bipolar planetary nebulae (Yungelson, Tutukov & Livio 1993; Han, Podsiadlowski, Eggleton 1995); the formation of blue stragglers (Pols & Marinus 1994); the formation of barium/CH stars (Han, Eggleton, Podsiadlowski et al. 1995) the evolution of globular clusters, N-body simulations (Tout, Aarseth & Pols 1997; Aarseth & Heggie 1998; Aarseth 1999). Webbink & Han (1998) obtained the spectrum of Galactic gravitational wave radiation background, which is important to LISA, from the BPS model of double degenerates of Han (1998).

Type Ia supernovae (SNe Ia) are the best distance indicator for cosmology. There are mainly two models for the progenitor of SN Ia. The first model is Nomoto's model, in which a carbon-oxygen white dwarf accretes mass from a main-sequence companion of a red giant companion and the mass of the WD increases to Chandrasekhar mass and the WD explodes as a SN Ia (Nomoto & Iben 1985, Hachisu, Kato & Nomoto 1996). The second model is a double degenerate model, in which two carbon-oxygen WDs coalesces and explodes if the merger is more massive than Chandrasekhar mass (Iben & Tutukov 1984; Webbink & Iben 1987). With a BPS technique, we find that the Galactic birth rate of SNe Ia is about 0.001 per year for Nomoto's model and 0.003 for the double degenerate model, while observation gives a birth rate of 0.003-0.004 per year in the Galaxy (van den Bergh & Tammann 1991). We see that both models agree with observations reasonably.

In the study of evolutionary population synthesis (EPS), UV upturn is used as an age indicator for giant elliptical galaxies (Yi et al. 1997). The determination of the age of the oldest galaxies can constrain the basic parameters of the cosmology model. The UV upturn is mainly from subdwarf B stars, which were assumed to be originated from single stellar evolution in the study of EPS. However, the latest observations show that most subdwarf B stars are in binaries, and Han et al. (2002) proposed a binary model for the formation of subdwarf B stars. If we apply the binary model to EPS, we can expect that the age of the oldest galaxies should be changed and some parameters in cosmological models should also be revised accordingly. Blue stragglers, which are resulted from binary evolution, contribute very much to some of the integrated colours in EPS.

However, no binary interaction has been included yet in EPS. The inclusion of binary evolution may greatly improve the study of EPS in the future.

4. Conclusions

The basic problems of binary evolution include the stability of Roche-lobe overflow (RLOF), the mass and the angular momentum loss during stable RLOF, the common-envelope (CE) ejection etc. In order to solve the problems, we need to carry out systematic study on binary evolution and compare theoretical models to observations with the binary population synthesis (BPS) approach. Binary evolution and BPS should be applied to the study of evolutionary population synthesis (EPS) in the future to get a better EPS model.

Acknowledgments. This work was in part supported by the Chinese National Science Foundation under Grant No. 19925312, 10073009 and NKBRSF No. 19990754. I thank the LOC for paying my travel expenses.

References

Aarseth, S. J. 1999, PASP, 111, 1333

Aarseth, S. J., Heggie, D. C. 1998, MNRAS, 297, 794

Bagot, P., Portegies, Z., Simon, F. et al. 1998, A&A, 332, L57

Belczynski, K., Bulik, T. 1999, A&A, 346, 91

Chen, X., Han, Z. 2002, MNRAS, in press

de Kool, M. 1990, ApJ, 358, 189

de Kool, M. 1992, A&A, 261, 188

de Kool, M., Ritter, H. 1993, A&A, 267, 397

Dewi, J. D. M., Tauris, T. M. 2000, A&A, 360, 1043

Eggleton, P. P. 1971, MNRAS, 151, 351

Eggleton, P. P. 1972, MNRAS, 156, 361

Eggleton, P. P. 1973, MNRAS, 163, 279

Eggleton, P. P., Fitchett, M. J., Tout, C. A. 1989, MNRAS, 347, 998

Hachisu, I., Kato, M., Nomoto, K. 1996, ApJ, 470, L97

Han, Z. 1995, PhD Thesis, University of Cambridge

Han, Z. 1998, MNRAS, 296, 1019

Han, Z., Eggleton, P. P., Podsiadlowski, Ph., Tout, C. A. 1995, MNRAS, 277, 1443

Han, Z., Eggleton, P. P., Podsiadlowski, Ph., Tout, C. A., Webbink, R.F. 2001, in Podsiadlowski Ph., Rappaport S., King A. R., D'Antona F., Burderi L., eds, Evolution of Binary and Multiple Star Systems, ASP Conf. Ser., Vol. 229, P. 205

Han, Z., Podsiadlowski, Ph., Eggleton, P. P. 1994, MNRAS, 270, 121

Han, Z., Podsiadlowski, Ph., Eggleton, P. P. 1995, MNRAS, 272, 800

Han, Z., Podsiadlowski, Ph., Maxted, P. F. L., Marsh, T. R., Ivanova, N. 2002a, MNRAS, in press

Han, Z., Podsiadlowski, Ph., Maxted, P. F. L., Marsh, T. R. 2002b, MNRAS, submitted

Han, Z., Tout, C. A., Eggleton, P. P. 2000, MNRAS, 319, 215

Han, Z., Webbink, R. F. 1999, A&A, 349, L17

Hjellming, M. S., Webbink, R. F. 1987, ApJ, 318, 794

Iben, I. Jr., Tutukov, A. V. 1984, ApJS, 54, 335

Iben, I. Jr., Tutukov, A. V., Yungelson, L. R. 1995, ApJS, 100, 217

Iben, I. Jr., Tutukov, A. V., Yungelson, L. R. 1997, ApJ, 475, 291

Kalogera, V. 1997, PASP, 109, 1394

Kalogera, V. 1998, ApJ, 493, 368

Kalogera, V., Kolb, U., King, A. R. 1998, ApJ, 504, 967

Kalogera, V., Webbink, R. F. 1996, ApJ, 458, 301

Kalogera, V., Webbink, R. F. 1998, ApJ, 493, 351

Kolb, U. 1993, A&A, 271, 149

Kolb, U., de Kool, M. 1993, A&A, 279, L5

Kolb, U., King, A. R., Ritter, H. 1998, MNRAS, 298, L29

Maxted, P. F. L., Heber, U., Marsh, T. R., North, R. C. 2001, MNRAS, 326, 1391

Nomoto, K., Iben, I. Jr. 1985, ApJ, 297, 531

Paczyński, B. 1976, in Structure and Evolution of Close Binary Systems, IAU Symp. No. 73, eds Eggleton P. P., Mitton S., Whelan J., Reidel, Dordrecht, p.75

Petrie, R. M. 1960, Ann. Astrophys., 23, 744

Podsiadlowski, Ph., Han, Z., Rappaport, S. 2002, MNRAS, in press

Pols, O. R., Marinus, M. 1994, A&A, 288, 475

Pols, O. R., Tout, C. A., Eggleton, P. P., Han, Z. 1995, MNRAS, 274, 964

Portegies, Z., Simon, F., Yungelson, L. R. 1998, A&A, 332, 173

Soberman, G. E., Phinney, E. S., van den Heuvel, E. P. J. 1997, A&A327, 620

Tout, C. A., Aarseth, S. J., Pols, O. R. 1997, MNRAS, 291, 732

Tutukov, A. V., Yungelson, L. 1996, MNRAS, 280, 1035

van den Bergh, S., Tammann, G. A. 1991, ARA&A, 29, 363

Webbink, R. F. 1988, in Mikolajewska J., Friedjung M., Kenyon S.J., Viotti R., eds, the Symbiotic Phenomenon. Kluwer, Dordrecht, p. 311

Webbink, R. F., Han, Z. 1998, AIP Conference Series, 456, 61

Webbink, R. F., Iben, I. Jr. 1987, in Philipp, A. G. D., Hayes, D. S., Liebert, J. W., eds, The Second Conference on Faint Blue Stars., IAU Colloq. No. 95, Davis Press, Schenectady, p.445

Yi, S., Demarque, P., Oemler, A. Jr. 1997, ApJ, 486, 201

Yungelson, L. R., Livio, M., Tutukov, A. V. 1997, ApJ, 481, 127

Yungelson, L. R., Tutukov, A. V., Livio, M. 1993, ApJ, 418, 794

IAU 8th Asia-Pacific Regional Meeting
ASP Conference Series, Vol. 289, 2003
S. Ikeuchi, J. Hearnshaw & T. Hanawa, eds.

Annual Parallax Measurements of Mira-Type Variables with Phase-Referencing VLBI Observations

Tomoharu Kurayama

Department of Astronomy, School of Science, the University of Tokyo, 7-3-1, Hongo, Bunkyo-ku, Tokyo, Japan
VERA Project office, National Astronomical Observatory, 2-21-1, Osawa, Mitaka, Tokyo, Japan

Tetsuo Sasao

VERA Project office, National Astronomical Observatory, 2-21-1, Osawa, Mitaka, Tokyo, Japan

Abstract. This study aims at a measurement of the annual parallax of a water maser around a Mira-type variable, AW Tau, with phase-referencing VLBI observations. This is the first detection of the annual parallax of water masers by VLBA.

1. Introduction

The period-luminosity relation of Mira-type variables (Miras) is first obtained in the observations of the Large Magellanic Cloud. Figure 1a is the period-luminosity relation of Mira-type variables in the Large Magellanic Cloud. The thickness of the Large Magellanic Cloud is much smaller than its distance, so we can consider all Miras in the LMC, and take apparent magnitudes as the vertical axes. A clear relation is seen.

On the other hand, Figure 1b is the period-luminosity relation of solar-neighborhood Miras. Unlike the case of Large Magellanic Cloud, we need the distances to Miras, in order to calculate the absolute magnitude of Miras. We can recognize large error bars in this diagram. The typical error is of the order of 1 mas in parallax.

So if we can measure annual parallaxes of Miras with enough accuracy, we can establish a good period-luminosity relation in solar-neighborhood Miras. This will lead to the distance modulus of the Large Magellanic Cloud and a restriction of physics of late-type stars.

2. What is Phase Referencing VLBI?

To measure the annual parallaxes of Miras, we have used the phase-referencing VLBI technique. First, we will explain this briefly. In interferometry systems, observational parameter is differences of two optical path lengths. The diffraction index of the earth's atmosphere varies with time by weather condition, so we must calibrate the path length by the Earth's atmosphere.

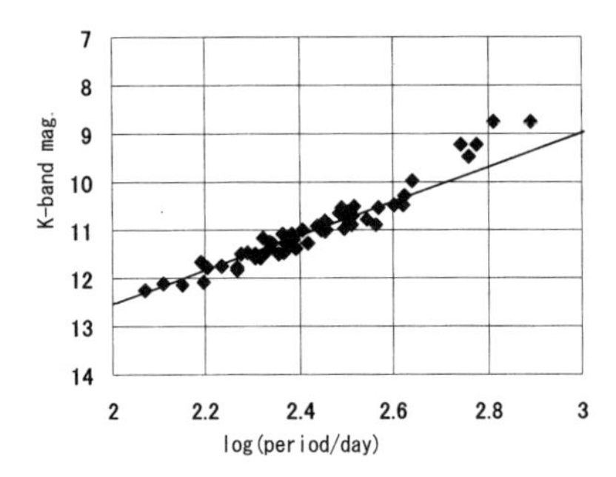

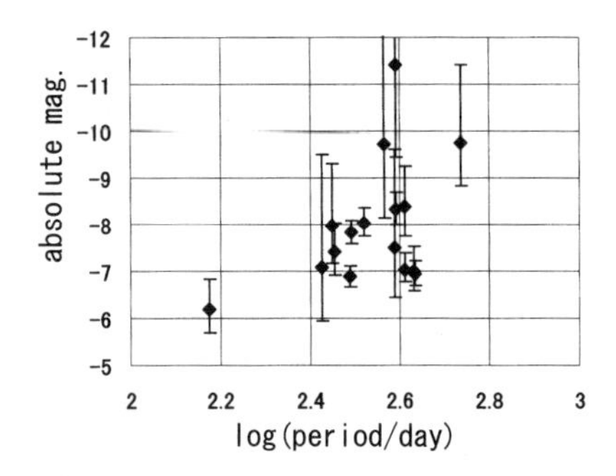

Figure 1. (a, left) Period luminosity relation of Miras in Large Magellanic Cloud. (b, right) Period luminosity relation of solar-neighborhood Miras used the parallax data of Hipparcos.

In phase-referencing VLBI observations, we observe two adjacent sources simultaneously. The effect of the atmosphere is almost same between these two sources, so we can cancel out the atmospheric fluctuation effectively by taking the difference of two optical path lengths. By taking one source a far source, this will be the reference point on celestial sphere to measure the annual parallaxes.

3. Observation and Analysis

Observation was carried out with following parameters (Table 1). Observed sources (Table 2) are selected according to the following criteria: (i) They have strong water maser emission. This information is given in a published catalogue, the Arcetri catalogue. (ii) They have at least one reference source within 2.0 degrees from the Miras. (iii) Variation periods and right ascensions are conveniently separated. We have finished the analysis of AW Tau, one of the five observed Miras. The procedure of analysis is shown in Table 3.

4. Movements of Masers around AW Tau

Movements of Maser spots around AW Tau is shown in Figure 2. Two components (A and B) are observed at adjacent frequency channels. The maser spot observed in the last epoch is separated from others, so we consider it is another maser spot and omit it from late discussion.

From the least square fitting method, we can fit these movements by parallax and proper motions. The result of fitting is

$$\begin{aligned}
\varpi &= 9.7 \pm 0.4 \text{ mas} \\
\mu_\alpha(A) &= 50.9 \pm 1.2 \text{ mas/yr} \\
\mu_\delta(A) &= 25.5 \pm 0.7 \text{ mas/yr} \\
\mu_\alpha(B) &= 64.1 \pm 1.2 \text{ mas/yr} \\
\mu_\delta(B) &= 1.2 \pm 0.8 \text{ mas/yr}
\end{aligned}$$

Array	VLBA 10 stations
Frequency	22 GHz
Nodding cycle	40 sec
Freq. set	32 Msps × 2 bit × 2 ch
Bandwidth	16 MHz
5 hours × 4 epochs × 5 sources	

Table 1. Observation setups.

Miras	Separa-tion	Period [day]	Hipparcos parallax [mas]	Calculated parallax [mas]	obs. date
VX UMa	0.58°	215		1.09	01.01.27 01.01.28 01.04.13 01.06.28
RT Aql	2.02°	327	0.00 ± 2.10	1.46	01.05.13 01.06.20 01.08.13 01.09.10
R Cas	1.64°	430	9.37 ± 1.10	5.37	02.03.30 02.04.25 02.05.23 02.06.26
UX Cyg	1.38°	565		0.875	01.02.22 01.10.12 01.11.11 02.02.04
AW Tau	0.24°	654		0.461	01.01.23 01.03.03 01.04.03 01.06.04

Table 2. Observed sources. Calculated parallaxes are calculated with the distance modulus of LMC (18.50) and K-magnitude.

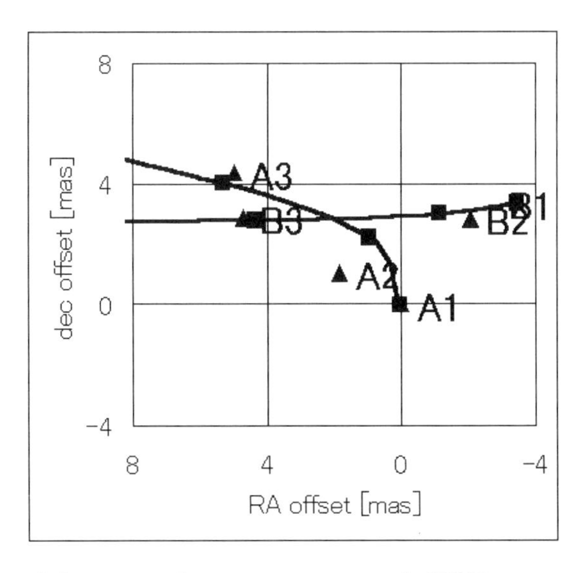

Figure 2. Movement of maser spots around AW Tau.

Calibration of sampler threshold level
Amplitude calibration
Calibration of clock lags between stations
Band-pass calibration
Calibration of Doppler shift by the earth's rotation
Subtraction of reference phases from maser phases (phase-referencing)
Mapping of masers
Measurements of positions on the celestial sphere
Tracing the position of 4 observation date

Table 3. Procedure of analysis.

5. Evaluation of Systematic Errors

For error estimation in a phase-referencing VLBI observation, here we consider three sources of systematic errors: (i) Error from reference source structure, (ii) Error from airmass correction uncertainties, (iii) Error from baseline error. (i) is almost negligible. This is seen from closure phases. (ii) is less than 0.1 mas at airmass estimation errors 1 cm and at 2000 km baselines. (iii) is 0.04 mas at baseline-length errors of 1 cm and at the 200 km baseline. Error estimations of (ii) and (iii) are in the worst case.

6. Discussion

The parallax value we got (9.7 ± 0.4mas) is much larger than that expected from period-luminosity relation and K-band magnitude (0.461 mas). One of this reason is the K-band magnitude uncertainty. AW Tau is long-period Mira, that is, an evolved star. Circumstellar extinction is expected to be large.

7. Conclusion and Future

We have carried out all the analysis of phase-referencing VLBI, and measured the positions of water masers. Although it is a preliminary one, we can get values of the annual parallax (9.7 ± 0.4 mas) and proper motions. This is the first time of parallax measurement in 22 GHz. We have evaluated systematic errors, and represented the possibility of the measurement of annual parallaxes at 0.1 mas error level. In order to establish the period-luminosity relation in solar neighborhood, we are going to analyze the other 4 sources, and willing to observe more sources by VLBA and VERA.

IAU 8th Asia-Pacific Regional Meeting
ASP Conference Series, Vol. 289, *2003*
S. Ikeuchi, J. Hearnshaw & T. Hanawa, eds.

Magnetic Reconnection in the Solar Lower Atmosphere

C. Fang, P. F. Chen, M. D. Ding

Department of Astronomy, Nanjing University, Nanjing 210093, China

Abstract. Accumulating observational evidence indicates that magnetic reconnection is a fundamental process in the solar lower atmosphere, which is responsible for many localized activities and the global maintenance of the hot dynamical corona. Meanwhile, qualitative theoretical considerations and quantitative numerical simulations demonstrate the applicability of the reconnection to a thin layer in the lower atmosphere. This paper reviews the research progress in the related observations, theories and numerical simulations.

1. Introduction

Magnetic reconnection is widely applied to explain various coronal eruptive and/or heating events, especially the solar flares with various length scales (Shibata 1999). Owing to both ground and space observations, many features are found indicative of the occurrence of magnetic reconnection in the corona, e.g., cusp-shaped soft X-ray loops, hard X-ray sources above the loop top, and so on. On the contrary, it is much more difficult to find out similar features which could show the evidence of the reconnection in the lower atmosphere (from photosphere to the transition region), although it is widely believed that this very thin layer is permeated with continuous small-scale magnetic reconnection processes. Nevertheless, a lot of efforts have been taken to understand the role and effect of the magnetic reconnection in the lower atmosphere. This paper reviews the progress in the research of such magnetic reconnection. §2 depicts the observational results, §3 is devoted to the theoretical and numerical approaches. Concluding remarks are discussed in §4.

2. Observations

2.1. Photospheric Magnetic Field

Photospheric magnetic measurement provides direct signatures of magnetic evolution. For the first time, Livi, Wang, & Martin (1985) discovered the gradual and mutual loss of magnetic flux with opposite polarities, which was proposed as the magnetic reconnection in the lower atmosphere. Despite diverse explanations for the observed magnetic cancellation, such as submergence of a loop or emergence of U-type flux tubes, Wang & Shi (1993) confirmed with the help of vector magnetograms that the observed flare-associated magnetic cancellation is really magnetic reconnection.

2.2. Activities in the Lower Atmosphere

There are many phenomena in different layers of the atmosphere, which are indicative of the occurrence of localized reconnection.

Photosphere: Solar white-light flares (WLFs) are very strong and rare flaring events. It was proposed that there are two types of WLFs which show distinctive emission features, i.e., type I WLFs reveal a Balmer or Paschen jump, while type II do not. Systematic studies by Fang & Ding (1995) indicated that the features for type I WLFs are well explained by the conventional flare model. However, for type II, since the known mechanisms of energy transport are no longer effective (Ding, Fang, & Yun 1999 and the references therein), an in situ heating mechanism deep in the chromosphere or the photosphere, e.g., magnetic reconnection, is required.

Chromosphere: Ellerman bombs (EBs), also known as moustaches, are small elongated brightening events which are observed in Hα wings around sunspots or under arch filament systems. They have a typical length of $\sim$1 arcsec (Kurokawa et al. 1982), and typical upward flow of $\sim$6 km s^{-1} in the chromosphere (Kitai 1983). EBs are found to be pushed away by expanding granules (Denker et al. 1995), where one polarity magnetic features may be driven to meet other opposite polarity features. It was suggested by many authors that the heating originates in the lower atmosphere (e.g., Georgoulis et al. 2002).

Transition Region: Observations with the High-Resolution Telescope and Spectrograph (HRTS) have shown that frequent explosive events are occurring mainly in or near the network lanes (Dere, 1994). The associated red and blue Doppler shifts are explained as the bidirectional reconnection jets. This proposal was confirmed by the SOHO spectroscopic observations (Innes et al. 1997).

3. Theoretical and Numerical Investigations

Unlike the fully ionized tenuous corona, the plasma in the lower atmosphere is partially ionized, and much denser. Therefore, both ionization and optically thick radiation become important in the dynamical evolution. Moreover, the strong stratification of the plasma and the drastic divergence of the magnetic field make the problem so complicated that it can be solved only after certain simplifications.

Zweibel (1989) studied the tearing instability in the partially ionized plasma, and pointed out that the reconnection rate becomes larger when the coupling between the ions and the neutral atoms gets weaker. This effect is in favor of the occurrence of fast reconnection in the lower atmosphere since the degree of ionization decreases rapidly from the transition region to the photosphere. Furthermore, Litvinenko (1999) showed that the temperature minimum region is the most favorite site for magnetic reconnection to occur, and the derived reconnection inflow velocity is consistent with the observed values, i.e., tens of meters per second.

Li et al. (1997) derived a combined set of MHD equations for the three-component (electrons, ions, and neutral atoms) plasma. They investigated the linear instability of partially ionized plasma from these equations, and found that the tearing-mode instability can account for many observational features of type II WLFs.

It has long been thought that magnetic reconnection in the lower atmosphere can release a spectrum of high-frequency Alfvén waves which can accelerate and heat the fast solar wind (e.g., Axford et al. 1999). Following this line of thought, Sturrock (1999) again illustrated that the chromosphere is a favorite site for reconnection to occur because of the large resistivity. The resulting reconnection can provide enough energy to explain the coronal heating, as well as inject sufficient mass to balance the steady downflow of the coronal mass.

Besides the qualitative results from the theoretical consideration, numerical simulations provided more detailed dynamical structures and evolution. By 2.5D numerical simulations, Karpen et al. (1995) reproduced the observed features of chromospheric activities as intermittency and large velocity, as well as the approximately concurrent appearance of oppositely directed flows. However, reconnection in their simulations is induced owing to numerical resistivity, which can not be quantitatively controlled.

Recently we investigated the effect of ionization and radiation on the magnetic reconnection in the lower atmosphere (Chen et al. 2001). It is found that due to the line-tying effect of the bottom boundary, the reconnection always saturates, and the lifetime of such reconnection is about 600–900 s, independent of the ionization and radiation. However, the ionization in the upper chromosphere consumes a large part of the released energy, making the heating effect significant only in the lower atmosphere. This feature is essential to explain the Hα spectra of Ellerman Bombs, i.e., emission in the wing and absorption in Hα center. The weak heating in the upper chromosphere may even account for the characteristic of blinkers (Harrison et al. 1997). The thermal quantities and the lifetime are also found in agreement with those of type II WLFs.

To explain the transition explosive events, Roussev and his colleagues (e.g., Roussev et al. 2001) have performed a series of numerical simulations, and synthesized the emission lines for the comparison with observations. Their researches also indicate that non-equilibrium ionization is important, and the direct comparison with observations needs more elaborate initial conditions.

Besides producing the localized activities, magnetic reconnection in the lower atmosphere can also change the magnetic connectivity of the coronal magnetic field. This led Wang & Shi (1993) to propose a two-step magnetic reconnection model for flares, i.e., the first reconnection in the lower atmosphere results in the loss of equilibrium of coronal field, which further triggers the flare-associated magnetic reconnection in the corona. This was confirmed by Chen & Shibata (2000), who illustrated that the low-layer reconnection leads to the large scale restructuring of the coronal field. The ensuing evolution is followed by the classic magnetic reconnection responsible for large flares.

The energy supply from the low-layer reconnection is studied by Takeuchi & Shibata (2001). Their research infers that the energy flux of the Alfvén waves might be enough to fuel the spicules, which are closely related to the pending mechanism for the coronal heating.

Part of the efforts are directed to the 3D magnetic reconnection. Several analytic solutions for the 3D magnetic reconnection in the partially ionized plasma are obtained for the first time by Ji & Song (2001).

4. Concluding Remarks

Observations have indicated that magnetic reconnection is occurring continuously especially along the chromospheric network. These small scale reconnection processes are not only related to some localized activities such as Ellerman bombs, type II WLFs, and bright points, but also may be important in large scale activities like the formation and eruption of filaments. At the same time, they could be essential for the maintenance of the global coronal heating and fast wind acceleration.

Whilst spectroscopic diagnostics will help to clarify the dynamical features of the activities in the lower atmosphere, numerical simulations, with the consideration of the canopy effect of the magnetic field, ionization, and radiative transfer, may provide deeper insight for our understanding of many phenomena as a unity.

Acknowledgments. This work was supported by NSFC (No. 4990451 and 19973009) and NKBRSF (G20000784). DMD was also supported by TRAPOYT.

References

Axford, W. I. et al. 1999, Space Sci.Rev., 87, 25
Chen, P. F., Fang, C. & Ding, M. D. 2001, Chin. J. Astron. Astrophys., 1, 176
Chen, P. F. & Shibata, K. 2000, ApJ, 545, 524
Denker, C., de Boer, C. R., Volkmer, R. & Kneer, F. 1995, A&A, 296, 567
Dere, K. P. 1994, Adv. Space Res., 14, 13.
Ding, M. D., Fang, C. & Yun, H. S. 1999, ApJ, 512, 454
Fang, C. & Ding, M. D. 1995, A&AS, 110, 99
Georgoulis, M. K. et al. 2002, ApJ, 575, 506
Harrison, R. A. et al. 1997, Solar Phys.170, 123
Innes, D. E. et al. 1997, Solar Phys., 175, 341
Ji, H. S. & Song, M. T. 2001, ApJ, 556, 1017
Karpen, J. T., Antiochos, S. K., & Devore, C. R. 1995, ApJ, 450, 422
Kitai, R. 1983, Solar Phys., 87, 135
Kurokawa, H. et al. 1982, Solar Phys., 79, 77
Li, X. Q., Song, M. T., Hu, F. M. & Fang, C. 1997, A&A, 320, 300
Livi, S. H., Wang, J. & Martin, S. F. 1985, Australian J. Phys., 38, 855
Litvinenko, Y. E. 1999, ApJ, 515, 435
Roussev, I. et al. 2001, A&A, 380, 719
Shibata, K. 1999, Ap&SS, 264, 129
Sturrock, P. A. 1999, ApJ, 521, 451
Takeuchi, A. & Shibata, K. 2001, ApJ, 546, L73
Wang, J. & Shi, Z. 1993, Solar Phys., 143, 119
Zweibel, E. G. 1989, ApJ, 340, 550

IAU 8th Asia-Pacific Regional Meeting
ASP Conference Series, Vol. 289, 2003
S. Ikeuchi, J. Hearnshaw & T. Hanawa, eds.

Future Microlensing Observations

Cheongho Han

Chungbuk National University, Korea

Abstract. Microlensing experiments were originally proposed and initiated to search for Galactic dark matter in the form of compact halo objects (MACHOs). Over the last decade, however, microlensing has been developed into a powerful tool in various aspects of astronomical research besides the original goal of searching for MACHOs. This progress was achieved mostly thanks to the improvement in the observational precision and the advent of new types of observations. In this paper, we discuss several additional fields for which lensing observations, through the use of a future generation of instruments, can provide useful information.

1. Introduction

A Galactic microlensing event occurs when a lensing object approaches very close to the line of sight towards a background source star. As a consequence of the lensing, the image of the source star appears to be distorted and split into two. The locations and magnifications of the individual images are

$$\boldsymbol{\theta}_\pm = \frac{1}{2}\left[\boldsymbol{u} \pm \sqrt{u^2+4}\frac{\boldsymbol{u}}{u}\right]\theta_{\rm E}, \tag{1}$$

and

$$A_\pm = \frac{u^2+2}{2u\sqrt{u^2+4}} \pm \frac{1}{2}, \tag{2}$$

where $\boldsymbol{u}$ is the projected lens-source separation vector normalized by the Einstein ring radius $\theta_{\rm E}$. The Einstein ring represents the effective lensing region around the lens within which the source star flux is magnified by more than $3/\sqrt{5}$ and it is related to the physical parameters of the lens by

$$\theta_{\rm E} \sim 0.72 \left(\frac{m}{0.5M_\odot}\right)^{1/2} \left(\frac{D_{\rm os}}{8\ {\rm kpc}}\right)^{-1/2} \left(\frac{D_{\rm os}}{D_{\rm ol}} - 1\right)^{1/2} {\rm mas}, \tag{3}$$

where m is the lens mass and $D_{\rm ol}$ and $D_{\rm os}$ are the distances to the lens and source star, respectively. The separation between the two images for a typical Galactic event is $\lesssim 1$ milli-arcsec, which is too small to be resolved even with the highest resolution achieved so far. However, a lensing event can be identified from its characteristic achromatic, smooth, and symmetric light curve, which is represented by

$$A = A_+ + A_- = \frac{u^2+2}{u\sqrt{u^2+4}}. \tag{4}$$

Since lensing events occur regardless of the lens brightness, microlensing provides an important tool to probe the nature of Galactic dark matter. For this reason, microlensing experiments were first proposed (Paczyński 1986) and initiated to search for Galactic dark matter in the form of compact halo objects (MACHOs) by observing millions of source stars located in the Large Magellanic Cloud (LMC) (MACHO, Alcock et al. 1993; EROS, Aubourg et al. 1993).

Over the last decade, microlensing has been developed into a powerful tool in various aspects of astronomical research besides its original use of searching for MACHOs. This progress was achieved thanks partly to intensive theoretical studies on various methods to extract extra information on lensed source stars and lensing objects, and more importantly to the improvement of observational precision and the advent of new types of observations. Some of the additional fields of lensing applicability include studies of stellar atmospheres, binary stars, extrasolar planets, the mass function of Galactic matter and its line-of-sight distribution (see the review of Paczyński 1996; Gould 2001). The most important progress made in the observational side includes the increase of the event detection rate by automatizing the data reduction process, the development of early warning systems to issue alerts of ongoing events (Udalski et al. 1994b; Alcock et al. 1996; Afonso et al. 2001; Bond et al. 2001), and the follow-up observational network for the frequent and precise observation of the alerted events (Rhie et al. 1999; Albrow et al. 1998). Judging from the progress achieved so far and the speed of progress, the extent of the lensing applicability will become even broader.

In this paper, we discuss several additional fields of astronomy to which lensing observations, by using future-generation instruments, can provide useful information. In § 2, we investigate the use of future high resolution space telescopes to directly resolve lenses from follow-up observations of source stars that have previously experienced lensing magnification. In § 3, we probe the feasibility of detecting and characterizing stellar spots from 1%-level photometry. In § 4, we show that lensing observations with very large telescopes (> 30m) will enable one to probe the detailed structures of distant extrasolar planets. We conclude our discussion in § 5.

2. Direct Lensing Imaging

Recently, from the *Hubble Space Telescope* (HST) images of one of the LMC events (MACHO LMC-5) taken 6.3 years after the original lensing measurement, Alcock et al. (2001) were able to resolve the lens from the lensed source star. By directly imaging the lens, they could identify that the event was caused by a nearby low-mass star located in the Galactic disk.

Besides the identification of the lens as a normal star, direct lens imaging is of scientific importance due for the following reasons. First, by directly and accurately measuring the lens proper motion with respect to the source, μ, one can better constrain the physical parameters of individual lenses. Secondly, if the lens is resolved for an event where the lens-source relative parallax, $\pi_{\rm rel} = 1/(D_{\rm ol}^{-1} - D_{\rm os}^{-1})$ AU, was previously measured during the lensing magnification, one can uniquely determine the lens mass from the known values of μ and $\pi_{\rm rel}$ (Gould 2001). Thirdly, if the source of an event is resolved via a caustic crossing and thus the source star radius normalized by the angular Einstein ring radius, $\rho_\star$, is measured, then one can determine the angular source star radius by $\theta_\star = \mu\, t_{\rm E}\, \rho_\star$, where the Einstein ring radius crossing time $t_{\rm E}$ is determined from the lensing light curve. By measuring $\theta_\star$, one can determine the effective temperature of the source star, which is important for the accurate construction of stellar atmosphere models (e.g., Alonso et al. 2000).

Although the first directly imaged lens was identified as an LMC event, much more numerous direct lens identifications are expected if high resolution follow-up observations are performed for the source stars of events detected towards the Galactic bulge. There are several reasons for this expectation. First, compared to the total number of LMC events, which is ~ 20, there are an overwhelmingly large number of bulge events ($\gtrsim 1\,000$). Secondly, while the majority of LMC events are suspected to be caused by dark (or very faint) objects, most bulge events are supposed to be caused by normal stars, for which imaging is possible. Thirdly, an important fraction of lenses responsible for bulge events are believed to be located in the Galactic disk with moderate distances, and thus are more likely to be imaged as a result of their tendency to be bright and have a large proper motion.

We estimate the fraction of Galactic bulge events whose lenses can be directly imaged. For this estimation, we first compute the expected distribution of the lens-source proper motions of the currently detected Galactic bulge events, based on standard models of the geometrical and dynamical distributions of lenses and their mass function. For the detailed models, see Han & Chang (2002). We then apply realistic detection criteria for lens resolution. The criteria we applied for lens resolution are described in detail in Fig. 1.

In Figure 2, we present the determined fractions of events with detectable lenses as a function of the time elapsed after the original lensing measurement, Δt. In the figure, the thick and thin curves represent the expected fractions when follow-up observations are conducted by using instruments with resolutions of $\theta_{\rm PSF} = 0''.05$ and $0''.1$, respectively. We note that the Advanced Camera for Surveys (ACS) recently installed on HST can achieve a resolution of $\theta_{\rm PSF} \sim 0''.1$. We also note that the Near Infrared Camera (NIRCam) of the *Next Generation Space Telescope* (NGST), which will have an aperture of 6–7 m, will be sensitive in the wavelength range from 0.6 to 5 microns, and thus can achieve $\theta_{\rm PSF} \sim 0''.05$. If the instrument has a resolution of $\theta_{\rm PSF} = 0''.1$, we estimate that lenses can be resolved for $\sim 3\%$ and 22% of disk-bulge events and for $\sim 0.3\%$ and 6% of bulge self-lensing events after $\Delta t = 10$ and 20 years, respectively. The fraction increases substantially with an increase in the resolving power. If observations are performed by using an instrument with $\theta_{\rm PSF} = 0''.05$, we estimate that lenses can be resolved for $\sim 22\%$ and 45% of disk-bulge events and for $\sim 6\%$ and 23% of bulge self-lensing events after $\Delta t = 10$ and 20 years, respectively. Therefore, we find that the proper choice of the instrument for lens resolution will be NGST. Bulge events have been reported since 1993 (Udalski et al. 1994a; Alcock et al. 1995). Under a rough assumption that disk-bulge and bulge self-lensing events equally contribute to the total Galactic bulge event rate and considering the life expectancy of HST, the fraction of events with resolvable lenses from HST observations will be just $\sim 5\%$ even if follow-up observations are performed at the end stage of the HST for the first generation of lensing events. However, by

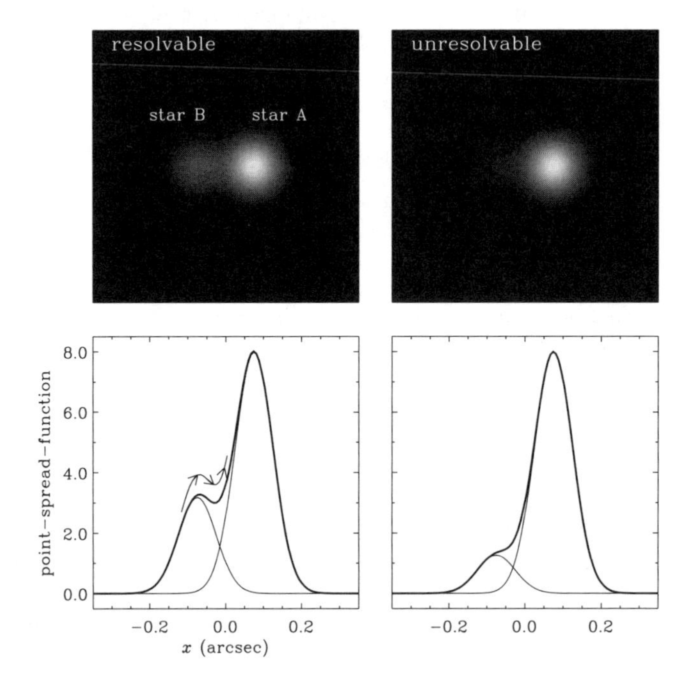

Figure 1. The criterion for resolving two closely located stars. Presented in each of the upper panels is the combined image of two stars. The thick solid curve in the corresponding lower panel is the one-dimensional PSF profile of the combined image. We assume that the individual stars are resolved if the sign of the combined image's PSF profile changes more than twice in the overlapping region between the centers of the individual stellar images. Under this criterion, the two stars in the left panel are resolved, while the stars in the right panel are not resolved. In addition to this restriction, we impose an additional restriction that detectable lenses should be brighter than $I = 22$.

using NGST, which is scheduled to be launched in 2009, it will be possible to resolve lenses for a significant fraction of events.

3. Stellar Surface Structure: Spots

One of the lensing applications in stellar astrophysics is the detection and characterization of stellar surface structures, such as spots. Spot detection via microlensing is possible for high magnification events produced by the source's crossing of the lens caustic. The caustic refers to the set of source positions on which the lensing magnification of a point source becomes infinite. For a single lens, the location of the caustic is that of the lens itself. For a binary lens, the set of caustics forms a closed curve in which each curve is composed of three or

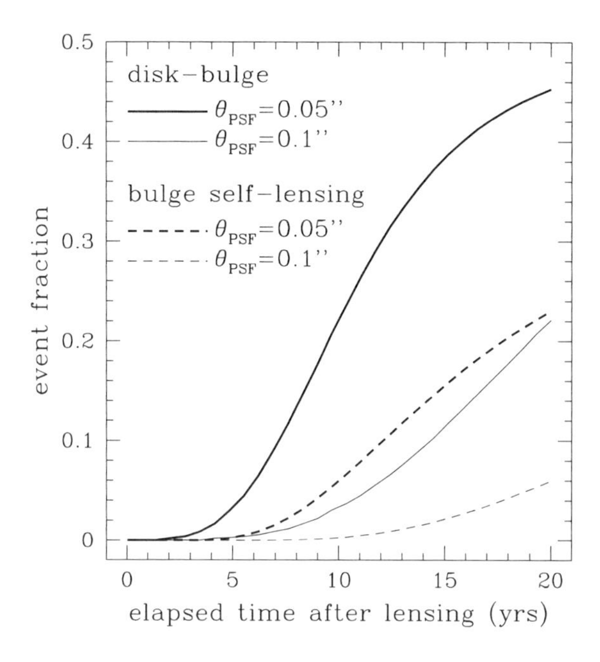

Figure 2. The fraction of Galactic bulge events with resolvable lenses as a function of the elapsed time after lensing magnification, Δt. The two pairs of curves drawn by solid lines and dashed lines correspond to the fractions expected when events are observed by using instruments with resolutions of $\Delta\theta_{\rm PSF} = 0''.05$ and $0''.1$, respectively. The solid curves are for disk-bulge events while the dashed curves are for bulge self-lensing events.

more concave line segments (fold caustic) that meet at cusps. Due to the larger cross-section of the fold caustic than that of the point caustic, caustic crossings occur more often for binary lens events. For caustic-crossing events, one can resolve the source star surface because different parts of the source are magnified by different amounts, because of the large gradient of magnification over the source during the caustic crossing (Gould 1994; Nemiroff & Wickramasinghe 1994; Witt & Mao 1994).

To investigate the patterns of spot-induced anomalies in lensing light curves, we perform simulations of caustic-crossing binary lens events occurring on source stars having spots on their surfaces. The physical state of a spot is characterized by various factors, including the size, shape, and the surface brightness contrast with respect to the unspotted stellar region. In addition, spots are likely to appear in groups and may have umbral/penumbral structures. We investigate the variation of the anomalies depending on these factors. Figure 3 presents an example of the simulations, showing the variations depending on the shape of the umbral/penumbral structures. The details of the spot model parameters

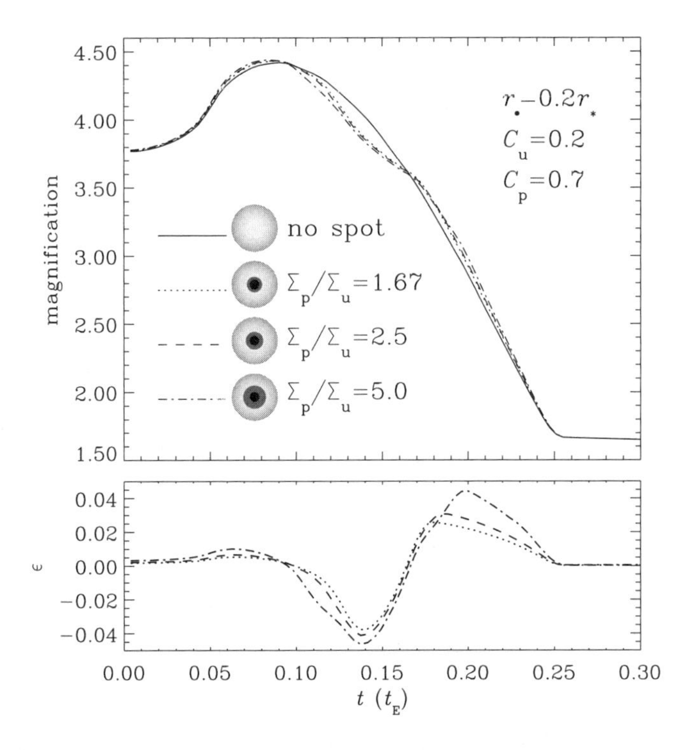

Figure 3. Light curves of caustic-crossing binary lens events occurring on source stars having spots with various umbra/penumbra structures. The lower panel shows the changes of the fractional deviation from the light curves of the unspotted source event. The source star has an angular radius of $0.1\theta_{\rm E}$. The area of the umbral part, $\Sigma_{\rm u}$, is fixed while the area of the penumbral region, $\Sigma_{\rm p}$, has three different values. The umbral region covers 4% of the total source surface area. The surface brightness contrasts are $C = 0.2$ and 0.7 for the umbral and penumbral parts, respectively.

are described in the figure caption. From this investigation, we find that a spot causes deviations at the level of a few per cent, which are readily detectable by using a 1-m class telescope at a good seeing site. However, to distinguish the subtle differences between different shapes of spots, photometric precision of 1% or less will be required.

4. Planetary Environment

Various methods have been proposed to search for extrasolar planets. These methods include pulsar timing analysis, direct imaging, accurate measurement of astrometric displacements, radial velocity measurement, planetary transit and microlensing [see the review of Perryman (2000)]. Planet detection by using microlensing is possible because the planet can induce noticeable anomalies in lensing light curves when the planet happens to be located close to one of the images produced by the primary lens (Mao & Paczyński 1991; Gould & Loeb 1992). The microlensing method has several important advantages over other methods. The most important advantage is that the strength of the planet's signal depends weakly on the planet/primary mass ratio and thus it is currently the only feasible method that is able to detect Earth-mass planets (Bennett & Rhie 1996). However, it also has disadvantages. One of the disadvantages is that, similar to other indirect methods based on the gravitational effect of the planet on the primary, the only useful information one can obtain is the planetary mass ratio. Thus, although one can identify the existence of planets, it is impossible to obtain information about the structure and environment of the planet.

Recently, as an additional channel to detect extrasolar planets by using microlensing, Graff & Gaudi (2000) and Lewis & Ibata (2000) proposed to monitor caustic-crossing binary lens events for the detections of close-in giant planets orbiting the source stars. In this method, the planet can be detected because the reflected light from the planet can be magnified during the caustic crossing highly enough to produce noticeable deviations in the lensing light curve of the primary. This method is important because the exquisite resolution afforded by caustics may allow one to study the features on and around planets in detail, and with large-aperture telescopes, one may be able to study the structure and environment of the detected planets by looking for small deviations in the normal light curve.

We estimate the expected magnitudes of deviations induced by various structures on and around planets, including satellites, rings, and atmospheric features (e.g., spots, zonal bands and Lambert scattering). For this estimation, we perform simulations of caustic-crossing events occurring on source stars having close-in planets with various structures. Figure 4 shows several light curves produced by a planet with rings of various widths. From this investigation, we find that, for reasonable assumptions of Galactic bulge events and close-in planets, rings can produce deviations on the order of 10% $\delta_{\rm p}$, where $\delta_{\rm p}$ is the magnitude of the deviations induced by the planet. For a planet with properties similar to the recently detected close-in planet of HD209458b (with a radius of $R_{\rm p} = 1.347 R_{\rm J}$ and a semimajor axis of $a = 0.0468$ AU, Brown et al. 2001), $\delta_{\rm p} \sim 1\%$. We also find that spots, zonal bands, and satellites produce substantially smaller deviations on the order of $\sim 1\% \delta_{\rm p}$. We asses the detectability of these planetary structures with current and future telescopes. We find that with 10-m-class telescopes, rings can be marginally detectable. For 30-m-class or larger telescopes, rings should be easily detectable, but it will still be very difficult to detect other features, even with these large telescopes.

5. Conclusion

We have discussed several new fields of astronomy to which lensing observations with future-generation instruments can provide useful information. Besides the ones mentioned in these proceedings, there already exists an important new type of lensing experiment that is scheduled to be carried out within the next

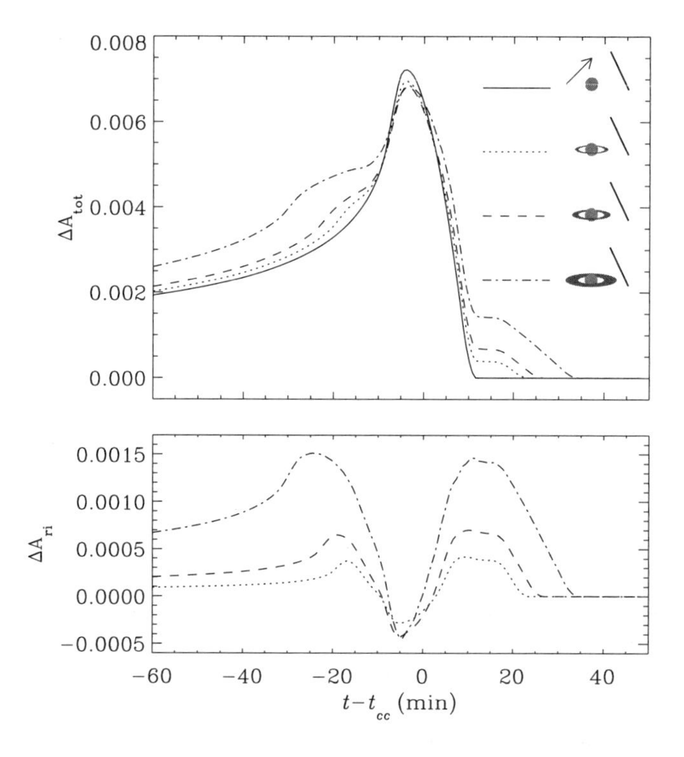

Figure 4. Microlensing light curves near the caustic crossings of planets having rings with various sizes. The rings of the tested planets have a common inner ring radius of $R_{\rm in} = 2.0$ in units of the planet radius, but have different outer ring radii of $R_{\rm out} = 2.6$, 3.0 and 4.0. The inclination of the ring is $i = 75°$. The arrow in the upper icon represents the direction of source motion with respect to the caustic (marked by a thick line).

decade. In this experiment, often called astrometric microlensing, one observes lensing-induced center-of-light motion of the lensed source star, by using high-precision interferometers. From this observation, the masses and locations of the individual lenses can be determined. Therefore, this experiment and others such as those proposed here will open a new era of lensing applications for stellar astrophysics.

References

Afonso, C. et al. 2001, A&A, 378, 1014

Alcock, C. et al. 1993, Nature, 365, 621

Alcock, C. et al. 1995, ApJ, 445, 133

Alcock, C. et al. 1996, ApJ, 463, 67
Alcock, C. et al. 1997, ApJ, 491, 436
Alcock, C. et al. 2001, Nature, 414, 617
Afonso, C. et al. 2001, 378, 1014
Albrow, M. et al. 1998, ApJ, 509, 687
Alonso, A. et al. 2002, A&A, 355, 1060
Aubourg, E. et al. 1993, Nat, 365, 623
Bennett, D. P., Rhie, S. H. 1996, ApJ, 472, 660
Bond, I. et al. 2001, MNRAS, 327, 868
Brown, T. M., Libbrecht, K. G., Charbonneau, D. 2002, PASP, 114, 826
Gould, A. 1994, ApJ, 421, L71
Gould, A. 2001, PASP, 113, 903
Gould, A., Loeb, A. 1992, ApJ, 396, 104
Graff D. S., Gaudi, B. S. 2000, ApJ, 538, L133
Han, C., Chang H.-Y. 2002, preprint (astro-ph/0209352)
Lewis G. F., Ibata R. A. 2000, ApJ, 539, L63
Mao, S., Paczyński, B. 1991, ApJ, 374, L37
Nemiroff, R. J., Wickramasinghe W. A. D. T. 1994, ApJ, 424, L21
Paczyński, B. 1986, ApJ, 304, 1
Paczyński, B. 1996, Ann. Rev. A. & A., 34, 419
Perryman, M. A. C. 2000, Rep. Prog. Phys., 63, 1209
Rhie S.H. et al. 1999, ApJ, 522, 1637
Udalski A. et al. 1994a, Acta Astron., 44, 1
Udalski A. et al. 1994b, Acta Astron., 44, 227
Witt H. J., Mao S. 1994, ApJ, 430, 505

IAU 8th Asia-Pacific Regional Meeting
ASP Conference Series, Vol. 289, 2003
S. Ikeuchi, J. Hearnshaw & T. Hanawa, eds.

A Perturbative Approach to Astrometric Microlensing Due to an Extrasolar Planet

Hideki Asada

Faculty of Science and Technology, Hirosaki University, Hirosaki 036-8561, Japan

Abstract. We develop a perturbative approach to microlensing due to an extrasolar planetary lens. For a weak lensing case, we show how the maximum angular size and the typical time scale of the anomalous shift of the light centroid are dependent on the mass ratio and angular separation between the star and the planet. Furthermore, it is shown that the lens equation for a binary gravitational lens being a set of two coupled real fifth-order algebraic equations can be reduced to a single real fifth-order algebraic equation, which provides a much simpler way to study lensing by binary objects.

1. Introduction

Extrasolar planets searches are successfully going on (Marcy and Butler 1998). In spite of the success, the Doppler method has some drawbacks: The inferred mass is the lower bound, since the inclination angle of the orbital plane is not determined except for the eclipse case. In addition, the radial velocity of a star becomes too small to detect its Doppler effect when separation between the star and the planet is of the order of 1 AU. Hence, it seems quite difficult to discover Earth-type planets by this method. Astrometry is considered as a supplementary method to find out planets by measuring the transverse motion of a star. Indeed, we can expect dramatic improvements in the precision of the future astrometric measurements by SIM, DIVA and GAIA.

Also by these future astrometry missions, the astrometric microlensing will become the third method for searching extrasolar planets: The typical scale in the gravitational lensing is the Einstein ring radius, which is of the order of 1AU for a lensing star within our galaxy. Hence, the microlensing is quite effective even for Earth-type planets (Mao and Paczynski 1991; Gould and Loeb 1992). In the planetary lens case, we can determine its true mass through observation of spikes in the light curve, which are produced by the magnification effect of the gravitational lensing. On the other hand, the astrometric microlensing is a consequence of a combination of the magnification and the position shift of the image. The numerical results (Safizadeh et al. 1999, Han and Lee 2002) have shown that the photo centroid shifts provide us a clue for extrasolar planets. The purpose of my talk is to present the analytic formulae and clarify the parameter dependence (Asada 2002a) and to show a new formulation for a binary gravitational lens (Asada 2002b).

2. Perturbative Approach

2.1. Microlensing Due to a Single Lens

A single lens is located at the distance $D_{\rm L}$ from the observer, and a source at $D_{\rm S}$. The distance between the lens and the source is denoted by $D_{\rm LS}$. Then, under the thin lens approximation, the lens equation for the single lens is written as

$$\boldsymbol{\beta} = \boldsymbol{\theta} - \frac{D_{\rm LS}}{D_{\rm S}}\boldsymbol{\alpha}, \tag{1}$$

where $\boldsymbol{\beta}$ and $\boldsymbol{\theta}$ are the angular positions of the source and the image, respectively, and $\boldsymbol{\alpha}$ is the deflection angle. In units of the Einstein ring radius, which is of the order of a milli-arcsecond (mas) for a solar mass lens within our galaxy, the lens equation is rewritten as $\boldsymbol{\beta} = \boldsymbol{\theta} - \boldsymbol{\theta}/\theta^2$, where θ denotes $|\boldsymbol{\theta}|$ and similar notations are taken below. We find out the two solutions $\boldsymbol{\theta}^{(\pm)}$.

Following the definition of the center of mass, we define the photo-center as (Walker 1995)

$$\boldsymbol{\theta}_{\rm C} = \frac{A^{(+)}\boldsymbol{\theta}^{(+)} + A^{(-)}\boldsymbol{\theta}^{(-)}}{A^{(+)} + A^{(-)}}. \tag{2}$$

With respect to the unlensed position, the location of the photo-center is $\Delta\boldsymbol{\theta}_{\rm C} = \boldsymbol{\theta}_{\rm C} - \boldsymbol{\beta}$, which expresses the deviation due to the lensing. For a single lens, it becomes an ellipse (Walker 1995; Jeong et al. 1999).

2.2. Microlensing Due to a Planetary Lens

Let us consider a planetary system with the stellar mass M_1, the planet mass M_2 and the projected separation vector $\boldsymbol{s}$ from the star to the planet. We adopt the frame of center of mass. Then, the lens equation is written as

$$\boldsymbol{\beta} = \boldsymbol{\theta} - \left(\nu_1 \frac{\boldsymbol{\theta} + \nu_2\boldsymbol{\epsilon}}{|\boldsymbol{\theta} + \nu_2\boldsymbol{\epsilon}|^2} + \nu_2 \frac{\boldsymbol{\theta} - \nu_1\boldsymbol{\epsilon}}{|\boldsymbol{\theta} - \nu_1\boldsymbol{\epsilon}|^2}\right), \tag{3}$$

where we defined $\nu_1 = M_1/(M_1 + M_2)$, $\nu_2 = 1 - \nu_1$ and $\boldsymbol{\epsilon} = \boldsymbol{s}/D_{\rm L}$.

In the planetary case, M_2 is much smaller than M_1; the Jupiter mass is about 10^{-3} of the solar mass. Hence, we introduce an expansion parameter as $\nu = \nu_2$ in our perturbation approach. Since we wish to consider the microlensing as a method supplementary to the Doppler technique, we concentrate ourselves on a large separation case $\epsilon > 1$, which is beyond the reach of the Doppler method. It is straightforward to extend our investigation to the case of $\epsilon < 1$. In addition, we consider a case of a large impact parameter $\beta \gg 1$, which is most probable because of the large cross section. In total, we consider the case of $\beta \gg \epsilon > 1$. Let us look for the solutions of the lens equation by taking a form of $\boldsymbol{\theta} = \boldsymbol{\theta}_0 + \delta\boldsymbol{\theta}$, for the zeroth and first order solutions $\boldsymbol{\theta}_0$ and $\delta\boldsymbol{\theta}$. This perturbation permits us to find a correction to the photo centroid as

$$\delta\boldsymbol{\theta}_{\rm C} = \frac{\mu}{\beta^6}\Big(-2\boldsymbol{\epsilon}(\boldsymbol{\beta}\cdot\boldsymbol{\epsilon})\beta^2 - \boldsymbol{\beta}\epsilon^2\beta^2 + 4\boldsymbol{\beta}(\boldsymbol{\beta}\cdot\boldsymbol{\epsilon})^2 + O(\beta^3)\Big) + O(\mu^2). \tag{4}$$

Denoting by $\boldsymbol{v}_\perp$ the transverse angular velocity of the source to the lens, we define

$$\boldsymbol{\epsilon}_\parallel = \frac{(\boldsymbol{\epsilon}\cdot\boldsymbol{v}_\perp)\boldsymbol{v}_\perp}{v_\perp^2}, \tag{5}$$

$$t_\mathrm{C} = \frac{\epsilon_\parallel}{v_\perp}. \tag{6}$$

For stellar cases in our Galaxy, the maximum angular size of the distortion is estimated as

$$\frac{\nu\theta_E}{\beta}\left(\frac{\epsilon}{\beta}\right)^2 \sim 1 \text{ micro-arcsec}\left(\frac{\nu}{10^{-3}}\right)\left(\frac{1}{\beta}\right)\left(\frac{\theta_E}{\text{mas}}\right), \tag{7}$$

where we assumed that β is comparable to ϵ. The maximal distortion occurs at

$$t_\mathrm{C} \sim 10^6\left(\frac{\epsilon_\parallel}{\text{AU}}\right)\left(\frac{100\text{ km/s}}{v_\perp}\right)\text{ s}, \tag{8}$$

about a few months before/after, depending on a location of the planet, the source passes the point closest to the lensing star.

3. Lens Equation for a Binary System

We consider a binary system of two bodies with mass M_1 and M_2 and angular separation vector $\boldsymbol{l}$ from the object 1 to 2. The lens equation is a set of two coupled real fifth-order equations for (θ_x, θ_y), equivalent to a single complex fifth-order equation for $\theta_x + i\theta_y$ (e.g. Witt 1990, 1993).

3.1. Off-axis Sources

Let us introduce polar coordinates: $(\theta_x, \theta_y) = (r\cos\phi, r\sin\phi)$ and $(\ell_x, \ell_y) = (\ell, 0)$, where r and $\ell \geq 0$. The lens equation is reduced to the fifth-order equation for $\tan\phi$ (Asada 2002b),

$$\sum_{i=0}^{5} a_i(\tan\phi)^i = 0, \tag{9}$$

where all of these coefficients $a_0, \cdots a_5$ are polynomials in ℓ, ν and $\boldsymbol{\beta}$, all of which are finite. We can also show that

$$r\cos\phi = F(\tan\phi). \tag{10}$$

As shown by Galois, a fifth-order equation cannot be solved in the algebraic manner (e.g. van der Waerden 1966). Hence, by solving numerically Eq. (9), the image position is obtained as $(\theta_x, \theta_y) = (r\cos\phi, r\cos\phi\tan\phi)$.

3.2. Sources on the Symmetry Axis

Let us consider the case of sources on the symmetry axis, for which analytic solutions can be obtained: For a binary with two equal masses, explicit solutions were found by Schneider and Weiß (1986). Analytic solutions for a binary with an arbitrary mass ratio, given by Asada (2002b), are useful for verification of numerical implementations, since numerical solutions in subsection 3.1 must approach analytic ones as $\beta_y \to 0$.

4. Conclusion

We have developed a perturbative approach to microlensing due to an extrasolar planetary lens. In particular, we have shown by Eqs. (7) and (8), how the light centroid shifts are dependent on the mass ratio and separation between the star and the planet. The typical time scale is of the order of months, depending strongly on $\epsilon_{\parallel}$, a projection of the separation vector onto the source motion.

We have also reexamined the lens equation for a binary system in the polar coordinates. Our formulation based on the one-dimensional equation (9) is significantly useful compared with previous two-dimensional treatments for which there are no well-established numerical methods; the new formulation enables us to study the binary lensing more precisely with saving time and computer resources.

Acknowledgments. The author would like to thank M. Bartelmann and M. Kasai for fruitful conversations. This work was supported by a Japanese Grant-in-Aid for Scientific Research from the Ministry of Education No. 13740137 and the Sumitomo Foundation.

References

Asada H. 2002a, ApJ 573, 825

Asada H. 2002b, A&A 390, L11

Gould, A. and Loeb, A. 1992, ApJ, 396, 104

Han, C. and Lee, C. 2002, MNRAS, 329, 163

Jeong, Y., Han, C. and Park, S. 1999, ApJ, 511, 569

Mao, S. and Paczynski, B. 1991, ApJ, 374, 37L ApJ, 522, 512

Marcy, G. W. and Butler, R. P. 1998, ARA&A, 36, 57

Safizadeh, N., Dalal N. and Griest, K. 1999, ApJ, 522, 512

Schneider P. and Weiß A. 1986, A&A 164, 237

van der Waerden B. L. 1966 *Algebra I* (Springer)

Walker, M. A. 1995, ApJ, 453, 37

Witt, H. J. 1990, A&A 236, 311

Witt, H. J. 1993, ApJ 403, 530

IAU 8th Asia-Pacific Regional Meeting
ASP Conference Series, Vol. 289, 2003
S. Ikeuchi, J. Hearnshaw & T. Hanawa, eds.

Detecting Astrometric Microlensing with VERA (VLBI Exploration of Radio Astrometry)

Mareki Honma[1,2,3]

[1] *VERA Project Office, NAOJ, 181-8588 Mitaka, Japan*
[2] *Earth Rotation Division, NAOJ, 023-0861 Mizusawa, Japan*
[3] *Graduate University for Advanced Study, 181-8588 Mitaka, Japan*

Tomoharu Kurayama[1,4]

[4] *Department of Astronomy, University of Tokyo, 113-0033 Tokyo, Japan*

Abstract. We investigate the possibility of detecting the astrometric microlensing of QSOs caused by Galactic stars and MACHOs using VERA (VLBI Exploration of Radio Astrometry). First we briefly introduce the VERA project, a ground-based radio astrometric mission which aims at astrometric accuracy at the 10 μas level. In the second part of this paper, we present model calculations of the optical depth and event duration of astrometric microlensing caused by Galactic stars and MACHOs, and we discuss the implications for VERA.

1. Introduction

Gravitational microlensing is a useful tool to study faint objects like MACHOs and low-mass stars in the Galaxy. Currently two types of microlensing events are known, depending on how they are detected: one is 'photometric' microlensing in which events are detected through photometric monitoring of source magnification, and the other is 'astrometric' microlensing which is detected by position shift of the lensed image. While hundreds of photometric microlensing events have already been detected by several massive photometry groups (e.g., MACHO, EROS, OGLE, MOA), astrometric microlensing has not been detected yet (except for those by the Sun and solar planets). However, recent studies have shown that astrometric microlensing events can be another tool to study faint objects in the Galaxy (e.g., Miralda-Escude 1996; Hosokawa et al. 1997; Dominik & Sahu 2000; Honma 2001; Honma & Kurayama 2002). The major point of those studies is that with astrometric accuracy at the 10 μas level, the probability of astrometric microlensing is much larger than that of photometric microlensing (e.g., Miralda-Escude 1996). For instance, Galactic astrometric lenses, at the 10 μas shift level, can be larger by 50 times than the Einstein-ring radius, which is the typical size of photometric microlensing (e.g., Honma 2001).

Although not yet achieved, we can expect that such a high astrometric accuracy will be available soon. For instance, there are two space astrometric missions which aim at an accuracy of 10 μas or higher, namely SIM and

GAIA. In addition to those space missions, there is a ground-based astrometric mission called VERA (VLBI Exploration of Radio Astrometry, e.g., Sasao 1996; Honma, Kawaguchi & Sasao 2000 and references therein), which utilize a phase-referencing VLBI technique for astrometry of radio sources. The most remarkable difference between VERA and space astrometric missions is that VERA can observe thousands of distant sources like QSOs and radio galaxies to trace the effect of astrometric microlensing. An advantage of using distant radio sources is that the column density of the lens can be much higher, leading to a higher event probability. Another advantage is that ground-based telescopes have much longer lifetimes than space satellites, and thus one can trace astrometric microlensing events with a long duration (typically 7 to 15 years). Thus, VERA is a potential tool to study the nature of MACHOs in the halo as well as low mass stars in the Galaxy's disk through astrometric microlensing. For these reasons, in the present paper we discuss the implications of astrometric microlensing for VERA.

2. The VERA Project

VERA (VLBI Exploration of Radio Astrometry), being promoted by the National Astronomical Observatory of Japan under collaboration with several Japanese universities (Sasao 1996; Honma, Kawaguchi & Sasao 2000), is a new VLBI array designed for phase referencing VLBI. VERA's dual-beam antenna enables us to observe a Galactic maser source and a nearby reference source simultaneously (Kawaguchi, Sasao & Manabe 2000), and with such a system we can cancel out the atmospheric fluctuations and measure the positions of Galactic maser sources relative to reference sources (QSOs and radio galaxies) with 10 microarcsec level accuracy. This accuracy allows us to determine the distance of an object D kpc away with uncertainty of D%, and the proper motion with uncertainty of $0.05D$ km/s. Therefore, VERA will be able to measure parallaxes and proper motions of maser sources in the whole of the Galaxy's disk. Hence, VERA will be one of most powerful tools for the study of the dynamics of the Milky Way Galaxy.

The VERA array consists of four 20-mϕ antennas spread over Japan with baseline lengths ranging from 1200 km to 2300 km. VERA's antenna is designed for observational frequencies of 2, 8, 22 and 43 GHz with a possible extension to 86 GHz. The main frequencies are 22 GHz for H_2O masers and 43 GHz for SiO masers, which are mainly emitted from star-forming regions and Mira-type variables. The other two frequencies (2 and 8 GHz) are mainly for geodetic observations. For phase-referencing observations, each antenna is installed with the dual-beam system on which two receivers are mounted, and simultaneously observes two sources separated by as much as 2.2 degrees to remove the atmospheric fluctuations effectively.

All the four stations had already been constructed by 2002 April, and are currently under system evaluation. We have already made test observations and detected fringes with all four stations, and also succeeded in dual-beam phase referencing. Astrometric accuracy is still under evaluation, but we expect that routine astrometric observations will be started by 2005.

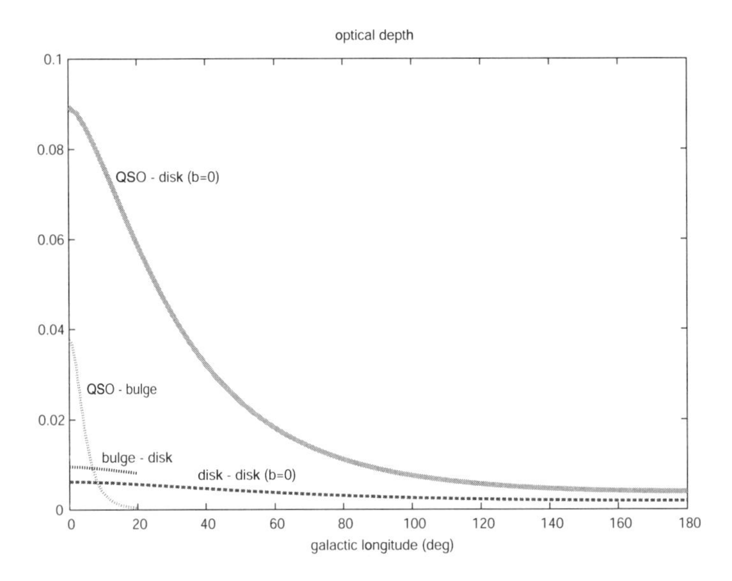

Figure 1. Distribution of optical depth for astrometric microlensing with galactic longitude.

3. Astrometric Microlensing Observation with VERA

Here we estimate the optical depth of astrometric microlensing events for VERA. The optical depth of astrometric microlensing events can be written as follows (e.g. Honma 2001; Honma & Kurayama 2002),

$$\tau_{\rm ast} = \frac{16\pi G^2 M}{c^4 \theta_{\rm min}^2} \int \rho \left(1 - \frac{D_{\rm d}}{D_{\rm s}}\right)^2 dD_{\rm d}. \tag{1}$$

Here, ρ is the mass density of the lens, $D_{\rm d}$ and $D_{\rm s}$ are the lens and source distances from the observer, and M is the mean lens mass. $\theta_{\rm min}$ denotes the minimum positional shift that can be detected based on astrometric observations, and in the present paper we assume $\theta_{\rm min}$ of 10 μas.

For the density distribution in the Galaxy's disk, we assume an exponential disk for both radial and vertical profiles. Here we take rather conservative values for disk parameters – the disk density at the Sun of $0.08 M_\odot$ pc^{-3}, and the radial and vertical scale lengths of 3.5 kpc and 300 pc. We also take into account the bulge star contributions based on a Plummer-type spherical bulge with a total mass of $0.8 \times 10^{10} M_\odot$ and a scale length of 1 kpc. For QSO-disk and QSO-bulge lensing, in which a distant QSO is being lensed by disk or bulge stars, the distance to the QSO is assumed to be infinite. On the other hand, for the disk-disk lensing case, we assume a typical source distance $D_{\rm s}$ of 8 kpc.

Figure 1 shows the optical depths calculated for the cases described above. First of all, for the QSO-disk/bulge lensing case, the maximum optical depth is obtained for sources behind the Galactic center (i.e., $(l, b) = (0^\circ, 0^\circ)$), being 8.9×10^{-2} for a disk lens, and 3.8×10^{-2} for a bulge lens, respectively. This result indicates that nearly one out of ten sources in this direction is always lensed by disk stars, and the optical depth of a bulge lens is nearly half that of a disk lens. For comparison, figure 1 shows that the maximum optical depth for disk-disk lensing is 6.2×10^{-3}, being smaller by a factor of 15 than the maximum optical depth for QSO-disk lensing. Also, the maximum optical depth for the bulge-disk lensing case is 9.3×10^{-3}, 10 times smaller than that for QSO-disk lensing. Therefore, observing distant QSOs with VERA gives a much higher optical depth than observing Galactic stars with a space astrometric satellite.

We also made similar calculations for the Galactic halo based on the standard halo models (see Honma 2001 for details). For astrometric microlensing by halo MACHOs, the optical depth is around 0.02 towards the Galactic center, assuming a lens mass of $0.5M_{\odot}$. Thus, to detect an astrometric microlensing event by MACHO, one has to monitor at least 50 sources. This is fairly large when compared to the case for disk star lensing, but still significantly smaller than the number of sources for a typical photometric microlensing search, which is 10^6 to 10^7.

In addition to optical depth, we also estimated the typical event durations for such astrometric events. The event duration of microlensing can be estimated by dividing the lens size by the lens proper motion relative to the sources. Hence, to estimate the duration, we need to assume the lens and source motions. For disk stars, we assume a circular rotating disk with a constant velocity of 200 km/s, independent of Galacto-centric distance (see Honma & Kurayama 2002 for details), and for halo MACHOs we assume a constant two-dimensional velocity dispersion of 200 km/s (see Honma 2001). We found that the typical event duration for disk star lensing is around 7 years, and 15 years for halo MACHO lensing. We note that these durations are longer than the lifetimes of typical space astrometric missions, and hence it may be difficult for space missions like SIM and GAIA to trace astrometric microlensing events for the full event duration. On the other hand, these event durations are less than the anticipated lifetime of VERA ($\sim$20 years), and hence one can trace the whole event using VERA.

In summary, detecting astrometric microlensing with VERA is fairly probable. Such an event, if detected, will provide useful information for constraining the mass and nature of lenses in the disk and the halo.

References

Dominik, M. & Sahu, K. C. 2000, ApJ, 534, 213

Honma, M., Kawaguchi, N. & Sasao, T. 2000, in Radio Telescopes, ed. H. R. Butcher, Proc. SPIE 4015, 624

Honma, M. 2001, PASJ, 53, 233

Honma, M. & Kurayama, T. 2002, ApJ, 568, 717

Hosokawa, M., Ohnishi, K. & Fukushima, T. 1997, AJ, 114, 1508

Kawaguchi, N., Sasao, T. & Manabe, S. 2000, in Radio Telescopes, ed. H. R. Butcher, Proc. SPIE 4015, 544

Miralda-Escude, J. 1996, ApJ, 470, L113

Sasao, T. 1996, in Proc. 4th Asia-Pacific Telescope Workshop, ed. E. A. King, 94 (Sydney, Austr. Tel. Nat. Facility)

IAU 8th Asia-Pacific Regional Meeting
ASP Conference Series, Vol. 289, 2003
S. Ikeuchi, J. Hearnshaw & T. Hanawa, eds.

The Dark Matter Halo of the Gravitational Lens Galaxy 0047-2808

Randall B. Wayth

School of Physics. University of Melbourne, 3010, Australia

Rachel L. Webster

School of Physics. University of Melbourne, 3010, Australia

Abstract. The location of the images in a multiple-image gravitational lens system are strongly dependent on the orientation angle of the mass distribution. As such, we can use the location of the images and the photometric properties of the visible matter to constrain the properties of the dark halo. We apply this to the optical Einstein Ring system 0047-2808 and find that the dark halo is almost spherical and is aligned in the same direction as the stars to within a few degrees.

1. Introduction

Numerical simulations of Cold Dark Matter (CDM) have been very successful in reproducing the observed large scale structure of the universe. The CDM model predicts that the dark matter (DM) haloes of today's galaxies are assembled through successive mergers of smaller haloes. Simulations using only dark matter predict that the haloes should be quite prolate. However, it is not clear how gas and/or stars interacting with the dark matter will change the shape of the halo. Studies have suggested that the DM halo can become more or less cuspy (El-Zant, Shlosman, & Hoffman, 2001; Tissera & Dominguez-Tenreiro, 1998) and rounder (Evrard, Summers, & Davis, 1994; Dubinski, 1994) after the interaction with stars and gas. An important test of galaxy formation and evolution models will be to compare the shape and profile of galaxy haloes with observed haloes. Thus, simple questions such as: "Do we expect the visible and dark matter to be aligned in elliptical galaxies?" and "Is the dark matter density in the central regions changed by the gravitational dominance of the stars?" must be answered with observations. For instance: the Milky Way, despite being a spiral galaxy, appears to have an almost spherical halo (Ibata et al., 2001).

Gravitational lensing offers a method to tightly constrain the shape of DM haloes in the population of medium redshift ($0.1 < z < 1.0$) lens galaxies. The image positions in a lens system are highly sensitive to the orientation of the overall mass profile. Keeton, Kochanek, & Falco (1998) showed that the *overall* mass distribution is typically aligned with the visible matter using a sample of lens galaxies and a simple SIE mass model. However, depending on the lens galaxy, the stellar mass can contribute a substantial fraction of the total mass inside the image. The extreme case is the lensed QSO 2237+0305 where the dark matter constitutes only 4% of the projected mass inside the images (Trott & Webster, 2002). In this case we expect the visible matter orientation and the total matter orientation derived from a lensing analysis to be very similar. The logical next step is to use a more complicated (stars + halo) model for the lens galaxy to determine the properties of the DM halo alone.

In this paper we use an implementation of the Lens MEM algorithm (Wallington, Kochanek, & Narayan, 1996) and a stars+halo lens model to study the optical Einstein Ring 0047-2808 (Warren et al., 1999, 1996) using data from the HST. This system is well suited for the study because it is an isolated lens galaxy, so we expect any external shear contributions to be small. The system is a $z = 0.485$ elliptical which is lensing a background starbursting galaxy at $z = 3.6$.

The algorithm we employ performs a non-parametric source reconstruction to match the observed data for a given lens model. The goodness-of-fit of the model is calculated using a χ^2 which takes into account the degrees of freedom used in the source. In this paper we assume $H_0 = 70$ kms^{-1}Mpc^{-1} and $(\Omega_{\rm m}, \Omega_\lambda) = (0.3, 0.7)$.

2. Method

The data were reduced as described in Wayth et al. (2002). The final image of the "ring" is 133×133 with $0.05''$ pixels as shown in Figure 1. The lens galaxy was best fit with a Sersic profile, where the surface brightness as a function of radius r is $\Sigma = \Sigma_{1/2} \exp\{-B(n)[(r/r_{1/2})^{1/n} - 1]\}$. The parameter n quantifies the shape of the profile: the values $n = 0.5$, $n = 1$, and $n = 4$ correspond to the Gaussian, exponential, and de Vaucouleurs profiles. Profiles with larger n are more cuspy. $B(n)$ is a constant for a particular n and we used the series asymptotic solution for $B(n)$ provided by Ciotti & Bertin (1999). Additional parameters used for the light profile are the axis ratio (q) and orientation angle (θ_s). The fitted parameters are shown in Table 1.

Table 1. Photometric parameters for the lens galaxy.

$R_{1/2}$ (pixels)	$\Sigma_{1/2}$ (counts)	q	θ_s (°)	n
21.69	0.7	0.693	125	3.115

We model the galaxy stellar component with fixed parameters from the photometry and allow only the M/L to vary. The halo is modelled as a Pseudo-Isothermal Elliptic Potential (PIEP) with a finite core. The PIEP model is defined by the lensing potential $\psi = b[r_{\rm c}^2 + (1-\epsilon)x^2 + (1+\epsilon)y^2]^{1/2}$ where $r_{\rm c}$ is the core radius, b is the mass scale (Einstein radius) and ϵ is the ellipticity. An additional parameter is used for the orientation angle ($\theta_{\rm h}$, being measured anti-clockwise from the horizontal). It is worth noting that the lens can be fit with the PIEP model alone (without a core) with the parameters $b = 1.165, \epsilon = 0.08$ and $\theta_{\rm h} = 129$. We use this mass scale for the halo model. The source in this system actually has two distinct components. The two-component model explains the

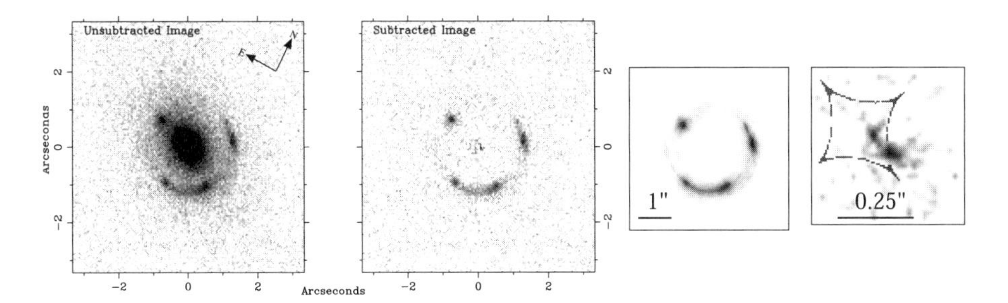

Figure 1. Data and model for the 0047-2808 lens. Left: the HST image and image with the galaxy subtracted. Right: Model image and reconstructed source for the PIEP model.

location and brightness of all features in the image with a standard lens model. Figure 1 shows the model source and corresponding image for the plain PIEP model.

The mass enclosed inside the image is tightly constrained by the Einstein radius. We use this constraint to normalize the stellar M/L for a halo of a given core radius. A large core is equivalent to a constant M/L mass model, whereas a small core will generate an unrealistically low M/L for the observed stellar component of the lens.

In preliminary tests, we found that we cannot fit the data for $r_{\rm c} \gtrsim 7''$ (42 kpc physical scale length) i.e. constant M/L models cannot fit the data. Therefore we have restricted our analysis to halo core radii $< 7''$. For the range of allowed core radius values, we have calculated the range of halo ellipticity and orientation angle which produce acceptable fits to the data.

3. Results

Figure 2 plots the acceptable range of halo ellipticity and orientation angle as a function of core radius. On the left, we see that the halo ellipticity is consistently less than the stellar ellipticity. For $1.5'' < r_{\rm c} < 2.5''$, the data permit a halo with projected mass density which is circular, although in all cases the best solution has a halo with non-zero ellipticity.

On the right of Figure 2, the plot shows that the halo orientation angle is independent of the core radius and is in the same direction as the projected stellar major axis (within errors). The acceptable range of orientation angles for $1.5'' < r_{\rm c} < 2.5''$ are for non-zero ellipticity.

4. Conclusion

By using a lens model which separates the stars from the halo, we have been able to determine some of the basic properties of the dark matter halo in the lens system 0047-2808. We find that the projected ellipticity of the halo is not circular, but is substantially rounder than the observed stellar ellipticity. A

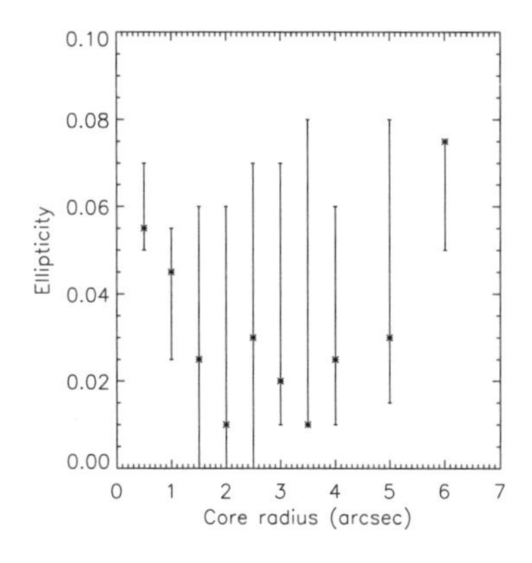

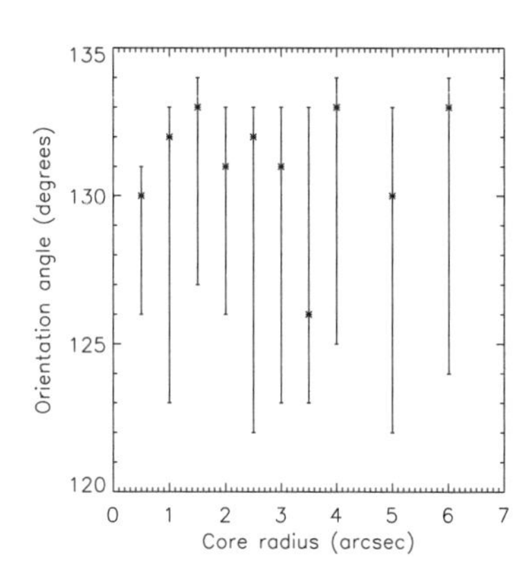

Figure 2. Constraints on the dark matter halo. Asterisks indicate the best fitting parameter values. The error bars indicate the range of the parameter which will produce acceptable (1σ) χ^2 values. Left: The acceptable range of ellipticity ($\epsilon_{\rm h}$) for various core radii. Right: the acceptable range of orientation angle ($\theta_{\rm h}$)

small range of halo core radii values ($1.5'' < r_{\rm c} < 2.5''$) allow the projected halo mass to be circular.

The halo's core, modelled as a constant density region, must be $< 7''$ to fit the observation. The core size could be further constrained by applying realistic limits to the stellar M/L which we intend to do in further work.

Finally, we find that although the halo is less elliptical than the stars, the orientation angle of the star's and halo's major axis are the same within errors.

References

Ciotti, L. & Bertin, G. 1999, A&A, 352, 447

Dubinski, J. 1994, ApJ, 431, 617

El-Zant, A., Shlosman, I. & Hoffman, Y. 2001, ApJ, 560, 636

Evrard, A. E., Summers, F. J. & Davis, M. 1994, ApJ, 422, 11

Ibata, R., Lewis, G. F., Irwin, M., Totten, E., & Quinn, T. 2001, ApJ, 551, 294

Keeton, C. R., Kochanek, C. S. & Falco, E. E. 1998, ApJ, 509, 561

Tissera, P. B. & Dominguez-Tenreiro, R. 1998, MNRAS, 297, 177

Trott, C. M. & Webster, R. L. 2002, MNRAS, 334, 621

Wallington, S., Kochanek, C. S. & Narayan, R. 1996, ApJ, 465, 64+

Warren, S. J., Hewett, P. C., Lewis, G. F., Moller, P., Iovino, A. & Shaver, P. A. 1996, MNRAS, 278, 139

Warren, S. J., Lewis, G. F., Hewett, P. C., Møller, P., Shaver, P. & Iovino, A. 1999, A&A, 343, L35

Wayth, R. B., Lewis, G. F., Warren, S. J. & Hewett, P. C. 2002, in prep.

IAU 8th Asia-Pacific Regional Meeting
ASP Conference Series, Vol. 289, 2003
S. Ikeuchi, J. Hearnshaw & T. Hanawa, eds.

The New Era of Precision Microlensing

Andrew Gould

Astronomy, Ohio State, 140 W. 18th Ave., Columbus, OH, USA

Abstract. In the past few years, microlensing has entered a new era of high-precision observations, which require equally precise interpretation. Very high quality data has enabled the measurement of the detailed surface structure of stars and even the mass of an individual microlens. Future observations using the *Space Interferometry Mission (SIM)*, with its 4 μas astrometric precision, will permit the measurement of hundreds of microlens masses, thus enabling for the first time a determination of the mass function of all objects in the Galactic bulge, both dark and luminous. *SIM* microlensing observations will also yield 1% mass measurements of non-binary stars, the first other than the Sun.

1. Introduction

When microlensing observations were initiated a decade ago by MACHO (Alcock et al. 1993), EROS (Aubourg et al. 1993) and OGLE (Udalski et al. 1993), the principal focus was on the quantity of the observations rather than their quality. The low optical depths, $\tau \lesssim 10^{-6}$ predicted for observations toward the Large Magellanic Cloud (Paczyński 1986) and Galactic bulge (Paczyński 1991; Griest et al. 1991) implied that millions of stars would have to be monitored to detect any microlensing events at all. Observing strategies naturally focused on obtaining this unprecedented data stream at sufficient precision to distinguish microlensing from other, much more common, forms of stellar variability.

However, almost immediately, theorists began suggesting additional applications of microlensing that had stricter observing requirements, either higher cadence than the roughly daily observations carried out by the survey teams, or higher precision than the $\sim 5\%$ photometry that was routinely achieved. These included the search for planetary companions of the microlens (Mao & Paczyński 1991; Gould & Loeb 1992), measurement of the microlens parallax $\pi_{\rm E}$ (Gould 1992) and of lens-source relative proper motion $\mu_{\rm rel}$ (Gould 1994; Nemiroff & Wickramasinghe 1994; Witt & Mao 1994). Of particular relevance here, Gould (1992) pointed out that if both the proper motion and the parallax could be measured, then so could the lens mass, M

$$M = \frac{{\rm AU}c^2}{4G}\frac{\theta_{\rm E}}{\pi_{\rm E}} \simeq \frac{M_\odot}{8}\frac{\theta_{\rm E}}{\rm mas}\frac{\tilde{r}_{\rm E}}{\rm AU}. \quad (1)$$

Here, $\tilde{r}_{\rm E} \equiv {\rm AU}/\pi_{\rm E}$ is the projected Einstein radius, $\theta_{\rm E} \equiv \mu_{\rm rel} t_{\rm E}$ is the angular Einstein radius, and $t_{\rm E}$ is the Einstein timescale, which last is routinely measured in microlensing events. Measurement of $\pi_{\rm E}$ and $\theta_{\rm E}$ also yields the lens-source relative parallax, $\pi_{\rm rel} = \pi_{\rm E}\theta_{\rm E}$.

Motivated primarily by the search for planets, but also to a certain extent these other higher-order effects, GMAN (Alcock et al. 1997), PLANET (Albrow et al. 1998), and MPS (Rhie et al. 2000) began in the mid 90s to obtain round-the-clock, high-cadence, high-precision observations of individual microlensing events from networks of observatories spanning the globe. Such observations became possible when OGLE and MACHO developed the ability to recognize microlensing events in real time and started to alert the community to these over the internet. EROS and a new survey group, MOA (Bond et al. 2002) also subsequently developed alert programs. Spurred on by these new observational capabilities, theorists developed even more ideas for higher order effects in microlensing, including limb-darkening (Witt 1995), xallarap (an "anti-parallax" effect induced by a binary source) (Griest & Hu 1992; Han & Gould 1997), star spots (Ignace & Hendry 1999; Han et al. 2000; Heyrovský & Sasselov 2000) as well as others. Almost as soon as these effects were proposed, some were detected (Udalski et al. 1994; Alcock et al. 1995; 1997).

2. Limb Darkening

While precision microlensing experiments have yielded important upper limits on the planetary companions to bulge lenses (Gaudi et al. 2002), the most robust detection of high-precision effects has been limb darkening (LD). Alcock et al. (1997) first detected this effect when a point lens transited the face of a very large giant-star source MACHO 95-BLG-30. However, all subsequent LD measurements have been made in binary-star events. Although binary events are about 15 times rarer than point-lens events (Alcock et al. 2000; Jaroszynski 2002), their cross section for a caustic crossing is much larger, of the order of the size of the Einstein ring, as opposed to the size of the source, which is typically several hundred times smaller. Furthermore, binary caustic crossings can often be reliably predicted, since once the source enters the caustic region, it must leave. Caustic crossings permit the resolution of the star's surface because they are contours of infinite magnification. Thus, even though microlens sources are extremely small, $100\,{\rm nas} \lesssim \theta_* \lesssim 10\,\mu{\rm as}$, they can still be resolved by taking a series of photometric measurements as the source crosses the caustic. Since typical proper motions are $\mu_{\rm rel} \sim 20\,{\rm km\,s^{-1}\,kpc^{-1}}$, the typical crossing times, $\Delta t = \theta_*/\mu_{\rm rel} \csc\phi$, are of order hours, where ϕ is the angle between the source-lens relative motion and the normal to the caustic. Hence, dozens to hundreds of high-precision measurements can be made. A major challenge is then to extract useful LD information from the resulting light curve.

The first such measurement was carried out by Martin Dominik of the PLANET collaboration (Albrow et al. 1999). The source in MACHO 97-BLG-28 passed over the cusp of a middle K giant. Such events are particularly difficult to model because not much of the caustic region is probed. However, they are especially sensitive to the stellar profile because the cusp magnification pattern is much more concentrated than that of a fold caustic. After more than a year of effort, Dominik succeeded in both determining the lens geometry and measuring two LD parameters in each of two bands, V and I. These profiles were shown to be in good agreement with stellar-model predictions for a middle K star. Moreover, the 2-parameter (linear + square-root) LD models were shown to be substantially closer to the theoretical predictions than the 1-parameter (linear) models.

The next three measurements lacked the favorable geometry of MACHO 97-BLG-41, and so yielded only linear LD measurements. Nevertheless, each marked an important advance in this subject. Afonso et al. (2000) combined data from five microlensing groups to solve the binary caustic crossing event MACHO 98-SMC-1. The main goal of this effort was to measure the $\mu_{\rm rel}$, whose low value proved that the lens was in the SMC itself. The LD measurement (in four different bandpasses) was a by-product, but was significant in part because the source was a metal-poor A star, which exist only outside the Milky Way – for such stars microlensing gives the only hope of obtaining an LD measurement – and in part because at 89 nas, the source is the smallest object that has been resolved by any technique. Albrow et al. (2000) solved the complex event MACHO 97-BLG-41 and in so-doing measured linear LD. Even though the source passed over two disconnected caustic structures, including a cusp crossing, short warning times and bad weather prevented coverage from being sufficient to obtain a more detailed source profile. However, the event was the first for which binary rotation was measured. Moreover, the rotating solution demonstrated that this complex event could be explained without recourse to a circumbinary planet (Bennett et al. 2000).

An important step on the road from detecting LD, to using it to challenge stellar models was achieved by Jin An of the PLANET collaboration in his analysis of OGLE 99-BUL-23, a fold-caustic crossing event (Albrow et al. 2001a). An made two important advances. First, he explored the role of errors in the LD parameters induced by correlations with the uncertainties in the lens geometry instead of calculating the error bars from the linear fit of the light curve to the photometric parameters while the geometric parameters are held fixed. Second, he developed a method to systematically compare the resulting V and I LD parameters (and covariance matrix) with the predictions of stellar models. Unfortunately, the geometry of this event was not particularly favorable, while the data were not of the highest precision, so this confrontation between data and stellar models could not reach the level required to really challenge the models.

3. EROS 2000-BLG-5: A Major Milestone

The spectacular event EROS 2000-BLG-5 (EB-2K-5) was a major turning point in this subject. As described below, it will allow by far the best LD measurement ever. It was the first event for which the lens mass was measured. Finally, high resolution spectra of the event during the crossing have brought the prospects of microlensing to probe the details of stellar atmosphere close to fruition.

The source of EB-2K-5 was a relatively large ($\theta_* \sim 7\,\mu$as) K3 giant and the geometry of the event was favorable. Heads up work by David Bennett of MPS recognized the first caustic crossing only a few hours after it started. This allowed the PLANET collaboration to obtain detailed photometry of this crossing, which in turn enabled Jin An accurately to predict both the time and duration of the unprecedentedly long 4-day second crossing, as well as a third peak due to a "cusp approach" four days later. Intensive round-the-clock observations of these structures by PLANET then set the stage for an astounding modelling effort by An et al. (2002). An & Gould (2001) had shown that triple-peak events of this type are susceptible to parallax measurements. In fact, the unusual length of this event ($t_{\rm E} \sim 100\,\text{days}$) also contributed strongly to the parallax signal, since robust parallax measurements usually require $t_{\rm E} \gtrsim \text{yr}/4$ (e.g. Bennett et al. 2002). Because caustic crossings resolve the source, they automatically yield $\rho_* \equiv \theta_*/\theta_{\rm E}$ as part of the solution. By locating the source relative to the clump on a CMD, it is then possible to determine θ_* from the empirically determined color/surface-brightness relation, and so extract $\theta_{\rm E}$. By combining the $\pi_{\rm E}$ and $\theta_{\rm E}$ measurements, An et al. (2002), were able to measure both the mass $M \sim 0.6\,M_\odot$ and relative parallax, $\pi_{\rm rel} = 0.4\,\text{mas}$, thus showing the lens was an M-star binary in the near disk. In a feat of incredible precision, An showed that after the source exited the caustic, it missed the cusp by $0.1\,\theta_* \sim 0.0005\,\theta_{\rm E}$, i.e., $< 1\,\mu\text{as}$.

An's real-time prediction of the caustic crossing enabled two groups to get spectra with 8-m class telescopes (Castro et al. 2001; Albrow et al. 2001b). On the basis of low-resolution spectra on each of four nights, Albrow et al. (2001b) argued that they had seen a dramatic decline of the Hα equivalent width. If real, this would imply that the K giant source has a chromosphere that is relatively much more important than that of the Sun (Afonso et al. 2001). More conservatively, Castro et al. (2001) showed from high-resolution spectra on two nights, that Hα was weaker on the stellar limb, in agreement with the general predictions of models. Currently, An et al. (2003) are systematically comparing all 240 nm of the two high-resolution spectra in order to probe the detailed differences in the stellar atmosphere of the limb and center of the star.

Finally, because of the excellent photometry, high cadence, and favorable geometry, EB-2K-5 offers a unique opportunity for highly precise measurements of LD. In fact, An et al. (2002) have already measured the 2-parameter I band LD as part of their solution of the event. However, they did not make a detailed analysis of the errors as An did for OGLE 99-BUL-23, nor did they attempt to confront models with their results. Moreover, PLANET obtained good coverage of the event in the V band, and for the first time for any caustic crossing event, in the H band as well. As An showed for the case of OGLE 99-BUL-23, combining several bands can place much more robust constraints on models. Finally, EROS obtained extremely good data for the second caustic crossing, so that by combining the EROS and PLANET data, it should be possible to obtain an incredibly detailed picture of the LD (Fields et al. 2003).

4. Bulge Mass Function

As impressive as An et al.'s (2002) accomplishment was in measuring the mass of EB-2K-5, the events for which this is possible will be rare and atypical. One would like to measure masses for a large and, very importantly, representative sample of bulge events so as to measure the bulge mass function (MF). Of course, MFs of various populations are frequently reported, but these are always derived from luminosity functions and so, by definition, refer only to luminous objects.

Microlensing is sensitive to mass, not luminosity, so this bulge MF would be of all objects, dark and luminous.

The key requirement here is to be able to *routinely* measure $\theta_{\rm E}$ and $\pi_{\rm E}$. Forty years ago, Refsdal (1966) showed that one could routinely measure microlens parallaxes (up to a 2-fold degeneracy) by simultaneously observing events from the ground and a satellite in solar orbit. The Earth and satellite see an event geometry that differs by $\pi_{\rm E}(d_{\rm sat}/{\rm AU})$, where $d_{\rm sat}$ is the distance to the satellite. Since $\pi_{\rm E}$ is typically a few tenths, this difference can be significant. Gould (1995) showed that the degeneracy could be resolved using higher order effects.

Building on the work of Hog, Novikov, & Polnarev (1995), Walker (1995), and Miyamoto & Yoshii (1995), Boden, Shao, & Van Buren (1998) and Paczyński (1998) then showed that the *Space Interferometry Mission (SIM)* could routinely measure $\theta_{\rm E}$. Even though microlensing events are by definition unresolved, so that the images cannot be separately observed, the centroid of the images moves by an amount of order $\theta_{\rm E}$ during the course of the event. Since it is much easier to centroid an image than resolve it, $\theta_{\rm E}$ can in principle be measured by precision centroiding. *SIM*, with its exquisite 4 μas astrometry should therefore be able to measure $\theta_{\rm E}$ with a precision of a few percent. Boden et al. (1998) argued that it should also be possible to measure $\pi_{\rm E}$ from astrometric effects, but in fact this is not practical (Gould & Salim 1999). However, since *SIM* will be in solar orbit, it could carry out photometric parallax measurements of the type advocated by Refsdal (1966), if it could only be equipped with a photometer. In fact, since *SIM* makes its astrometric measurements by *counting photons* as a function of delay time, it automatically performs photometry during its astrometric observations. Gould & Salim (1999) recognized this and showed that only about five hours of *SIM* time are required to make 5% mass measurements for $I = 15$ sources; 1200 hours of *SIM* time have now been awarded for this purpose, so it should be possible to measure 200 masses. At present, one expects that about 20% of bulge events are due to black holes, neutron stars, white dwarfs, and brown dwarfs (Gould 2000b). We have not the least idea which 20% these are because the full-width half maximum (FWHM) of the $t_{\rm E}$ distribution at fixed mass is about a factor 10. Since $M \propto t_{\rm E}^{1/2}$, this implies a FWHM of 100 in mass, roughly the full extent of the MF. With *SIM* we will for the first time be able to probe these dark and dim populations.

5. Masses of Nearby Stars

Refsdal (1964) first suggested that astrometric microlensing could be used to measure the masses of nearby stars by their deflection of the light of more distant sources. The deflection is given by

$$\Delta\theta = 8\,{\rm mas}\,\frac{M}{M_\odot}\left(\frac{\beta}{1''}\right)^{-1}\left(\frac{d}{\rm pc}\right)^{-1}, \tag{2}$$

where β is the lens-source angular separation and d is the distance to the lens (more accurately, $d = {\rm AU}/\pi_{\rm rel}$). No candidate source-lens pairs were identified for more than three decades, for two reasons. First, the available astrometric precision was inadequate to measure this effect, which could plausibly reach an amplitude of 1 mas at most. Second, there was no way to sort out from among the half-billion $V \lesssim 19$ sources in the sky which dozen or so would show significant deflection. Paczyński (1995, 1999) revived this idea, noting that *SIM* now made it practical. In particular, he suggested that the neighborhoods of the future trajectories of high proper motion stars from the Hipparcos catalogue should be checked for candidate sources. Gould (2000a) investigated this possibility in greater detail, and Salim & Gould (2000) carried out such a search and identified about a dozen candidates for which 1% mass measurements could be possible with about 10 hours of *SIM* time. Whether these candidates will in fact prove viable will depend on the precise *SIM* design. The major potential problem is that Hipparcos stars are bright ($V \lesssim 11$). While the *SIM* baseline is 10 m (so that the central fringe is ~ 10 mas), the mirrors are only of order 30 cm, so the point-spread function is several arcsec. This sets the scale of the stop if *SIM* is to recover the majority of source photons. Unfortunately, viable events typically have impact parameters $\beta \lesssim 1''$ (Salim & Gould 2000), which means that it may be very difficult to observe the source in the presence of a bright lens. The obvious solution to this problem is to seek fainter lenses.

To understand why this is difficult, one should first consider the advantages of the Hipparcos stars. First, the trajectories of Hipparcos stars are known extremely well: the errors in their 2010 positions are typically tens of mas and almost always less than 200 mas, well within the range of what is required. The astrometric problem for Hipparcos stars is determining the positions of the sources, whose 1950 positions are known to 250 mas from USNO-A (Monet 1998), but which may have moved in the meantime. Additional ground-based observations are required to determine the source-lens relative positions circa 2000. Second, the distances to Hipparcos stars are generally quite well determined. Since $\Delta\theta \propto d^{-1}$, the amount of *SIM* time required scales as $T \propto d^2$. The distance to a random star in the sky is generally not known to a factor of 10, so without better distance information, it is impossible to assess which candidates are viable. Finally, the accurate distances, colors, and fluxes of Hipparcos stars make it possible to give reasonable estimates of their masses, which also enter the *SIM* time for 1% measurements as $T \propto M^{-2}$.

The best source of high proper motion stars is the *New Luyten Two Tenths* (NLTT) catalogue (Luyten 1979, 1980). Indeed, essentially all of the high proper motion stars in Hipparcos were input directly from NLTT. In comparison to Hipparcos, however, both the astrometry and the photometry of the 59 000 NLTT stars are poor. The 1950 positions are recorded only to $6''$ (even though in the majority of cases, Luyten actually measured the positions accurate to $1''$!). With some effort, it is possible to locate more than half the NLTT stars in USNO-A, but because of the NLTT proper motion errors of $\sim 20\,\mathrm{mas\,yr^{-1}}$, the positions deteriorate by $> 1''$ by 2010. Salim & Gould (2000) were thus able to identify proto-candidates that might plausibly pass near enough to a source in 2010 to be truly viable candidates. However, they had to observe several hundred such proto-candidates to find the genuine candidates. There had to be a better way, and there was!

Samir Salim first recognized that once the approximate 2000 position had been determined from the USNO-A position and NLTT proper motion, it would be straight forward to refine this using 2MASS (Skrutskie et al. 1997). Next he realized that even when NLTT positions were so bad that the USNO-A counterpart could not be located directly, it was still possible to find USNO-A/2MASS (1950/2000) counterpart pairs with separations predicted by NLTT proper motions, and so sift through literally millions of stars to locate the NLTT stars. Eventually Gould & Salim (2003) and Salim & Gould (2003) undertook a massive project to identify the great majority ($\sim 97\%$) of NLTT stars using this and related techniques. As a bonus, they were able to construct a $V - J$ reduced proper motion (RPM) diagram. In contrast to the RPM diagram based on Luyten's original photographic B and R, the $V - J$ RPM clearly separates out white dwarfs, subdwarfs, and main-sequence stars (Salim & Gould 2002). Once the stars are typed in this way, it becomes possible to make photometric distance estimates accurate to half a magnitude, which means that viable candidates can easily be separated from those that would require excessive *SIM* time.

References

Afonso, C. et al. 2000, ApJ, 532, 340

Afonso, C. et al. 2001, A&A, 378, 1014

Albrow, M. D. et al. 1998, ApJ, 509, 687

Albrow, M. D. et al. 1999, ApJ, 512, 1022

Albrow, M. D. et al. 2000, ApJ, 534, 894

Albrow, M. D. et al. 2001a, ApJ, 549, 759

Albrow, M. D. et al. 2001b, ApJ, 550, L173

Alcock, C. et al. 1993, Nature, 365, 621

Alcock, C. et al. 1995, ApJ, 454, L125

Alcock, C. et al. 1997, ApJ, 491, 436

Alcock, C. et al. 2000, ApJ, 541, 270

An, J. H. & Gould, A. 2001, ApJ, 63, L111

An, J. H. et al. 2002, ApJ, 572, 521

An, J. H. et al. 2003, in preparation

Aubourg, E. et al. 1993, Nature, 365, 623

Bennett, D. P. et al., 2000, Nature, 402, 57

Bennett, D. P. et al. 2002, ApJ, submitted (astro-ph/0109467)

Boden, A.F., Shao, M. & Van Buren, D., 1998 ApJ, 502, 538

Bond, I. A. et al. 2002, MNRAS, 333, 71

Castro, S. M., Pogge, R. W., Rich, R. M., DePoy, D. L. & Gould, A. 2001, ApJ 548, L197

Fields, D. et al. 2003, in preparation

Gaudi, B. S. et al. 2002, ApJ, 566, 463

Gould, A. 1992, ApJ, 392, 442

Gould, A. 1994, ApJ, 421, L71

Gould, A. 1995, ApJ, 441, L21

Gould, A. 2000a, ApJ, 532, 936
Gould, A. 2000b, ApJ, 535, 928
Gould, A. & Loeb, A. 1992, ApJ, 396, 104
Gould, A. & Salim, S. 1999, ApJ, 524, 794
Gould, A. & Salim, S. 2003, ApJ, in press
Griest, K. et al. 1991, ApJ, 372, L79
Griest, K. & Hu, W. 1992, ApJ, 397, 362
Han, C., & Gould, A. 1997, ApJ, 480, 196
Han, C., Park, S.-H., Kim, H.-I., & Chang, K. 2000, MNRAS, 316, 665
Heyrovský, D., & Sasselov, D. 2000, ApJ, 529, 69
Hog, E., Novikov, I. D. & Polnare v, A.G. 1995 A&A, 294, 287
Ignace, R. & Hendry, M. A. 1999, A&A, 341, 201
Jaroszynski, M. 2002, Acta Astronomica, 52, 39
Luyten, W. J. 1979, 1980, New Luyten Catalogue of Stars with Proper Motions Larger than Two Tenths of an Arcsecond (Minneapolis: University of Minnesota Press)
Mao, S. & Paczyński, B. 1991, ApJ, 374, L37
Miyamoto, M. & Yoshii, Y. 1995, AJ, 110, 1427
Monet, D. 1998, BAAS, 193, 112003
Nemiroff, R.J. & Wickramasinghe, W. A. D. T. 1994, ApJ, 424, L21
Paczyński, B. 1986, ApJ, 304, 1
Paczyński, B. 1991, ApJ, 371, L63
Paczyński, B. 1995, Acta Astronomica, 45, 345
Paczyński, B. 1998, ApJ, 494, L23
Refsdal, S. 1964, MNRAS, 128, 295
Refsdal, S. 1966, MNRAS, 134, 315
Rhie, S. H., et al., 2000 ApJ, 535, 378
Salim, S. & Gould, A. 2000, ApJ, 539, 241
Salim, S. & Gould, A. 2002, ApJ, 575, L83
Salim, S. & Gould, A. 2003, ApJ, submitted
Skrutskie, M. F. et al. 1997, in The Impact of Large-Scale Near-IR Sky Survey, ed. F. Garzon et al (Kluwer: Dordrecht), p. 187
Udalski, A., Szymanski, M., Kaluzny, J., Kubiak, M., Krzeminski, W., Mateo, M.,Preston, G.W. & Paczynski, B. 1993, Acta Astronomica, 43, 289
Udalski, A., Szymanski, M., Mao, S., di Stefano, R., Kaluzny, J., Kubiak, M., Mateo, M. & Krzeminski, W. 1994, ApJ, 436, L103
Walker, M. A. 1995, ApJ, 453, 37
Witt, H. J. 1995, ApJ, 449, 42
Witt, H. J. & Mao, S. 1994, ApJ, 429, 66

IAU 8th Asia-Pacific Regional Meeting
ASP Conference Series, Vol. 289, 2003
S. Ikeuchi, J. Hearnshaw & T. Hanawa, eds.

Secular Component of Apparent Proper Motion of QSOs Induced by Gravitational Lens of the Galaxy

Kouji Ohnishi[1,2], Mizuhiko Hosokawa[3], & Toshio Fukushima[2]

[1]*Nagano National College of Technology, Nagano, 381-8550, Japan*

[2]*National Astronomical Observatory, Mitaka, Tokyo 181-8588, Japan*

[3]*Communications Research Laboratory, Tokyo 184-8795, Japan*

Abstract. The observed position of QSOs may vary due to the gravitational deflection by foreground objects. This effect is eminent in the direction of the Galactic Center. To measure the distance to the Galactic Center with the relative error of a few %, in addition to the astrometric microlensing by individual stars in the Galaxy, we have to consider the macro lens effect, the collective gravitational deflection by the core and the bulge of the Galaxy. This effect is important because it has a secular component. Its magnitude is 0.4 μas/yr in the case of internal motion of QSOs by the core and reaches 0.6 μas/yr in the case of the collective motion produced by both the core and the bulge. The measurement of these effects will provide us valuable information on the density and mass function of the Galactic Center.

1. Introduction

The *macro lens* is the gravitational deflection by a group of stars. Of course, the macro lens by distant galaxies is well known and studied intensively (e.g. Schneider et al. 1992). However, that caused by the Galaxy, which is eminent in the direction to the Galactic Center, is not examined so well.

Recently, the motion of Sgr A* has been measured by referring to QSOs in the direction of Galactic Center. The observed magnitude is 6.0 mas/yr (Reid et al. 1999; Backer and Sramek 1999). This is consistent with the expected value computed from the galactic rotation. More accurate observation makes it possible to measure the annual parallax of Sgr A*, that is the distance to the Galactic Center. The QSOs referred in the observations, W56, W109, and GC441, are close to Sgr A* on the celestial sphere; with about 0.7 degree separation. This means that we observe these QSOs through the core (i.e. the nuclear bulge), the bulge, and the disk of the Galaxy where stars are very densely distributed. In such case, the gravitational lens should be taken into account (Hosokawa, Ohnishi, & Fukusima 1997; Hosokawa et al. 2002).

Table 1 summarizes various effects in the apparent motion of Sgr A* relative to QSOs. As for Sgr A* itself, the kinematic effect of galactic rotation is most dominant. This looks like a constant motion. The next is the annual parallax, which is periodic of course. As for QSOs, the gravitational effects of individual

Table 1. Various effects in the apparent motion

	Sgr A*		QSOs	
Nature	Secular	Periodic	Secular	Random
Magnitude	6 mas/yr	250μas/yr	0.6μas/yr	10μas/yr
Cause	Galactic Rotation	Annual Parallax	Macro Lens	Microlensing

stars, namely the astrometric microlensing, is large but random. While the macro lens is secular and less than 1μas/yr.

In determining the distance to Sgr A* with the accuracy of a few %, we have to consider the macro lens to separate the fluctuation due to the microlensing from the proper motion and the annual parallax. On the other hand, the detection and separation of the macro lens effect will provide the information on the density and mass function of the Galactic Center. As for the astrometric microlensing effect toward the Galactic Center, we have already discussed (Hosokawa et al. 2002). In this paper, we will report the macro lens effect toward the Galactic Center.

2. Model

We assumed that the mass distribution near the Galactic Center is almost spherical symmetric to the Galactic Center. Thus the column density $\Sigma(L)$ is axis-symmetric to the axis of galactic rotation, where L is the impact parameter of QSO's light ray and the Galactic Center. The line integration with the distance from the observer, D is expressed as $\Sigma(L) = \int \rho(L)\mathrm{d}D$, where ρ is the mass density. The total column mass of the stars in a cylinder of the radius of impact parameter L, $m(L)$ is expressed as

$$m(L) = \int_0^L \Sigma(L')2\pi L'\mathrm{d}L' \tag{1}$$

Then, the total gravitational deflection of the images of the referenced QSOs by the stars in the Galaxy is approximated as $\theta = m(L)/L$. Though it amounts to $1''$ if $L < 100$pc, this deflection itself is not observable. Rather the time variation due to the galactic rotation is measurable;

$$\mathrm{d}\theta = \left(2\pi\Sigma(L) - \frac{1}{L^2}\int_0^L \Sigma(L')2\pi L'\mathrm{d}L'\right)\mathrm{d}L\,. \tag{2}$$

For example, let us evaluate the gravitational deflection due to a thick sheet of a constant column density, Σ_0. In this case, the column mass within the cylinder is $m(L) = \pi L^2\Sigma_0$. Then the deflection angle becomes $\theta = \pi L\Sigma_0$. Thus an apparent proper motion $\mu = \mathrm{d}\theta/\mathrm{d}t$ is constant of being the order of sub-μas/yr. In this manner, we estimate the magnitude of the macro lens effect by the core and the bulge. The disk contribution is negligible due to the symmetrical distribution with respect to the axis of galactic rotation.

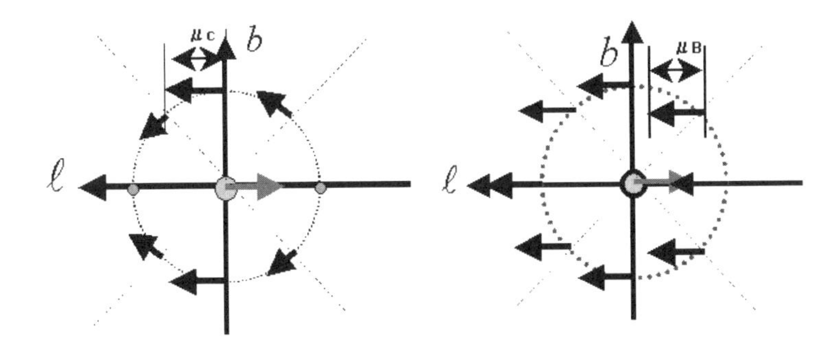

Figure 1. Effect of core motion (left) and bulge motion (right)

As for the density distribution near the galactic center, we adopt the model of Alexander & Sternberg (1999), which gives the mass density distribution of the core and the bulge as

$$\rho_{\rm core}(r) = \frac{\rho_{\rm c}}{1+3(r/r_{\rm c})^2}, \qquad \rho_{\rm bulge}(r) = \rho_{\rm b} K_0(\frac{r}{r_{\rm b}}), \tag{3}$$

where $K_0(x)$ is a modified Bessel function. The numerical values of the parameters are $\rho_{\rm c} = 4 \times 10^6 M_\odot {\rm pc}^{-3}$, $r_{\rm c} = 0.38{\rm pc}$ (Genzel et al. 1996), $\rho_{\rm b} = 3.53 M_\odot {\rm pc}^{-3}$, and $r_{\rm b} = 667$ pc (Kent 1992).

3. Estimates

Hereafter we will discuss the effects of the core and the bulge separately. Before going further, we remark that the impact parameter L is much smaller than the scale length of bulge $r_{\rm b}$. Then we treat the bulge as a thick sheet of a constant column density. While the size of core $r_{\rm c}$ is much smaller than L. Therefore we have to consider the radial density distribution carefully.

First, the apparent proper motion due to the core becomes

$$(\mu_\ell, \mu_b) = (\mu_{\rm core}(1 - \cos 2\varphi) \ , -\mu_{\rm core} \ \sin 2\varphi) \ , \tag{4}$$

where μ_ℓ and μ_b is the proper motion in galactic longitude and in galactic latitude, respectively, and

$$\mu_{\rm core} = 0.2\mu{\rm as/yr} \left(\frac{L}{100{\rm pc}}\right) \left(\frac{\rho_{\rm c}}{4 \times 10^6 m_\odot/{\rm pc}^3}\right) \left(\frac{r_{\rm c}}{0.38{\rm pc}}\right)^2 \left(\frac{v}{220{\rm km/s}}\right). \tag{5}$$

The angle φ is the position angle measured from the galactic plane. Note that the direction of apparent shift is opposite to the direction of the motion of the core relative to the observer. The shift reduces to zero when the QSO is on the galactic plane. The maximum is around 0.4 μas/yr and is reached on the galactic meridian. On the other hand, the effect of the bulge is constant everywhere;

$$(\mu_\ell, \mu_b) = (\mu_{\rm bulge}, \ 0) \ ; \ \mu_{\rm bulge} = 0.2\mu{\rm as/yr} \left(\frac{\Sigma(100{\rm pc})}{6 \times 10^6 m_\odot/{\rm pc}^2}\right) \left(\frac{v}{220{\rm km/s}}\right). \tag{6}$$

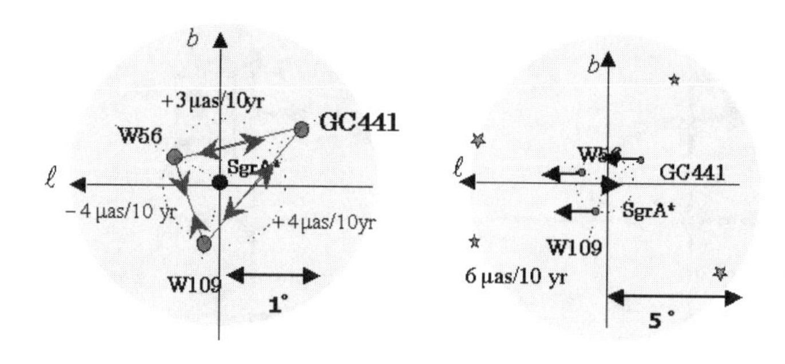

Figure 2. Internal (left) and collective (left) motions of QSOs

This time, the direction of apparent shift is opposite to the direction of the motion of the bulge. Note that μ_{bulge} and μ_{core} are almost the same.

4. Discussion

Combining the contributions of the core and the bulge of the Galaxy, we obtained the internal motion of the apparent places of QSOs closed to the direction of the Galactic Center as illustrated Fig.2. Some baselines contract and some others expand. This depends on their locations in a complicated manner. The magnitudes of contraction/expansion are 0.3 to 0.4 μas/yr. They depend on the parameters of the core, ρ_{c} and r_{c}, namely the density and the manner of concentration. Therefore, by measuring the magnitudes, we will obtain the information on the core.

On the other hand, the collective motion of the apparent places of QSOs is in the direction ℓ increases. The magnitude is about 0.6μas/yr. This will be detected by measuring the relative motion of QSOs near Sgr A* to those far away from the galactic center. Apart from the magnitude, the most important nature of this effect is being secular. This enhances the possibility of its detection when we observe continuously in the long term.

References

Alexander, T., Sternberg, A. 1999, ApJ, 520, 137

Backer, D. C., & Sramek, R. A. 1999, ApJ, 524, 805

Genzel, R., Thatte, N., Krabbe, A., Kroker, H. & Tacconi-Garman, L. E. 1996, ApJ, 472,153

Hosokawa, M., Ohnishi, K. & Fukushima, T. 1997, AJ, 114, 1508

Hosokawa, M., Jauncey, D., Reynolds, J., Tzioumis, A., Ohnishi, K., & Fukushima, T. 2002, ApJL, in press

Kent, S. 1992, ApJ, 387,181

Reid, M. J., Readhead, A. C. S., Vermeulen, R. C., & Treuhaft, R.N. 1999, ApJ, 524, 816

Schneider, P., Ehlers, J., Falco, E. E. 1992, Gravitational Lenses (Springer, Berlin)

IAU 8th Asia-Pacific Regional Meeting
ASP Conference Series, Vol. 289, 2003
S. Ikeuchi, J. Hearnshaw & T. Hanawa, eds.

Chandra Spectroscopy and Mass Estimation of the Lensing Cluster of Galaxies CL0024+17

Naomi Ota

Tokyo Metropolitan University, 1-1 Minami-osawa, Hachiouji, Tokyo 192-0397, Japan

Makoto Hattori, Etienne Pointecouteau

Tohoku University, Aoba Aramaki, Sendai 980-8578, Japan

Kazuhisa Mitsuda

ISAS, 3-1-1 Yoshinodai, Sagamihara, Kanagawa 229-8510, Japan

Abstract. We present the X-ray analysis and the mass estimation of the lensing cluster of galaxies CL0024+17 with *Chandra*. We found that the temperature profile is consistent with being isothermal and the average X-ray temperature is $4.47^{+0.83}_{-0.54}$ keV. The X-ray surface brightness profile is represented by the sum of emissions associated with the central three bright elliptical galaxies and the emission from intracluster medium (ICM), which can be well described by a spherical β-model. Assuming the ICM to be in hydrostatic equilibrium, we estimated the X-ray mass and found it is significantly smaller than the strong lensing mass by a factor of 3.

1. Introduction

CL0024+17 is one of the most extensively studied lensing clusters of galaxies, located at $z = 0.395$. Since the discovery of the multi-lensed arc system, several authors modelled the matter distributions in the cluster. Tyson, Kochanski, & Dell'Antonio (1998) constructed a very detailed mass map and suggested that the dark matter profile has a soft core. Broadhurst et al. (2000) measured the arc redshift to be 1.675 and also built a lens model in a simplified manner.

On the other hand, the X-ray emitting gas is an excellent tracer of the dark matter potential. Soucail et al. (2000) performed a combined analysis of the *ROSAT* and *ASCA* data and estimated the cluster mass within the arc radius (hereinafter the X-ray mass). They found that there is a factor of ~ 3 discrepancy between the X-ray mass and the strong lensing mass (Tyson et al. 1998; Broadhurst et al. 2000). Because the *ROSAT* HRI image suggested the elongated gas distribution, they considered that the discrepancy may be caused by the irregular mass distribution.

However, there were still large measurement uncertainties in both the X-ray temperature and the image morphology. It was mainly because of the heavy contamination from the bright Seyfert galaxy. Thus for the cluster mass estimation the temperature determination is crucial. In this paper, we report on the accurate measurements of the temperature and the morphology with *Chandra*, from which we discuss whether there is an inevitable mass discrepancy between the X-ray and the strong lensing. We use $H_0 = 50$ km/s/Mpc and $\Omega_0 = 1$. At $z = 0.395$, $1' = 383\, h_{50}^{-1}$ kpc.

2. Observation

We observed CL0024+17 with the *Chandra* ACIS-S detector on 2000 September 20. The net exposure time was 37 121 s. In Figure 1a, we show the ACIS-S3 image. The strongest X-ray peak is at (00:26:36.0, +17:09:45.9) (J2000) and the extended emission is detected out to $\sim 2'$ in radius. The point sources detected in the field were removed in the following analysis.

3. Spectral Analysis

We extracted the cluster spectrum from a circular region of $r = 1'.5$, centered at the X-ray peak (the dashed circle in Figure 1a). The background was estimated from the $2'.5 < r < 3'.2$ ring region. We fitted the spectrum to the MEKAL thin-thermal plasma model with the Galactic absorption (Figure 2a) and determined the temperature to be $kT = 4.47^{+0.83}_{-0.54}$ keV (90% error). This is consistent with our previous result with *ASCA* (Soucail et al. 2000; Ota & Mitsuda 2002). We detected the strong redshifted Fe-K line from the cluster for the first time. The iron abundance is $0.76^{+0.37}_{-0.31}$ solar.

In order to investigate the radial temperature profile, we accumulated spectra from four ring regions with various radii and fitted them with the MEKAL model. The radius ranges were chosen so that the each spectrum contains more than 400 photons. We found that there is not any meaningful temperature variation against radius (Figure 2b); the gas is consistent with being isothermal.

4. Image Analysis

4.1. X-ray Surface Brightness and Galaxy Distribution

Though the original ACIS CCD has a pixel size of $0''.5$, we rebinned the image by a factor of four. We restricted the energy range to 0.5 – 5 keV in the image analysis. We find that there is a second X-ray peak at (00:26:35.1,+17:09:38.0) (J2000). From comparison with the galaxy catalogue by Czoske et al. (2001), we recognized that the three central bright elliptical galaxies are located at the positions consistent with the fist and second X-ray peaks (G1 and G2, hereinafter). Note that G1 contains two of the three elliptical galaxies (Figure 1a).

4.2. 2-D Surface Brightness Distribution

In order to determine the X-ray emission profile of the ICM, we fitted the 2-dimensional surface brightness distribution with a model consisting of three β profiles which we consider to represent emissions from two elliptical-galaxy components and ICM component; $S(r) = \Sigma_{i=1}^{3} S_i(1+(r/r_{c,i}))^{-3\beta_i+1/2}$. We fitted the image of $3'.3 \times 3'.3$ region with the maximum-likelihood method. The center positions of the two elliptical-galaxy components were fixed at the G1 and G2 peaks, respectively, while for the third component, which we consider describes the ICM emission, the position was allowed to vary. The results of the fits are shown in Table 1 and Figure 1b. In order to check the goodness of the fit, we rebinned the image into two single dimensional profiles of two perpendicular directions and calculated the χ^2 values between the model and data profiles to find they are enough small ($\chi^2 < 112$ for 99 degrees of freedom). The best-fit cluster center position is $80\,h_{50}^{-1}$ kpc away from the G1 peak. We consider that the emission of G1 and G2 can be attributed to the elliptical galaxies because of the small luminosities. Furthermore, we tested the significance of the ellipticity of the cluster image and found it is not significant ($\epsilon < 0.2$).

Table 1. Results of the 2-D image fitting with the three β-models

Model component	Center position RA, dec. in J2000	β	r_c h_{50}^{-1} kpc	$L_{\rm X,bol}$ erg/s
G1	00:26:36.0,+17:09:45.9 (F)	1 (F)	52^{+11}_{-9}	5.5×10^{43}
G2	00:26:35.1,+17:09:38.0 (F)	1 (F)	10 (F)	3×10^{42}
Cluster	00:26:35.6,+17:09:35.2 †	$0.71^{+0.07}_{-0.06}$	210^{+33}_{-30}	4.5×10^{44}

(F) Fixed parameters. † The 90% errors are $\pm 1''.3$ for RA and $\pm 1''.5$ for dec.

5. Mass Estimation and Comparison

From the spectral and spatial analysis mentioned above, we found that the gas is isothermal and can be described with the spherical β-model. Assuming the gas is hydrostatic, we obtained the projected X-ray mass within the arc radius to be $M_{X,\beta}(< r_{\rm arc} = 220\,{\rm kpc}) = 0.85^{+0.12}_{-0.09} \times 10^{14} h_{50}^{-1}\, M_{\odot}$. On the other hand, the strong lensing mass was estimated to be $M_{\rm lens}(< r_{\rm arc}) = (3.117 \pm 0.004) \times 10^{14} h_{50}^{-1}\, M_{\odot}$ by Tyson et al. (1998). Therefore the discrepancy of a factor of 3 is evident. This is consistent with our previous result (Soucail et al. 2000). On the other hand, the X-ray surface brightness profile of the NFW potential is similar to that of the β-model and can be converted from the β-model parameters through the relations of $r_s = r_c/0.22$ and $B = 15\beta$ (Makino, Sasaki, & Suto 1998). We thus derived the gas profile for the NFW case based on the results of β-model fitting and estimated the X-ray mass to be $M_{\rm X,NFW}(< r_{\rm arc}) = 0.75^{+0.11}_{-0.05} \times 10^{14} h_{50}^{-1}\, M_{\odot}$. Thus the discrepancy still remains even in this case.

Since the gas is isothermal and spherical, we consider that the gas is relaxed in the cluster potential and the hydrostatic equilibrium is a good approximation in the X-ray mass estimation. This suggests that the lens mass is significantly overestimated. Czoske et al. (2001, 2002) measured the redshift distribution of galaxies in the direction of CL0024+17 and revealed the presence of foreground and background groups of galaxies. This may possibly enhance the strong lensing mass. Thus updated lens modelling of the cluster is urged.

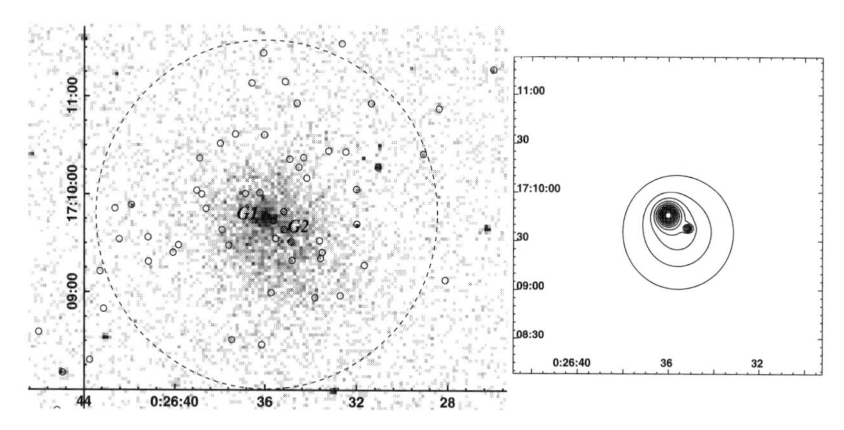

Figure 1. (a) *Chandra* ACIS-S3 image of CL0024+17 in the 0.5 – 5 keV. The first and second X-ray peaks are labelled as G1 and G2. The small circles are the positions of the galaxies with $0.38 < z < 0.41$(Czoske et al. 2001). (b) Contours of the best-fit 2D image of the three β-models.

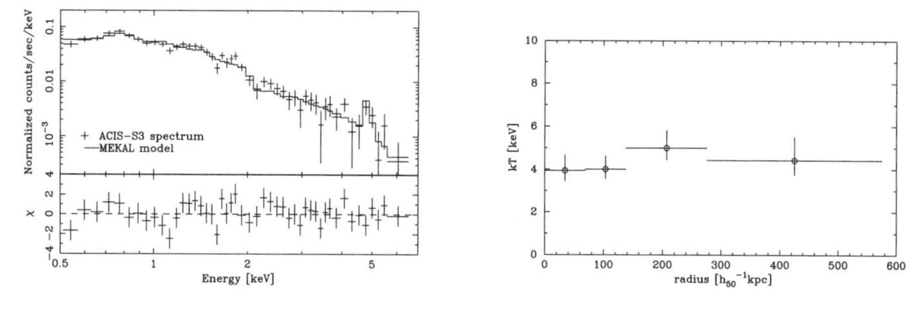

Figure 2. (a) *Chandra* average spectrum of CL0024+17 fitted with the MEKAL model and (b) radial temperature profile.

References

Tyson, J. A., Kochanski, G. P., & Dell'Antonio, I. P. 1998, ApJ, 498, L107

Broadhurst, T., Huang, X., Frye, B., & Ellis, R. 2000, ApJ, 534, L15

Soucail, G., Ota, N., Böhringer, H., Czoske, O., Hattori, M., & Mellier, Y. 2000, A&A, 355, 433

Ota, N.,& Mitsuda, K. 2002, ApJ, 567, L23

Makino, N., Sasaki, S. & Suto, Y. 1998, ApJ, 497, 555

Czoske, O., Kneib, J.-P., Soucail, G., Bridges, T.J., Mellier, Y. & Cuillandre, J.-C. 2001, A&A, 372, 391

Czoske, O., Moore, B., Kneib, J.-P. & Soucail, G. 2002, A&A, 386, 31

IAU 8th Asia-Pacific Regional Meeting
ASP Conference Series, Vol. 289, 2003
S. Ikeuchi, J. Hearnshaw & T. Hanawa, eds.

Towards Direct Detection of Substructure around Galaxies – Quasar Mesolensing

Atsunori Yonehara, Masayuki Umemura

Center for Computational Physics, University of Tsukuba, Tennoudai 1-1-1, Tsukuba, Ibaraki, 305-8577, Japan

Hajime Susa

Institute of Theoretical Physics, Rikkyo University, Nishi-Ikebukuro 3-34-1, Toshima-ku, Tokyo, 171-8501, Japan

Abstract. We discuss about detectability of substructure around galaxies by utilizing a quasar which is gravitationally lensed by an intervening galaxy. In this situation, background quasars have already shown multiple images. If a substructure in the lens galaxy is superposed on one of the images, additional gravitational lensing by the substructure will occur. The typical image separation due to the substructures with mass $10^8 M_\odot$ is on the order of micro-arcseconds. Even if we cannot resolve these images, the total flux variations of these images are expected to be observed, both as its own flux variation plus the "echo"-like variation due to time delay between the images. The time delay of this "echo"-like variation is estimated to be several tens of minutes.

1. Introduction

The cold dark matter scenario for structure formation meets a serious crisis; overproduction of substructure around the Galaxy in N-body simulations compared with observation (Klypin et al. 1999, Moore et al. 1999). If the scenario is correct, a large fraction of such substructures should be invisible or too faint to detect. Then, how can we observationally probe such substructures? One of the most powerful approaches can be realized by gravitational lensing, since the gravitational lens effect of objects is determined by mass, whereas brightness of the matter plays no role.

Recently, Chiba (2002), Dalal & Kochanek (2002), and Metcalf & Madau (2001) have investigated a nice idea for this problem in the weak lensing regime. However, weak effects due to gravitational lensing are sometimes confusing, because a flux anomaly in one of the multiple quasar images can be explained by other reasons, e.g., quasar microlensing with a long time scale. Therefore, in this paper, we will discuss another possibility to detect the substructures via strong lensing. This should present more direct evidence for the existence of large numbers of substructures around galaxies.

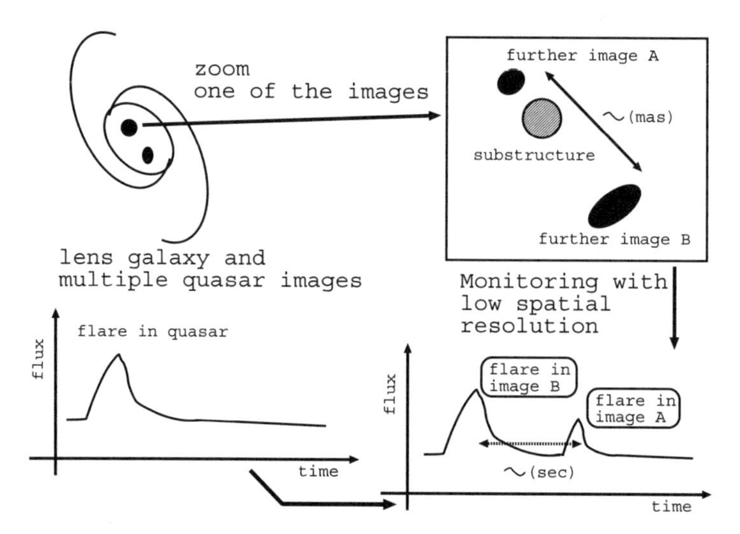

Figure 1. Schematic picture of gravitational lensing due to substructure around the lens galaxy.

2. Basic Idea

The best target for our purpose should be gravitationally lensed quasars with multiple images. The reason is that the apparent Einstein ring radius, the effective size of a lens, is larger in a distant lens system and there is a larger fraction of space in front of the source, so one of the multiple quasar images may be covered with the Einstein ring of the substructure.

Here, we consider the situation that one of the substructures around the lens galaxy is located beside one of the multiple quasar images. In such a case, additional gravitational lensing effects appear in the image, i.e., medium-scale quasar lensing will occur. This is "quasar mesolensing". Expected phenomena due to such gravitational lensing are schematically presented in Figure 1, and simple estimations compared with other quasar lensing effects are presented in Table 1.

Table 1. Typical values of gravitational lens effects in quasars for three mass scales of the lens: galaxies, substructures, and stellar objects.

Name	Lens mass	Image separation	Time delay
Macrolensing	$\sim 10^{12} M_\odot$	~ 1 arcsec	~ 1 month
Mesolensing	$\gtrsim 10^{6} M_\odot$	$\gtrsim 1$ mas	$\gtrsim 1$ s
Microlensing	$\sim 1 M_\odot$	~ 1 μas	~ 1 μs

The most popular effect of gravitational lensing may be magnification. Unfortunately, magnification does not reflect the mass scale of the lens, and it may be difficult to extract the information about the lens mass from magnification. Thus, proper effects for our purpose can be the image splitting and the time delay.

One of the multiple quasar images should have more than two images caused by gravitational lens effects of the substructure in the vicinity of the image. The separation between such images is comparable to the Einstein ring radius of the lens object, i.e., the substructure with $\gtrsim 10^6 M_\odot$, and if we use observational instruments with high spatial resolution, we will be able to find multiple images with $\gtrsim 1$ mas separation in one of the multiple images.

Even if we cannot directly resolve such multiple images, monitoring observations with short intervals enable us to find any evidence for numerous substructures around the lens galaxy. In the case where the substructure is superposed on a quasar image and gravitational lensing occurs, multiple images are not observable but exist within the image. Additionally, a quasar as the source of the lens system has intrinsic variabilities or flares due to some physical instabilities in the accretion disk of the quasar (Kawaguchi et al. 2000). Thus, when we monitor such a quasar image, consequently, we will hunt an "echo"-like variation that comes from the relatively trailing image, and the main variation comes from the relatively leading image. Such an "echo"-like event occurs recurrently, and we can easily discriminate the substructure-origin flux variation from quasar microlensing, which is an accidental event.

3. Estimations

Following a rough estimation and presentation of the basic idea, we perform a more realistic estimation for the expected image separations and the time delays between multiple images in one of the multiple quasar images.

We do not know the mass or density profile of the substructures, and we assume two types of lens model, a point lens and a singular isothermal sphere (SIS) lens as the substructures. Furthermore, substructure lens(es) in a lens galaxy should be affected by a lens galaxy itself. Thus, we include external convergence (κ) and shear (γ) to take this effect into account. κ and γ values in a lens galaxy beside quasar images are of the order of unity (e.g., Schmidt, Webster & Lewis 1998), and we set $\kappa = \gamma = 0.4$. Therefore, we adopt the "Chang & Refsdal lens" model (Chang & Refsdal 1984) and approach to the estimations.

In this lens model, the number of images is sometimes larger than two, and we only pick-up the two brightest images. Moreover, if the second brightest image is too faint compared with the brightest one, the detection of the image or echo of the second brightest image may be practically difficult, and we exclude the case in which the flux ratio between these two images is smaller than 0.1. By using the lens models, assumptions and limitations described above, we performed Monte-Carlo simulations to calculate the probability distributions for image separations and the time delay. Here, we set the lens and the source redshifts as 1.0 and 3.0, respectively.

The results for a $10^8 M_\odot$ substructure or corresponding velocity dispersion for the SIS lens model (Chiba 2002) are shown in Figure 2. Probable values (more than 90%) for a more realistic SIS lens model are $\gtrsim 100$ s for the time delay and $\gtrsim$ 1 mas, and somewhat smaller than that for the point lens model. Here, we only consider a single mass lens, though N-body simulations show more massive substructures. If we include a mass spectrum of substructures, the expected values can be larger than these values in actual cases. The lensing probability may be estimated roughly by the fraction of the total size of the Einstein ring radius of the substructures to the size of the host galaxy, and it is of the order of 1%. The event probability is not so frequent, but in this regime the "Chang & Refsdal lens" treatment is guaranteed and our estimation should be correct.

The direct detection of the substructure is not an easy task, but it will be realized in the near future using our proposed method.

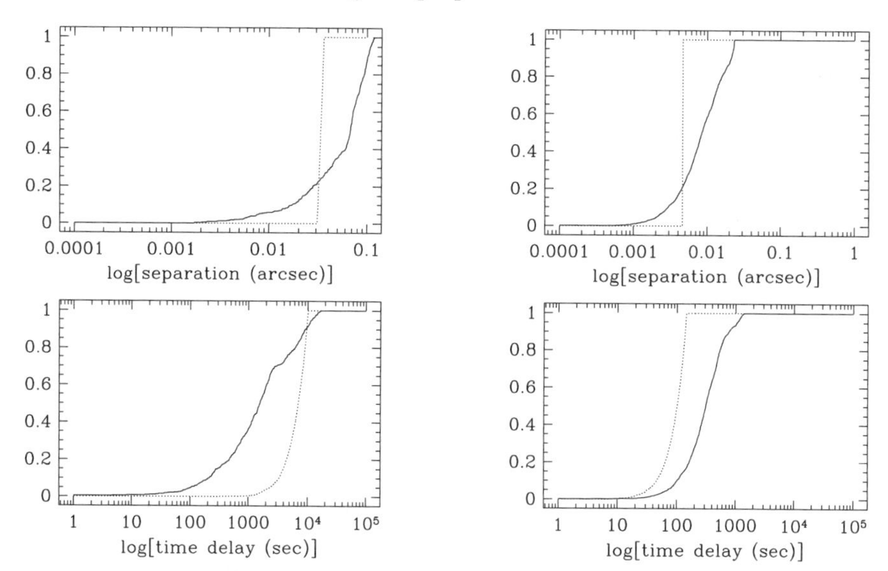

Figure 2. Cumulative probability distributions of the expected image separation (upper) and the time delay (lower) in the case of a point lens (left) and an SIS lens (right) are presented with solid lines. The case without external κ and γ is also plotted with dotted lines.

References

Chang, K. & Refsdal, S. 1984, A&A, 132, 168

Chiba, M. 2002, ApJ, 565, 17

Dalal, N. & Kochanek, C. S. 2002, ApJ, 572, 25

Kawaguchi, T. et al. 2000, PASJ, 52, L1

Klypin, A. et al. 1999, ApJ, 516, 530

Metcalf, R. B. & Madau, P. 2001, ApJ, 563, 9

Moore, B. et al. 1999, ApJ, 524, L19

Schmidt, R., Webster, R. L. & Lewis G. F. 1998, MNRAS, 295, 488

IAU 8th Asia-Pacific Regional Meeting
ASP Conference Series, Vol. 289, 2003
S. Ikeuchi, J. Hearnshaw & T. Hanawa, eds.

An Analytical Model for the Distribution of Image Separations in Gravitational Lensing

Premana W. Premadi

Department of Astronomy, Institut Teknologi Bandung, Indonesia

Hugo Martel

Department of Astronomy, University of Texas at Austin, Austin, TX, USA

Abstract. Using an analytical model, we compute the distribution of image separations resulting from gravitational lensing of distant sources, for 3 *COBE*-normalized CDM models. The model assumes that multiple imaging results only from lensing by individual galaxies, which are modelled as nonsingular isothermal spheres. The contribution of the background matter to the lensing is neglected. We found that, while the number of multiple-imaged sources can put strong constraints on the cosmological parameters, the distribution of image separations does not constrain the cosmological models in any significant way.

1. Introduction

The distribution of image separations resulting from the gravitational lensing of distant sources can be used to study the properties of the lenses and the underlying cosmological background. In Premadi et al. (2001, hereinafter PMMF), we have studied the distribution of image separations in CDM universes, and their dependence upon the cosmological parameters, using a multiple lens-plane algorithm. To gain more insight into these results, we have designed a simple analytical model. We consider three different CDM models: an Einstein-de Sitter (EdS) model ($\Omega_0 = 1$, $\lambda_0 = 0$), an open model ($\Omega_0 = 0.2$, $\lambda_0 = 0$), and a flat, Λ model ($\Omega_0 = 0.2$, $\lambda_0 = 0.8$), and combine the galaxy distribution of PMMF with our analytical model to compute the distributions of image separations.

2. The Effect of the Background Matter

In the analytical model, gravitational lensing is caused by individual galaxies, with negligible contribution from the background matter. To check the validity of this approximation in the context of a CDM universe, we compare the results of ray tracing experiments with and without background matter for the three cosmological models considered. The left panels of Figure 1 show the distributions of image separations. The only noticeable effect of the background matter is the presence of a high-separation tail. This is consistent with claims that the effect of the background matter on image separation is of order 20% or

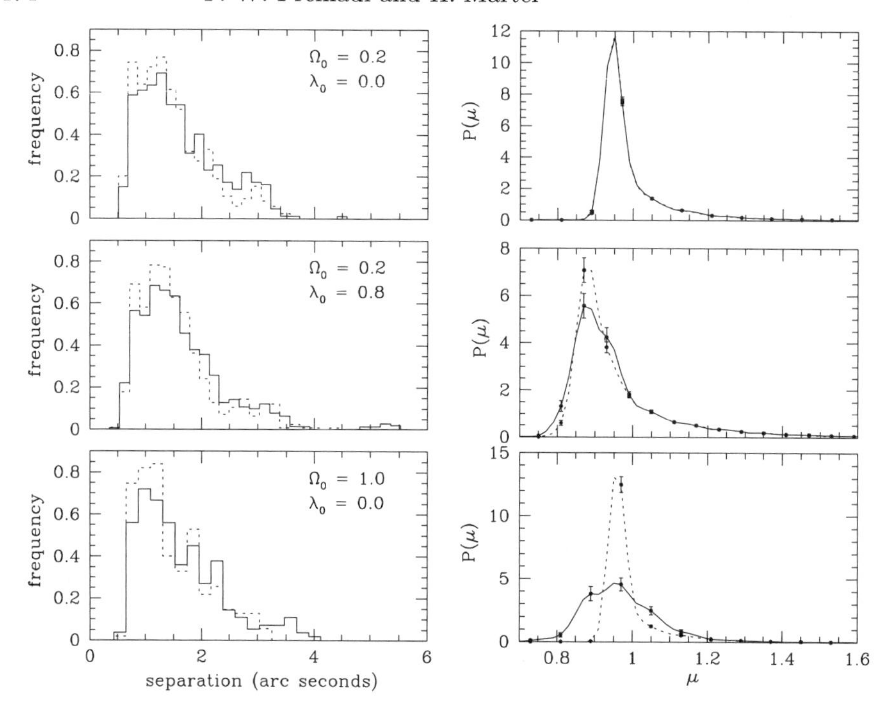

Figure 1. Distributions of image separations (left), and magnifications (right). Solid and dotted curves correspond to ray-tracing experiments with and without background matter, respectively.

less (Bernstein & Fischer 1999; Romanowsky & Kochanek 1999). Notice that the effect is similar for all three cosmological models. Overall, the effect of the background matter is always small, and we are justified in ignoring it in our analytical model. The right panels show the distributions of magnifications. While the effect of the background matter on the distribution of image separations is small and independent of the model, the effect on the magnification distribution can be very important, and strongly depends on thc cosmological model.

3. Analytical Model of the Distribution of Image Separations

Our analytical model is described in detail in Martel, Premadi, & Matzner (2002). It is based on the following assumptions: (1) Lensing is entirely caused by galaxies and each galaxy acts alone. (2) Galaxies are modelled as nonsingular isothermal spheres. Each galaxy is parameterized by a core radius $r_{\rm c}$ and a rotation velocity v, which are functions of the the luminosity. (3) We neglect any spatial correlation between the sources and the lens. The probability that a particular galaxy will produce multiple images is then proportional to its angular cross section for multiple imaging. (4) We impose limits on the smallest image separation that allows each image to be resolved by introduc-

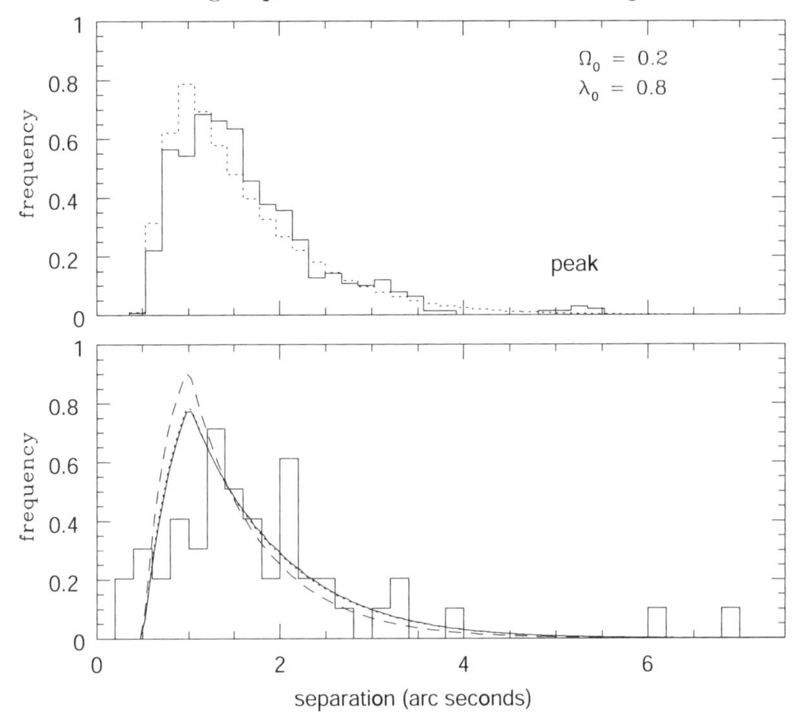

Figure 2. Distributions of angular separation. Top: predictions of the analytical model (dotted) compared with results of ray-tracing experiments (solid), for a particular cosmological model. Bottom: predictions of the analytical model for the Einstein-de Sitter model (solid), the open model (dashed) and the Λ model (dotted). The histogram shows the distribution from the CASTLES database.

ing a probability of "resolvability" which depends on the source size and image separation. With these assumptions, the problem is reduced to studying lensing by isolated, nonsingular isothermal spheres. We define x_c and y to be the core radius of the galaxy and the position of the source, respectively, in units of $\xi_0 \equiv 2\pi(v/c)^2 D_L D_{LS}/D_S$, where D_L, D_{LS}, and D_S are the angular diameter distance between observer and lens, lens and source, and observer and source, respectively. We then define, for $x_c < 1$, a critical radius $y_r = (1 - x_c^{2/3})^{3/2}$. The nonsingular isothermal sphere has the following properties: (i) If $x_c \geq 1$, the source has only one image; (ii) If $x_c < 1$ the source has one image if $y \geq y_r$, and three images if $y < y_r$. Hence, each lens which satisfies the condition $x_c < 1$ has an angular cross section for multiple imaging $\sigma_{\mathrm{m.i.}}$ given by $\sigma_{\mathrm{m.i.}} = \pi(y_r\xi_0/D_S)^2$ (In practice, we increase this cross-section to account for the extended sources). If a galaxy modelled as a nonsingular isothermal sphere produces multiple images, the angular separation s between the two outermost images is always close to its maximum possible value $s_{\max} = (2\xi_0/D_L)(1 - x_c^2)^{1/2}$ (Hinshaw & Krauss 1987). By combining these equations with the distributions of galaxies taken from PMMF, we can directly compute the distribution of images separations.

4. Distributions of Image Separations

The top panel of Figure 2 shows, for the Λ model, a comparison of the distributions of image separations predicted by the analytical model and obtained with ray-tracing experiments. The agreement is excellent. The ray-tracing experiments show a secondary peak at large separations which is likely caused by the background matter, whose effect is not included in the analytical model. The bottom panel shows the distributions predicted by the analytical model, for the EdS, open, and Λ models. The distributions are quite similar for the EdS and open models and totally indistinguishable for the EdS and Λ models. For comparison, we also show the distribution for 49 known gravitational lenses from the CASTLES database (Kochanek et al. 1998). The overall agreement between the models and the observations is quite good. The main differences are that the peaks of the distributions are located at $s = 1''$ for the models, and $s = 1.3''$ for the observations, and two known lenses (RX J0921+4528 and Q0957+561) have separations $s > 6''$, in conflict with the analytical model.

5. Conclusion

Our main results are the following: (1) The presence of the background matter tends to increase the image separations produced by lensing galaxies. (2) Simulations with galaxies and background matter occasionally produce a secondary peak in the distribution of image separations at large separations. (3) The effect of the background matter on the magnification distribution strongly depends on the adopted cosmology. (4) The analytical model shows that the distributions of image separations are indistinguishable for flat, non-zero Λ models with various values of Ω_0. For models without Λ, the distributions of image separations depend weakly upon Ω_0.

Acknowledgments. We thank Richard Matzner and Karl Gerhardt (University of Texas), and Rachel Webster (Melbourne University) for fruitful discussions. PWP thanks the APRM2002 organizing committee for travel support. HM acknowledges the support of NASA grants NAG5-10825 and NAG5-10826.

References

Bernstein, G. & Fischer, P. 1999, AJ, 118, 14

Hinshaw, G. & Krauss, L. M. 1987, ApJ, 320, 468

Kochanek, C. S., Falco, E. E., Impey, C., Lehár, J., McLeod, B. & Rix, H.-W. 1998, CASTLE Survey Gravitational Lens Data Base (Cambridge: CfA)

Martel, H., Premadi, P. & Matzner, R. 2002, ApJ, 570, 17

Premadi, P., Martel, H., Matzner, R. & Futamase, T. 2001, ApJ Suppl., 135, 7 (PMMF)

Romanowsky, A. J., & Kochanek, C. S. 1999, ApJ, 516, 18

Business Sessions

IAU 8th Asia-Pacific Regional Meeting
ASP Conference Series, Vol. 289, *2003*
S. Ikeuchi, J. Hearnshaw & T. Hanawa, eds.

Summary of Business Sessions

Satoru Ikeuchi
Department of Astrophysics, Nagoya University, Chikusa-ku, Nagoya 464-8602, Japan

1. Introduction

During IAU 8th APRM, we had two business sessions: BS1 on future arrangement of APRM and BS2 on the regional publication and network. This paper gives summary of these bussiness sessions. On the behalf of the SOC, I would like to thank the chair persons, Professor M. Raharto (BS1) and Professor Hyung Mok Lee (BS2) for their all the efforts. The following sections are based on their summary reports.

2. BS1: Future Arrangement of APRM

reported by M. Raharto

Scientific activities in Astronomy have existed in several Asian-Pacific countries. By the nature, it is embedded in long scientific tradition of the country in this region. The development of scientific activities may be also influenced by important conditions such as education, scientific tradition, economy and politics of each country. It is noticed that there is markedly different in the speed of development of astronomy in each country. It is also noticed that the interest in the field of astronomy is getting broader and deeper. We listened to the presentation of the result on many executed and new plan to observe astronomical objects through multiwavelength windows. Astronomers in Asia-Pacific region contribute the progress in observational and theoretical astronomy and astrophysics.

APRM is a chance to meet each other among astronomer in this Asian-Pacific region. Some reasons behind the APRM:

1. to broadening the new horizon in astronomy and astrophysics
2. to learn of astronomical development in different country in a Asia-Pacific region
3. to exchange the results of current astronomical research
4. to promote the exchanges of astronomers and collaborations on observational technique in the Asia-Pacific region
5. to pusuade young astronomers' activity for further development in astronomy
6. to discuss on the way of driving a high quality astronomical research activities in the region and popularizing astronomy

7. A chance to meet astronomers in the Asia-Pacific region. It will create an opportunity or inspiration to collaborate in developing astronomy. IAU gives more support for the future APRM (this view was given by elected President of IAU, Prof. Ecker during discussion in BS1).

8. Others in relation to the problems in astronomy in this region.

Asian-Pacific Regional meeting was initiated in Wellington, New Zealand in 1978, it has been regularly exist every three years since 1978, with an unfortunately the meeting in 1999 should happened in Indonesia, but it did not executed due to the worse political, economical and security situation in Indonesia. Thanks to the Japanese astronomers to take over as host country for the APRM 2002. It is assumed that there is no change for the periodicity of APRM then the next APRM will be held on 2005, 2008 and 2011. The discussions in BS 1 APRM 2002 4 July 2002 was limited to APRM 2005 and 2008. Indonesian astronomers accepted and committed as host country for the next APRM meeting in 2005 after about a quarter of century organized the second APRM in 1981. Australia, China/Taiwan and New Zealand were proposed as host country for the APRM-2008 (BS 1 APRM 2002, 4 July 2002).

NOTE ON APRM 2002, 2005 and 2008

(a) APRM 2002

- Historically some information that Indonesia was a candidate a host country for APRM-1999, it was proposed and discussed in Seoul for APRM-1996.
- Due to the worse political situation in 1998 then the APRM-1999 in Indonesia was canceled.
- In 2000 during GA in Manchester, Japan take over for the APRM-2002 and Indonesia was proposed to prepare the next APRM after APRM-2002

(b) APRM 2005

- In general Indonesia astronomers accepted to host the APRM-2005 (meeting in the Department of Astronomy July 2002). During business session in APRM-2002 it was openly talked and there was no objection and it had been announced in closing talked of the APRM-2002.
- A remarked: In case something happened in preparing APRM-2005 then Australia will take over the APRM-2005 meeting.

(c) APRM 2008

- Australia, China/Taiwan and New Zealand were proposed as host country for the APRM-2008 (BS 1 APRM 2002, 4 July 2002).

(d) Standing Committee (SC)

- Standing committee was proposed by Prof. Ikeuchi, the practical purpose to anticipate that for some reasons the host country will not able to organize the APRM, the decision of new host country for the APRM will be made by standing committee as soon as possible. The existence of SC was accepted in the BS1 APRM 2002. Membership of Standing Committee of APRM, one from each country, the chairman is the chairman of SOC and the co-chairman is the next SOC. Member of SC consist of member of SOC of the APRM plus proposed delegate from each country in Asia Pacific. Main function of SC is to give a guarantee that IAU-APRM regularly held on the A-P Region and SC propose member of SOC for the next APRM. A set of membership of SC will be discussed and renewed on each APRM. Furthermore, the role of SC is to discuss about the future arrangement of regional publication, network, creation of international institution and others in order to strengthen the connection among countries of this region.

Australia	R. Webster
Azerbaijan	E. S. Babayev (proposed)
Canada	R. Taylor
Chile	L. Bronfman
China Nanjing	C. Fang, K. J. Lo or G. Zhao
China Taipei	K. Y. Lo
Egypt	
Georgia	
India	S.M. Chitre or J. V. Narlikar
Indonesia	M. Raharto
Iran	Y. Soubouti (proposed)
Iraq	
Japan	S. Ikeuchi (Chair of SC)
Jordan	
Kazakhstan	
Korea	K. Chang or H. M. Lee
Malaysia	
New Zealand	J. Hearnshaw
Philipines	
Russia	I. S. Kim
Singapore	
South Africa	
North Korea	
Thailand	
USA	V. Rubin
Uzbekistan	A. S. Hojaev (proposed)
Vietnam	

Table 1. Tentative members of the standing committee of APRM (term 2002 - 2005)

3. BS2: The Regional Publication and Network

reported by H. M. Lee

The business session 2 was held on July 4, 2002. This session was attended by about 20 people from several different countries. We have discussed two topics: setting up of an international institute for the Pacific Rim countries and creation of a new international journal for Asian-Pacific region. Here is the summary of the discussions and conclusions:

3.1. Creation of an International Institute

Prof. Colin Norman of Space Telescope Science Institute has delivered an e-mail proposing Asian-Pacific Rim Institute for Astrophysics (APRIA). The main function of this institute is to run of topical programs of substantial lengths (three to four months) among astrophysicists from Asian-Pacific Rim area, including South American countries. This institute also provides a forum for the discussion of the large scale collaborative projects in this region.

However, the participants of the business session were somewhat cautious in pushing such an institute right away because of the difficulties in creating a new international institute without much experience of collaboration within this region. The participants agreed to set a long-term goal of forming an international organization to carry out collaborative projects and run topical programs within this region by following more realistic steps. The interim short projects could include the construction of a 2-m class telescope such as the one now being pursued by Chinese and Japanese colleagues. Utilization of the pre-existing establishment such as the Asia-Pacific Center for Theoretical Physics (APCTP) should also be emphasized. APCTP is currently located in Pohang, Korea.

The Taiwanese deligation noted that a new institute for astrophysics at Chinghwa University will soon take off under the leadership of Prof. Frank Shu (currently the president of the university). The experiences through these institutes and other small projects will enable us to realize more ambitious goal such as the creation of APRIA.

3.2. Regional Publication

As the size of the Asian-Pacific astronomical community grows we should seriously consider the creation of a new high standard journal in this region. At the moment, the major astronomical works are published mostly in American, British, or European Journals. Some works are published in the national journals of individual countries. Except for the Publication of Astronomical Society of Japan (PASJ) most of the national journals experience difficulties in maintaining the quality and quantity. The participants agreed that the regional journal, if created, will be of great benifit to the astronomical community in this region. However, it was also realized that there are subatantial difficulties in creating such a journal. First, the journal has to be created under the enthusiastic support among the participating countries. Second, we need financial resources and manpower to launch a new journal. The electronic publication would reduce the financial burden, and we also should seriously consider such possbilities. Therefore we concluded that we should take a cautionary steps forward. The specific action items discussed in this meeting include:

1. Take a poll among possible participating societies. This should be done under the name of SOC chair of this meeting. The BS2 Chair (H. M. Lee) will write a draft.

2. Form a standing committee to discuss the matters like this. The first standing committee members will be the SOC members of this meeting.

3.3. Conclusion

The business session 2 was quite a fruitful meeting. Although the number of participants was not large, we had rather intense and candid discussions. Although there was no firm conclusion on the two issues, we were able to set up the standing committee and action items. We sincerely hope that important issues discussed in this session will be seriously considered and followed up by the standing committee. In this way, the international institute and the regional journal can be finally realized in the near future.

First Author Index

Asada, H. 441
Athar, H. 323
Babayev, E. S. 157
Bagla, J. 251
Black, D. 77
Blain, A. 247
Cao, X. 341
Celebre, C. 145
Chen, W.-P. 181
Chen, Y. 295
Couch, W. 235
Ekers, R. 21
Fahlman, G. 3
Fang, C. 425
Goto, M. 189
Gould, A. 453
Gulyaev, S. 151
Han, C. 431
Han, Z. 413
Hearnshaw, J. 11, 55
Hidas, M. 65
Hiei, E. 137
Hirabayashi, H. 375
Honma, M. 445
Hunstead, R. 329
Ikeuchi , S. 479
Imanishi, M. 345
Inutsuka, S. 195
Ishitsuka, M. 49
Ji, J. 89
Jing, Y. P. 259
Jones, P. 173
Kawasaki, M. 207
Kohmura, T. 285
Kohno, K. 349
Koide, S. 319
Koo, B.-C. 199
Kurayama, T. 421
Kuwabara, J. 389
Lee, H.-W. 281
Li, X. 273
Malasan, H. L. 45
Matsuhara, H. 113
Matsumoto, H. 291
Mineshige, S. 301
Miyoshi, M. 33
Momose, M. 85
Murakami, T. 305
Nakajima, Y. 177
Nakamura, O. 243
Nityananda, R. 29
Ohnishi, K. 461
Ojha, D. K. 121
Oreshina, A . 315
Ota, N. 465
Parker, Q. 165
Peterson, B. A. 219
Podgorny, A. 397
Premadi, P. 473
Ratag, M. A. 211
Rattenbury, N. 69
Ryu, D. 371
Sadler, E. 105
Shibai, H. 117
Shibata, K. 381
Shigeyama, T. 263
Skuljan, J. 17
Sobouti, Y. 409
Sofue, Y. 129
Somov, B. 393
Suematsu, Y. 37
Tamura, M. 73
Taniguchi, Y. 353
Totani, T. 227, 311
Tsuboi, M. 255
Uchida, Y. 367
Umemura, M. 363
Wakamatsu, K. 97
Wang, Y. 267
Wayth, R. 449
Yan, Y. 401
Yang, J. 185
Yin, Q. F. 125
Yonehara, A. 469
Yoshida, M. 337

A LIST OF THE VOLUMES

Published
by

THE ASTRONOMICAL SOCIETY OF THE PACIFIC
(ASP)

An international, nonprofit, scientific and educational organization founded in 1889

All book orders or inquiries concerning

THE ASTRONOMICAL SOCIETY OF THE PACIFIC
CONFERENCE SERIES
(ASP - CS)

and

INTERNATIONAL ASTRONOMICAL UNION VOLUMES
(IAU)

should be directed to the:

The Astronomical Society of the Pacific Conference Series
390 Ashton Avenue
San Francisco CA 94112-1722 USA

Phone: 800-335-2624 (Within USA)
Phone: 415-337-2126
Fax: 415-337-5205

E-mail: service@astrosociety.org
Web Slte: http://www.astrosociety.org

Complete lists of proceedings of past IAU Meetings are maintained at the IAU Web site at the URL: http://www.iau.org/publicat.html

Volumes 32 - 189 in the IAU Symposia Series may be ordered from:

Kluwer Academic Publishers
P. O. Box 117
NL 3300 AA Dordrecht
The Netherlands

Kluwer@wKap.com

ASP CONFERENCE SERIES VOLUMES

Published by the Astronomical Society of the Pacific

PUBLISHED: 1988 (* asterisk means OUT OF STOCK)

Vol. CS -1 PROGRESS AND OPPORTUNITIES IN SOUTHERN HEMISPHERE OPTICAL ASTRONOMY: CTIO 25TH Anniversary Symposium
eds. V. M. Blanco and M. M. Phillips
ISBN 0-937707-18-X

Vol. CS-2 PROCEEDINGS OF A WORKSHOP ON OPTICAL SURVEYS FOR QUASARS
eds. Patrick S. Osmer, Alain C. Porter, Richard F. Green, and Craig B. Foltz
ISBN 0-937707-19-8

Vol. CS-3 FIBER OPTICS IN ASTRONOMY
ed. Samuel C. Barden
ISBN 0-937707-20-1

Vol. CS-4 THE EXTRAGALACTIC DISTANCE SCALE:
Proceedings of the ASP 100th Anniversary Symposium
eds. Sidney van den Bergh and Christopher J. Pritchet
ISBN 0-937707-21-X

Vol. CS-5 THE MINNESOTA LECTURES ON CLUSTERS OF GALAXIES AND LARGE-SCALE STRUCTURE
ed. John M. Dickey
ISBN 0-937707-22-8

PUBLISHED: 1989

Vol. CS-6 * SYNTHESIS IMAGING IN RADIO ASTRONOMY: A Collection of Lectures from the Third NRAO Synthesis Imaging Summer School
eds. Richard A. Perley, Frederic R. Schwab, and Alan H. Bridle
ISBN 0-937707-23-6

PUBLISHED: 1990

Vol. CS-7 PROPERTIES OF HOT LUMINOUS STARS: Boulder-Munich Workshop
ed. Catharine D. Garmany
ISBN 0-937707-24-4

Vol. CS-8 * CCDs IN ASTRONOMY
ed. George H. Jacoby
ISBN 0-937707-25-2

Vol. CS-9 COOL STARS, STELLAR SYSTEMS, AND THE SUN: Sixth Cambridge Workshop
ed. George Wallerstein
ISBN 0-937707-27-9

Vol. CS-10 * EVOLUTION OF THE UNIVERSE OF GALAXIES:
Edwin Hubble Centennial Symposium
ed. Richard G. Kron
ISBN 0-937707-28-7

Vol. CS-11 CONFRONTATION BETWEEN STELLAR PULSATION AND EVOLUTION
eds. Carla Cacciari and Gisella Clementini
ISBN 0-937707-30-9

Vol. CS-12 THE EVOLUTION OF THE INTERSTELLAR MEDIUM
ed. Leo Blitz
ISBN 0-937707-31-7

PUBLISHED: 1991

Vol. CS-13 THE FORMATION AND EVOLUTION OF STAR CLUSTERS
ed. Kenneth Janes
ISBN 0-937707-32-5

ASP CONFERENCE SERIES VOLUMES

Published by the Astronomical Society of the Pacific

PUBLISHED: 1991 (* asterisk means OUT OF STOCK)

Vol. CS-14 — ASTROPHYSICS WITH INFRARED ARRAYS
ed. Richard Elston
ISBN 0-937707-33-3

Vol. CS-15 — LARGE-SCALE STRUCTURES AND PECULIAR MOTIONS IN THE UNIVERSE
eds. David W. Latham and L. A. Nicolaci da Costa
ISBN 0-937707-34-1

Vol. CS-16 — Proceedings of the 3rd Haystack Observatory Conference on ATOMS, IONS, AND MOLECULES: NEW RESULTS IN SPECTRAL LINE ASTROPHYSICS
eds. Aubrey D. Haschick and Paul T. P. Ho
ISBN 0-937707-35-X

Vol. CS-17 — LIGHT POLLUTION, RADIO INTERFERENCE, AND SPACE DEBRIS
ed. David L. Crawford
ISBN 0-937707-36-8

Vol. CS-18 — THE INTERPRETATION OF MODERN SYNTHESIS OBSERVATIONS OF SPIRAL GALAXIES
eds. Nebojsa Duric and Patrick C. Crane
ISBN 0-937707-37-6

Vol. CS-19 — RADIO INTERFEROMETRY: THEORY, TECHNIQUES, AND APPLICATIONS, IAU Colloquium 131
eds. T. J. Cornwell and R. A. Perley
ISBN 0-937707-38-4

Vol. CS-20 — FRONTIERS OF STELLAR EVOLUTION: 50th Anniversary McDonald Observatory (1939-1989)
ed. David L. Lambert
ISBN 0-937707-39-2

Vol. CS-21 — THE SPACE DISTRIBUTION OF QUASARS
ed . David Crampton
ISBN 0-937707-40-6

PUBLISHED: 1992

Vol. CS-22 — NONISOTROPIC AND VARIABLE OUTFLOWS FROM STARS
eds. Laurent Drissen, Claus Leitherer, and Antonella Nota
ISBN 0-937707-41-4

Vol CS-23 * — ASTRONOMICAL CCD OBSERVING AND REDUCTION TECHNIQUES
ed. Steve B. Howell
ISBN 0-937707-42-4

Vol. CS-24 — COSMOLOGY AND LARGE-SCALE STRUCTURE IN THE UNIVERSE
ed. Reinaldo R. de Carvalho
ISBN 0-937707-43-0

Vol. CS-25 — ASTRONOMICAL DATA ANALYSIS, SOFTWARE AND SYSTEMS I - (ADASS I)
eds. Diana M. Worrall, Chris Biemesderfer, and Jeannette Barnes
ISBN 0-937707-44-9

Vol. CS-26 — COOL STARS, STELLAR SYSTEMS, AND THE SUN: Seventh Cambridge Workshop
eds. Mark S. Giampapa and Jay A. Bookbinder
ISBN 0-937707-45-7

Vol. CS-27 — THE SOLAR CYCLE: Proceedings of the National Solar Observatory/Sacramento Peak 12th Summer Workshop
ed. Karen L. Harvey
ISBN 0-937707-46-5

ASP CONFERENCE SERIES VOLUMES

Published by the Astronomical Society of the Pacific

PUBLISHED: 1992 (asterisk means OUT OF STOCK)

Vol. CS-28 AUTOMATED TELESCOPES FOR PHOTOMETRY AND IMAGING
eds. Saul J. Adelman, Robert J. Dukes, Jr., and Carol J. Adelman
ISBN 0-937707-47-3

Vol. CS-29 Viña del Mar Workshop on CATACLYSMIC VARIABLE STARS
ed. Nikolaus Vogt
ISBN 0-937707-48-1

Vol. CS-30 VARIABLE STARS AND GALAXIES
ed. Brian Warner
ISBN 0-937707-49-X

Vol. CS-31 RELATIONSHIPS BETWEEN ACTIVE GALACTIC NUCLEI AND STARBURST GALAXIES
ed. Alexei V. Filippenko
ISBN 0-937707-50-3

Vol. CS-32 COMPLEMENTARY APPROACHES TO DOUBLE AND MULTIPLE STAR RESEARCH, IAU Colloquium 135
eds. Harold A. McAlister and William I. Hartkopf
ISBN 0-937707-51-1

Vol. CS-33 * RESEARCH AMATEUR ASTRONOMY
ed. Stephen J. Edberg
ISBN 0-937707-52-X

Vol. CS-34 ROBOTIC TELESCOPES IN THE 1990's
ed. Alexei V. Filippenko
ISBN 0-937707-53-8

PUBLISHED: 1993

Vol. CS-35 * MASSIVE STARS: THEIR LIVES IN THE INTERSTELLAR MEDIUM
eds. Joseph P. Cassinelli and Edward B. Churchwell
ISBN 0-937707-54-6

Vol. CS-36 PLANETS AROUND PULSARS
ed. J. A. Phillips, S. E. Thorsett, and S. R. Kulkarni
ISBN 0-937707-55-4

Vol. CS-37 FIBER OPTICS IN ASTRONOMY II
ed. Peter M. Gray
ISBN 0-937707-56-2

Vol. CS-38 NEW FRONTIERS IN BINARY STAR RESEARCH: Pacific Rim Colloquium
eds. K. C. Leung and I.-S. Nha
ISBN 0-937707-57-0

Vol. CS-39 THE MINNESOTA LECTURES ON THE STRUCTURE AND DYNAMICS OF THE MILKY WAY
ed. Roberta M. Humphreys
ISBN 0-937707-58-9

Vol. CS-40 INSIDE THE STARS, IAU Colloquium 137
eds. Werner W. Weiss and Annie Baglin
ISBN 0-937707-59-7

Vol. CS-41 ASTRONOMICAL INFRARED SPECTROSCOPY: FUTURE OBSERVATIONAL DIRECTIONS
ed. Sun Kwok
ISBN 0-937707-60-0

ASP CONFERENCE SERIES VOLUMES

Published by the Astronomical Society of the Pacific

PUBLISHED: 1993 (* asterisk means OUT OF STOCK)

Vol. CS-42 GONG 1992: SEISMIC INVESTIGATION OF THE SUN AND STARS
ed. Timothy M. Brown
ISBN 0-937707-61-9

Vol. CS-43 SKY SURVEYS: PROTOSTARS TO PROTOGALAXIES
ed. B. T. Soifer
ISBN 0-937707-62-7

Vol. CS-44 PECULIAR VERSUS NORMAL PHENOMENA IN A-TYPE AND RELATED STARS, IAU Colloquium 138
eds. M. M. Dworetsky, F. Castelli, and R. Faraggiana
ISBN 0-937707-63-5

Vol. CS-45 LUMINOUS HIGH-LATITUDE STARS
ed. Dimitar D. Sasselov
ISBN 0-937707-64-3

Vol. CS-46 THE MAGNETIC AND VELOCITY FIELDS OF SOLAR ACTIVE REGIONS, IAU Colloquium 141
eds. Harold Zirin, Guoxiang Ai, and Haimin Wang
ISBN 0-937707-65-1

Vol. CS-47 THIRD DECENNIAL US-USSR CONFERENCE ON SETI -- Santa Cruz, California, USA
ed. G. Seth Shostak
ISBN 0-937707-66-X

Vol. CS-48 THE GLOBULAR CLUSTER-GALAXY CONNECTION
eds. Graeme H. Smith and Jean P. Brodie
ISBN 0-937707-67-8

Vol. CS-49 GALAXY EVOLUTION: THE MILKY WAY PERSPECTIVE
ed. Steven R. Majewski
ISBN 0-937707-68-6

Vol. CS-50 STRUCTURE AND DYNAMICS OF GLOBULAR CLUSTERS
eds. S. G. Djorgovski and G. Meylan
ISBN 0-937707-69-4

Vol. CS-51 OBSERVATIONAL COSMOLOGY
eds. Guido Chincarini, Angela Iovino, Tommaso Maccacaro, and Dario Maccagni
ISBN 0-937707-70-8

Vol. CS-52 ASTRONOMICAL DATA ANALYSIS SOFTWARE AND SYSTEMS II - (ADASS II)
eds. R. J. Hanisch, R. J. V. Brissenden, and Jeannette Barnes
ISBN 0-937707-71-6

Vol. CS-53 BLUE STRAGGLERS
ed. Rex A. Saffer
ISBN 0-937707-72-4

PUBLISHED: 1994

Vol. CS-54 * THE FIRST STROMLO SYMPOSIUM: THE PHYSICS OF ACTIVE GALAXIES
eds. Geoffrey V. Bicknell, Michael A. Dopita, and Peter J. Quinn
ISBN 0-937707-73-2

Vol. CS-55 OPTICAL ASTRONOMY FROM THE EARTH AND MOON
eds. Diane M. Pyper and Ronald J. Angione
ISBN 0-937707-74-0

Vol. CS-56 INTERACTING BINARY STARS
ed. Allen W. Shafter
ISBN 0-937707-75-9

ASP CONFERENCE SERIES VOLUMES

Published by the Astronomical Society of the Pacific

PUBLISHED: 1994 (* asterisk means OUT OF STOCK)

Vol. CS-57 STELLAR AND CIRCUMSTELLAR ASTROPHYSICS
eds. George Wallerstein and Alberto Noriega-Crespo
ISBN 0-937707-76-7

Vol. CS-58 * THE FIRST SYMPOSIUM ON THE INFRARED CIRRUS AND DIFFUSE INTERSTELLAR CLOUDS
eds. Roc M. Cutri and William B. Latter
ISBN 0-937707-77-5

Vol. CS-59 ASTRONOMY WITH MILLIMETER AND SUBMILLIMETER WAVE INTERFEROMETRY,
IAU Colloquium 140
eds. M. Ishiguro and Wm. J. Welch
ISBN 0-937707-78-3

Vol. CS-60 THE MK PROCESS AT 50 YEARS: A POWERFUL TOOL FOR ASTROPHYSICAL INSIGHT, A Workshop of the Vatican Observatory --Tucson, Arizona, USA
eds. C. J. Corbally, R. O. Gray, and R. F. Garrison
ISBN 0-937707-79-1

Vol. CS-61 ASTRONOMICAL DATA ANALYSIS SOFTWARE AND SYSTEMS III - (ADASS III)
eds. Dennis R. Crabtree, R. J. Hanisch, and Jeannette Barnes
ISBN 0-937707-80-5

Vol. CS-62 THE NATURE AND EVOLUTIONARY STATUS OF HERBIG Ae/Be STARS
eds. Pik Sin Thé, Mario R. Pérez, and Ed P. J. van den Heuvel
ISBN 0-9837707-81-3

Vol. CS-63 SEVENTY-FIVE YEARS OF HIRAYAMA ASTEROID FAMILIES: THE ROLE OF COLLISIONS IN THE SOLAR SYSTEM HISTORY
eds. Yoshihide Kozai, Richard P. Binzel, and Tomohiro Hirayama
ISBN 0-937707-82-1

Vol. CS-64 * COOL STARS, STELLAR SYSTEMS, AND THE SUN:
Eighth Cambridge Workshop
ed. Jean-Pierre Caillault
ISBN 0-937707-83-X

Vol. CS-65 * CLOUDS, CORES, AND LOW MASS STARS:
The Fourth Haystack Observatory Conference
eds. Dan P. Clemens and Richard Barvainis
ISBN 0-937707-84-8

Vol. CS-66 * PHYSICS OF THE GASEOUS AND STELLAR DISKS OF THE GALAXY
ed. Ivan R. King
ISBN 0-937707-85-6

Vol. CS-67 UNVEILING LARGE-SCALE STRUCTURES BEHIND THE MILKY WAY
eds. C. Balkowski and R. C. Kraan-Korteweg
ISBN 0-937707-86-4

Vol. CS-68 * SOLAR ACTIVE REGION EVOLUTION: COMPARING MODELS WITH OBSERVATIONS
eds. K. S. Balasubramaniam and George W. Simon
ISBN 0-937707-87-2

Vol. CS-69 REVERBERATION MAPPING OF THE BROAD-LINE REGION IN ACTIVE GALACTIC NUCLEI
eds. P. M. Gondhalekar, K. Horne, and B. M. Peterson
ISBN 0-937707-88-0

Vol. CS-70 * GROUPS OF GALAXIES
eds. Otto-G. Richter and Kirk Borne
ISBN 0-937707-89-9

ASP CONFERENCE SERIES VOLUMES

Published by the Astronomical Society of the Pacific

PUBLISHED: 1995 (* asterisk means OUT OF STOCK)

Vol. CS-71 TRIDIMENSIONAL OPTICAL SPECTROSCOPIC METHODS IN ASTROPHYSICS, IAU Colloquium 149
eds. Georges Comte and Michel Marcelin
ISBN 0-937707-90-2

Vol. CS-72 MILLISECOND PULSARS: A DECADE OF SURPRISE
eds. A. S Fruchter, M. Tavani, and D. C. Backer
ISBN 0-937707-91-0

Vol. CS-73 AIRBORNE ASTRONOMY SYMPOSIUM ON THE GALACTIC ECOSYSTEM: FROM GAS TO STARS TO DUST
eds. Michael R. Haas, Jacqueline A. Davidson, and Edwin F. Erickson
ISBN 0-937707-92-9

Vol. CS-74 PROGRESS IN THE SEARCH FOR EXTRATERRESTRIAL LIFE: 1993 Bioastronomy Symposium
ed. G. Seth Shostak
ISBN 0-937707-93-7

Vol. CS-75 MULTI-FEED SYSTEMS FOR RADIO TELESCOPES
eds. Darrel T. Emerson and John M. Payne
ISBN 0-937707-94-5

Vol. CS-76 GONG '94: HELIO- AND ASTERO-SEISMOLOGY FROM THE EARTH AND SPACE
eds. Roger K. Ulrich, Edward J. Rhodes, Jr., and Werner Däppen
ISBN 0-937707-95-3

Vol. CS-77 ASTRONOMICAL DATA ANALYSIS SOFTWARE AND SYSTEMS IV - (ADASS IV)
eds. R. A. Shaw, H. E. Payne, and J. J. E. Hayes
ISBN 0-937707-96-1

Vol. CS-78 ASTROPHYSICAL APPLICATIONS OF POWERFUL NEW DATABASES: Joint Discussion No. 16 of the 22nd General Assembly of the IAU
eds. S. J. Adelman and W. L. Wiese
ISBN 0-937707-97-X

Vol. CS-79 * ROBOTIC TELESCOPES: CURRENT CAPABILITIES, PRESENT DEVELOPMENTS, AND FUTURE PROSPECTS FOR AUTOMATED ASTRONOMY
eds. Gregory W. Henry and Joel A. Eaton
ISBN 0-937707-98-8

Vol. CS-80 * THE PHYSICS OF THE INTERSTELLAR MEDIUM AND INTERGALACTIC MEDIUM
eds. A. Ferrara, C. F. McKee, C. Heiles, and P. R. Shapiro
ISBN 0-937707-99-6

Vol. CS-81 LABORATORY AND ASTRONOMICAL HIGH RESOLUTION SPECTRA
eds. A. J. Sauval, R. Blomme, and N. Grevesse
ISBN 1-886733-01-5

Vol. CS-82 * VERY LONG BASELINE INTERFEROMETRY AND THE VLBA
eds. J. A. Zensus, P. J. Diamond, and P. J. Napier
ISBN 1-886733-02-3

Vol. CS-83 * ASTROPHYSICAL APPLICATIONS OF STELLAR PULSATION, IAU Colloquium 155
eds. R. S. Stobie and P. A. Whitelock
ISBN 1-886733-03-1

ATLAS INFRARED ATLAS OF THE ARCTURUS SPECTRUM, 0.9 - 5.3 μm
eds. Kenneth Hinkle, Lloyd Wallace, and William Livingston
ISBN: 1-886733-04-X

ASP CONFERENCE SERIES VOLUMES

Published by the Astronomical Society of the Pacific

PUBLISHED: 1995 (* asterisk means OUT OF STOCK)

Vol. CS-84 THE FUTURE UTILIZATION OF SCHMIDT TELESCOPES, IAU Colloquium 148
eds. Jessica Chapman, Russell Cannon, Sandra Harrison, and Bambang Hidayat
ISBN 1-886733-05-8

Vol. CS-85 * CAPE WORKSHOP ON MAGNETIC CATACLYSMIC VARIABLES
eds. D. A. H. Buckley and B. Warner
ISBN 1-886733-06-6

Vol. CS-86 FRESH VIEWS OF ELLIPTICAL GALAXIES
eds. Alberto Buzzoni, Alvio Renzini, and Alfonso Serrano
ISBN 1-886733-07-4

PUBLISHED: 1996

Vol. CS-87 NEW OBSERVING MODES FOR THE NEXT CENTURY
eds. Todd Boroson, John Davies, and Ian Robson
ISBN 1-886733-08-2

Vol. CS-88 * CLUSTERS, LENSING, AND THE FUTURE OF THE UNIVERSE
eds. Virginia Trimble and Andreas Reisenegger
ISBN 1-886733-09-0

Vol. CS-89 ASTRONOMY EDUCATION: CURRENT DEVELOPMENTS, FUTURE COORDINATION
ed. John R. Percy
ISBN 1-886733-10-4

Vol. CS-90 THE ORIGINS, EVOLUTION, AND DESTINIES OF BINARY STARS IN CLUSTERS
eds. E. F. Milone and J. -C. Mermilliod
ISBN 1-886733-11-2

Vol. CS-91 BARRED GALAXIES, IAU Colloquium 157
eds. R. Buta, D. A. Crocker, and B. G. Elmegreen
ISBN 1-886733-12-0

Vol. CS-92 * FORMATION OF THE GALACTIC HALO .... INSIDE AND OUT
eds. Heather L. Morrison and Ata Sarajedini
ISBN 1-886733-13-9

Vol. CS-93 RADIO EMISSION FROM THE STARS AND THE SUN
eds. A. R. Taylor and J. M. Paredes
ISBN 1-886733-14-7

Vol. CS-94 MAPPING, MEASURING, AND MODELING THE UNIVERSE
eds. Peter Coles, Vicent J. Martinez, and Maria-Jesus Pons-Borderia
ISBN 1-886733-15-5

Vol. CS-95 SOLAR DRIVERS OF INTERPLANETARY AND TERRESTRIAL DISTURBANCES: Proceedings of 16th International Workshop National Solar Observatory/Sacramento Peak
eds. K. S. Balasubramaniam, Stephen L. Keil, and Raymond N. Smartt
ISBN 1-886733-16-3

Vol. CS-96 HYDROGEN-DEFICIENT STARS
eds. C. S. Jeffery and U. Heber
ISBN 1-886733-17-1

Vol. CS-97 POLARIMETRY OF THE INTERSTELLAR MEDIUM
eds. W. G. Roberge and D. C. B. Whittet
ISBN 1-886733-18-X

ASP CONFERENCE SERIES VOLUMES

Published by the Astronomical Society of the Pacific

PUBLISHED: 1996 (* asterisk means OUT OF STOCK)

Vol. CS-98 FROM STARS TO GALAXIES: THE IMPACT OF STELLAR PHYSICS ON GALAXY EVOLUTION
eds. Claus Leitherer, Uta Fritze-von Alvensleben, and John Huchra
ISBN 1-886733-19-8

Vol. CS-99 COSMIC ABUNDANCES:
Proceedings of the 6th Annual October Astrophysics Conference
eds. Stephen S. Holt and George Sonneborn
ISBN 1-886733-20-1

Vol. CS-100 ENERGY TRANSPORT IN RADIO GALAXIES AND QUASARS
eds. P. E. Hardee, A. H. Bridle, and J. A. Zensus
ISBN 1-886733-21-X

Vol. CS-101 ASTRONOMICAL DATA ANALYSIS SOFTWARE AND SYSTEMS V – (ADASS V)
eds. George H. Jacoby and Jeannette Barnes
ISBN 1080-7926

Vol. CS-102 THE GALACTIC CENTER, 4th ESO/CTIO Workshop
ed. Roland Gredel
ISBN 1-886733-22-8

Vol. CS-103 THE PHYSICS OF LINERS IN VIEW OF RECENT OBSERVATIONS
eds. M. Eracleous, A. Koratkar, C. Leitherer, and L. Ho
ISBN 1-886733-23-6

Vol. CS-104 PHYSICS, CHEMISTRY, AND DYNAMICS OF INTERPLANETARY DUST, IAU Colloquium 150
eds. Bo Å. S. Gustafson and Martha S. Hanner
ISBN 1-886733-24-4

Vol. CS-105 PULSARS: PROBLEMS AND PROGRESS, IAU Colloquium 160
ed. S. Johnston, M. A. Walker, and M. Bailes
ISBN 1-886733-25-2

Vol. CS-106 THE MINNESOTA LECTURES ON EXTRAGALACTIC NEUTRAL HYDROGEN
ed. Evan D. Skillman
ISBN 1-886733-26-0

Vol. CS-107 COMPLETING THE INVENTORY OF THE SOLAR SYSTEM:
A Symposium held in conjunction with the 106th Annual Meeting of the ASP
eds. Terrence W. Rettig and Joseph M. Hahn
ISBN 1-886733-27-9

Vol. CS-108 M.A.S.S. -- MODEL ATMOSPHERES AND SPECTRUM SYNTHESIS:
5th Vienna - Workshop
eds. Saul J. Adelman, Friedrich Kupka, and Werner W. Weiss
ISBN 1-886733-28-7

Vol. CS-109 COOL STARS, STELLAR SYSTEMS, AND THE SUN: Ninth Cambridge Workshop
eds. Roberto Pallavicini and Andrea K. Dupree
ISBN 1-886733-29-5

Vol. CS-110 BLAZAR CONTINUUM VARIABILITY
eds. H. R. Miller, J. R. Webb, and J. C. Noble
ISBN 1-886733-30-9

Vol. CS-111 MAGNETIC RECONNECTION IN THE SOLAR ATMOSPHERE:
Proceedings of a Yohkoh Conference
eds. R. D. Bentley and J. T. Mariska
ISBN 1-886733-31-7

ASP CONFERENCE SERIES VOLUMES

Published by the Astronomical Society of the Pacific

PUBLISHED: 1996 (* asterisk means OUT OF STOCK)

Vol. CS-112 THE HISTORY OF THE MILKY WAY AND ITS SATELLITE SYSTEM
eds. Andreas Burkert, Dieter H. Hartmann, and Steven R. Majewski
ISBN 1-886733-32-5

PUBLISHED: 1997

Vol. CS-113 EMISSION LINES IN ACTIVE GALAXIES: NEW METHODS AND TECHNIQUES, IAU Colloquium 159
eds. B. M. Peterson, F.-Z. Cheng, and A. S. Wilson
ISBN 1-886733-33-3

Vol. CS-114 YOUNG GALAXIES AND QSO ABSORPTION-LINE SYSTEMS
eds. Sueli M. Viegas, Ruth Gruenwald, and Reinaldo R. de Carvalho
ISBN 1-886733-34-1

Vol. CS-115 GALACTIC CLUSTER COOLING FLOWS
ed. Noam Soker
ISBN 1-886733-35-X

Vol. CS-116 THE SECOND STROMLO SYMPOSIUM: THE NATURE OF ELLIPTICAL GALAXIES
eds. M. Arnaboldi, G. S. Da Costa, and P. Saha
ISBN 1-886733-36-8

Vol. CS-117 DARK AND VISIBLE MATTER IN GALAXIES
eds. Massimo Persic and Paolo Salucci
ISBN-1-886733-37-6

Vol. CS-118 FIRST ADVANCES IN SOLAR PHYSICS EUROCONFERENCE: ADVANCES IN THE PHYSICS OF SUNSPOTS
eds. B. Schmieder. J. C. del Toro Iniesta, and M. Vázquez
ISBN 1-886733-38-4

Vol. CS-119 PLANETS BEYOND THE SOLAR SYSTEM AND THE NEXT GENERATION OF SPACE MISSIONS
ed. David R. Soderblom
ISBN 1-886733-39-2

Vol. CS-120 LUMINOUS BLUE VARIABLES: MASSIVE STARS IN TRANSITION
eds. Antonella Nota and Henny J. G. L. M. Lamers
ISBN 1-886733-40-6

Vol. CS-121 ACCRETION PHENOMENA AND RELATED OUTFLOWS, IAU Colloquium 163
eds. D. T. Wickramasinghe, G. V. Bicknell, and L. Ferrario
ISBN 1-886733-41-4

Vol. CS-122 FROM STARDUST TO PLANETESIMALS: Symposium held as part of the 108th Annual Meeting of the ASP
eds. Yvonne J. Pendleton and A. G. G. M. Tielens
ISBN 1-886733-42-2

Vol. CS-123 THE 12th 'KINGSTON MEETING': COMPUTATIONAL ASTROPHYSICS
eds. David A. Clarke and Michael J. West
ISBN 1-886733-43-0

Vol. CS-124 DIFFUSE INFRARED RADIATION AND THE IRTS
eds. Haruyuki Okuda, Toshio Matsumoto, and Thomas Roellig
ISBN 1-886733-44-9

Vol. CS-125 ASTRONOMICAL DATA ANALYSIS SOFTWARE AND SYSTEMS VI
eds. Gareth Hunt and H. E. Payne
ISBN 1-886733-45-7

ASP CONFERENCE SERIES VOLUMES

Published by the Astronomical Society of the Pacific

PUBLISHED: 1997 (* asterisk means OUT OF STOCK)

Vol. CS-126 FROM QUANTUM FLUCTUATIONS TO COSMOLOGICAL STRUCTURES
eds. David Valls-Gabaud, Martin A. Hendry, Paolo Molaro, and Khalil Chamcham
ISBN 1-886733-46-5

Vol. CS-127 PROPER MOTIONS AND GALACTIC ASTRONOMY
ed. Roberta M. Humphreys
ISBN 1-886733-47-3

Vol. CS-128 MASS EJECTION FROM AGN (Active Galactic Nuclei)
eds. N. Arav, I. Shlosman, and R. J. Weymann
ISBN 1-886733-48-1

Vol. CS-129 THE GEORGE GAMOW SYMPOSIUM
eds. E. Harper, W. C. Parke, and G. D. Anderson
ISBN 1-886733-49-X

Vol. CS-130 THE THIRD PACIFIC RIM CONFERENCE ON
RECENT DEVELOPMENT ON BINARY STAR RESEARCH
eds. Kam-Ching Leung
ISBN 1-886733-50-3

PUBLISHED: 1998

Vol. CS-131 BOULDER-MUNICH II: PROPERTIES OF HOT, LUMINOUS STARS
ed. Ian D. Howarth
ISBN 1-886733-51-1

Vol. CS-132 STAR FORMATION WITH THE INFRARED SPACE OBSERVATORY (ISO)
eds. João L. Yun and René Liseau
ISBN 1-886733-52-X

Vol. CS-133 SCIENCE WITH THE NGST (Next Generation Space Telescope)
eds. Eric P. Smith and Anuradha Koratkar
ISBN 1-886733-53-8

Vol. CS-134 BROWN DWARFS AND EXTRASOLAR PLANETS
eds. Rafael Rebolo, Eduardo L. Martin, and Maria Rosa Zapatero Osorio
ISBN 1-886733-54-6

Vol. CS-135 A HALF CENTURY OF STELLAR PULSATION INTERPRETATIONS:
A TRIBUTE TO ARTHUR N. COX
eds. P. A. Bradley and J. A. Guzik
ISBN 1-886733-55-4

Vol. CS-136 GALACTIC HALOS: A UC SANTA CRUZ WORKSHOP
ed. Dennis Zaritsky
ISBN 1-886733-56-2

Vol. CS-137 WILD STARS IN THE OLD WEST: PROCEEDINGS OF THE 13th NORTH
AMERICAN WORKSHOP ON CATACLYSMIC VARIABLES
AND RELATED OBJECTS
eds. S. Howell, E. Kuulkers, and C. Woodward
ISBN 1-886733-57-0

Vol. CS-138 1997 PACIFIC RIM CONFERENCE ON STELLAR ASTROPHYSICS
eds. Kwing Lam Chan, K. S. Cheng, and H. P. Singh
ISBN 1-886733-58-9

Vol. CS-139 PRESERVING THE ASTRONOMICAL WINDOWS:
Proceedings of Joint Discussion No. 5 of the 23rd General Assembly of the IAU
eds. Syuzo Isobe and Tomohiro Hirayama
ISBN 1-886733-59-7

ASP CONFERENCE SERIES VOLUMES

Published by the Astronomical Society of the Pacific

PUBLISHED: 1998 (* asterisk means OUT OF STOCK)

Vol. CS-140 SYNOPTIC SOLAR PHYSICS --18th NSO/Sacramento Peak Summer Workshop
eds. K. S. Balasubramaniam, J. W. Harvey, and D. M. Rabin
ISBN 1-886733-60-0

Vol. CS-141 ASTROPHYSICS FROM ANTARCTICA:
A Symposium held as a part of the 109th Annual Meeting of the ASP
eds. Giles Novak and Randall H. Landsberg
ISBN 1-886733-61-9

Vol. CS-142 THE STELLAR INITIAL MASS FUNCTION: 38th Herstmonceux Conference
eds. Gerry Gilmore and Debbie Howell
ISBN 1-886733-62-7

Vol. CS-143 * THE SCIENTIFIC IMPACT OF THE GODDARD HIGH RESOLUTION SPECTROGRAPH (GHRS)
eds. John C. Brandt, Thomas B. Ake III, and Carolyn Collins Petersen
ISBN 1-886733-63-5

Vol. CS-144 RADIO EMISSION FROM GALACTIC AND EXTRAGALACTIC COMPACT SOURCES, IAU Colloquium 164
eds. J. Anton Zensus, G. B. Taylor, and J. M. Wrobel
ISBN 1-886733-64-3

Vol. CS-145 ASTRONOMICAL DATA ANALYSIS SOFTWARE AND SYSTEMS VII – (ADASS VII)
eds. Rudolf Albrecht, Richard N. Hook, and Howard A. Bushouse
ISBN 1-886733-65-1

Vol. CS-146 THE YOUNG UNIVERSE GALAXY FORMATION
AND EVOLUTION AT INTERMEDIATE AND HIGH REDSHIFT
eds. S. D'Odorico, A. Fontana, and E. Giallongo
ISBN 1-886733-66-X

Vol. CS-147 ABUNDANCE PROFILES: DIAGNOSTIC TOOLS FOR GALAXY HISTORY
eds. Daniel Friedli, Mike Edmunds, Carmelle Robert, and Laurent Drissen
ISBN 1-886733-67-8

Vol. CS-148 ORIGINS
eds. Charles E. Woodward, J. Michael Shull, and Harley A. Thronson, Jr.
ISBN 1-886733-68-6

Vol. CS-149 SOLAR SYSTEM FORMATION AND EVOLUTION
eds. D. Lazzaro, R. Vieira Martins, S. Ferraz-Mello, J. Fernández, and C. Beaugé
ISBN 1-886733-69-4

Vol. CS-150 NEW PERSPECTIVES ON SOLAR PROMINENCES, IAU Colloquium 167
eds. David Webb, David Rust, and Brigitte Schmieder
ISBN 1-886733-70-8

Vol. CS-151 COSMIC MICROWAVE BACKGROUND
AND LARGE SCALE STRUCTURES OF THE UNIVERSE
eds. Yong-Ik Byun and Kin-Wang Ng
ISBN 1-886733-71-6

Vol. CS-152 FIBER OPTICS IN ASTRONOMY III
eds. S. Arribas, E. Mediavilla, and F. Watson
ISBN 1-886733-72-4

Vol. CS-153 LIBRARY AND INFORMATION SERVICES IN ASTRONOMY III -- (LISA III)
eds. Uta Grothkopf, Heinz Andernach, Sarah Stevens-Rayburn, and Monique Gomez
ISBN 1-886733-73-2

ASP CONFERENCE SERIES VOLUMES

Published by the Astronomical Society of the Pacific

PUBLISHED: 1998 (* asterisk means OUT OF STOCK)

Vol. CS-154 COOL STARS, STELLAR SYSTEMS AND THE SUN: Tenth Cambridge Workshop
eds. Robert A. Donahue and Jay A. Bookbinder
ISBN 1-886733-74-0

Vol. CS-155 SECOND ADVANCES IN SOLAR PHYSICS EUROCONFERENCE: THREE-DIMENSIONAL STRUCTURE OF SOLAR ACTIVE REGIONS
eds. Costas E. Alissandrakis and Brigitte Schmieder
ISBN 1-886733-75-9

PUBLISHED: 1999

Vol. CS-156 HIGHLY REDSHIFTED RADIO LINES
eds. C. L. Carilli, S. J. E. Radford, K. M. Menten, and G. I. Langston
ISBN 1-886733-76-7

Vol. CS-157 ANNAPOLIS WORKSHOP ON MAGNETIC CATACLYSMIC VARIABLES
eds. Coel Hellier and Koji Mukai
ISBN 1-886733-77-5

Vol. CS-158 SOLAR AND STELLAR ACTIVITY: SIMILARITIES AND DIFFERENCES
eds. C. J. Butler and J. G. Doyle
ISBN 1-886733-78-3

Vol. CS-159 BL LAC PHENOMENON
eds. Leo O. Takalo and Aimo Sillanpää
ISBN 1-886733-79-1

Vol. CS-160 ASTROPHYSICAL DISCS: An EC Summer School
eds. J. A. Sellwood and Jeremy Goodman
ISBN 1-886733-80-5

Vol. CS-161 HIGH ENERGY PROCESSES IN ACCRETING BLACK HOLES
eds. Juri Poutanen and Roland Svensson
ISBN 1-886733-81-3

Vol. CS-162 QUASARS AND COSMOLOGY
eds. Gary Forland and Jack Baldwin
ISBN 1-886733-83-X

Vol. CS-163 STAR FORMATION IN EARLY-TYPE GALAXIES
eds. Jordi Cepa and Patricia Carral
ISBN 1-886733-84-8

Vol. CS-164 ULTRAVIOLET–OPTICAL SPACE ASTRONOMY BEYOND HST
eds. Jon A. Morse, J. Michael Shull, and Anne L. Kinney
ISBN 1-886733-85-6

Vol. CS-165 THE THIRD STROMLO SYMPOSIUM: THE GALACTIC HALO
eds. Brad K. Gibson, Tim S. Axelrod, and Mary E. Putman
ISBN 1-886733-86-4

Vol. CS-166 STROMLO WORKSHOP ON HIGH-VELOCITY CLOUDS
eds. Brad K. Gibson and Mary E. Putman
ISBN 1-886733-87-2

Vol. CS-167 HARMONIZING COSMIC DISTANCE SCALES IN A POST-HIPPARCOS ERA
eds. Daniel Egret and André Heck
ISBN 1-886733-88-0

Vol. CS-168 NEW PERSPECTIVES ON THE INTERSTELLAR MEDIUM
eds. A. R. Taylor, T. L. Landecker, and G. Joncas
ISBN 1-886733-89-9

ASP CONFERENCE SERIES VOLUMES

Published by the Astronomical Society of the Pacific

PUBLISHED: 1999 (* asterisk means OUT OF STOCK)

Vol. CS-169 11th EUROPEAN WORKSHOP ON WHITE DWARFS
eds. J.-E. Solheim and E. G. Meištas
ISBN 1-886733-91-0

Vol. CS-170 THE LOW SURFACE BRIGHTNESS UNIVERSE, IAU Colloquium 171
eds. J. I. Davies, C. Impey, and S. Phillipps
ISBN 1-886733-92-9

Vol. CS-171 LiBeB, COSMIC RAYS, AND RELATED X- AND GAMMA-RAYS
eds. Reuven Ramaty, Elisabeth Vangioni-Flam, Michel Cassé, and Keith Olive
ISBN 1-886733-93-7

Vol. CS-172 ASTRONOMICAL DATA ANALYSIS SOFTWARE AND SYSTEMS VIII
eds. David M. Mehringer, Raymond L. Plante, and Douglas A. Roberts
ISBN 1-886733-94-5

Vol. CS-173 THEORY AND TESTS OF CONVECTION IN STELLAR STRUCTURE: First Granada Workshop
ed. Álvaro Giménez, Edward F. Guinan, and Benjamín Montesinos
ISBN 1-886733-95-3

Vol. CS-174 CATCHING THE PERFECT WAVE: ADAPTIVE OPTICS AND INTERFEROMETRY IN THE 21st CENTURY,
A Symposium held as a part of the 110th Annual Meeting of the ASP
eds. Sergio R. Restaino, William Junor, and Nebojsa Duric
ISBN 1-886733-96-1

Vol. CS-175 STRUCTURE AND KINEMATICS OF QUASAR BROAD LINE REGIONS
eds. C. M. Gaskell, W. N. Brandt, M. Dietrich, D. Dultzin-Hacyan, and M. Eracleous
ISBN 1-886733-97-X

Vol. CS-176 OBSERVATIONAL COSMOLOGY: THE DEVELOPMENT OF GALAXY SYSTEMS
eds. Giuliano Giuricin, Marino Mezzetti, and Paolo Salucci
ISBN 1-58381-000-5

Vol. CS-177 ASTROPHYSICS WITH INFRARED SURVEYS: A Prelude to SIRTF
eds. Michael D. Bicay, Chas A. Beichman, Roc M. Cutri, and Barry F. Madore
ISBN 1-58381-001-3

Vol. CS-178 STELLAR DYNAMOS: NONLINEARITY AND CHAOTIC FLOWS
eds. Manuel Núñez and Antonio Ferriz-Mas
ISBN 1-58381-002-1

Vol. CS-179 ETA CARINAE AT THE MILLENNIUM
eds. Jon A. Morse, Roberta M. Humphreys, and Augusto Damineli
ISBN 1-58381-003-X

Vol. CS-180 SYNTHESIS IMAGING IN RADIO ASTRONOMY II
eds. G. B. Taylor, C. L. Carilli, and R. A. Perley
ISBN 1-58381-005-6

Vol. CS-181 MICROWAVE FOREGROUNDS
eds. Angelica de Oliveira-Costa and Max Tegmark
ISBN 1-58381-006-4

Vol. CS-182 GALAXY DYNAMICS: A Rutgers Symposium
eds. David Merritt, J. A. Sellwood, and Monica Valluri
ISBN 1-58381-007-2

Vol. CS-183 HIGH RESOLUTION SOLAR PHYSICS: THEORY, OBSERVATIONS, AND TECHNIQUES
eds. T. R. Rimmele, K. S. Balasubramaniam, and R. R. Radick
ISBN 1-58381-009-9

ASP CONFERENCE SERIES VOLUMES

Published by the Astronomical Society of the Pacific

PUBLISHED: 1999 (* asterisk means OUT OF STOCK)

Vol. CS-184 THIRD ADVANCES IN SOLAR PHYSICS EUROCONFERENCE: MAGNETIC FIELDS AND OSCILLATIONS
eds. B. Schmieder, A. Hofmann, and J. Staude
ISBN 1-58381-010-2

Vol. CS-185 PRECISE STELLAR RADIAL VELOCITIES, IAU Colloquium 170
eds. J. B. Hearnshaw and C. D. Scarfe
ISBN 1-58381-011-0

Vol. CS-186 THE CENTRAL PARSECS OF THE GALAXY
eds. Heino Falcke, Angela Cotera, Wolfgang J. Duschl, Fulvio Melia, and Marcia J. Rieke
ISBN 1-58381-012-9

Vol. CS-187 THE EVOLUTION OF GALAXIES ON COSMOLOGICAL TIMESCALES
eds. J. E. Beckman and T. J. Mahoney
ISBN 1-58381-013-7

Vol. CS-188 OPTICAL AND INFRARED SPECTROSCOPY OF CIRCUMSTELLAR MATTER
eds. Eike W. Guenther, Bringfried Stecklum, and Sylvio Klose
ISBN 1-58381-014-5

Vol. CS-189 CCD PRECISION PHOTOMETRY WORKSHOP
eds. Eric R. Craine, Roy A. Tucker, and Jeannette Barnes
ISBN 1-58381-015-3

Vol. CS-190 GAMMA-RAY BURSTS: THE FIRST THREE MINUTES
eds. Juri Poutanen and Roland Svensson
ISBN 1-58381-016-1

Vol. CS-191 PHOTOMETRIC REDSHIFTS AND HIGH REDSHIFT GALAXIES
eds. Ray J. Weymann, Lisa J. Storrie-Lombardi, Marcin Sawicki, and Robert J. Brunner
ISBN 1-58381-017-X

Vol. CS-192 SPECTROPHOTOMETRIC DATING OF STARS AND GALAXIES
ed. I. Hubeny, S. R. Heap, and R. H. Cornett
ISBN 1-58381-018-8

Vol. CS-193 THE HY-REDSHIFT UNIVERSE: GALAXY FORMATION AND EVOLUTION AT HIGH REDSHIFT
eds. Andrew J. Bunker and Wil J. M. van Breugel
ISBN 1-58381-019-6

Vol. CS-194 WORKING ON THE FRINGE: OPTICAL AND IR INTERFEROMETRY FROM GROUND AND SPACE
eds. Stephen Unwin and Robert Stachnik
ISBN 1-58381-020-X

PUBLISHED: 2000

Vol. CS-195 IMAGING THE UNIVERSE IN THREE DIMENSIONS: Astrophysics with Advanced Multi-Wavelength Imaging Devices
eds. W. van Breugel and J. Bland-Hawthorn
ISBN 1-58381-022-6

Vol. CS-196 THERMAL EMISSION SPECTROSCOPY AND ANALYSIS OF DUST, DISKS, AND REGOLITHS
eds. Michael L. Sitko, Ann L. Sprague, and David K. Lynch
ISBN: 1-58381-023-4

Vol. CS-197 XVth IAP MEETING DYNAMICS OF GALAXIES: FROM THE EARLY UNIVERSE TO THE PRESENT
eds. F. Combes, G. A. Mamon, and V. Charmandaris
ISBN: 1-58381-24-2

ASP CONFERENCE SERIES VOLUMES

Published by the Astronomical Society of the Pacific

PUBLISHED: 2000 (* asterisk means OUT OF STOCK)

Vol. CS-198 EUROCONFERENCE ON "STELLAR CLUSTERS AND ASSOCIATIONS: CONVECTION, ROTATION, AND DYNAMOS"
eds. R. Pallavicini, G. Micela, and S. Sciortino
ISBN: 1-58381-25-0

Vol. CS-199 ASYMMETRICAL PLANETARY NEBULAE II: FROM ORIGINS TO MICROSTRUCTURES
eds. J. H. Kastner, N. Soker, and S. Rappaport
ISBN: 1-58381-026-9

Vol. CS-200 CLUSTERING AT HIGH REDSHIFT
eds. A. Mazure, O. Le Fèvre, and V. Le Brun
ISBN: 1-58381-027-7

Vol. CS-201 COSMIC FLOWS 1999: TOWARDS AN UNDERSTANDING OF LARGE-SCALE STRUCTURES
eds. Stéphane Courteau, Michael A. Strauss, and Jeffrey A. Willick
ISBN: 1-58381-028-5

Vol. CS-202 * PULSAR ASTRONOMY – 2000 AND BEYOND, IAU Colloquium 177
eds. M. Kramer, N. Wex, and R. Wielebinski
ISBN: 1-58381-029-3

Vol. CS-203 THE IMPACT OF LARGE-SCALE SURVEYS ON PULSATING STAR RESEARCH, IAU Colloquium 176
eds. L. Szabados and D. W. Kurtz
ISBN: 1-58381-030-7

Vol. CS-204 THERMAL AND IONIZATION ASPECTS OF FLOWS FROM HOT STARS: OBSERVATIONS AND THEORY
eds. Henny J. G. L. M. Lamers and Arved Sapar
ISBN: 1-58381-031-5

Vol. CS-205 THE LAST TOTAL SOLAR ECLIPSE OF THE MILLENNIUM IN TURKEY
eds. W. C. Livingston and A. Özgüç
ISBN: 1-58381-032-3

Vol. CS-206 HIGH ENERGY SOLAR PHYSICS – *ANTICIPATING HESSI*
eds. Reuven Ramaty and Natalie Mandzhavidze
ISBN: 1-58381-033-1

Vol. CS-207 NGST SCIENCE AND TECHNOLOGY EXPOSITION
eds. Eric P. Smith and Knox S. Long
ISBN: 1-58381-036-6

ATLAS VISIBLE AND NEAR INFRARED ATLAS OF THE ARCTURUS SPECTRUM 3727-9300 Å
eds. Kenneth Hinkle, Lloyd Wallace, Jeff Valenti, and Dianne Harmer
ISBN: 1-58381-037-4

Vol. CS-208 POLAR MOTION: HISTORICAL AND SCIENTIFIC PROBLEMS, IAU Colloquium 178
eds. Steven Dick, Dennis McCarthy, and Brian Luzum
ISBN: 1-58381-039-0

Vol. CS-209 SMALL GALAXY GROUPS, IAU Colloquium 174
eds. Mauri J. Valtonen and Chris Flynn
ISBN: 1-58381-040-4

Vol. CS-210 DELTA SCUTI AND RELATED STARS: Reference Handbook and Proceedings of the 6th Vienna Workshop in Astrophysics
eds. Michel Breger and Michael Houston Montgomery
ISBN: 1-58381-043-9

ASP CONFERENCE SERIES VOLUMES

Published by the Astronomical Society of the Pacific

PUBLISHED: 2000 (* asterisk means OUT OF STOCK)

Vol. CS-211 MASSIVE STELLAR CLUSTERS
eds. Ariane Lançon and Christian M. Boily
ISBN: 1-58381-042-0

Vol. CS-212 FROM GIANT PLANETS TO COOL STARS
eds. Caitlin A. Griffith and Mark S. Marley
ISBN: 1-58381-041-2

Vol. CS-213 BIOASTRONOMY '99: A NEW ERA IN BIOASTRONOMY
eds. Guillermo A. Lemarchand and Karen J. Meech
ISBN: 1-58381-044-7

Vol. CS-214 THE Be PHENOMENON IN EARLY-TYPE STARS, IAU Colloquium 175
eds. Myron A. Smith, Huib F. Henrichs and Juan Fabregat
ISBN: 1-58381-045-5

Vol. CS-215 COSMIC EVOLUTION AND GALAXY FORMATION:
STRUCTURE, INTERACTIONS AND FEEDBACK
The 3rd Guillermo Haro Astrophysics Conference
eds. José Franco, Elena Terlevich, Omar López-Cruz, and Itziar Aretxaga
ISBN: 1-58381-046-3

Vol. CS-216 ASTRONOMICAL DATA ANALYSIS SOFTWARE AND SYSTEMS IX
eds. Nadine Manset, Christian Veillet, and Dennis Crabtree
ISBN: 1-58381-047-1 ISSN: 1080-7926

Vol. CS-217 IMAGING AT RADIO THROUGH SUBMILLIMETER WAVELENGTHS
eds. Jeffrey G. Mangum and Simon J. E. Radford
ISBN: 1-58381-049-8

Vol. CS-218 MAPPING THE HIDDEN UNIVERSE: THE UNIVERSE BEHIND THE MILKY WAY
THE UNIVERSE IN HI
eds. Renée C. Kraan-Korteweg, Patricia A. Henning, and Heinz Andernach
ISBN: 1-58381-050-1

Vol. CS-219 DISKS, PLANETESIMALS, AND PLANETS
eds. F. Garzón, C. Eiroa, D. de Winter, and T. J. Mahoney
ISBN: 1-58381-051-X

Vol. CS-220 AMATEUR - PROFESSIONAL PARTNERSHIPS IN ASTRONOMY:
The 111th Annual Meeting of the ASP
eds. John R. Percy and Joseph B. Wilson
ISBN: 1-58381-052-8

Vol. CS-221 STARS, GAS AND DUST IN GALAXIES: EXPLORING THE LINKS
eds. Danielle Alloin, Knut Olsen, and Gaspar Galaz
ISBN: 1-58381-053-6

PUBLISHED: 2001

Vol. CS-222 THE PHYSICS OF GALAXY FORMATION
eds. M. Umemura and H. Susa
ISBN: 1-58381-054-4

Vol. CS-223 COOL STARS, STELLAR SYSTEMS AND THE SUN:
Eleventh Cambridge Workshop
eds. Ramón J. García López, Rafael Rebolo, and María Zapatero Osorio
ISBN: 1-58381-056-0

Vol. CS-224 PROBING THE PHYSICS OF ACTIVE GALACTIC NUCLEI
BY MULTIWAVELENGTH MONITORING
eds. Bradley M. Peterson, Ronald S. Polidan, and Richard W. Pogge
ISBN: 1-58381-055-2

ASP CONFERENCE SERIES VOLUMES

Published by the Astronomical Society of the Pacific

PUBLISHED: 2001 (* asterisk means OUT OF STOCK)

Vol. CS-225 VIRTUAL OBSERVATORIES OF THE FUTURE
eds. Robert J. Brunner, S. George Djorgovski, and Alex S. Szalay
ISBN: 1-58381-057-9

Vol. CS-226 12th EUROPEAN CONFERENCE ON WHITE DWARFS
eds. J. L. Provencal, H. L. Shipman, J. MacDonald, and S. Goodchild
ISBN: 1-58381-058-7

Vol. CS-227 BLAZAR DEMOGRAPHICS AND PHYSICS
eds. Paolo Padovani and C. Megan Urry
ISBN: 1-58381-059-5

Vol. CS-228 DYNAMICS OF STAR CLUSTERS AND THE MILKY WAY
eds. S. Deiters, B. Fuchs, A. Just, R. Spurzem, and R. Wielen
ISBN: 1-58381-060-9

Vol. CS-229 EVOLUTION OF BINARY AND MULTIPLE STAR SYSTEMS
A Meeting in Celebration of Peter Eggleton's 60th Birthday
eds. Ph. Podsiadlowski, S. Rappaport, A. R. King, F. D'Antona, and L. Burderi
IBSN: 1-58381-061-7

Vol. CS-230 GALAXY DISKS AND DISK GALAXIES
eds. Jose G. Funes, S. J. and Enrico Maria Corsini
ISBN: 1-58381-063-3

Vol. CS-231 TETONS 4: GALACTIC STRUCTURE, STARS, AND THE INTERSTELLAR MEDIUM
eds. Charles E. Woodward, Michael D. Bicay, and J. Michael Shull
ISBN: 1-58381-064-1

Vol. CS-232 THE NEW ERA OF WIDE FIELD ASTRONOMY
eds. Roger Clowes, Andrew Adamson, and Gordon Bromage
ISBN: 1-58381-065-X

Vol. CS-233 P CYGNI 2000: 400 YEARS OF PROGRESS
eds. Mart de Groot and Christiaan Sterken
ISBN: 1-58381-070-6

Vol. CS-234 X-RAY ASTRONOMY 2000
eds. R. Giacconi, S. Serio, and L. Stella
ISBN: 1-58381-071-4

Vol. CS-235 SCIENCE WITH THE ATACAMA LARGE MILLIMETER ARRAY (ALMA)
ed. Alwyn Wootten
ISBN: 1-58381-072-2

Vol. CS-236 ADVANCED SOLAR POLARIMETRY: THEORY, OBSERVATION, AND INSTRUMENTATION, The 20th Sacramento Peak Summer Workshop
ed. M. Sigwarth
ISBN: 1-58381-073-0

Vol. CS-237 GRAVITATIONAL LENSING: RECENT PROGRESS AND FUTURE GOALS
eds. Tereasa G. Brainerd and Christopher S. Kochanek
ISBN: 1-58381-074-9

Vol. CS-238 ASTRONOMICAL DATA ANALYSIS SOFTWARE AND SYSTEMS X
eds. F. R. Harnden, Jr., Francis A. Primini, and Harry E. Payne
ISBN: 1-58381-075-7

Vol. CS-239 MICROLENSING 2000: A NEW ERA OF MICROLENSING ASTROPHYSICS
ed. John Menzies and Penny D. Sackett
ISBN: 1-58381-076-5

ASP CONFERENCE SERIES VOLUMES

Published by the Astronomical Society of the Pacific

PUBLISHED: 2001 (* asterisk means OUT OF STOCK)

Vol. CS-240 GAS AND GALAXY EVOLUTION,
A Conference in Honor of the 20th Anniversary of the VLA
eds. J. E. Hibbard, M. P. Rupen, and J. H. van Gorkom
ISBN: 1-58381-077-3

Vol. CS-241 CS-241 THE 7TH TAIPEI ASTROPHYSICS WORKSHOP ON
COSMIC RAYS IN THE UNIVERSE
ed. Chung-Ming Ko
ISBN: 1-58381-079-X

Vol. CS-242 ETA CARINAE AND OTHER MYSTERIOUS STARS:
THE HIDDEN OPPORTUNITIES OF EMISSION SPECTROSCOPY
eds. Theodore R. Gull, Sveneric Johannson, and Kris Davidson
ISBN: 1-58381-080-3

Vol. CS-243 FROM DARKNESS TO LIGHT:
ORIGIN AND EVOLUTION OF YOUNG STELLAR CLUSTERS
eds. Thierry Montmerle and Philippe André
ISBN: 1-58381-081-1

Vol. CS-244 YOUNG STARS NEAR EARTH: PROGRESS AND PROSPECTS
eds. Ray Jayawardhana and Thomas P. Greene
ISBN: 1-58381-082-X

Vol. CS-245 ASTROPHYSICAL AGES AND TIME SCALES
eds. Ted von Hippel, Chris Simpson, and Nadine Manset
ISBN: 1-58381-083-8

Vol. CS-246 SMALL TELESCOPE ASTRONOMY ON GLOBAL SCALES, IAU Colloquium 183
eds. Wen-Ping Chen, Claudia Lemme, and Bohdan Paczyński
ISBN: 1-58381-084-6

Vol. CS-247 SPECTROSCOPIC CHALLENGES OF PHOTOIONIZED PLASMAS
eds. Gary Ferland and Daniel Wolf Savin
ISBN: 1-58381-085-4

Vol. CS-248 MAGNETIC FIELDS ACROSS THE HERTZSPRUNG-RUSSELL DIAGRAM
eds. G. Mathys, S. K. Solanki, and D. T. Wickramasinghe
ISBN: 1-58381-088-9

Vol. CS-249 THE CENTRAL KILOPARSEC OF STARBURSTS AND AGN:
THE LA PALMA CONNECTION
eds. J. H. Knapen, J. E. Beckman, I. Shlosman, and T. J. Mahoney
ISBN: 1-58381-089-7

Vol. CS-250 PARTICLES AND FIELDS IN RADIO GALAXIES CONFERENCE
eds. Robert A. Laing and Katherine M. Blundell
ISBN: 1-58381-090-0

Vol. CS-251 NEW CENTURY OF X-RAY ASTRONOMY
eds. H. Inoue and H. Kunieda
ISBN: 1-58381-091-9

Vol. CS-252 HISTORICAL DEVELOPMENT OF MODERN COSMOLOGY
eds. Vicent J. Martínez, Virginia Trimble, and María Jesús Pons-Bordería
ISBN: 1-58381-092-7

PUBLISHED: 2002

Vol. CS-253 CHEMICAL ENRICHMENT OF INTRACLUSTER AND INTERGALACTIC MEDIUM
eds. Roberto Fusco-Femiano and Francesca Matteucci
ISBN: 1-58381-093-5

ASP CONFERENCE SERIES VOLUMES

Published by the Astronomical Society of the Pacific

PUBLISHED: 2002 (* asterisk means OUT OF STOCK)

Vol. CS-254 EXTRAGALACTIC GAS AT LOW REDSHIFT
eds. John S. Mulchaey and John T. Stocke
ISBN: 1-58381-094-3

Vol. CS-255 MASS OUTFLOW IN ACTIVE GALACTIC NUCLEI: NEW PERSPECTIVES
eds. D. M. Crenshaw, S. B. Kraemer, and I. M. George
ISBN: 1-58381-095-1

Vol. CS-256 OBSERVATIONAL ASPECTS OF PULSATING B AND A STARS
eds. Christiaan Sterken and Donald W. Kurtz
ISBN: 1-58381-096-X

Vol. CS-257 AMiBA 2001: HIGH-*Z* CLUSTERS, MISSING BARYONS, AND CMB POLARIZATION
eds. Lin-Wen Chen, Chung-Pei Ma, Kin-Wang Ng, and Ue-Li Pen
ISBN: 1-58381-097-8

Vol. CS-258 ISSUES IN UNIFICATION OF ACTIVE GALACTIC NUCLEI
eds. Roberto Maiolino, Alessandro Marconi, and Neil Nagar
ISBN: 1-58381-098-6

Vol. CS-259 RADIAL AND NONRADIAL PULSATIONS AS PROBES OF STELLAR PHYSICS, IAU Colloquium 185
eds. Conny Aerts, Timothy R. Bedding, and Jørgen Christensen-Dalsgaard
ISBN: 1-58381-099-4

Vol. CS-260 INTERACTING WINDS FROM MASSIVE STARS
eds. Anthony F. J. Moffat and Nicole St-Louis
ISBN: 1-58381-100-1

Vol. CS-261 THE PHYSICS OF CATACLYSMIC VARIABLES AND RELATED OBJECTS
eds. B. T. Gänsicke, K. Beuermann, and K. Reinsch
ISBN: 1-58381-101-X

Vol. CS-262 THE HIGH ENERGY UNIVERSE AT SHARP FOCUS: CHANDRA SCIENCE, held in conjunction with the 113th Annual Meeting of the ASP
eds. Eric M. Schlegel and Saeqa Dil Vrtilek
ISBN: 1-58381-102-8

Vol. CS-263 STELLAR COLLISIONS, MERGERS AND THEIR CONSEQUENCES
ed. Michael M. Shara
ISBN: 1-58381-103-6

Vol. CS-264 CONTINUING THE CHALLENGE OF EUV ASTRONOMY: CURRENT ANALYSIS AND PROSPECTS FOR THE FUTURE
eds. Steve B. Howell, Jean Dupuis, Daniel Golombek, Frederick M. Walter, and Jennifer Cullison
ISBN: 1-58381-104-4

Vol. CS-265 ω CENTAURI, A UNIQUE WINDOW INTO ASTROPHYSICS
eds. Floor van Leeuwen, Joanne D. Hughes, and Giampaolo Piotto
ISBN: 1-58381-105-2

Vol. CS-266 ASTRONOMICAL SITE EVALUATION IN THE VISIBLE AND RADIO RANGE, IAU Technical Workshop
eds. J. Vernin, Z. Benkhaldoun, and C. Muñoz-Tuñón
ISBN: 1-58381-106-0

Vol. CS-267 HOT STAR WORKSHOP III: THE EARLIEST STAGES OF MASSIVE STAR BIRTH
ed. Paul A. Crowther
ISBN: 1-58381-107-9

Vol. CS-268 TRACING COSMIC EVOLUTION WITH GALAXY CLUSTERS
eds. Stefano Borgani, Marino Mezzetti, and Riccardo Valdarnini
ISBN: 1-58381-108-7

ASP CONFERENCE SERIES VOLUMES

Published by the Astronomical Society of the Pacific

PUBLISHED: 2002 (* asterisk means OUT OF STOCK)

Vol. CS-269 THE EVOLVING SUN AND ITS INFLUENCE ON PLANETARY ENVIRONMENTS
eds. Benjamín Montesinos, Álvaro Giménez, and Edward F. Guinan
ISBN: 1-58381-109-5

Vol. CS-270 ASTRONOMICAL INSTRUMENTATION AND THE BIRTH AND GROWTH OF ASTROPHYSICS: A Symposium held in honor of Robert G. Tull
eds. Frank N. Bash and Christopher Sneden
ISBN: 1-58381-110-9

Vol. CS-271 NEUTRON STARS IN SUPERNOVA REMNANTS
eds. Patrick O. Slane and Bryan M. Gaensler
ISBN: 1-58381-111-7

Vol. CS-272 THE FUTURE OF SOLAR SYSTEM EXPLORATION, 2003-2013
Community Contributions to the NRC Solar System Exploration Decadal Survey
ed. Mark V. Sykes
ISBN: 1-58381-113-3

Vol. CS-273 THE DYNAMICS, STRUCTURE AND HISTORY OF GALAXIES
eds. G. S. Da Costa and H. Jerjen
ISBN: 1-58381-114-1

Vol. CS-274 OBSERVED HR DIAGRAMS AND STELLAR EVOLUTION
eds. Thibault Lejeune and João Fernandes
ISBN: 1-58381-116-8

Vol. CS-275 DISKS OF GALAXIES: KINEMATICS, DYNAMICS AND PERTURBATIONS
eds. E. Athanassoula, A. Bosma, and R. Mujica
ISBN: 1-58381-117-6

Vol. CS-276 SEEING THROUGH THE DUST:
THE DETECTION OF HI AND THE EXPLORATION OF THE ISM IN GALAXIES
eds. A. R. Taylor, T. L. Landecker, and A. G. Willis
ISBN: 1-58381-118-4

Vol. CS 277 STELLAR CORONAE IN THE CHANDRA AND XMM-NEWTON ERA
eds. Fabio Favata and Jeremy J. Drake
ISBN: 1-58381-119-2

Vol. CS 278 NAIC–NRAO SCHOOL ON SINGLE-DISH ASTRONOMY:
TECHNIQUES AND APPLICATIONS
eds. Snezana Stanimirovic, Daniel Altschuler, Paul Goldsmith, and Chris Salter
ISBN: 1-58381-120-6

Vol. CS 279 EXOTIC STARS AS CHALLENGES TO EVOLUTION, IAU Colloquium 187
eds. Christopher A. Tout and Walter Van Hamme
ISBN: 1-58381-122-2

Vol. CS 280 NEXT GENERATION WIDE-FIELD MULTI-OBJECT SPECTROSCOPY
eds. Michael J. I. Brown and Arjun Dey
ISBN: 1-58381-123-0

Vol. CS 281 ASTRONOMICAL DATA ANALYSIS SOFTWARE AND SYSTEM XI
eds. David A. Bohlender, Daniel Durand, and Thomas H. Handley
ISBN: 1-58381-124-9 ISSN: 1080-7926

Vol. CS 282 GALAXIES: THE THIRD DIMENSION
eds. Margarita Rosado, Luc Binette, and Lorena Arias
ISBN: 1-58381-125-7

Vol. CS 283 A NEW ERA IN COSMOLOGY
eds. Nigel Metcalfe and Tom Shanks
ISBN: 1-58381-126-5

ASP CONFERENCE SERIES VOLUMES

Published by the Astronomical Society of the Pacific

PUBLISHED: 2002 (* asterisk means OUT OF STOCK)

Vol. CS 284 AGN SURVEYS
eds. R. F. Green, E. Ye. Khachikian, and D. B. Sanders
ISBN: 1-58381-127-3

Vol. CS 285 MODES OF STAR FORMATION AND THE ORIGIN OF FIELD POPULATIONS
eds. Eva K. Grebel and Walfgang Brandner
ISBN: 1-58381-128-1

PUBLISHED: 2003

Vol. CS 286 CURRENT THEORETICAL MODESL AND HIGH RESOLUTION SOLAR OBSERVATIONS: PREPARING FOR ATST
eds. Alexei A. Pevtsov and Han Uitenbroek
ISBN: 1-58381-129-X

Vol. CS 287 GALACTIC STAR FORMATION ACROSS THE STELLAR MASS SPECTRUM
eds. J.M. De Buizer and N.S. van der Bliek
ISBN:1-58381-130-3

Vol. CS 288 STELLAR ATMOSPHERE MODELING
eds. I. Hubeny, D. Mihalas and K. Werner
ISBN: 1-58381-131-1

Vol. CS 289 THE PROCEEDINGS OF THE IAU 8TH ASIAN-PACIFIC REGIONAL MEETING, VOLUME 1
eds. Satoru Ikeuchi, John Hearnshaw and Tomoyuki Hanawa
ISBN: 1-58381-134-6

A LISTING OF IAU VOLUMES MAY BE FOUND ON THE NEXT PAGE

INTERNATIONAL ASTRONOMICAL UNION (IAU) VOLUMES

Published by the Astronomical Society of the Pacific

PUBLISHED: 1999 (* asterisk means OUT OF STOCK)

Vol. No. 190 NEW VIEWS OF THE MAGELLANIC CLOUDS
eds. You-Hua Chu, Nicholas B. Suntzeff, James E. Hesser, and David A. Bohlender
ISBN: 1-58381-021-8

Vol. No. 191 ASYMPTOTIC GIANT BRANCH STARS
eds. T. Le Bertre, A. Lèbre, and C. Waelkens
ISBN: 1-886733-90-2

Vol. No. 192 THE STELLAR CONTENT OF LOCAL GROUP GALAXIES
eds. Patricia Whitelock and Russell Cannon
ISBN: 1-886733-82-1

Vol. No. 193 WOLF-RAYET PHENOMENA IN MASSIVE STARS AND STARBURST GALAXIES
eds. Karel A. van der Hucht, Gloria Koenigsberger, and Philippe R. J. Eenens
ISBN: 1-58381-004-8

Vol. No. 194 ACTIVE GALACTIC NUCLEI AND RELATED PHENOMENA
eds. Yervant Terzian, Daniel Weedman, and Edward Khachikian
ISBN: 1-58381-008-0

PUBLISHED: 2000

Vol. XXIVA TRANSACTIONS OF THE INTERNATIONAL ASTRONOMICAL UNION REPORTS ON ASTRONOMY 1996-1999
ed. Johannes Andersen
ISBN: 1-58381-035-8

Vol. No. 195 HIGHLY ENERGETIC PHYSICAL PROCESSES AND MECHANISMS FOR EMISSION FROM ASTROPHYSICAL PLASMAS
eds. P. C. H. Martens, S. Tsuruta, and M. A. Weber
ISBN: 1-58381-038-2

Vol. No. 197 ASTROCHEMISTRY: FROM MOLECULAR CLOUDS TO PLANETARY SYSTEMS
eds. Y. C. Minh and E. F. van Dishoeck
ISBN: 1-58381-034-X

Vol. No. 198 THE LIGHT ELEMENTS AND THEIR EVOLUTION
eds. L. da Silva, M. Spite, and J. R. de Medeiros
ISBN: 1-58381-048-X

PUBLISHED: 2001

IAU SPS ASTRONOMY FOR DEVELOPING COUNTRIES
Special Session of the XXIV General Assembly of the IAU
ed. Alan H. Batten
ISBN: 1-58381-067-6

Vol. No. 196 PRESERVING THE ASTRONOMICAL SKY
eds. R. J. Cohen and W. T. Sullivan, III
ISBN: 1-58381-078-1

Vol. No. 200 THE FORMATION OF BINARY STARS
eds. Hans Zinnecker and Robert D. Mathieu
ISBN: 1-58381-068-4

Vol. No. 203 RECENT INSIGHTS INTO THE PHYSICS OF THE SUN AND HELIOSPHERE: HIGHLIGHTS FROM SOHO AND OTHER SPACE MISSIONS
eds. Pål Brekke, Bernhard Fleck, and Joseph B. Gurman
ISBN: 1-58381-069-2

Vol. No. 204 THE EXTRAGALACTIC INFRARED BACKGROUND AND ITS COSMOLOGICAL IMPLICATIONS
eds. Martin Harwit and Michael G. Hauser
ISBN: 1-58381-062-5

INTERNATIONAL ASTRONOMICAL UNION (IAU) VOLUMES

Published by the Astronomical Society of the Pacific

PUBLISHED: 2001 (* asterisk means OUT OF STOCK)

Vol. No. 205 GALAXIES AND THEIR CONSTITUENTS AT THE HIGHEST ANGULAR RESOLUTIONS
eds. Richard T. Schilizzi, Stuart N. Vogel, Francesco Paresce, and Martin S. Elvis
ISBN: 1-58381-066-8

Vol. XXIVB TRANSACTIONS OF THE INTERNATIONAL ASTRONOMICAL UNION REPORTS ON ASTRONOMY
ed. Hans Rickman
ISBN: 1-58381-087-0

PUBLISHED: 2002

Vol. No. 12 HIGHLIGHTS OF ASTRONOMY
ed. Hans Rickman
ISBN: 1-58381-086-2

Vol. No. 199 THE UNIVERSE AT LOW RADIO FREQUENCIES
eds. A. Pramesh Rao, G. Swarup, and Gopal-Krishna
ISBN: 58381-121-4

Vol. No. 206 COSMIC MASERS: FROM PROTOSTARS TO BLACKHOLES
eds. Victor Migenes and Mark J. Reid
ISBN: 1-58381-112-5

Vol. No. 207 EXTRAGALACTIC STAR CLUSTERS
eds. Doug Geisler, Eva K. Grebel, and Dante Minniti
ISBN: 1-58381-115-X

PUBLISHED: 2003

Vol. XXVA TRANSACTIONS OF THE INTERNATIONAL ASTRONOMICAL UNION REPORTS ON ASTRONOMY 1999-2002
ed. Hans Rickman
ISBN: 1-58381-137-0

Vol. No. 208 ASTROPHYSICAL SUPERCOMPUTING USING PARTICLE SIMULATIONS
eds. Junichiro Makino and Piet Hut
ISBN: 1-58381-139-7

Vol. No. 211 BROWN DWARFS
ed. Eduardo Martín
ISBN: 1-58381-132-X

Vol. No. 212 A MASSIVE STAR ODYSSEY: FROM MAIN SEQUENCE TO SUPERNOVA
ed. Karel A. van der Hucht, Artemio Herrero and César Esteban
ISBN: 1-58381-133-8

Ordering information is available at the beginning of the listing